INSIDE HISTORY of the USAF LIGHTWEIGHT FIGHTERS

1900 to 1975

Herbert A. Hutchinson

Chief System Engineer
USAF LWF Program Office

INSIDE HISTORY OF THE USAF LIGHTWEIGHT FIGHTERS, 1900 to 1975

(Second Edition)

Herbert A. Hutchinson

Chief System Engineer
USAF LWF Program Office

Durham, North Carolina
2025

SECOND EDITION CHANGES

This is the second edition of "INSIDE HISTORY OF THE USAF LIGHTWEIGHT FIGHTERS,1900 to 1975". The major changes from the first edition are:

1.) Increasing the dimensions of the original book size from 6x9 to 8x11; and

2.) Converting several photographs, illustrations, and tabular figures from black and white to their original color.

Published in the United States of America

ISBN 979-8-89395-284-1 (SC)
ISBN 979-8-89395-285-8 (HC)
ISBN 979-8-89395-286-5 (Ebook)

Library of Congress Control Number: 2025925470

Verity Publications
41 Grandwood Cir
Durham, NC 27712
www.stellarliterary.com

Order Information and Rights Permission:

Quantity sales. Special discounts might be available on quantity purchases by corporations, associations, and others. For details, contact the publisher at the address above.

For Book Rights Adaptation and other Rights Permission,
Call toll-free 1-888-945-8513 or send an email at
admin@stellarliterary.com.

TABLE OF CONTENTS

DISCLAIMER

Much of the material contained in this book – information, dates, photographs, diagrams, charts, etc. – have been obtained from documents and correspondence generated by US Government personnel during the planning for and/or during technology development activities conducted to advance the capabilities of military flight weapon systems. There are also illustrations, diagrams, charts, tables which were generated by the two USAF Lightweight Fighter Program contractors (General Dynamics and Northrop) which were prepared for briefings to the USAF Lightweight Fighter Program Office revealing the status of their development activities required to be performed per the terms of their respective contracts with the USAF. Regardless of the source of those Government documents in the technology information database, if the document (in any format) was developed using US Government funds, the control of dissemination of the information is maintained by US Government regulatory organizations and with regulations over the content of the information to adhere to any security classifications and any release restrictions to the general public. Government documents are not controlled by the same copyright restrictions and regulations as for non-Government (i.e. commercially or privately created) documents or products, but are controlled under the US Government Freedom of Information Act of 1966 with all subsequent interpretations and modifications.

Further, much of the material is taken from the Author's personal career memorabilia files and holdings while as a civilian employee of the US Government in the USAF at Wright-Patterson AFB, Dayton, Ohio. That material which at one time was classified up to a high level of security classification has all since been downgraded by the US Government to be Unclassified and Approved for Release to the Public under the provisions of the US Freedom of Information Act. All such material originated by US Government activities and contained in this book is provided only for educational or informational purposes but expressly not for the purpose of selling that material or information for financial profit.

It is known that previously published reports by authors "outside" Government employment about the historical development of the USAF Lightweight Fighters are not entirely factually correct and/or does not reveal a critical event happening without which the F-16 Fighting Falcon Weapon System might never have been created. The Author has published this book as a previous "insider" Government employee to clarify and correct that erroneous, misleading, or missing information in the historical technology development database on which the F-16 Fighting Falcon and other military flight systems relied during their respective developments.

Some of the material herein has been extracted from reference sources on the internet but only those identified as having an origin as a government document are used in this book. All of the material in that latter category that is contained herein includes photographs of military aircraft used by the United States Armed Forces and taken by personnel on active military duty. Where photographs of military aircraft are created by US military photographers and are shown, they have been copied herein for educational and visual informational transfer purposes only. The name of the producing company of the aircraft (known to be in the military inventory) shown in a photograph is acknowledged for its contribution to the technology database comprising military aviation history.

The Author stresses that publication of this book be considered as a not-for personal-profit venture so as to achieve maximum distribution without copyright limitations restricting it from being distributed openly to the educational/scholastic public. This book of encyclopedic nature is copyrighted by the Publisher as an entity, i.e., the book is prohibited from being copied as a whole and resold for financial profit.

As this book is not intended to be sold for personal profit, once the Author has been re-compensated by royalties from its sale for the costs expended in its publication, any further royalties from sales will be gifted to the not-for-profit Alumni Foundation of the Brooklyn Technical High School on Fort Greene Place in Brooklyn, New York, for its discretionary use toward the teaching of the technology development process in the commercial, military, or public aviation production or testing domain. The Author was a student at Brooklyn Tech in the College Preparatory Course from 1945 to 1949. It is hoped that any monetary profit will be applied by the Alumni Foundation of Brooklyn Tech to inspire and further the technical education of young students aspiring to pursue some field of science or engineering – just as Brooklyn Tech did for the Author.

There is no intentional infringement of any copyrighted material from any source and contained herein to enhance profitable gain. Any improbable infringement occurring herein is unintentional and the Author pre-apologizes for any such infringement if claimed and verified to infringe on valid copyright protection measures. The Publisher of this book is absolved of any and all proven copyright infringements that may be unintentionally contained in this book.

The Author, Herbert A. Hutchinson

DEDICATION

This book is dedicated most specially to:

USAF Col. John R. Boyd (and his "soldiers" in the "Light Weight Fighter Mafia") for his creative ingenuity and persistence without which the USAF Lightweight Fighter Program may never have been brought to happen; and to

Harry J. Hillaker for his inspired and dedicated configuration development talents applied to the USAF YF-16 Lightweight Fighter as Principal Chief Designer at the General Dynamics Corporation in Fort Worth, Texas. Harry had earlier designed the B-58 Hustler and F-111 Aardvark configurations. With professional and dedicated support from the entire General Dynamics YF-16 Lightweight Fighter development team, an aircraft labeled as "the fighter pilot's fighter" was created with many major flight technology "firsts". Most notable was the YF-16 being the first aircraft to fly with an all-electric (no mechanical back-up), fly-by-wire flight control system.

John Boyd and Harry Hillaker are rightfully forever enshrined in aviation history for having "bent the twig" of the USAF Lightweight Fighters in the right direction. A copy of their obituaries which appeared in their local news media, are provided in Appendix 1 at the end of this book. Those obituaries pay tribute to the many accomplishments both achieved in their respective careers which followed two very diverse paths until they intersected and merged to create the USAF Lightweight Fighters. The Author had the honor of having worked closely with them both to transform their conceptual ideas and design thoughts about a "lightweight, highly maneuverable, affordable cost, multi-mode capable fighter aircraft as a 'low end mix' with the 'high end' F-15 Eagle" into the flight-worthy USAF Lightweight Fighters.

Every F-16 Fighting Falcon in flight was, is, and continues to be a flying monument to their genius and dedicated creativity.

PREFACE

As the twig is bent, so shall the tree grow.

That metaphor applies to the planting of sapling trees, to people and their careers, to the design of new machinery or vehicles, and most especially to each of the numerous diverse technologies embodied in military weapon systems. This book describes the "As the twig is bent" portion of that metaphor. A second book by the Author "The F-16 Multi-National Weapon System (1967 to 2017)" describes the "...so shall the tree grow." portion of the metaphor.

The overall objective of this book is to inspire anyone contemplating whether or not to embark on a scientific or engineering career, within private industry or anywhere within the US Department of Defense, to strongly consider doing so after completion of their college education. This book is written by an Author having a more than 40 years professional career divided equally as a Government employee and as an employee in private industry. The Author has the highest of hopes that this book will prove to be interesting, informative, and useful to all who may read it.

The figures and photographs shown in this book are in varying format and degree of detail, but most are presented in the exact form in which they appeared in original within Government briefings, reports, and news media during the early years of the creation of a small "advanced technology demonstrator" aircraft. The small aircraft later went on to become "The arms deal of the 20th century" as cited in the news media accounts of its successful development. A parallel hope is that this book will inspire those newly embarking on their careers in science and engineering to passionately "THINK"– outside, and not just inside, the mythical "box" - throughout their careers.

There are two statues in front of the National Archives in Washington DC. The statue on the left at the front entrance has the inscription "Study the past" and the statue on the right has the inscription "What is past is prologue". Shown below is a photograph of each of the two statues which the Author observed while on one of many trips to the Pentagon and HQ AFSC at Andrews AFB during his many years of service with the US Government.:

Statues at the National Archives, Washington, D.C. (Photos by the Author)

To better understand how a new or improved technology comes into use, it is always enlightening to look back at the prior history of that technology. In a sense, it is much like developing the genealogy "tree" of a family ancestral history. The efforts Man has made to fly like the birds is a fascinating story that has been told in numerous books dating back to ancient Greece with the story of Daedalus and his son Icarus attempting to fly with wings made from feathers embedded in wax. This book begins much later in that aviation technology "tree" with a short history of the earliest activities by Orville and Wilbur Wright that led to their first successful (at least adequately if not fully) controlled flights at Kill Devil Hill near Kitty Hawk, North Carolina, on December 13, 1904. The development of aviation took a historic leap when the Wright brothers removed the barrier to understanding of flight control of an aircraft. The singular accomplishment of flight control by the Wright brothers led to technology advancements in the five primary technology domains of the five critical elements acting to determine the flight path and controlled motion of airplanes (lift, drag, propelling force, weight, and flight control). Military airplanes of a myriad of designs were subsequently developed for, during, and after the several major wars around the World in which airplanes were a key weapon of war.

In order to keep this book to a manageable size and to focus on primarily those technology developments which had profound influence on the design of the US Air Force Lightweight Fighters, it is of necessity to not detail other technologies (e.g., those that led to the carrier-capable fighter aircraft developed for the US Navy and US Marine Corps aircraft inventory). That does not infer that those technologies for naval aircraft carrier-based fighter aircraft were less important than those of land-based fighter aircraft, but only to reflect that those branches of atmospheric flight technology were not of high significance in the "technology tree" for that of the USAF Lightweight Fighters. It is factual that the development of the Northrop YF-17 (later developed by the USN as the F/A-18 Hornet) was in fact derived from the same USAF Advanced Day Fighter Study done in the mid-1960s which spawned the USAF YF-16 and YF-17 Lightweight Fighter prototypes.

The two-engine Northrop YF-17 Lightweight Fighter competed in a fly-off comparison with the single engine General Dynamics YF-16 Lightweight Fighter for the USAF Air Combat Fighter Program in the mid-1970s. This competition clearly added vital database information addressing long-standing questions as to whether a two-engine fighter design is better (or not) than a single engine fighter aircraft design such as the YF-16. Accordingly, there was no "loser" in the USAF Lightweight Fighter fly-off competition. The General Dynamics YF-16 went on to become the USAF F-16 Air Combat Fighter, and the Northrop YF-17 went on to become the USN/USMC F/A-18 Hornet. Both programs must be regarded as being "winners".

Also, in keeping within the scope of this book and to avoid security classification requirements, there is no information included herein on the technologies of stealth, super-cruise, thrust vectoring, advanced sensor fusion, and helmet mounted display systems that are embodied in various fifth generation jet powered fighters of the United States or other countries. Suffice it to say that the YF-16 Lightweight Fighter was the first jet fighter to fly with an all fly-by-wire flight control system and proved its advanced capabilities in flight control responsiveness for fighter aircraft agility and improved survivability when the aircraft is potentially subjected to battle damage. Flight proven fly-by-wire flight control technology is incorporated from the fourth generation F-16 fighter aircraft into integration within most fifth-generation jet propelled fighter aircraft.

For those readers, especially those newcomers to the world of aerospace engineering who wish more details about the technological history of manned atmospheric flight, The Author most strongly recommends that a book recently published *"The Airplane: A History Of Its Technology"* by the eminent aviation historian John D. Anderson, Jr. be read in its entirety. Dr. Anderson is a Curator with the National Air and Space Museum at the Smithsonian Institute and Professor Emeritus with the Department of Aerospace Engineering at the University of Maryland. His literary award-winning book is published through the auspices of the American Institute of Aeronautics and Astronautics in which he is an Honorary Fellow. [The Author is an Associate Fellow Emeritus in the AIAA.] In parallel, the internet contains a more extensive historical electronically-accessible database on military aviation technology than can be shown on paper in any book. It is from all of that military aviation history as compiled so magnificently by Dr. John Anderson and augmented with government documents used in compiling the internet database that describes the technology development paths of the first four generations of jet powered military fighter aircraft. This book selects only example technologies from each of those first four generations of US military fighter aircraft that shaped the technology database of the USAF Lightweight Fighters – the General Dynamics YF-16 and the Northrop YF-17.

A primary purpose of this book is to describe that, as in any technology development, it is the personalities who pursue the idea for advancement of an existing technology or the creation of a radically new application for a set of emerging technologies. The major technologies of military fighter aircraft are shown to also have advanced as a result of needs and capabilities to become effective against military enemy threats in the air and on the ground as demonstrated in the several wars of the past 100 years. Interwoven with the historical geopolitical events and inspired technology developments in the fascinating and "unorthodox" development process is the creative thinking "outside the box" done by a small group of professionals whose scientific and analytical DNA were merged into developing the USAF Lightweight Fighters.

The contents of this book place emphasis on the "genealogy" of the technologies that were selected for inclusion in the initial designs for the Lightweight Fighter configurations as "advanced technology demonstrator" aircraft. Accordingly, much emphasis was placed by both General Dynamics and Northrop on those technologies improving the state of the art in maneuverability of the fighter aircraft and "desirements" (i.e., desired capabilities rather than contractually specified required capabilities) which a few different organizations within the USAF had expressed to the aircraft industry during many USAF IR&D (Independent Research and Development) programs. Northrop focused their efforts on a fighter that would replace their F-5 series of fighter aircraft, noted for being small (light weight), low-cost, and for being a contender for the Foreign Military Sales acquisitions by other free-world nations. General Dynamics read the tea leaves slightly differently than did Northrop and focused their IR&D activities on the several technologies in the areas of configuration aerodynamics and flight control systems that were emerging from the technology developments fostered by the USAF laboratories at Wright-Patterson AFB. That database of information was augmented with cooperative wind tunnel tests and data analysis provided by the wind tunnel test experts at the several NASA research centers (at Langley Research Center, Virginia; at Ames Research Center, Mountain View, California; and at Lewis Research Center, Cleveland, Ohio.)

Both General Dynamics and Northrop had also applied Lessons Learned from air combat and ground support activities in the Korean War and the Vietnam War against the principal threat aircraft – those created by the Soviet Air Force for its inventory as well as those of Communist bloc countries aligned with the Soviet political and military

"Cold War" objectives. Those Lessons Learned all pointed to the need for future fighter aircraft to have a high degree of Survivability and low Vulnerability built into the design from the start – not added as later upgrades (as was done for the F-4s being lost at an alarming rate to ground gunfire in Vietnam). The same need was foreseen in the areas of Reliability, Maintainability, Reparability, and Serviceability to maintain the inventory of fighter aircraft at a high state of readiness and availability when based at far-flung operational bases. Those capabilities also must be included as initial design parameters, and not relied upon for being accomplished as upgrades.

In Appendix 1 at the end of this book can be found the biographies and obituaries of two key individuals without whom the USAF Lightweight Fighters may not have proceeded beyond being another interesting flight demonstration of advanced flight technologies embodied in a prototype platform. Only after a weapon system has entered the operational inventory and performed well for a few years in hazardous operational environments do the talents of the initial team developing the weapon system from its inception become recognized and commemorated. Aspiring young engineers reading those biographies should note that no one career path is the same as another. They should also note that when those several diverse careers are woven into a team effort, it is that team effort that determines the degree of success of a new aircraft design having outstanding performance characteristics.

There have been several outstanding books and numerous aviation magazine articles written during the past 40 years describing the successful prototype approach taken during the Lightweight Fighter phase of the F-16 Program, and would make this book seem to be redundant. However, those books were written by authors referencing official program documents in USAF reports to Congress which typical show only achieved progress milestones and schedules. This book is written by the USAF Chief System Engineer who "was there and did that" every step of the way for 6 fascinating years in the Lightweight Fighter and F-16 Air Combat Fighter Program Offices. While the role of the USAF Chief System Engineer is not as vital to the creation of an advanced technology demonstrator version of a weapons system concept, the responsibilities for transitioning that advanced technology demonstrator prototype into a fully operational capable weapon system is a huge challenge and responsibility given to the entire development team. A number of unusual problems surfaced during the USAF Lightweight Fighter Program, and many of those more significant problems and resolutions are described herein to portray the types of unusual circumstances that arise when integrating a set of new technological advances into any new aircraft design.

It should be noted to newcomers to technology development activities that it is of paramount importance when solving a technological problem that it is necessary to not only include the dominant independent and dependent parameters within that technology, but important to also consider the interfaces (physical and functional) of that technology with adjacent technologies and design features. Another most important factor to be considered is the exposure of the technology to very harsh operational environments much different from that in controlled laboratory environments. Woven into the technical and engineering aspects of the USAF Lightweight Fighters design is the human influence on requirements definition, setting of goals and schedules, management approaches, creating sources for funding, and a multitude of considerations inherently involved in every new design development.

Every new flight system development begins with "baselining" the state of the art in the technologies that exist at the start of the new development and that baseline is heavily influenced by the experience with the most recent preceding aircraft used extensively in a preceding war. The baseline of the technologies described herein are understandably not taught in significant depth and detail in engineering colleges and universities, and are included herein primarily to educate and inform those who surely will be confronted with similar surprises and challenging problems in their new career paths for advancing flight technologies of the next generations of turbojet-propelled military aircraft.

It is commonly accepted that the next war will not be fought just like the previous war. It is the wisdom and hindsight/foresight of program managers, requirements definition personnel, technology creators and developers, creative designers, and engineering personnel acting as a team in developing the next weapon system for that next unknown war. Teamwork is hard to define, but it is much like the water in a mix of concrete. It is water which produces the strength of the mix – without it the mix crumbles. Also commonly accepted is that there is no crystal ball which will prognosticate where, when, and how that next war will be fought. Based on the history of fighter aircraft aviation, the only assurance is that regrettably there will be a next war, and the losses of life and equipment on both sides are inevitable. In general, the winner of a war is that country which suffers the least losses of human lives and war equipment.

The readers of this book should bear in mind that when John Boyd created the concept of a multi-mode lightweight low-cost fighter in the early 1960s, he had no information predicting that there would be a war in Bosnia, or Iraq, or Afghanistan. It was the foresight of the General Dynamics Chief Designer Harry Hillaker who developed the set of advanced technologies (that were emerging from USAF Research Laboratories based on needs and changes in capabilities derived from battle damage assessments of the F-4 Phantom II in the Vietnam War) to be applied to the GD Model 401 design identified later as the YF-16 in the USAF Lightweight Fighter Program. The prime example of Harry Hillaker's foresight was in his selection of the Survivable Flight Control System technology developed to reduce the losses of fighter aircraft to enemy ground gunfire. Those survivability needs led to implementation of the YF-16 all-electric flight control system which became more commonly known as the fly-by-wire system. Fly-by-wire dictated the use of a central flight control computer with quadraplex (4) channels for redundancy (i.e., Survivability) – and that in turn provided the flexibility for integrating flight control algorithms with advanced targeting and weapons delivery controls and displays. More details of this appear later in this book and are surfaced here summarily to provide readers with the need to recognize that there is much need to "think outside the box" when selecting a technology for one purpose and finding it to have beneficial (synergistic) effects when combined/integrated with other technologies in the same flight platform. Conversely, there must be constant analysis and testing done to assure that incorporation of one technology does not degrade or nullify the benefits of an allied technology on the same platform.

As the chapters to be read hereafter describe, development of a typical new technology goes through phases of development before being integrated with allied technologies into the system design. The readers of this book who have little interest in this process should optionally skip to the last page of this Preface.

This was the same general process which was heuristically developed by the Wright brothers for their Wright Flyer Prototype in the early 1900s and which has been used as a management template in almost every aircraft-related technology development up through that of the F-16 Fighting Falcon Weapon System.

1) The initial phase consists of conducting a survey of that technology domain to ascertain whether it is a technology development with merely an academic-driven basis (the hardest kind to find sources of funding to continue), or to determine if there is a potential high payoff justifying taking the risk for that technology to be applied to a much-needed system (which typically enjoy high priority and receive adequate funding for development).

2) The next logical phase is to analyze the past efforts, if any, in the development of that technology to ascertain the development status of the technology as well as to apply the Lessons Learned from previous development failures experienced with earlier attempts in its development. This phase also includes planning of laboratory testing to establishing the baseline status of the potential new technology or improvements to a previously existing but not fully developed technology. This phase is commonly referred to as the Baselining Phase, defining what is essentially the starting point (often referred to as "a scratch in the mud") from which the technology to be developed progresses.

3) The third phase is to conduct exploratory laboratory tests and analytical studies leading to development of a more comprehensive ground test and analysis plan for obtaining a quantified initial estimate of the relative importance of the independent and dependent variables associated with the technology. It is common to test the unknowns ("Unks") in the technology, but it is also common to encounter unknown unknowns ("Unk Unk"s) inherent with the new technology. Neither Unks nor Unk Unks are ever to be ignored.

4) The ground testing phase (wind tunnel tests, structural tests, electrical and electronics functional tests, software validation tests, etc.) is conducted next to define the envelope within which the technology is capable of operating and to identify those regions where the potential technology is either highly advantageous or disadvantageous to have when applied to a system prototype design;

5) The flight test phase of a prototype (if schedule and funding permits) is the preferred next phase for flight demonstrating that technology (or set of allied technologies) with the objective much like that of earlier ground tests. That objective is to define the envelope limits and constraints (if any) for applying that technology, and to verify with flight data the degree to which that technology has on the advantages and disadvantages when integrated as a subsystem into a flight demonstrator or prototype.

6) Regardless of whether all or only some of the previous phases are conducted, the definition of the risk status and payoff potential of the technology must be thoroughly documented and analyzed to include identification of residual unknowns, risk potential, payoff potential, and with cost estimates as realistically as can be forecasted. These results are then presented as options to the design team management and program management for a) applying and integrating that technology with other allied technologies, or b) deferring that technology to a future system design, or c) at worst, abandoning further development and consideration of that technology;

7) The preferable last technical phase is to conduct flight testing of a pre-production prototype platform, with allied technologies integrated into the platform as a system, with a goal of verifying to what degree the estimated full payoff of that technology development activity has been achieved. Often an isolated technology will have been demonstrated fully in ground testing, but when integrated with other allied technologies in the demonstration platform may net an exceedance of the sum of the benefits of the individual technologies (i.e., a synergism). Similarly, there may be realized a net reduction in the sum of the allied technologies when integrated and evaluated as a total system (i.e., a detriment).

To determine when each phase has been adequately accomplished for readiness to proceed with the next phase, it is common practice to establish a Progress Milestone for the completion of each phase. A Progress Milestone differs from a Schedule Milestone in that a Progress Milestone is no defined to be completed by a specific calendar date, but instead is a milestone indicating accomplishment of a successful test regardless of the date of conduct of the test. The completion criteria for each Progress Milestone should be complete, be realistic, and be well defined. In practice, it is common for the Procuring Agent and the Contract Performing Agent to use those Progress Milestones also as a basis for incremental payment of funds for the contract effort, such that the Progress Milestones are important indicators of the technical progress, the budgetary/expenditures aspects of the contract, as well as for overall program progress when the status of the program is reported to the Contracting Program Office.

When presented with only an artist's rendition or a photograph of a test model but not a photograph of the end product incorporating that technology, viewers should not assume that the technology development process as outlined above has been developed to a totally successful degree of completeness by that design team. The urgency of a technology improvement needs and a set of design requirements obtained as feedback from dynamically changing event(s), such as those that are frequently identified from operational combat battles and/or foreign technology updates, may require modification in planning. Unfortunately, some steps in the process may be short-cut, and the schedule and costs for some steps in the process to be accelerated, altered, or perhaps some steps may be deferred because of program urgency requirements and/or funding availability. That is when technology developers do the best that can be done with available resources – leading to the development team having to use much resourcefulness". The Contracting Program Office must always be kept informed and in the active loop when planning and schedules/costs changes are being considered.

For those Readers who are planning to be employed as engineering interns or junior engineers and to be assigned to an on-going technology or system of technologies development activity, their value to the program will be determined by the timing of their employment relative to which phase identified above and at a time relative to a completed Progress Milestone. Regardless of which phase exists or is underway at the time of his/her employment, the new employee (engineer or technician) is well advised to rapidly become familiar as to what is the status (i.e., what has already been accomplished) in that technology or system of technologies.

A diagram (commonly referred to as a "Learning Curve") will often be generated to graphically orient those readers who may be interested as to how Program Management tracks progress as each phase of the program unfolds. This type of diagram, supplemented with dozens of other management plans and schedule, is commonly kept in a Program Control Room and kept up to date with periodic (bi-weekly or monthly) updates. Milestones which are slipping are marked in red. An actual achieved Milestone is marked in green on the chart when the Phase is completed. The Program Management at both the Performing Contractor facility and at the Contracting Program Office can better visualize how each of the major elements of the Program is progressing in their development according to the planned schedule. When "hot spots" are identified they are typically given closer attention, especially if it requires additional tests of some type to address an unknown (an "Unk") which has surfaced unexpectedly during a test. The more troublesome type surprises to accommodate are the emergence of unknown unknowns (the "Unk Unks") for which planning schedules and funding changes must be adjusted in a logical and affordable work-around manner.

An example of what a typical Progress Tracking and Planning Chart looks like in the technology development and program management process appears next:

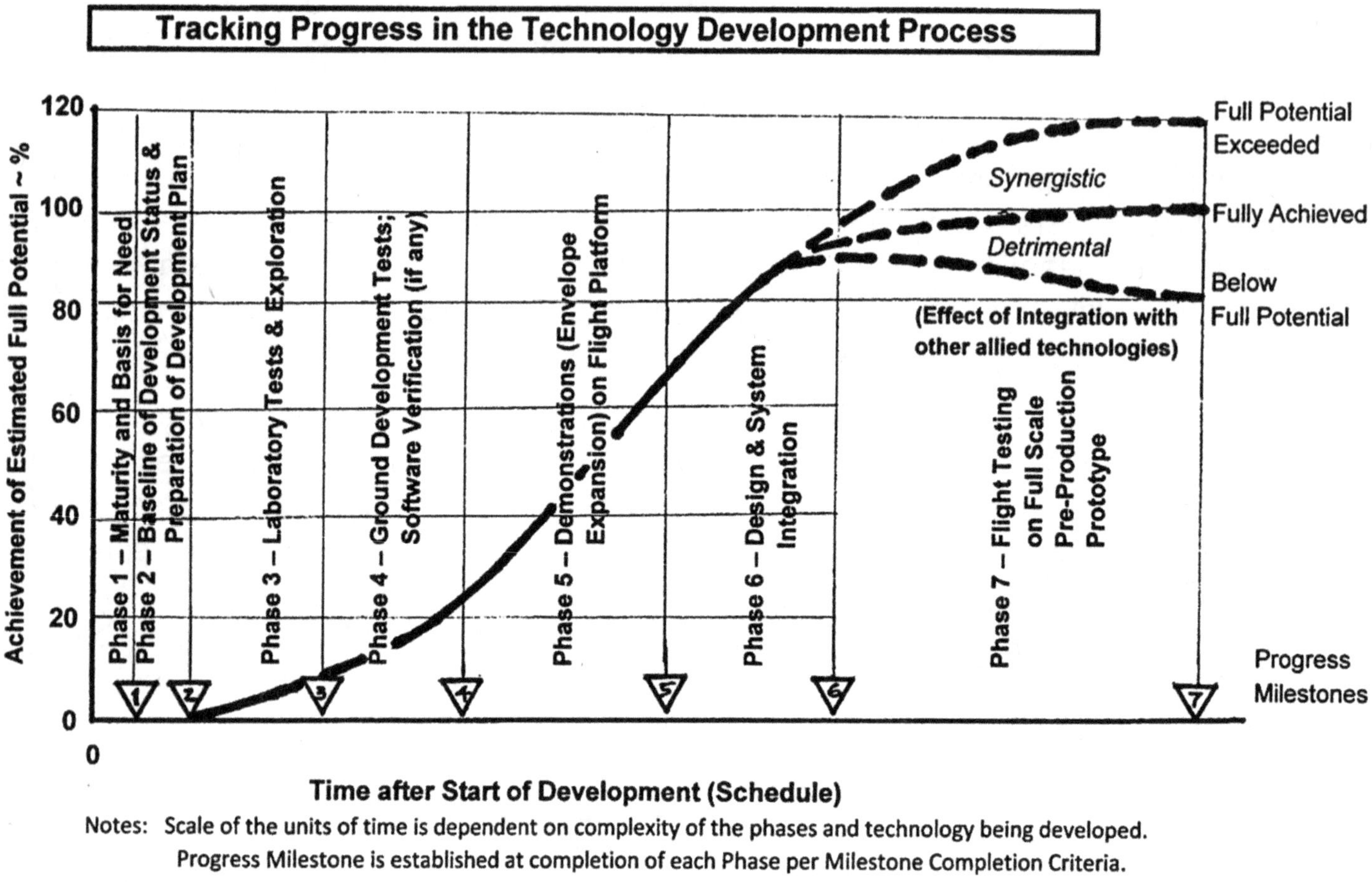

Tracking Progress in Technology Development (Example)
(Source: The Author)

As shown in the diagram above, not until the latter part of Phase 6 when the technology is detail designed and integrated with allied technologies are design conflicts identified. Not until Phase 7 Flight Tests on a pre-production prototype will those conflicts be quantified with actual flight test data. On the positive side, there may be synergistic results identified during the design integration that also are quantified by flight test data on the pre-production prototype. Seldom does it occur that there are neither detrimental nor synergistic effects emerging during the design integration and flight testing.

Having previewed as to what is, and what is not, in this book and why, it is now time to read a book which describes the technology advancements achieved and the unusual events that took place in the development of many exceptionally well-designed aircraft and weapon systems responding to major events that have occurred in the intriguing history of military aviation since the Wright brothers. Not all weapons systems programs meet with success when integrated with other aircraft in the same operational inventory and are "sent back to the drawing board". The experience base and pattern of thought used by a few highly creative and analytical individuals are cited in this book as creating major new technology developments and/or its application to a new aircraft design that had significant high pay-off despite high risks taken during their development.

Readers should note that there is much photographic content in this book. Anyone developing a new technology should photographically record test set-ups, instrumentation type and installations, models (of all scales) being tested, and test facilities being used such that when the inevitable surprise in the test data recorded is revealed, or failure of a component of the test model should happen, it can be more rapidly investigated to determine the root cause. Numerous photographs, augmented by other visual exhibits, will be used to not only visually describe the progress made in the development of the USAF Lightweight Fighters but to also visually describe the actual problems encountered and surprise results of tests conducted while developing its capabilities.

The first chapters of this book (Chapter 1 through Chapter 4) describe fighter aircraft development history – from the Wright brothers' achievements in aircraft flight technologies, World War I, World War II, the onset of the "Cold War" with the Soviet Union, the Korean War 1950-1953, and through the end of the 1950s decade. Those Readers who are familiar with that aviation history and wish not to repeat that history should feel free to skip over that history in Chapters 1, 2, 3, and 4 and should start reading at Chapter 5 to be exposed to those few key events and identification of key personnel which combined and created the concept of the Lightweight Fighter for the USAF. Chapters 5 through 7 are dedicated to describing in substantial detail how the multi-mode General Dynamics YF-16 and Northrop YF-17 Lightweight Fighters came to be, but almost did not (for non-technological reasons).

Appendix 1 at the end of this book describes the biographies (and obituaries) for two key individuals credited with creating the USAF Lightweight Fighters. This book has been dedicated to them for their professionalism and loyalty to their country without seeking or expecting any large monetary reward for their dedicated efforts. Readers who are newly embarking on a career in technology development, and especially a development activity in which other professionals of varying levels of experience and specialties are teamed in developing an entirely new, or major modification of an existing, military weapon system, will gain much from observing how the team works together. As Col. John Boyd's career was entirely within the USAF and Harry Hillaker's career was entirely within the aircraft design, test, and manufacturing environment of Government Contractors, their obituaries describe what are the variety of assignments and challenges as well as opportunities that will be presented during the career of a professional engineer within Government or private industry employment.

It would be difficult to portray this book with a one- or two-word descriptor such as a "documentary", a "non-fictional novel", a "technical report", or a "historical summary" – it is a mélange of all of those and with a few Author's Notes injected randomly to clarify an event or to tell a humorous anecdote about the personality of a person. All of the personalities mentioned in this book are no longer alive to challenge those quotes, but since the quotes are all of a complimentary nature The Author felt they would not object.

The Author chose his career to be half within the employ of large airframe manufacturing companies (designing and/or testing aircraft to meet Government contract requirements), and half within the employ of the US Air Force as a civilian engineer – in Federal Civil Service. Having served in the US Army in the Korean War time period added a third dimension to The Author's background in gaining a hands-on appreciation of how and why the military services generate the need for an item in their weapons inventories.

The developments and technology advances described in this book occurred mostly before The Author became an active participant in those earlier developments, although some of the events such as the Japanese aircraft attacking the US Naval warships at Pearl Harbor on December 7, 1941, are still stored in the memory of a pre-teenager forming a calibration of what the World is really like – not as ideally as taught in elementary education textbooks. The contents in the early chapters (1-4) of this book are mostly obtained from historical records in military aviation history, and are as interpreted in perspective of their influence on technologies emerging as "Lessons Learned" from previous wars.

The contents in the latter chapters (5-9) of this book were obtained primarily from the two contractors – General Dynamics and Northrop – providing material to the USAF Lightweight Fighter Program files. That material was supplemented with personal memorabilia from the Author's Pack Rat file. Some material was taken from Government documents (technical reports, briefings, weekly status reports, etc.) generated either by the two Contractors and provided to the USAF Lightweight Fighter Program for tracking progress of design, analysis, or test activities, or generated by USAF personnel for presentations to be given by the USAF Lightweight Fighter Program to higher headquarters or to other USAF organizations (e.g., the USAF Laboratories. Within the USAF Research Laboratories, flight technologies are developed to maintain a constant awareness of how foreign and domestic technology developments are progressing. All such references and all photographs, documents, charts, diagrams, etc. have been declassified (if originally classified), have been Approved for Public Release, or are in the photographic files of the military services on the internet and released under the provisions of the Freedom of Information Act FOIA.

This book is intended to be only an educational, informational transfer of material and information describing the status of technologies of military fighter aircraft design, as well as to provide career guidance based on actual highly successful careers. The Disclaimer at the front of this book should be read carefully before any information is extracted from this not-for-personal-profit book.

CHAPTER 1
PRE-WW I AVIATION TECHNOLOGY DEVELOPMENTS (1900 TO 1935)

For Readers to better understand and more fully appreciate what actually takes place in the development of any new aircraft, the "genealogy" (documented history) of the technologies being considered for application to and integration in the new aircraft development should be very thoroughly researched. The fewer the "Unks" and the more the knowns are discovered, the prospects are increased for developing the new technology. Unlike the genealogies of humans however, it is not a backward tracing process starting with the birth parents of the individual being traced and determining from Public Record documents their parents, grandparents, great grandparents, and so forth until the document trail can be traced no further. Technology "genealogy" is better described by identifying the original idea for the technology as a concept and then describing the events and personalities involved in developing that technology from its birth as a concept and then describing that concept as it matures into hardware (and software) in practical, useful applications.

This book has the ultimate goal of showing in chronological order the evolution of the multiple technologies that were developed individually and then merged later into development of the USAF Lightweight Fighters.

Initially, it was considered to start this book at the point where jet engines had just emerged in the aviation technology data base; however, it was then felt that Readers might derive additional benefit by starting this book at an even earlier period in time (i.e., for the ancestral "roots" of the technologies that are in the USAF Lightweight Fighter "genealogical" history). It was then decided that a better starting point for this book would be two decades prior to the start of the 20th century. It was in the 1880s that other scientists and inventors were performing scientifically credible experiments to "learn how to fly like the birds". It is from that foundation in knowledge and experiences that the Wright brothers, Orville and Wilbur, were inspired to conduct their own exploratory tests with glider aircraft to understand how birds were able to control their flight path to get from Point A to a desired Point B as birds do in search of life-sustaining food. By necessity, this book starts at that point in time when early rockets and hot-air balloons first started to evolve in Man's understanding for how physical masses move in the 3-

dimensional atmosphere above Earth's surface between Point A and Point B. The long-standing challenge was to understand and replicate the way uneducated birds were routinely doing so and in a fully controlled manner (except in gale force winds through which no animal or human should dare to fly).

As mentioned in the Preface, the book *"The Airplane: A History of its Technology"* by aviation historian Dr. John D. Anderson, a Curator at the Smithsonian National Air and Space Museum, describes in detail what its title advertises – the history of airplane technologies. All Readers newly embarking on a career in any branch of aviation should take the time to read Dr. Anderson's book. The Anderson book is replete with many diagrams and photographs describing the earliest days of aviation technologies. So as to not get entangled in copyright issues, none of Dr. Anderson's book is copied (a few sentences are paraphrased in the text of this book) and graphical diagrams and photographs from Government documents and records on file at the National Archives and/or Smithsonian Institution National Aviation and Space Museum are used instead.

Although many Readers may perceive in error otherwise, the Wright brothers were *not* the first humans to lift themselves off the ground in a "flying machine" and travel from Point A to Point B as that had been accomplished by many before in hot-air balloons and gliders – and also in powered aero planes. However, in the case of hot-air balloons and gliders there was total reliance on the forces of Nature – wind near the ground and wind currents aloft – to determine the direction in which the balloon or glider would travel from Point A to an unplanned Point B.

As Dr. Anderson describes, early attempts in Europe to fly in self-powered airplanes, a decade or two before the Wright brothers first adequately controlled flight in 1903, were disputed as being "successful" but were considered to be merely "powered leaps" and with no very large distance between the start and end point of the leap. Most of these attempts to "fly like the birds" are kindly referred to as being "cut and try" experiments. The scientific principles of aerodynamics, efficient light-weight propulsion, structural construction materials that were strong and yet lightweight, means for controlling the "flying machine", were all in various stages of knowledge and technology being considered for application.

It is well recognized that except for "controlling the flight path" of the flying machine, most of the other elements crucial to flight were undergoing development by several scientists and theoreticians and mostly in European countries – principally in Germany, France, and Great Britain. Two of those were names that are not recognized as being as vital as they actually were in aviation development as follows:

The first deserving recognition is Ernst Mach of Austria who earned his doctorate in Physics in 1860. While as a Professor of Experimental Physics at the Charles University in Prague in 1887, Dr. Mach established many of the basic scientific principles of supersonic aerodynamic flow of air. It is from this work that the velocity of an object moving through air when expressed in a ratio to the velocity of sound at those air temperature and pressure density conditions at which the object was moving could be expressed as the Mach ratio (later in common use to be known as the "Mach number"). These foundations in supersonic flow of air and the formation of shock waves gave rise to the entire technology area of transonic and supersonic, compressible aerodynamics. The work by Mach also spawned the generation of ground test facilities, most notably transonic and supersonic wind tunnels, in use in every developed country having aircraft in its commercial transportation and /or military aviation inventories. Ground test facilities are used to prove that the flight vehicle is capable of safe, controlled flight from Mach 0.1 to Mach 3 and beyond.

The second prominent name in the development of aviation technologies in the 1880s time period is that of Otto Lilienthal in Germany. Otto and his 1-year younger brother Gustav, were both born in the mid-1800s in Prussia (Germany). Unlike Dr. Ernst Mach, the Lilienthal brothers were not as highly educated and were not theoreticians – they were experimentalists with an extraordinary gift of organization and persistence. They both, especially Otto, believed and dedicated their lives to proving that if uneducated birds could fly with complete control of their flight path, there was no reason why humans also could not learn from the Aves (birds) how to control Man-made devices/machines that also could fly with adequate control of their flight path. In 1889, Otto Lilienthal published the book "Bird flight as the Basis of Aviation"

The Lilienthal brothers conducted their systematic experiments with unpowered gliders and which were manned mostly by Otto Lilienthal. The wings of the glider were strapped to Otto's shoulders and his glider was launched from atop a 50 feet high conical shaped hill known as the "Fliegeberg" – translated into English as the "Flight Hill" – which the Lilienthal brothers constructed as their "test facility". As gliders rely on wind currents to gain relative wind over the wings which creates lift, and inasmuch as wind currents shift direction devilishly frequent and suddenly, Otto could always launch his manned glider from the top of conical Fliegeberg headed into the wind. Otto Lilienthal had also developed the technique of shifting his legs and lower body as a means of controlling the center of gravity of the "flight system" relative to the center of lift of the wings strapped to his shoulders. A restored Lilienthal Glider (on display at the Smithsonian National Air and Space Museum) is shown next:

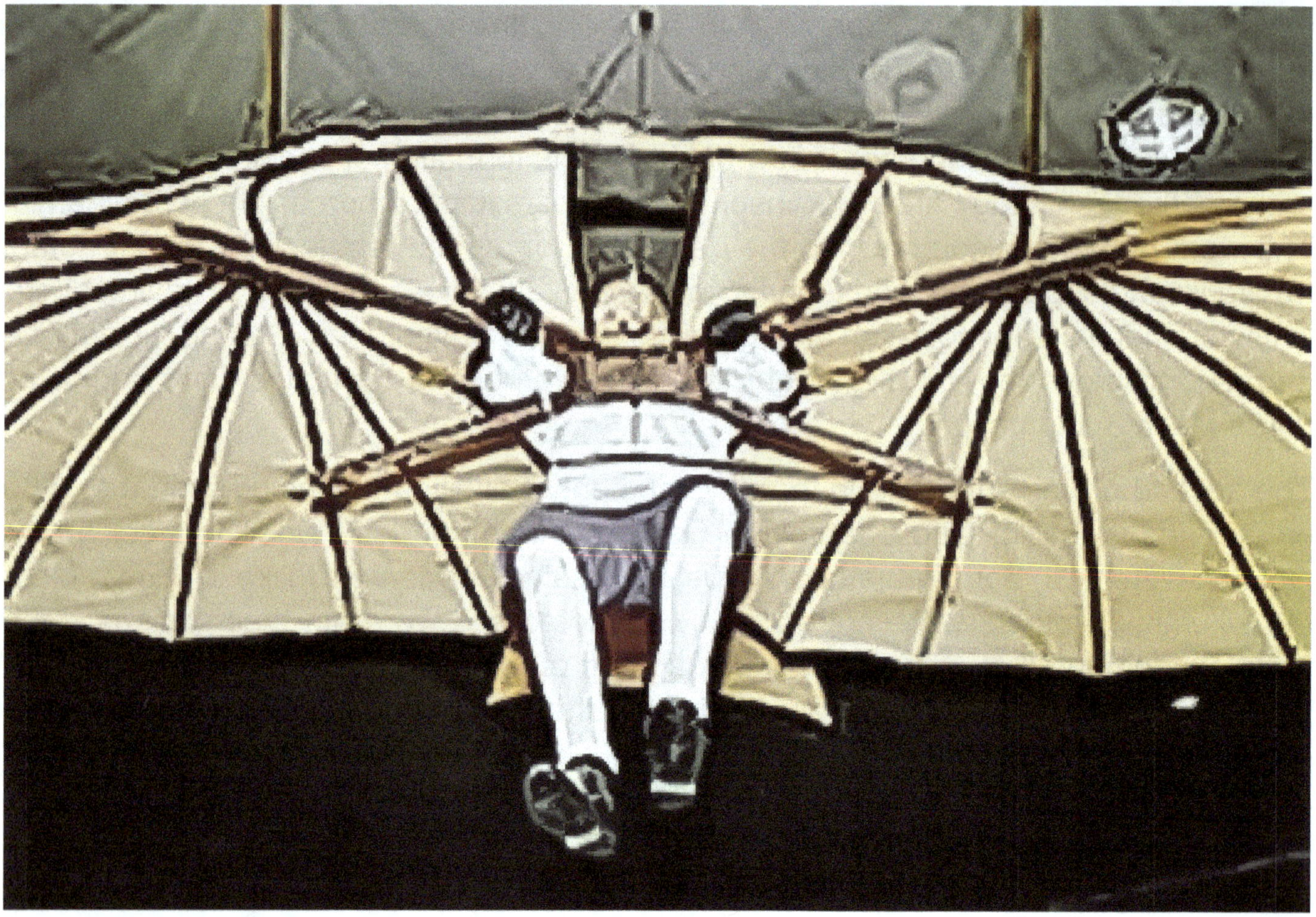

Restored Lilienthal Glider (1 of 5) at the National Air and Space Museum, Washington DC (Photo taken by the Author)

In modern “hang-gliding”, “pilots” grasp a bar and can move their entire body mass relative to the glider (parasail) wings.

On August 9, 1896, strong wind gust caused Otto Lilienthal to lose control of his manned glider (attributed to his inability to shift enough of his body mass to correct for the sudden change in longitudinal stability – “negative static margin”) and he suffered major injury to cervical vertebrae. As there was no medical treatment developed for cervical vertebrae injuries at that point in medicine, Otto Lilienthal died the next day August 10, 1896 at the age of only 48 years. His last words were to his brother Gustav (in German) “*Opfer müssen gebracht werden*!” which translates in English to: “Sacrifices must be made!”

[Author’s note to Readers: On occasion the developer(s) of a technology may sometimes, knowingly or unknowingly, disregard the hazards to his/her health which are suspected of being inherent in development of a technology feature for which little or no precedence exists. Probably the most famous instance of that is that of Madame Marie Curie who developed the element radium. She handled the radium without adequate hand protection and without being aware that she would suffer severe radiation burns to her hands. Similarly, Otto Lilienthal risked the hazard of a glider crashing and injuring him fatally as it did. In more modern time, test pilots of newly developed aircraft accept the risk of crash of a new airplane being tested in order that whatever was determined as the "root cause" of the crash could be avoided when the airplane enters operational inventory and is flown by lesser qualified and experienced pilots. This book is inherently dedicated to all flight test pilots who heroically sacrificed their lives in order to protect others – that is the definition of a hero.]

The contributions Otto Lilienthal made to aviation technology were never fully successful, but the brilliant work of Otto (and Gustav) Lilienthal inspired the Wright brothers to develop their flying machine with the ability to adequately control the flight path of their Flying Machine which successfully flew on December 17, 1903, seven years after Otto Lilienthal's tragic death. Lilienthal was fatally injured while exploring the variables permitting manned flight in a heavier-than-air machine.

In 1912 at a ceremony honoring the Wright brothers' first successful flight, Wilbur Wright most graciously attributed Otto Lilienthal as having been a major inspiration for their (Wright brothers) pursuit of manned flight. Wilbur is quoted as having stated: *"Of all the men who attacked the flying problem in the 19th century, Otto Lilienthal was easily the most important. It is true that attempts at gliding had been made hundreds of years before him, and that in the 19th century, Cayley, Spencer, Wenham, Mouillard, and others were reported to have made feeble attempts to glide, but their failures were so complete that nothing of (technological) value resulted."*

One of the more credible attempts to fly, from which the Wright brothers undoubtedly also drew many of their ideas for their own "Flying Machine" in the United States, was that of the Russian inventor Alexander Mozhaysky in the early 1880s. The Russian Government issued a postage stamp in 1974 commemorating Mozhaysky's aircraft design and an extracted copy of the design from the postage stamp (shown next) reveals the most interesting aircraft design. Fine details of the design are somewhat "fuzzy" as they become enlarged from the small postage stamp, but it is the overall configuration which is more important to note.

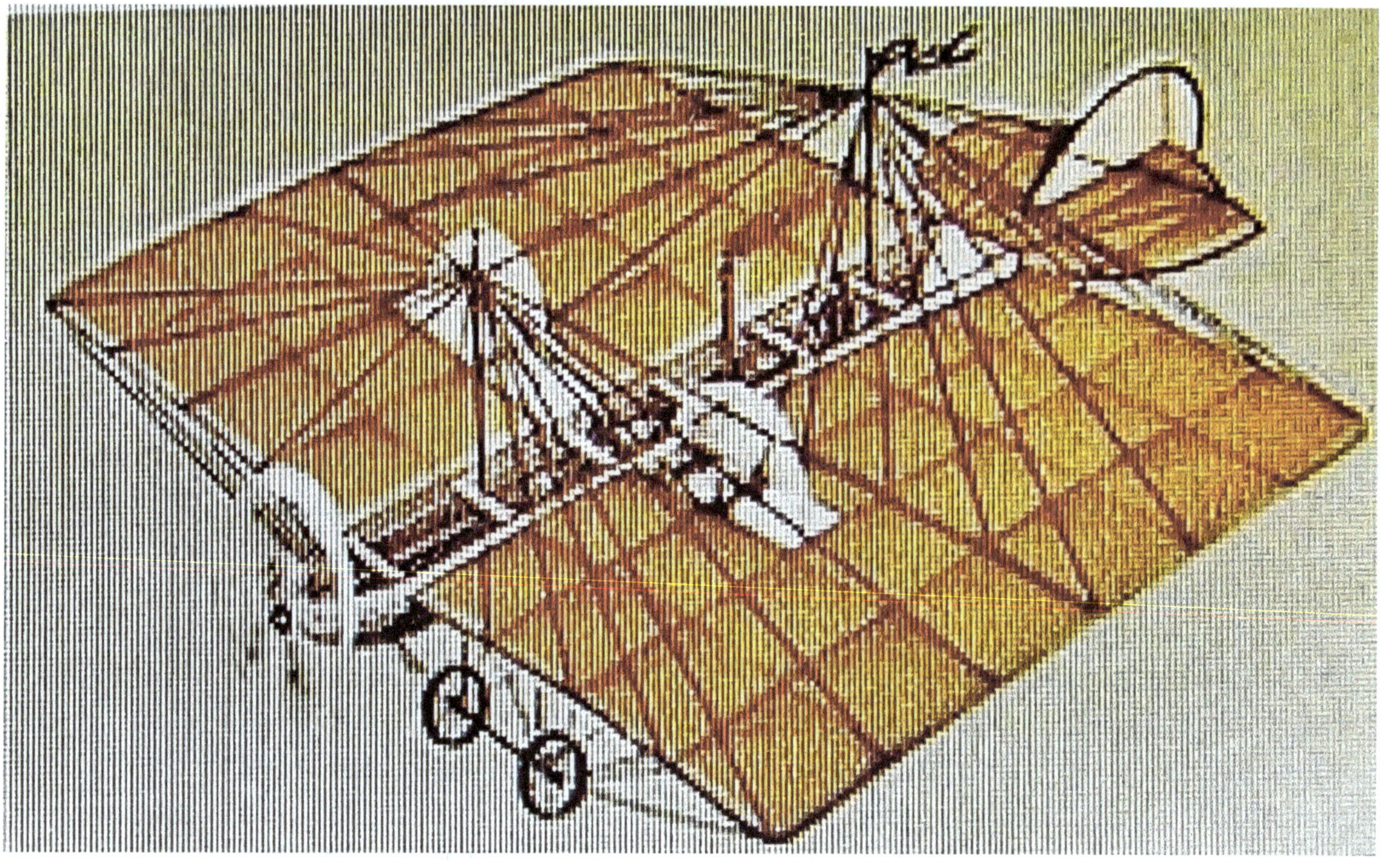

Alexander Mozhaysky aeroplane configuration, Russia (circa 1884)
(Source: Russian postage stamps issued circa 1974)

Shown are all of the major "subsystems" for the Mozhaysky configuration – the fuselage containing the propulsion system, the propelling force generated by a propeller (then referred to as an "air screw"), the large flat plate planar wing, the posts and support rods to the wing to maintain its structural shape when loaded with lift forces, the empennage with both vertical tail and horizontal tail surfaces, and a wheeled landing gear. The single pilot (shown in a white enclosure to the left of the engine in the fuselage) is positioned at about the center of gravity of the airplane "system". The pilot was connected by cables to the empennage to allow him to create angle of attack when at flight speed to have the large planar wing create flat plate lift, and with a small vertical tail to provide directional stability to the apparatus.

Conceptually, all of the elements necessary for flight were present in the design. Realistically, the source of power for propelling the airplane to gain forward speed was from a coal-burning steam engine. Seen in the small illustration in the fuselage just aft of the pilot is a smoke stack (a chimney pipe) with black smoke trailing from the top of the stack. The thrust to weight ratio for steam engines is abyssmal and not very well-suited for propelling an airplane to gain sufficient airspeed for lifting the massive weight of the airplane off the ground and for overcoming the aerodynamic drag associated with flat plates at angle of attack.

As extracted from Dr. Anderson's book: "*In 1884, with I. N. Golubev as the pilot, Mozhaiski's flying machine was launched down an inclined ramp at Krasnoye Selo, near St. Petersburg, Russia. Building momentum, it became airborne for a distance of about 80 feet".* In the Russian book "Meet Aerospace Vehicles" (by Mir Publishers, Moscow, 1974) is found the following statement: *"A. Mozharsky (sic) is rightfully claimed to have created the world's first aeroplane . . . This was the world's first successful flight of an aeroplane."* This claim is not accepted by most historians of aviation technology.

In his book *"The World's First Aeroplane Flights"* (1965), the dean of historians of aeronautics, C. H. Gibbs-Smith, states the following criteria used by aviation historians to judge a successful powered flight:

"In order to qualify for having made a simple powered and sustained flight, a conventional airplane should have sustained itself freely in a horizontal or rising flight path – without loss of airspeed – beyond a point where it could be influenced by any momentum built up before it left the ground: otherwise, its performance can only be related as a powered leap, i.e., it will not have made a fully self-propelled flight, but will only have followed a ballistic trajectory modified by the thrust of its propeller and by the aerodynamic forces acting upon its aero foils. Furthermore, it must be shown that the machine can be kept in satisfactory equilibrium. Simple sustained flight obviously need not include full controllability, but the maintenance of adequate equilibrium in flight is part and parcel of sustentation."

On this basis, Mozhaiski's flying machine accomplished the second powered hop in history; neither his machine nor that of Felix du Temple before him satisfied the described criteria for a successful piloted flight. After his attempt in 1884, Mozhaiski bowed out of the flying machine arena."

The first recognized successful flight (self-powered, manned, adequately controlled for significant distance) took place on December 17, 1903, at Kill Devil Hill near Kitty Hawk, North Carolina. The historic event was captured on film by a local Beach Guardsman to whom the Wright brothers had loaned their camera to document the successful flight that had eluded so many others prior to that date.

First Flight by Wright Brothers at Kitty Hawk, NC (December 17, 1903)
(Source: National Archives)

In the "genealogy" of aviation, this was the moment of birth for aviation in a self-propelled, manned, safely constructed airplane with an "adequately controlled" flight path over a significant distance at equal or greater height above the ground level launch point. It is seen that Orville Wright is in a prone position piloting (i.e., controlling) the flying machine and his brother Wilbur standing to the side in the event assistance was needed to keep the wings in a level orientation relative to the sand. The flight traveled a distance of 120 feet from the end of the launch rail to the point of touchdown of the skid and with a flight duration of 12 seconds. Ground speed therefore averaged about 10 feet per second, but taking into consideration the headwind (of about 35 feet per second (25 mph) and steady at the time of the flight), the average airspeed was about 45 feet per second or about 30 miles per hour.

A sharp-eyed Reader will observe that the forward canard surfaces on the Wright brothers "aero plane" are at a several degrees angle of attack (to lift the nose of the flying machine off the launch rail), and that the wing is not being warped (to command a correction in roll motion). At the first pilot debriefing in aviation history, pilot Orville Wright commented after the first flight: "I found the control of the front rudder quite difficult". At an airspeed of about 30 miles per hour, the aerodynamic responsiveness of control surfaces to pilot command would be small indeed. The brothers conducted three subsequent flights later that same day, and Wilbur flew the fourth and longest flight for a distance of 852 feet with a flight duration of 59 seconds (averaging 15 feet per sec ground speed and about 50 miles per hour airspeed). On the fifth flight attempt, with Orville Wright at the controls, the Wright Flying Machine crashed into the sands when hit by a sudden strong crosswind gust and the aircraft testing was terminated for the next few months.

There were three significant "Lessons Learned" by the Wright brothers from the glider technology advancements by Otto Lilienthal in Germany and from the clever external aircraft configuration work of Alexander Mozhaiski in Russia:

The first "Lesson Learned" by the Wright brothers was that from the tragic crash that killed their mentor, Otto Lilienthal. When conducting tests of a flying machine or glider, in which the loss of control for any reason has a high possibility of resulting in a crash, the tests would be safer if conducted over the soft sand of a beach or on sand dunes adjacent to many beaches.

The second "Lesson Learned" is shown by noting that there was no wheeled landing gear embodied in the first successful Wright brothers' flying machine. This was a very conscious decision the Wright brothers made as they realistically assessed that their test flights with gliders at Kitty Hawk could land in the soft sand with relatively minor and easily reparable damage. The possibility that the first flying machine would undergo pre-flight trials and would have landings in soft sand causing less extensive damage to the one-of-a-kind machine landing on a harder surface (including grass-covered compact soil) induced the Wright brothers to select a rail and catapult type system for their earliest manned flights. The wheel type landing gear system used by the Russian Mozhaiski in 1882 was considered by the Wright brothers to be an advancement in landing gear technology in their later series of flying machines in test flights above harder ground surfaces such as at the Huffman Prairie in Dayton, Ohio, near the Area A portion of Wright-Patterson Air Force Base.

The third, and far more technologically significant, "Lesson Learned" by the Wright brothers was that the approach of shifting the mass of the "pilot" to achieve control stability of the machine being flown was not a correct one. The tragic death of their mentor Otto Lilienthal had convinced them that the control must be aerodynamically achieved and by some manner of adjustment of an external configuration feature of the machine by the "pilot". Further, the Wright brothers reasoned that the external feature(s) to achieve control of the flight path must be in all 3 axes of freedom of flight.

Their reasoning led, after many wind tunnel tests in their own small wind tunnel, to the use of a wing-like structure being designed into the external configuration and adjusted by the "pilot" so as to produce a force and moment in each axis to control the flight path motion in that axis – and simultaneously in all 3 axes. The concept of having a vertical rudder plane such as used on boats to change direction of the boat in a "yaw" (horizontal) direction, and similarly turning the rudder 90^0 so as to change the "pitch" (vertical) direction of the airplane were both relatively straight-forward thinking. What type of surface could be used to generate a motion of the airplane to "roll" about its central axis remained as a challenge. So as not to build Readers' suspense, the concept of warping the wing to achieve control in the "roll" direction of motion was a fall-out from the Wright brothers testing of box shaped kites' years before in their development of their Flying

Machine". By warping the wing of their very flexible wing, which resembles the construction of a box kite, the control technology for the elusive third axis of controlled motion was introduced.

One might think that the Wright brothers would apply for a US Patent to protect their work in developing the entire set of components that composed their first successful aeroplane, but the Wright brothers were fully aware that their contribution to having advanced the technology of manned powered flight was primarily in the mechanism they had developed for control of in-flight motion of their otherwise rudimentary configured aeroplane.

The Attorney for the Wright brothers had indeed filed an Application for Patent to the US Patent and Trademark Office on March 23, 1903, approximately 9 months before their first successful flight, but it was only to cover the control system for maintaining positive flight control with the 3 primary aerodynamic surfaces – the wings (developing lift and control forces about the longitudinal x-axis to produce angular roll by warping the wings asymmetrically); the horizontal rudder or canard (developing forces to control the pitch angle relative to the lateral y-axis of the aero plane); and the vertical rudder (developing forces to control the direction of the aero plane in yaw – the angle relative to the vertical z-axis). All that remained was to size the vertical and horizontal rudders and how to achieve the warping of the wings with a system of cables all combined into a "control stick" manually manipulated by the pilot. Their Application for Patent No. 821,393 contained 5 figures and these are shown next exactly as presented in their patent application describing the mechanization of the first successful flight control subsystem.

FIG. 1.

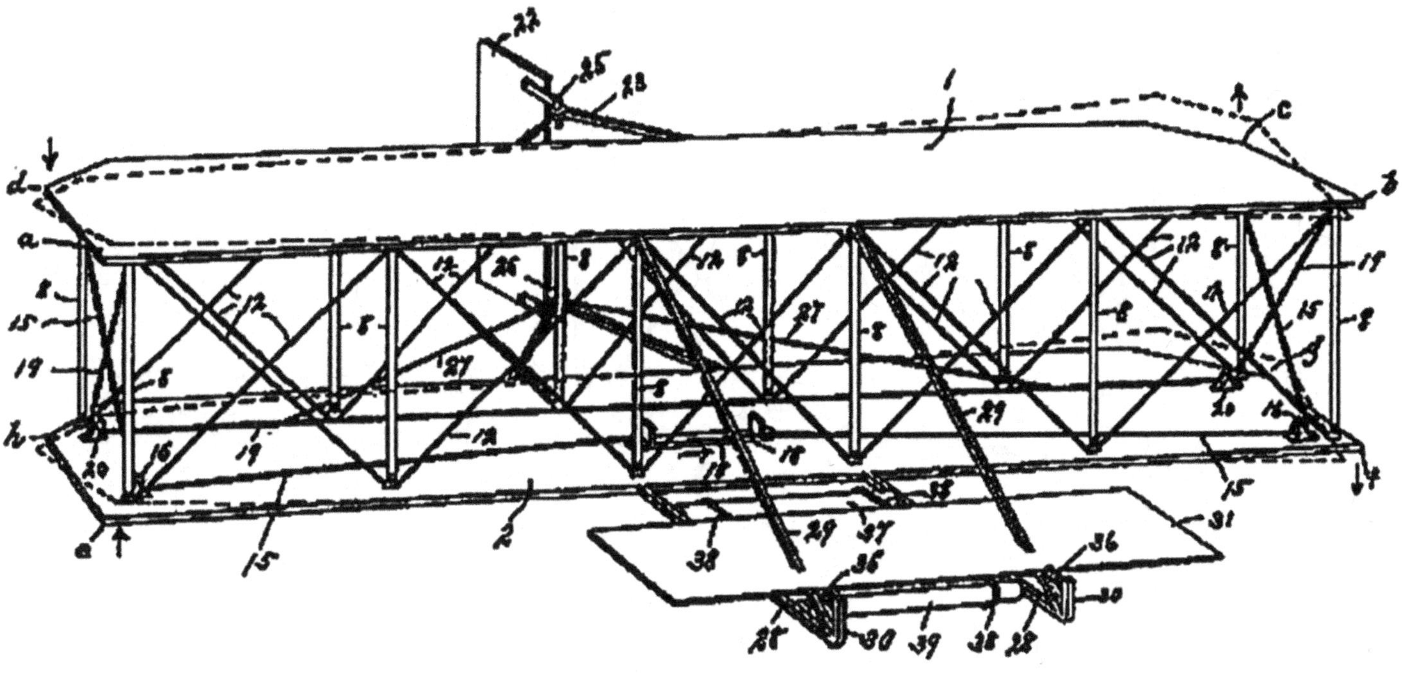

Fig 1 of Patent No. 821,393
(Source: USPTO)

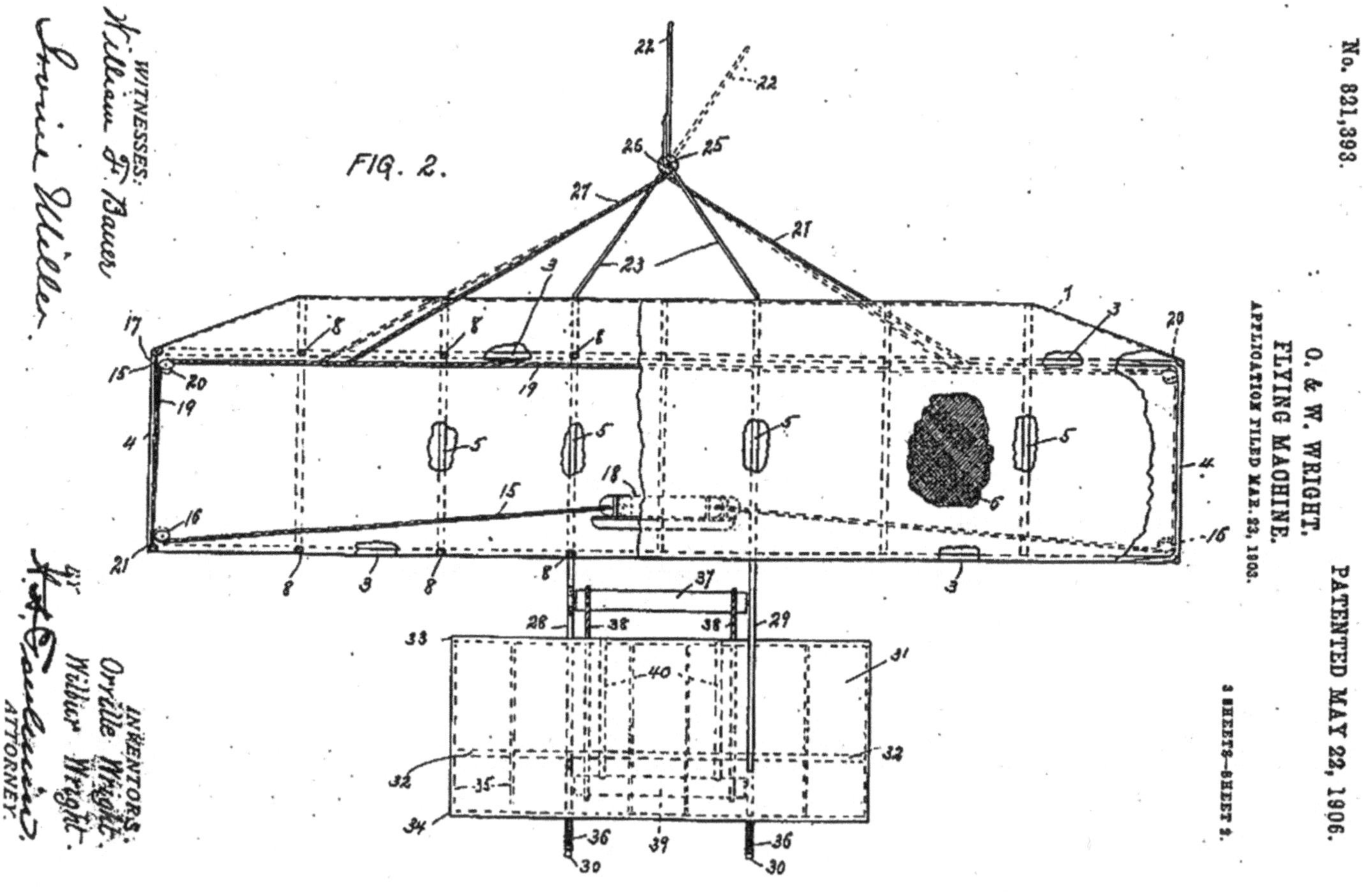

Fig. 2 of Patent No. 821,393
(Source: USPTO)

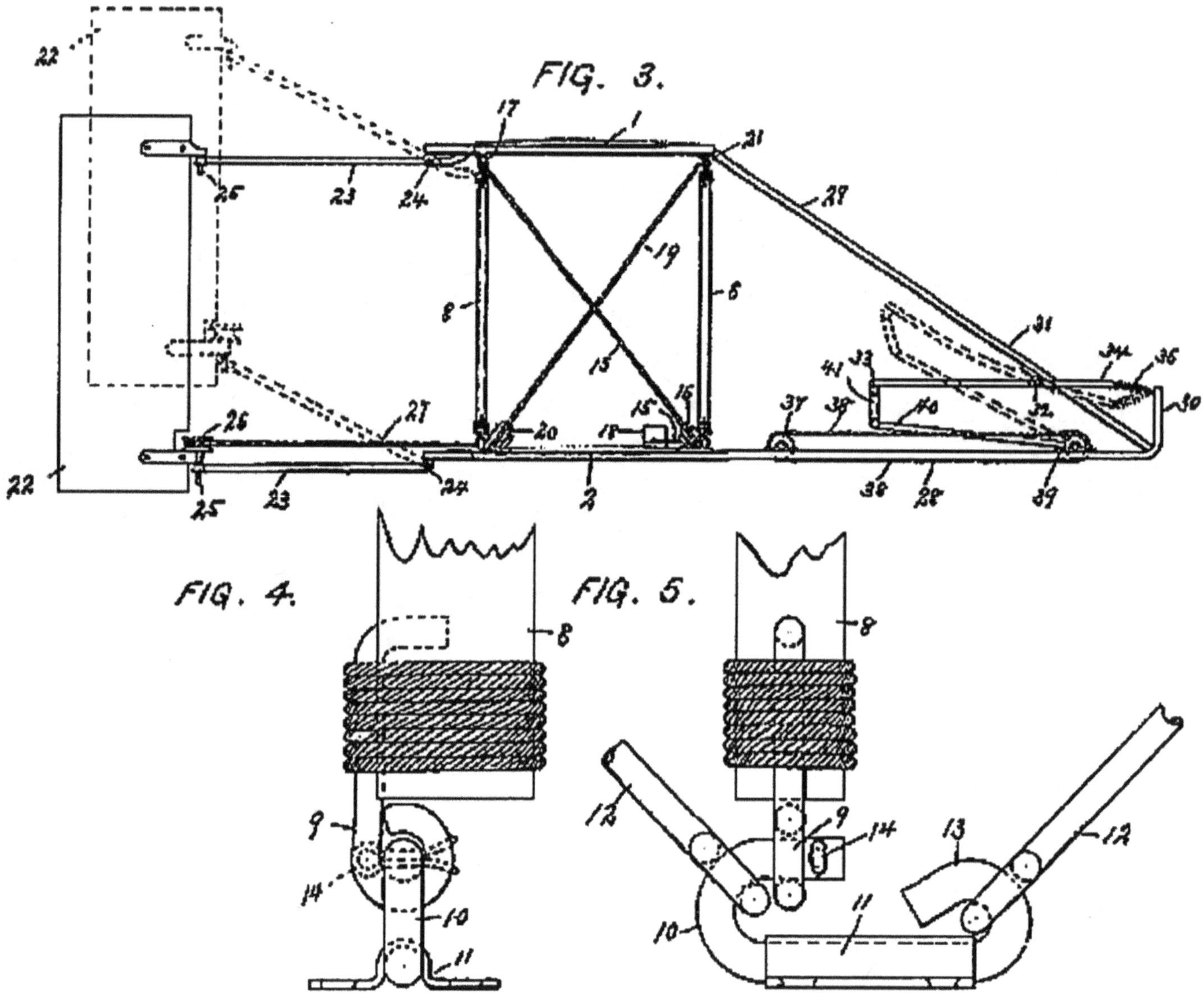

Figs. 3, 4 and 5 from Patent No. 821,393
(Source: USPTO)

It should be noticed from the drawings submitted in support of the claims in their Application for Patent of their Flying Machine invention that no mention or feature on the drawing shows the means for how the Flying Machine would be self-propelled or that it would even be manned by a human manipulating the levers that would cause, and control, the wings to warp, to control the angle of the forward canard in pitch, and/or to control the angle of the aft rudder to control the angle of yaw relative to the flow of air over the Flying Machine when in forward motion. Despite having only demonstrated success of their unique flight control system on manned glider flights, the USPTO granted Patent No. 821,393 to O. & W. Wright on May 22, 1906, a full 2 ½ years after their official first flight on December 20, 1903. Regardless of the lack of official description for their complete Flying Machine that made that official first successful flight, aviation history recognizes that the Wright brothers, had indeed made the flight of significant horizontal distance between two points on Earth, with the ending point being at least of equal or higher height than the starting point in a manned, self-propelled and in an adequately controlled flying machine without singular dependence on natural wind to cause sufficient lift on the wings for the flight or be the primary force controlling the direction of the flight.

A lesser recognized feature in the Wright brother's development of their airplane achieving their successful flight was in the propulsion system technology developed for their Flying Machine. The Wright brothers undoubtedly were aware that the steam engine, which was in the Mozhaysky failed attempt at sustained flight with his airplane design in Russia in 1884 described earlier, was not the propulsion technology needed for practical efficient aviation. The technology of internal combustion engines being developed by others in both Europe and the United States for automobile transportation offered a much better prospect for adaptation in aviation. Calculations by the Wright brothers for the horsepower needed for an aircraft engine at minimal weight (a much higher "power to weight ratio" of the engine ("thrust to weight" in today's turbojet propulsion system parlance) indicated that a tailored engine design was needed for their Wright Flying Machine.

While the Wright family bicycle shop in Dayton, Ohio, gained a reputation for high quality manufacture of their bicycles, the credit for that is largely due to the machinist Charlie Taylor who actually hand-made the parts of the bicycles which were not otherwise manufactured by other companies (tires, spoked wheels, handlebars, etc. Charlie Taylor was about the same age as Orville and Wilbur Wright, and had an equal passion as they did for inventing a machine that could fly. Taylor machined almost all of the various instruments and machine parts the Wright brothers used in their gliders for testing at Kitty Hawk, but lamented that the Wright brothers would never teach him how to fly as a pilot of a Wright Flyer aircraft. As he later told newspaper interviewers: "I always wanted to learn to fly, but I never did. The Wrights refused to teach me and discouraged the idea. They said they needed me in the shop and to service their machines, and if I learned to fly, I'd be gadding about the country and maybe become an exhibition pilot, and then they'd never see me again."

[The Author would like to inject here a few thoughts for Readers to consider, especially as they read later sections of this book. First is that the success of development of any new airplane is not currently attributed to one or two individuals, it is due to the teamwork of many others at all levels of experience and with widely varying talents. Second is that much of that success is due to the "homework" done before the first line is created on a drawing (or on a Computer Aided Design (CAD) app of a digital computer) by setting down fairly detailed goals and requirements for the airplane to achieve as a flight system. Third is that the pilot of that airplane relies on those who design and build the airframe he/she is piloting to safely propel him/her self and any other humans aboard from Point A to a pre-planned destination Point B through all possible unusual environments enroute. Fourth (and last for this note) is that the ground crews who service (replenish expendables – fuel, oil, hydraulic fluid, armament on military airplanes, etc.) and repair any wear and tear damage on the airplane during and after a flight. Great credit should be awarded to each and every member of the team associated with the creation of flying machines and every phase of their operation in the real-world environments. As Readers continue reading about other airplanes and propulsion systems that have been produced over the 115 years since the Wright brothers first successful flight, they should look behind the scenes at the team/personalities involved in the creation of each different aircraft and system.]

Returning attention to that of the engine developed by the Wright brothers and manufactured for their successful venture in aviation history, the entire 4-man team (Wilbur and Orville Wright, Charlie Taylor, and the "Program Manager" father Milton Wright, a church Bishop) worked closely on candidate designs for an "8 Horse Power engine that weighed not more than 200 pounds" and used fuel of ordinary gasoline being used in contemporary automobile designs. Conceptual sketches were made of the primary structural and functional parts, and Charlie Taylor took it from there. A one-piece aluminum casting was made of the engine block and was the first-time aluminum material had been incorporated in the design of airplanes and their subsystems. The block was cast by the Buckeye Iron and Brass Works using the raw aluminum obtained from the nearby Pittsburgh Reduction Company (renamed Alcoa in 1907). Brass castings were also made for some of the engine parts at Miami Brass Foundry near Dayton. A photograph of those basic engine parts is shown next:

Castings for the Wright Brothers Engine (1903)
(Source: National Archives)

Charlie Taylor personally machined the internal components to be assembled into the engine as shown next:

Steel Parts of Engine Machined by Charlie Taylor (1903)
(Source: National Archives)

A cutaway drawing shown next illustrates the assembly of the engine as used by the Wright brothers on their first successful flight:

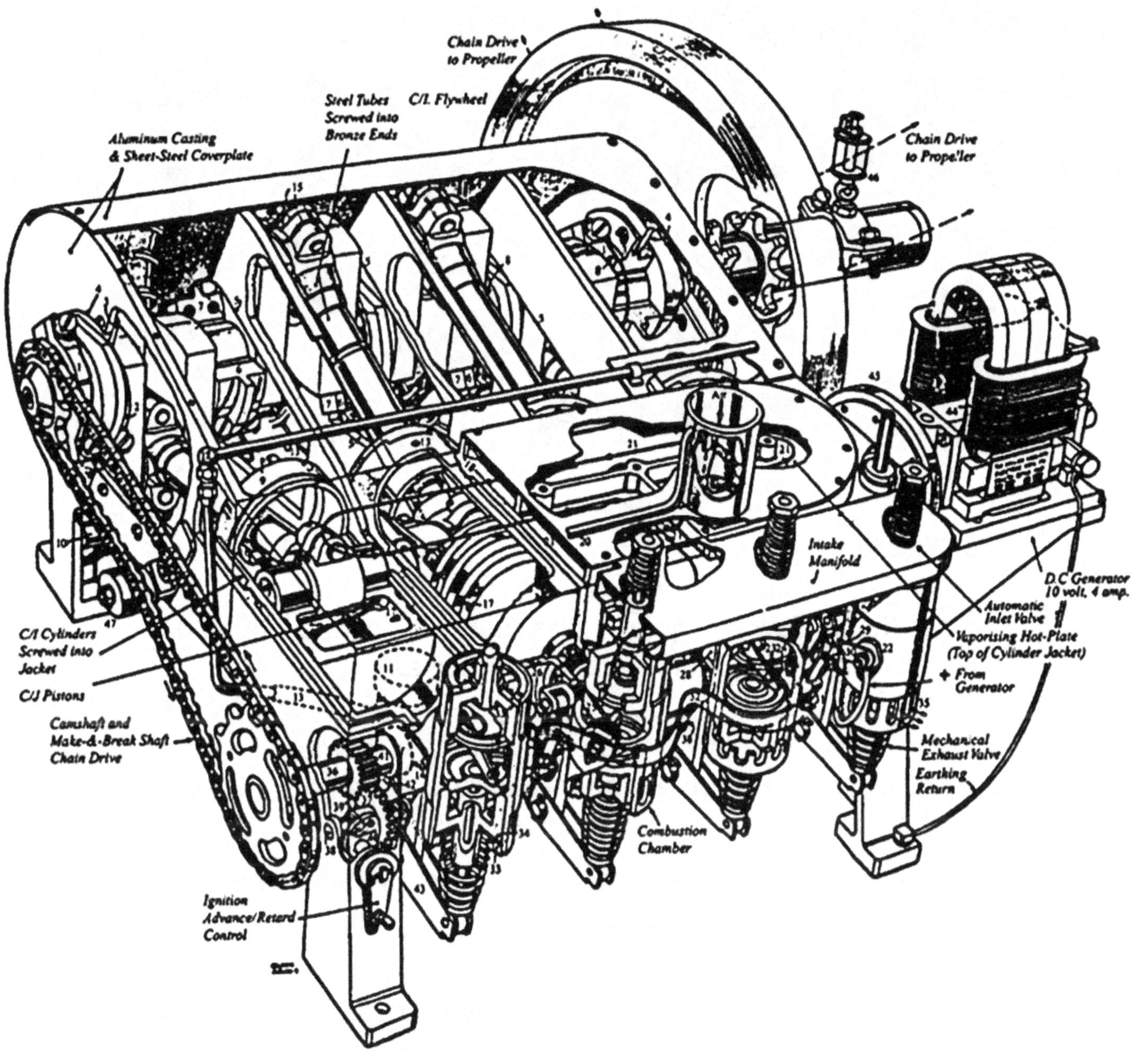

Cutaway Drawing of the Assembled Engine Wright Brothers Used
(Source: National Archives)

The final configuration of Charlie Taylor's engine is shown next in a photograph of his creation:

The Charlie Taylor engine in the Wright Flying Machine (1904)
(Source: National Archives)

The final engine exceeded the 8 Horse Power "desirement" (i.e., not a formal "requirement") of the Wright brothers and actually generated 12 Horse Power in its installed configuration and at under 200 pounds.

Details of the combustion process and cooling scheme for the Charlie Taylor engine are remarkable in themselves but are left for Readers to research for themselves on the internet.

The experimental approach taken by the ingenuous Wright brothers, Orville and Wilbur, of Dayton, Ohio, to develop a flying machine prototype in the early 1900s has often been characterized as being a good model for the development of subsequent military weapon systems developed during the ensuing 11 decades. While developing the key technologies for Man to fly like the birds and have positive control of their flight path, the ingenuous development of a mechanical means for Man to control the flight path of their powered, manned gliders took place. Readers should be also inquisitive as to what the Wright brothers had re-set for their next goal after their historic flights in December 1903 on the sand dunes of the Atlantic Ocean coast at Kitty Hawk, North Carolina.

Soon after the Wright brothers successful first flight, there were claims and lawsuits submitted from several European countries, most notably Germany, as well as in the

United States claiming that the Wright brothers had taken their work and denied them the ability to have made the first flight. The diaries and technical notes, recording failures and successes in wind tunnel tests and kite and glider (manned and unmanned) exploratory flight testing, were meticulously kept and dated by the Wright brothers for the few years preceding their first successful flight. The meticulously kept and dated records and photographs caused all such claims and lawsuits to be rejected.

The experimental approach taken by the ingenuous Wright brothers, Orville and Wilbur, of Dayton, Ohio, to develop a flying machine prototype in the early 1900s has often been characterized as being a good model for the development of subsequent military weapon systems developed during the ensuing 11 decades. While developing the key technologies for Man to fly like the birds and have positive control of their flight path, the ingenuous development of a mechanical means for Man to control the flight path of their powered, manned gliders took place. Readers should be also inquisitive as to what the Wright brothers had re-set for their next goal after their historic flights in December 1903 on the sand dunes of the Atlantic Ocean coast at Kitty Hawk, North Carolina.

Orville and Wilbur Wright kept prolific and detailed diaries, notes, diagrams, and letters, and also photographic record of certain key events in their exploratory testing (ground and flight). Those Readers who are entering into the first steps of a career in the development, advancement, or application of any new technology are advised to keep the same type of records as did the Wright brothers (although perhaps much more concise and without personal notes and correspondences to relatives about the progress) for whatever it is they are designing or testing. Of prime importance is the taking of photographs to record test set-ups and perhaps high-speed photography to capture the dynamics of the test article as it enters into the ever-possible failure mode and unexpectedly disintegrates within a flash.

Speaking from experience, this Author can attest that during the testing of a test article of any nature, especially high-cost aircraft, there may be tests to determine the safe strength and functional limits of the article being tested. In aircraft early development flight testing, this is generally referred to as "establishing the flight envelope" and the only method that will best define what are the safe limits for that test are to go beyond the limit and into a set of conditions that may in fact cause failure of the test article. It is far better for a test article to fail in a test facility (ground or flight) and confirm its operating limit, than later in production when that article is being inadvertently operated beyond that safe limit and results in not only costly damages to the unique test article but worse if it fails and results in human injury or fatality.

The Wright brothers' records (and voluminous other extraneous material) are on file at the U.S. National Archives. Readers are strongly encouraged to read the papers of the Wright brothers, especially those embarking on a career in engineering of any field related to aeronautics, as they serve as a good model for how development testing in a technology domain for which there may be no precedence should be conducted and documented. For convenience, the papers of Wilbur and Orville Wright were compiled in a two-volume book by Marvin W. McFarland of the Aeronautics Division of the Library of Congress to commemorate the 100th Anniversary of the Wright brother's world-changing first successful *adequately controlled* flight at Kitty Hawk. [The Author had the inspirational experience of reading the Wright brothers papers and diaries at the U.S. National Archives. Readers are strongly encouraged to read the Wright brothers papers and detailed flight records and photographs, especially those embarking on a career in engineering, whether related to aeronautics or not. They serve as a good model for how development testing in a technology domain (for which there may be no precedence or database) should be conducted and the level of detail which should be documented throughout the development process].

By today's standards, the research, analysis, manufacture, and test (ground and flight) resources used by the Wright brothers may seem primitive. However, one must realize that 115 years ago there were no wind tunnels locally available for testing, there was no reliable and complete database for airfoil characteristics of aerodynamic surfaces, there were no computers to assist in analyzing and designing the structural integrity of the airframe, there was no large numerically-controlled machinery to accurately fabricate the physical components (the airframe, the engine, and the propeller), there were no Government build-to specifications and no schedule (not entirely a detriment to their progress) for defeating an enemy threat, and there was no precedence for the ability of a human to sustain adequate control of a flight vehicle. Further, since all funds expended were out of pocket, there were no audits to be done by the Government Accounting Office. Everything the Wright brothers needed to develop their flying machine was necessarily of their own ingenuity, creativity, resourcefulness, and persistence.

From a system perspective, the original Wright brothers design had an inefficient landing gear, that deficiency being the absence of a wheel type landing gear system. Because the Wright brothers were doing their development tests on the sand dunes and sandy hills of Kitty Hawk at the North Carolina Atlantic Ocean coastline, they elected to use a launch rail system on which the airplane would move while gaining flight speed. A wheeled landing gear would be impractical for use on the soft sand of the Kill Devil Hill sand dunes at Kitty Hawk. The Wright brothers relied on a skid type landing gear when the inevitable landing took place. In addition, a better method for take-off and landing became necessary for testing and proving their unique wing-warping technique for obtaining adequate flight control while in flight.

After their first successful flight at Kitty Hawk, the Wright brothers resumed development flights at the Huffman Prairie outside Dayton, Ohio, and not far from their bicycle shop. For takeoff from the Huffman Prairie, the same rail launch track that was used at Kitty Hawk was used at the Huffman Prairie (now on the grounds within the boundary lines of Wright-Patterson Air Force Base). To further assist in take-offs in no/low-wind conditions, the Wright brothers devised a tower with a weight and cable attached to the underside of their Wright Flyer to achieve basically catapult launches from the rail. Shown below is a photograph, documenting the flight test launch apparatus used.

Wright Development Flight Tests on Huffman Prairie, Dayton, Ohio (1905)
(Source: National Archives)

The question which quickly arose as the post first flight development tests were being conducted was how can Man make practical use of the aero plane? It was much the same as in the old quip: "When a dog catches the tire of a moving car, what does the dog do with it?" It is, or should be, apparent that this was a situation where a new technology had been developed to provide a new capability. But that illustrates the development of a technology for technology's sake. Most technologies are developed or advanced to satisfy a defined need for that capability. Nevertheless, the ability for Man to fly from Point A to a distant Point B opened up many doors of opportunity – e.g., for entertainment and sport flight purposes, for public transportation purposes, for commercial airfreight transportation purposes, and as history records for a variety of military offensive and defensive purposes.

As the Wright brothers quickly realized, it was the military door that seemed to offer the most immediate opportunity for use of their aero plane in the United States, with the carriage of US Mail as the next best alternate or parallel opportunity. The Wright brothers contacted the US Army and suggested the aero plane as a better option for observing the placement, movements and battlefield equipment resources of the enemy rather than by sending observers in tethered balloons above a battlefield - as was done in the US Civil War for reconnaissance purposes with often disastrous results.

The US Army initially saw little value in the original Wright brother's airplane which had very limited range (the fuel tank on the original Wright brothers aero plane held about 0.4 gal of fuel), had no provisions for a second crew member to act as the Observer while the pilot was totally occupied in flying the airplane, and had no wheels to allow it to be repeatedly operated from near the FEBA (Forward Edge of the Battle Area), etc. Once the Army expressed interest in the overall concept and basic capabilities, the operational features desired in the Wright Flying Machine were better defined and were incorporated in the Wright brothers Wright Flyer III. A second person, to act as an Observer above a battlefield, was added as was a much larger fuel tank than the 0.4-gallon fuel tank on the Wright brothers first successful flight airplane. For greater radius of action. The original propeller was also changed to a more propulsive-efficient "bent tip" configuration.

Incorporating most of the above changes first noted by the US Army, the modified Wright Flyer III model was demonstrated to the U.S. Army Signal Corps at Fort Myers, Virginia, in September 1908. Its external configuration had been modified from the first 1903 Wright Flying Machine to that of the Wright Flyer III shown in the drawing below:

Wright Flyer III (seen with 2 Man Flight Crew) (1908)
(Source: National Archives)

After making additional modifications to the Wright Flyer III design and demonstrating its ability to meet US Army desires, the Wright brothers sold the design of the Wright Flyer III to the US Army for approximately $30,000 in 1908. This officially became the first military airplane procurement – for reconnaissance. The ability to drop hand grenades and to launch bombs, and the addition of a machine gun for air-to-air engagements did not come until a few years later.

The US Army also noted that the Wright Flyer III airframe was so structurally flimsy that one gunshot tearing through the canvas covered wings might cause the airplane to crash. Although it was not emphasized as such at that point in time (circa 1906), the technology area in military airplane design of Survivability and Vulnerability to enemy ground gunfire had not yet been created.

Readers may ask why so much emphasis as to what the Wright brothers did? Detailing the approach the Wright brothers took, the ingenuity they used, the obstacles and failures they overcame, and the persistence they applied to achieve the goal of advancing flight technology they envisioned is presented to inspire and be emulated by newcomers to the field of technology development. There is little question in the minds of aviation historians that it was the technologies of the flight control system and propulsion system of aircraft which the Wright brothers developed which were the critical technology advances needed for the further development of aviation – military and commercial.

To improve upon the operational and design deficiencies of the Wright Flyer III of 1908, major advances in the new flight technologies were made in Europe in the years shortly after the Wright brothers' successful first flight. Political and military tensions between Germany and France, too complex to be totally detailed here in this book, resulted in the start of World War I in 1914. It quickly became apparent to the aviation world of the importance of major advances in the multiple technologies for efficient and maneuverable flight. The critical technology areas were a) aerodynamics of the external configuration; b) aerodynamically efficient wing and tail surfaces design; c) propulsion systems (piston engines and propellers) producing a large ratio of propulsive thrust to the weight of the airplane; d) structural integrity of the airframe when subjected to maneuvering flight loads (and to ground gunfire); and e) flight control mechanisms for controlling the flight permitting high maneuvering to be achieved. Careful integration of these aircraft design features led to the development of the airplane as a weapon of war. To those basic flight technologies was rapidly added the technology of airborne weaponry - machine guns for destroying the enemy in an air-to-air engagement and the ability to drop munitions from the airplane onto enemy fortified positions.

[Author's note: The Author desires that those Readers who have had no training or service in one of the military forces, of the US or any other country, be made aware that the basic purpose of engaging an enemy force (on the ground, in the air, or at sea) is to "Destroy the enemy's ability to wage war". It is far more preferable to destroy the enemy's weapons used to engage in war operations than it is to kill the human who is operating that weapon system. Fighter pilots do not shoot guns or launch missiles at enemy aircraft trying to hit the pilot – they shoot at the enemy aircraft with the intention of striking a vital mechanical or functional part of the aircraft itself and downing the aircraft. In hand-to-hand combat between foot soldiers there is no honorable alternative except to mortally wound the other human. In war, "destroying the enemy's ability to wage war" is the primary goal to be achieved by any manner.]

The two best airplanes of World War I in the view of many military aircraft historians were the French SPAD XIII and the German Albatross D.Va. Another factor which contributed to the effectiveness of those fighter aircraft was that the pilots who flew them quickly developed attack and maneuvering tactics which were self-taught by innate reasoning. Two of the most famous "aces" of World War I were Captain Eddie Rickenbacker of the US Army in the French SPAD XIII and Baron Manfred von Richthofen (of "Red Baron" fame) in the German Albatross D.Va and Fokker DVi. However, Douglas Campbell was the first American Air Ace with 6 kills in World War I while flying earlier in the same "Hat in The Ring" squadron in France in which Eddie Rickenbacker served and scored his 26 kills.

Photos of the "aces" of World War 1 and their airplanes are shown on the next pages:

Lt. David Campbell, first American air Ace (1917)
(Source: National Archives)

Capt. Eddie Rickenbacker and French SPAD XIII (1917)
(Source: National Archives)

Albatros D.Va bi-wing (in which the Red Baron had 65 kills) (1917)
(Source: USAF Foreign Technology)

A third excellent fighter aircraft for its time period was the Fokker D.VI tri-wing and a photograph of the Red Baron (Baron Manfred von Richthofen) and with two security guards is shown next:

Fokker D.Vi tri-wing (in which the Red Baron had15 kills)) (1917)
(Source: USAF Foreign Technology)

Baron Manfred von Richthofen (the "Red Baron") had 80 kills in two aircraft – 65 kills in the bi-wing Albatros D.Va and 15 kills in the tri-wing Fokker D.Vi – reflecting that he had masterful flying and tactical skills regardless of the type aircraft he had flown. The Albatros D model had a structural problem with the lower portion of its bi-wing failing in pullups during steep descents. The "Red Baron" was then provided with the tri-wing Fokker D.Vi aircraft which developed more wing lift and had a significantly larger horizontal tail, both of which gave the Red Baron more effective maneuverability for his last 15 aircraft "kills".

Readers should also note that unlike the Wright brothers design which used wing-warping to achieve roll control, both the Albatros D.Va and the Fokker D.Vi were designed with cable-operated mechanically-linked flight control systems and to have more conventional "ailerons" near the outer portion of the top wing trailing edge for higher roll control authority and responsiveness to pilot maneuvering commands. The technology of flight control, which was the reason for the Wright Brothers success at Kitty Hawk, rapidly advanced from the wing-warping technique with exposed ropes and wires to that of the all enclosed metal fuselage and with a mechanical system of cables and pulleys attached to the bottom of a control column (or "stick") and mechanized such that pilot commands for motion of a control surface (aileron, elevator, and rudder by pedal controls) were conducted by wire cable to those control surfaces.

The technology of the standard mechanical flight control system made important advancements by improvement of mechanical advantages and reduction of frictional force losses within the system of linkages during the period 1920 to about 1938. At that point in time, the use of hydraulic systems (high pressure pumps, hydraulic reservoirs, rapidly responsive hydraulic actuators, etc.) became adapted into the mechanization of the flight control systems in almost all aircraft. Pilot control of military aircraft became more positive and responsive and the reliability and survivability aspects also were greatly improved as a result of applying "Lessons Learned" from earlier aircraft operational uses during World War I.

A more obvious change to be noted in the above photographs of the French SPAD XIII, the Fokker D.Va and the Fokker D.Vi, is the difference in the means of propulsion from that used by the Wright brothers in 1903. There are several reasons for the difference in propulsion technologies being used by the Wright brothers and by those in Europe. The Wright brothers initially had no plan for developing their airplane as a military weapon of war, and were well aware that internal combustion engines were being developed for the emerging automobile industry both in the United States and in Europe. However, all of the candidate automobile engines were either too protected under patent regulations for use or weighed far too much for the Wright brothers' application.

France and Germany were at geopolitical odds in the early 1900s and it appeared inevitable that war would break out between France and Germany within the next 10-15 years. Recognizing that the airplane could be an effective weapon in a future war, but also recognizing that the Wright brothers' system of flight technologies was too flimsy for operating in the harsh environment of the battlefield. the European countries, principally Germany and France, began developing their own weapons they estimated were needed for offensive and defensive war purposes. Foremost being rapidly developed were the technologies of the fighter type airplane, and of those, it was the propulsion systems in the fighter aircraft which saw the largest technology advances.

The French (Leon Levavasseur, 1905) developed a uniquely configured rotary engine, which was developed to overcome the conventional air-cooled engine problem of not enough airflow passing over the engine to cool the heat produced from the internal combustion process within its cylinders. In the rotary engine, the crankshaft is fixed and the engine rotates around the fixed camshaft. A propeller is attached in front of the engine cylinders and the entire cylinders and propeller rotate within the engine compartment, with increased airflow around the rotating cylinders. While the first rotary engine was developed for use on speedboats, its application to airplanes was quickly recognized. A major disadvantage however, was that there was a high centrifugal force associated with the rotating heavy mass of the engine cylinders and accessories and reduced its reliability.

The Le Rhone 9C had 9 cylinders (hence the 9C designation), was air-cooled, and developed 80 Horse Power – almost 7 times as much as the Wright brother's engine made by Charlie Taylor which produced 12 Horse Power. A photograph of the Le Rhone 9C rotary engine, developed in the 1911 time period, is shown next. Visitors at the Aerospace Museum of California, adjacent to the McLellan Air Base in Sacramento can view this engine on display:

French Le Rhone 9C rotary engine (1914)
(Source: Photo by the Author)

Companies in Europe also began producing the rotary engine starting as early as in 1911 - among them Le Rhone and Clerget in France and Oberusel in Germany. Rotary engine technology in various forms powered 80 percent of the airplanes built in Europe for operation earlier in World War I. However, around 1918 rotary engines were replaced with water-cooled engines and were more reliable with improved cooling and since the centrifugal forces associated with the rotating high mass of the rotary engine cylinders and propeller mass was eliminated.

An advantage of the circular geometry of the rotary engine made the fit into the near–circular forward fuselages of contemporary aircraft a relatively easy design consideration, but this did not offset the disadvantages in its reliability and centrifugal forces. A photograph of a rotary engine installed in a British Sop with Pup is shown next.

Rotary Engine installed in British Sop with Pup (circa 1920)
(Source: Photo by the Author)

Adding to the reliability problems inherent in rotary engines was the fact that being air-cooled, the front of the installed engine had a doughnut shaped cowling cover with an open face to allow the cooling air to enter over the rotating cylinders. Operating military aircraft in war theaters are subjected to the harshest of environments. It should be apparent that the open face of the rotary engine is not only exposed to the cooling airflow but it is also exposed to any materials contained in that entering cooling airflow (rain, snow, dust, sand, etc.) which may have an adverse effect on the rotating cylinders. There could be frictional damage to the internal moving parts of the engine, and perhaps worse could cause a dynamic imbalance of one or more cylinders. The adverse factors associated with the rotary engines on military aircraft caused a reassessment of using the engine technology being generated in the automobile production assembly lines developing in most countries of the World.

In 1915, Mercedes of Germany produced a 6-cylinder liquid-cooled engine, derived from their automobile production developments, and which developed 180 Horse Power. That was 15 times as much horsepower than that of Charlie Taylor's engine in the Wright brothers first engine. A drawing of that Mercedes engine is shown next.

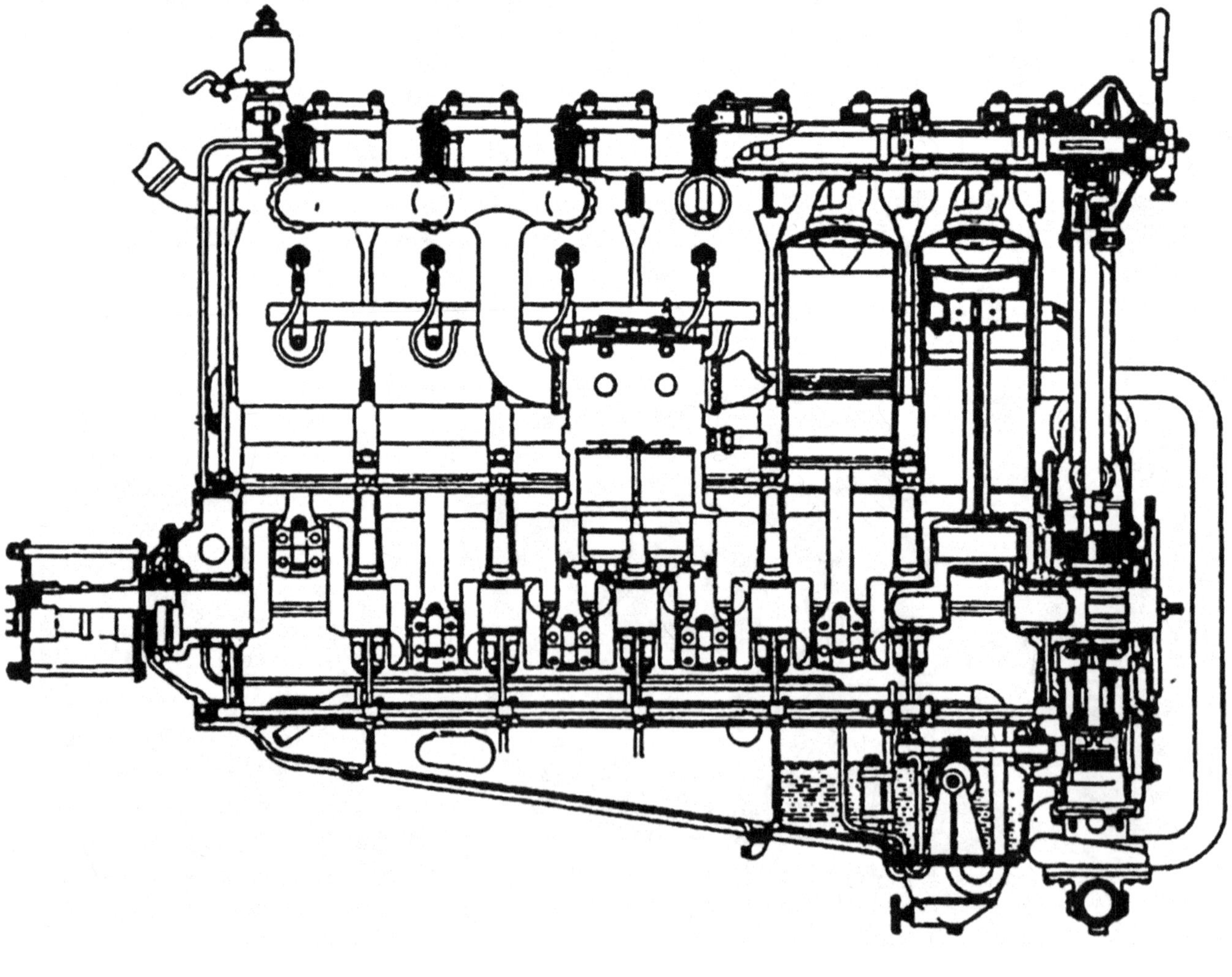

Mercedes 6-cylinder in-line, liquid-cooled engine,1915
(Source: USAF Foreign Technology)

The leading water-cooled engine was that of the Hispano-Suiza technology design, and which powered the French SPAD VII and SPAD XIII aircraft used nearer the end of World War I. The Hispano-Souza engine was also a derivation from automobile engine production technology in Spain and Switzerland (hence the Hispano-Suiza descriptor) and adapted to use in aircraft, principally the allied fighter aircraft in World War I. This was the same engine used in the SPAD VII in which Capt. Eddie Rickenbacker achieved his most air-kills to gain "Ace" rating. Its V-8 configuration made it relatively easy to be designed in to the near-circular forward fuselage of aircraft.

The Hispano-Suiza 200 was a V-8 configuration and generated 200 Horse Power. It was highly powerful, light weight, reliable and with high efficiency, causing it to be sought for purchase by numerous companies in Europe as well as in the United States. In 1916, a group of US investors merged the Wright Company with the Glen L. Martin Company and the Simplex Automobile Company to form the Wright-Martin Corporation. They obtained an order and license from France for the construction of 450 of the Hispano-Suiza Model A engines. The Simplex Division of Wright-Martin manufactured the Hispano-Suiza engines in New Brunswick, New Jersey.

An impressive display model of the Hispano-Suiza 150 engine (150 Horse Power version) is on display at the Aerospace Museum of California located adjacent to McClellan Air Base in the Sacramento area. A photograph (front view) of the classic Hispano-Suiza engine is shown below.

Front view of the Hispano-Suiza 150, V-8 built in 1914-1918) (Source: Photo by the Author)

Every engine requires a set of "accessories" that permit the engine to function efficiently, to remove the excess heat generated by the internal combustion process (by air cooling, by water cooling, or by oil cooling), to ingest (by injection or by carburetion) the fuel for combustion into the combustion component, to provide the electric power for both the combustion ignition and the engine control functions, and to have as small an external "envelope" so as to fit into the usually tight internal space allocated to the propulsion system in the airplane configuration. That is no small challenge for engine designers and propulsion engineers.

To portray better how the multiple accessory components are located relative to the basic block of the classic Hispano-Suiza engine, a photograph (aft view) of the engine, taken by this Author at the Aerospace Museum of California, is shown next.

Rear view of the Hispano-Suiza Model 150, V-8 built in 1914-1918
(Photo by the Author)

World War I ended on November 1, 1919, and was optimistically referred to as "The war to end all wars", and as a result the post-war military need for a large inventory of fighter aircraft was sharply reduced as well as the drive to advance the technologies for military fighter aircraft. The peace in Europe and the vision of airplanes having peaceful applications rather than solely military also prevailed in the United States.

The left-over bi-wing aircraft ("biplanes") from World War I were privately operated for entertainment (stunt-flying with wing-walkers), for agricultural use (as crop dusters), for pilot training schools (with two-seat configurations), and for transporting US Post Office mail on distant city-pairs. Aircraft were also designed with larger enclosed fuselages capable of transporting people on business and personal travel for longer distances – and much faster than by train. By virtue of carrying larger useful payloads, the need for larger engines and more efficient propeller design saw significant advances in those two technology areas especially from 1920 to 1935 and continued well after as commercial aircraft grew in size.

A provision of the Versailles Treaty which ended World War I in 1919 prohibited Germany from developing any war machinery and that included aircraft development. In 1922 that Versailles Treaty was relaxed and allowed Germany to proceed with development of aircraft for commercial air transportation, for movement of cargo by air, and for movement of postal material by air. In a still further relaxation of the use of aircraft for commercial and commercial transportation purposes, the Messerschmitt Company of Germany developed an aircraft which was a large advance in piston engine propeller propulsion, in the smoother external fuselage configuration for improved aerodynamics, in a retractable landing gear system to avoid that drag component at high speeds, had an aerodynamically efficient low wing somewhat blended into the fuselage, and had an enclosed cockpit. That aircraft was the Messerschmitt Me Bf 108 Taifun (Typhoon) in design at the start of the 1930 decade and which first flew in 1934. Shown next is a photograph of that Me Bf 108 configuration.

Messerschmitt Me Bf 108,1934
(Source: USAF Foreign Technology)

The period from 1900 to 1935 covered in this Chapter 1 is intended to portray how the technologies of aviation rapidly changed in the United States in a short 30 years. It has been shown that world events drive the direction and scope of technologies more so than just the expected progress to be made when only technological problems and discoveries are being addressed without any external influence. The world was in a relatively peaceful state as the Wright brothers systematically flight tested with their gliders and then progressed into developing what was to become the first successful flying machine. Airplane technologies made advancements using the Wright brothers' level of technologies as a baseline, but geopolitical events in Europe caused France and Germany to steer technology developments toward those needed to support military operations of the impending World War I. The United States slowly recognized in the early 20th century that the Wright brothers had opened up the door to a new form of transportation to augment that of railroad trains and motor vehicles. However, world events again entered into the picture as "the war to end all wars" (World War I) did not achieve that end to war.

The time period of 1900 to 1935 is inarguably the origin of military fighter aviation and gives rise to the metaphor: "As the twig is bent, so shall the tree grow." The Dedication section at the front of this book identifies 2 of the "forefathers" credited with the development of the USAF YF-16 Lightweight Fighter, both born during the time period 1900 to 1935. A second book by the Author "The F-16 Multi-National Weapon System, 1967 to 2017" gives credit to all 5 of the founding forefathers of the F-16 Multi-National Weapon System.

CHAPTER 2
PRE-WW II MILITARY AIRCRAFT TECHNOLOGY DEVELOPMENTS (1931 TO 1941)

Readers' attention is now redirected to the 1931 time period in the United States, and about the time of onset of significant geopolitical unrest stirring both in Europe and the Far East. In Europe, the aggressor nation threatening its otherwise peace-loving neighbors was that of Germany, seeking retaliation for its humbling defeat in World War I. German leaders also had a quest for "Lebensraum" – space for living – and were developing weapons of war as a means of forcefully obtaining additional physical land area for its growing population. Germany had a third factor in its aggressive planning which was to obtain regions which could support the petroleum, oil and lubricants (POL) needs for its growing industrial activities and capabilities. In the Far East, the Japanese Empire had neither sufficient minerals nor POL resources for its growing population and sought to obtain land areas in China and Manchuria which had abundant mineral resources, and in Indonesia which had abundant natural oil resources. The United States was well aware of what was happening in Europe to threaten outbreak of another World War, but was less aware that Japan would also be wanting to go to war in the Far East to capture its needed vital resources.

In a company- funded development activity, but with US Army Air Corps participation (supplying GFAE engines – Pratt & Whitney R-1340 WASP - and instruments), the Boeing Company designed a prototype of a militarily-configured aircraft designated by Boeing as Model 248. This prototype was the first American all-metal construction aircraft replacing the technology of a World War I vintage bi-wing aircraft of wood structure and cloth covered skin. It flew for the first time on 20 March 1932. It later was acquired by the US Army Air Corps and designated as the P-26 Peashooter, becoming the first aircraft to be produced in "large quantity" (151 being produced) for military usage as a pursuit fighter aircraft. The primary users of the P-26 Peashooter were to be initially the US Army Air Corps, the Republic of China Air Force, the Philippine Army Air Corps, and the Guatemalan Air Force.

A photograph of the unusual appearing external shape of the P-26 Peashooter is shown next:

Boeing P-26 "Peashooter" for the US Army Air Corps,1932)
(Source: USAF Photography files)

With the oversize radial engine cowling, and with the massive fairings over the main landing gear strut, and not to mention the external braces between the landing gear structure to support the wing lift loads, the P-26 was not an aerodynamically well-designed aircraft – but that was the design technology "state of the art" at the time. To experienced aerodynamicists: "If it looks good, it will fly good".

In conventional subsonic aircraft design, effort is expended to result in a wing placement about 1/3 of the fuselage length behind the nose of the aircraft in order to have the fully-equipped aircraft have the smallest static margin (distance between the aerodynamic center of pressure location and the aircraft center of gravity). This is done to avoid causing the aircraft to be very "stiff" to maneuver. It is also conventional in design of an aircraft with a nose landing gear to place the main landing gear wheels also just aft of the aircraft center of gravity to prevent the aircraft from tipping over tail down. The P-26 configuration must be considered as falling well outside the bounds of conventional design. The photograph above suggests that the main landing gear placed so far forward drove the design of the overall configuration and is there to support the weight of the large diameter radial engine. The engine was the Pratt & Whitney R-1340-33 Wasp 9-cylinder engine, which weighed 930 pounds and generated 600 horsepower.

The Wasp engine initially was a fuel injected engine but was soon thereafter modified to be a carbureted combustion system when installed on the P-26 airplane. A photograph of the uninstalled Wasp engine is shown below:

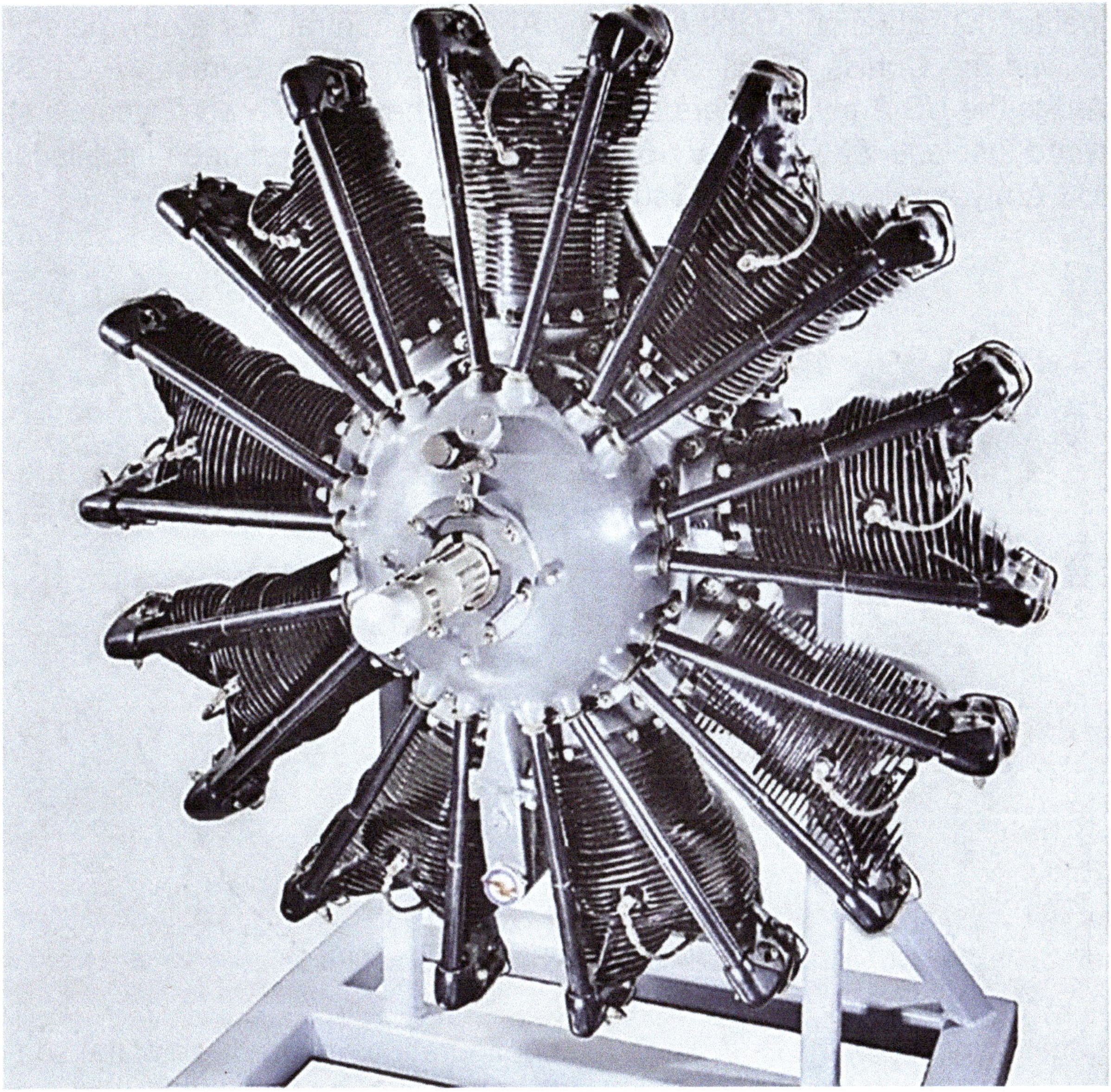

Pratt & Whitney R-1340-33 Wasp engine,1932.
(Source: Photo by the Author)

The P&W R-1340 was the first engine built by the newly-established Pratt & Whitney Company (started up in 1920.) The engine was a fine piece of radial engine technology as witnessed by the fact that 34,966 were produced and were applied in at least 55 different aircraft, including the P-26, over a production span of more than 25 years. The high reliability R-1340 Wasp engine was designated as a National Engineering Landmark by the American Society of Mechanical Engineers for the advancement it made in propulsion technology.

The performance of the P-26 as a military aircraft for the US Army Air Corps was dismal. The best that can be said is that the P-26 set the base-line for military aircraft configuration technology from which there could only be improvements made.

A competitor configuration for the US Army Air Corps contract for a new pursuit fighter in 1932 was the Curtiss XP-31 Swift. A photograph of the Curtiss XP-31 Swift, as proposed to the US Army, appears next, reflecting the far more aerodynamic shape of the forward fuselage if a Curtiss V-1570 Conqueror engine had been installed in place of the US Army required P&W R-1340 Wasp engine:

Curtiss XP-31 Swift, 1933.
(Source: USAF Photography files)

In the original version of the P-31 Swift, the configuration featured use of the same large diameter P&W R-1430 Wasp radial engine as designed into the P-26 Peashooter. With the original Wasp engine, there was little difference in the external configuration of both the Boeing P-26 Peashooter and the Curtiss XP-31 Swift, creating an efficient means of propulsion for the new USAAC pursuit fighter. Due to the XP-31 being of heavier weight than the P-26, its performance was somewhat lower than that of even the base-line technology P-26. This was the discerning factor for the US Army Air Corps selection of the P-26 for production as the better performer of the two bad performers.

To try and improve its performance and its appearance, the Curtiss Company retrofitted the one prototype XP-31 aircraft with a higher horsepower V-1570 Conqueror 12-cylinders (6 per bank) engine and which not only improved the XP-31 appearance but resulted in much better aerodynamic performance. The Curtiss V-1570-59 Conqueror 12-cylinder V engine was a liquid-cooled engine, which generated 675 Horsepower at

2,450 rpm. It weighed 770 pounds. A photograph of the alternate engine Curtiss V-1570-59 engine appears next:

Curtiss V-1570 12-Cylinder V engine, 1928.
(Source: Photo by the Author)

It is easily seen that in a comparison of the photographs of the Pratt & Whitney R-1340 Wasp engine with the Curtiss V-1570 Conqueror engine was that the latter had a much smaller diameter but a much longer length. The ungainly main landing gear fairings and the larger underwing braces shown in the photograph of the XP-31 Swift still however resulted in higher weight and lower performance of what the US Army Air Corps hoped to achieve and the USAAC did not opt to revise its selection of the Boeing P-26 to be its new pursuit fighter.

[Author's note: Dick Whitcomb (a brilliant practical aerodynamicist at the NASA Langley wind tunnel facility in Hampton, Virginia) later in 1970 expressed his opinion to this Author that whomever at the US Army Air Corps at Wright Field in Dayton, Ohio, had not selected the more aerodynamically efficient Curtiss XP-31 Swift and "didn't clean up the clunky XP-31 configuration was 9 cents short of a dime". Dick Whitcomb in his mid-career was the creator of the Whitcomb "Area Rule" which is often referred to as the "Coke bottle" shaping of aircraft fuselages to minimize transonic wave drag, but that is left for discussion in a later chapter of this book.

For those readers not familiar with the name Dick Whitcomb, the readers might appreciate, as this Author did tremendously, being exposed to Dick Whitcomb's practical "fill and file" approach to improving aerodynamics technologies and applying them to candidate aircraft configurations being wind tunnel tested at his facility. While the Author learned the principles of theoretical aerodynamics at Georgia Tech, it must be admitted that a more useful education in practical applied aerodynamics was obtained from informal discussions with the renowned aerodynamicist Dr. Richard T. Whitcomb at the NASA Wind Tunnel facilities at Langley Air Force Base in Virginia. It was Dick Whitcomb who taught the Author to "think like a molecule". At the core of that thinking was that a molecule of air is at peaceful rest (scientifically termed as being at equilibrium) in the air, and when an aircraft approaches on that same collision path containing the air molecule, the air molecule would be very abruptly (and "most rudely so" per Dick Whitcomb) displaced from its place of rest.

When the now-agitated air molecule flows with its equally agitated neighbor molecules over a surface of the airplane it is least further disturbed when flowing over smoothly varying contours until the gang of molecules departs the surface and returns to a state of rest (equilibrium). However, if the molecule must abruptly change direction or is made to spin abruptly by a sharp change in the pressure distribution on the surface contour, the molecule and its cohorts then become turbulent, like an angry mob. The energy expended to displace one air molecule may be miniscule, but when a large mass of those air molecules is displaced, made turbulent, or made to spin, the summation of all those miniscule energy losses adds up to be wasted energy, equivalent to increased aerodynamic drag. The propulsion system of the aircraft must overcome the additional drag to produce continuous forward motion. Therein lies the lesson of Applied Aerodynamics 101 as expounded by one of the giants of aerodynamics and aircraft design. Dick Whitcomb was fond of the short piece of aerodynamics prose: "Little whirls breed smaller whirls that feed on their velocity. Small whirls breed smaller whirls, and so on, to viscosity".

Later in this book (during the Advanced Day Fighter configuration development phase), the major influence is described which Dick Whitcomb had on the understanding of the aerodynamics of supersonic aircraft having shock-boundary layer interactions on the top surface of wings when transitioning from subsonic flow into/ through the transonic flight region and beyond. Whitcomb's "fingerprints" are all over the smoothly blended wing-body contouring of the F-16 aircraft.]

In the mid-1930s, Germany began rebuilding its war machinery despite being in violation of the Versailles Treaty and met opposition from the United States and the League of Nations. This book cannot adequately describe all of the geopolitical events that took place in the mid-1930s. Suffice it to say that those events produced a surge in the aviation technologies for the purposes of creating war machinery needed to support the misguided offensive intentions of only a few countries. In reciprocation, several other countries in Europe, Asia, and in North America saw the need to develop defensive war machinery to protect those countries from being overtaken by the offensive countries.

The philosophy, "An enemy of my friend is my enemy also", pervaded in many countries of the World, some for-trade considerations but most from moral considerations.

In 1934, the German Luftwaffe (Air Force) issued what was essentially a Request for Proposals (RFPs in the United States Defense Department parlance) for a new fighter aircraft for inclusion in the Luftwaffe military aircraft inventory. Four German manufacturing companies responded, and the leading two contenders produced prototype aircraft for that competition requested in the Luftwaffe RFP.

Capitalizing on the technology advancements developed for the Messerschmitt Me Bf 108 *Taifun* (Typhoon), shown at the end of Chapter 1, Messerschmitt offered the Me Bf 109 fighter aircraft, not very dissimilar to their Me Bf 108, a photo of which is shown next.

Messerschmitt Me Bf 109, circa 1936.
(Source: USAF Foreign Technology files)

Although the main landing gear of the Me-109 was of the retractable type, the tail wheel was left extended as there was little structural room for its stowage in the aft fuselage without compromise of the aft fuselage structure to accommodate the aerodynamic loads on the tail surfaces, especially those generated by the horizontal tail while maneuvering heavily when in air-to-air engagements. Vertical tail rudder loads were comparatively lower in maneuvering.

Most notable was the improvement in the design of the Me Bf 109 cockpit canopy to give the Luftwaffe pilot then- unparalleled visibility to the upper forward hemisphere except for the small sector directly aft of his head (in the "6" direction). The higher scoring aces of World War II (several having greater than 50 kills) were predominantly German pilots who had learned from World War I and pre-World War II warfare that most kills were easier to achieve by attacking from above and from the rear, and capitalizing on the glare of the Sun at their back to make visible detection of them nearly impossible. (As described later in this book, the technology development of advanced airborne radar detection sensors removes that tactical advantage edge in combat tactics.) The importance of excellent visibility through the canopy was clearly established and persists in modern fighter aircraft.

The engine in the pre-World War II Messerschmitt Me Bf 109 was the Daimler-Benz DB 601 V12 Piston Engine built and first run in 1935. It was a liquid-cooled, 12-cylinder inverted V piston engine. A photograph of a preserved Daimler-Benz DB 601 engine is shown next.

Daimler-Benz DB-601,1936.
(Source: USAF Foreign Technology files)

The Heinkel aircraft manufacturing company designed and entered their design into the Luftwaffe fighter aircraft competition as the Heinkel He 112. That configuration was as shown in the next photograph:

Heinkel 112, circa 1935.
(Source: USAF Foreign Technology files)

The Luftwaffe flight evaluated the Me-109 and the He 112 and selected the Me-109 as the fighter to be produced in quantity (and which later was used very effectively in the air war over Europe against Allied fighter aircraft). To be noted is that the Heinkel 112 had an elliptically shaped wing quite similar to that of the British Supermarine Spitfire. Aviation historians are uncertain if one country had capitalized on a technology advancement from another or whether it was by sheer coincidence that the wing designs were so similar.

Given that most major countries producing aircraft designs have a Foreign Technology organization within their purview and have many field agents performing clandestine activities. That known Foreign Technology activity reduces the probability of two technology developments resulting in almost the same shape wing being a sheer coincidence – but which one was the first to apply the theoretical elliptical wing pressure distribution to an operational aircraft wing design is not certain.

Another German aircraft design, the Junkers Ju 87 Stuk" (*Sturzkampfflugzeug* – dive bomber), which had its first flight in 1935, was one of the most feared aircraft when it flew. The Stuka was actually a close air support and ground attack aircraft and was not intended for use in air-to-air combat. It is being included here because it very much of several other fighters in other country's fighter inventory, but was without the high performance as a fighter due to its fixed extended landing gear.

A photograph of the Junkers Ju 87 Stuka is shown next.

Germany Junkers Ju 87 Stuka, 1935.
(Source: USAF Foreign Technology files)

The Stuka is seen to have a large tear-drop style canopy which gave the pilot exceptionally good visibility in the forward hemisphere of sight, most important when in a steep dive and strafing or dropping bombs on a ground target. The dominant external feature identifying the Stuka was its "inverted gull" wing, similar to that which was designed into the US Navy F4U Corsair later in World War II. It is a high probability that the USN Vought F4U gull wing design is an application drawn from the German Stuka gull wing. Both the Stuka and Vought F-4U aircraft had larger-diameter propellers than contemporary fighters and in order to keep the main landing gear as short as possible, but yet long enough to give the propeller sufficient clearance when moving on the ground, the inner portion of the wings canted downward from the junction at the fuselage to the landing gear structural mount span-wise on the wing. The outer portion of the wing from that landing gear mounting point was canted upward.

The landing gear on the Stuka was non-retractable which is not as aerodynamically efficient as a retractable one. As a psychological weapon, the fairings ("spats") over the top portion of the Stuka main landing gear wheels were designed to give a very audible acoustic shriek when in a steep fast dive toward a target on the ground. To anyone on the ground near a potential target and hearing that Jericho-like trumpet sound getting louder, it meant that the target was in imminent threat of being bombed by the approaching Stuka. (When later used in World War II by the Italian Air Force, that shriek emanated by the Stuka diving in an attack was referred to as the *picchiatello*, a piccolo musical instrument).

To the Stuka pilot, a pull-out from that steep dive (60-90 degrees) at high speed (350-370 mph) meant pulling high Gs (about 6G max), and a system of automatic pull-up dive brakes were installed under both wings in case the Stuka pilot blacked out during recovery from the high-G dive maneuver.

Due to the original Stuka's otherwise poor turning maneuverability and lack of defensive armament, the target area was cleared from defending fighter aircraft cover by the German fighters serving as escorts for the Stuka dive bombers. The Stuka sank or badly damaged numerous small ships being used to evacuate the allied troops concentrated on the beach at Dunkirk in May 1940. Readers should note that military aircraft can be designed for a ground-attack mission if there are other aircraft with better air-air combat capability to provide cover for the aircraft in the ground-attack mode.

In the mid-1930s, Great Britain was keenly aware of the pursuit of the *Lebensraum* movement in Germany and was well aware of the German advances being made in their aircraft technologies – all being of militaristic nature and all for obvious use in the German occupation of neighboring countries in Europe. Germany had obvious intentions of expanding not only toward the East through Poland and into Russia, but also toward the West and to occupy France. Great Britain is situated about 25 miles across the English Channel from France and feared that they too would be invaded and occupied by Nazi Germany.

Accordingly, Great Britain developed fighter aircraft configurations with the capabilities for defending against the German armed forces that might invade Great Britain or at least to inflict major damage on Great Britain for having defeated the Germans in World War I. Three of the more recognizable and capable British fighter aircraft were the Hawker Hunter, which was first flown in 1937, the Supermarine Spitfire, which was flown a year later in 1938, and the Hawker Hurricane which was introduced in 1939.

The Hawker Hunter configuration is shown in the next photograph.

Hawker Hunter,1937.
(Source: USAF Foreign Technology files)

The Hawker Hunter became one of the most used aircraft in the next 7 years – over 15,000 were built during that period – and was more sturdily built than was the Supermarine Spitfire. However, because of its heavier weight, the Hurricane was not quite as maneuverable as the contemporary Spitfire. Another serious g design flaw was that due to the design of the cockpit Environmental Control System very close to the engine compartment and without an adequate protective seal present, there was a constant presence of carbon monoxide gas in the Hurricane cockpit. The problem took a few years to be rectified, and the solution required the pilot to wear his oxygen mask from the time of engine start until engine shutdown.

The British Supermarine Spitfire was perhaps the most aerodynamically- configured aircraft of that era, and what else should it be named than "Spitfire"? [The Author had thought that it was named for the fact that it had four machine guns in each wing which "spit" a spread of gun rounds at the target, greatly increasing the probability of hitting the target. The name Spitfire however was dubbed because, initially, the Spitfire was powered by a steam-cooled engine (to be described later) and which caused the engine to literally spit fire from the engine exhaust ports.]

Shown next is a photo of the Supermarine Spitfire and its uniquely shaped elliptical wing planform:

Supermarine Spitfire, 1938.
(Source: USAF Foreign Technology files)

It should be noted that there are no linear edges anywhere on the external configuration of the Spitfire. Readers may recall the observation described earlier as being made by Dick Whitcomb at the NASA Langley wind tunnel complex in 1970 when comparing the external configurations of military aircraft: "Air molecules enjoy having the freedom to flow in smoothly curved directions when displaced by a passing aircraft. It minimizes their anger at being so rudely disturbed by an approaching aircraft".

In aerodynamics, the distribution of lift on a wing is proportional to the geometric planform characteristics of the wing (taper ratio and aspect ratio) as well as to its camber and airfoil section. The Spitfire wing was designed with an elliptical planform so that it would produce the near ideally aerodynamic elliptical lift distribution. In the photograph, readers should note that there are no linear edges anywhere on the external configuration of the Spitfire.

The original engine around which several British aircraft were designed in the early 1930s was the Rolls-Royce Kestrel, V-12, 22-liter (1,342 cubic inches) engine generating initially 450 horsepower. With gradual improvements in combustor and cooling design and type of fuel used, the generated power of the Kestrel was increased to 700 Horsepower. The Kestrel was manufactured and first operated in late 1926 by Rolls-Royce, which had gained a reputation for high-quality and reliable engines. Not only was the Kestrel the engine of choice in the early 1930s for British aircraft but it was also used by several other countries in Europe for their aircraft propulsion needs. Ironically, the Kestrel was used to power the prototypes of a few German military aircraft which saw service against Britain in the Battle of Britain in the late 1930s. A photograph of a Rolls-Royce Kestrel XVI engine is shown next.

Rolls-Royce Kestrel XVI V-12 Engine,1929.
(Source: Photo by the Author)

When the Rolls-Royce Kestrel for the Spitfire original design was replaced with the highly efficient and powerful Rolls-Royce Merlin engine, the airframe/engine match was nearly 100 percent optimally obtained and gave the Spitfire its outstanding maneuverability in air- to-air combat against all of its adversaries. The Merlin engine was a liquid-cooled engine, and a photograph of the Merlin is shown next:

Rolls-Royce Merlin Engine, 1934.
(Source: Photo by the Author)

A measure of success for propulsion systems is the number of the engines built for use in the aviation industry in a variety of different airplanes. Rolls-Royce produced 4,750 of the very fine Kestrel engines from late 1926 to approximately 1930 but produced 149,659 of the superior Merlin engines used from late1933, through the Battle of Britain (1940), through World War II (1941-1945), and well into the late 1940s. The Rolls-Royce Merlin engine is regarded by most as having set the technology standard of piston combustion engines for high performance military aircraft worldwide.

With a supercharged Merlin engine, the USAAC P-51 Mustang later was to join the P-38 Lightning in becoming the two most-feared piston-powered fighter aircraft of World War II. With its high altitude supercharged propulsive capability and combat radius obtained by external fuel tanks, the P-51 Mustangs dominated the air war in the European theater. The P-38 Lightning dominated the Pacific theater due to its high-speed advantage over much slower aircraft flown by Japan. More about their kill record will be described in Chapter 3.

Returning to the Spitfire external configuration, the inner-most guns (of the four guns mounted in each wing) on the early Spitfire aircraft fired rounds (.303 caliber initially) that adequately cleared the propeller circle, and which eliminated the tricky technology challenge of the guns being mounted behind the propeller circle. This was achieved by relying on an engine shaft-driven synchronized came to interrupt the firing of rounds which otherwise would intersect a propeller blade and self-destroy the propeller blade (as many aircraft in World War I had experienced). In the photo of the Supermarine Spitfire shown earlier, two larger cannons can be seen protruding from the wings instead of the four smaller guns in each wing leading edge.

To capitalize on the Hawker Hunter ruggedness (i.e., survivability to battle damage) and the agility of the Supermarine Spitfire, Great Britain whenever possible employed the Hawker Hurricane more for close air support of ground troops and attacks on enemy ground forces and reserved the Spitfires for achieving and maintaining air superiority over land masses under British control. During the historically famed Battle of Britain in mid-1940, the British Royal Air Force initially had a total of 527 Hurricanes and 321 Spitfires to deal with the German Luftwaffe incoming force of 2,700 aircraft of which 1,887 (70 percent) of the Luftwaffe force was destroyed. The Battle of Britain spanned a period of about 3 months, during which a total of 1,547 British aircraft also were destroyed (most of the original Hurricanes and Spitfires and aircraft of other makes).

At the end of the Battle of Britain, then-Prime Minister Winston Churchill issued his historical proclamation that: “Never have so many owed so much to so few” referring to those gallant Hurricane and Spitfire pilots. Their heroic achievement caused Adolf Hitler and the Luftwaffe (directed by Air Minister Hermann Goering) to disengage from air-to-air combat over Great Britain and from sending bombing raids on its populous. The latter two reasoned that the losses of aircraft over Great Britain were much too heavy to warrant those losses at that rate, and that the remainder of the Luftwaffe aircraft inventory was to be employed in missions to support the Nazi military operations toward the East and against Russia.

In Japan, the build-up of their war machinery was not as focused as it was in Germany and was due to the opposing missions and equipment inventory needs of the powerful Japanese military departments. There was intense rivalry between the Japanese Army generals and the Japanese Navy admirals as to what types of fighter aircraft were needed for their conquest of other Asian-South Pacific countries. As a consequence, both the Japanese Army and Navy produced fighter aircraft for their separate needs but which, in this Author’s opinion, were much less technologically advanced compared to those being manufactured in Germany.

In 1937, Nakajima developed and rushed into production the Nakajima Ki-27 (Nate) primarily for the support of the Japanese Army forces invading the Chinese mainland for its mineral resources. In 1938, Mitsubishi developed and also rushed into production the Mitsubishi A6M (Zeke or Zero) primarily for use by the Japanese Navy operating from Japanese aircraft carriers to capture the oil resources in the South Pacific region. In general, the Japanese put higher emphasis of keeping their fighter aircraft as light as possible for maneuverability because their piston engine propeller technology had not advanced quite as well as had the more industrial machine design nation of Germany. A photograph of the Nakajima Ki-27 "Nate" appears next.

Nakajima Ki-27 "Nate", 1937.
(Source: USAF Foreign Technology files)

As can be seen in the photograph of the Nakajima Ki-27, the forward fuselage features a rather blunt engine cowling and which results in reduced net propulsive power from that blunt engine installation. The small horizontal tail is seen also to have rather small area elevators and the wings have rather small area ailerons which are not conducive to achieving high maneuverability, but the overall light weight of the aircraft structural skeleton compensated for the maneuverability of the entire aircraft as a system. As another example of that emphasis on light weight, it can be seen in the photograph that the Nakajima Ki-27 aircraft did not have a retractable landing gear which was not favorable to aerodynamics efficiency, although there were wheel fairings incorporated over the main landing gear wheels to minimize that external drag.

The Japanese government, driven by the military departments, requested the Mitsubishi Heavy Industries Company to design a competitive alternate design for a fighter aircraft, and Mitsubishi responded with its A6M design. A photograph of the Mitsubishi A6M "Zeke" (or more commonly called "Zero", as the circular-shaped Rising Sun suggests) is shown in the next photograph:

Mitsubishi A6M "Zero", 1938.
(Source: USAF Foreign Technology files)

As the war clouds in Europe and in Southeast Asia further darkened, the US Army Air Corps saw the configuration of the German Heinkel He 112 (which first flew in 1935) and Messerschmitt Bf 109 (which first flew in 1937), as well as the Japanese Nate and Zero and concluded there was an urgent need for more potentially capable fighter aircraft than the ungainly looking P-26 Peashooter. At least four of the US aircraft manufacturers also perceived the need for a more maneuverable fighter aircraft for the US Army Air Corps, as well as for the US Navy and Marine Corps carrier-based fleet. All began design and development of fighter aircraft designs having more advanced propulsion systems and more aerodynamically efficient external configurations plus including an enclosed canopy for pilot protection from the elements as well as to improve visibility during air-to-air engagements. In addition, serious efforts were begun to replace the stationary extended landing gear with a retractable one, the large reduction in aerodynamic drag off-setting the small difference in weight between a landing gear with a hydro-mechanical retraction system versus a stationary landing gear having heavy struts and large fairings. Structural construction of the wings to not require underwing structural braces was also a structural technology domain being actively pursued for structural efficiency, safety, and of growing importance, producibility.

Building on its technology baseline developed for the XP-31, Curtiss Aircraft Company designed a Curtiss Hawk Model 75 as a company-sponsored development and which improved on all of the technology deficiencies that were exhibited in the Boeing P-26 Peashooter and its XP-31 Swift. A photo of the Curtiss P-36 Hawk is shown next:

Curtiss P-36 Hawk,1935.
(Source: USAF photographic files)

Although there still was little interest expressed by the US Army Air Corps in the Curtiss P-36 Hawk, France, the Netherlands, Norway, China, and India acquired a large number of them for defense of their airspace against the German and Japanese and later the Italians in various parts of the European, Asian, and African continent. When the advantages became apparent from P-36 Hawk flight experience of having a V-type engine (as was the liquid-cooled, supercharged Allison V-1710 V-12 engine) rather than the blunt faced Pratt& Whitney R-1830 (Twin Wasp) 14-cylinder air cooled engine, the Curtiss Chief Engineer (Donovan R. Berlin) directed the replacement of the radial engine in the P-36 Hawk with an in-line V engine. At least three various V-type engines were given consideration and had been designed for installation in the Curtiss P-36 Hawk. It was a design decision made by Don Berlin that the P-36 Hawk would be fitted with the Allison V-1710 engine.

The Allison V-1710 engine was a single overhead cam V-12 (60-degree angle between valves) with 4 valves per cylinder. It was liquid-cooled (radiator) and oil-cooled. The V-1710 had a Normal Power rating of 910 horsepower at 2,600 rpm at Sea Level, and 1,060 Horsepower at 2,950 rpm for Takeoff. The V-1710-19 was 8 feet 1.75 inches long, was 3 feet 4.75 inches high, and 2 feet 4.94 inches wide. It weighed 1,320 pounds, and required use of 100/130 octane aviation gasoline.

A cutaway drawing of the Allison V-1710 is shown next:

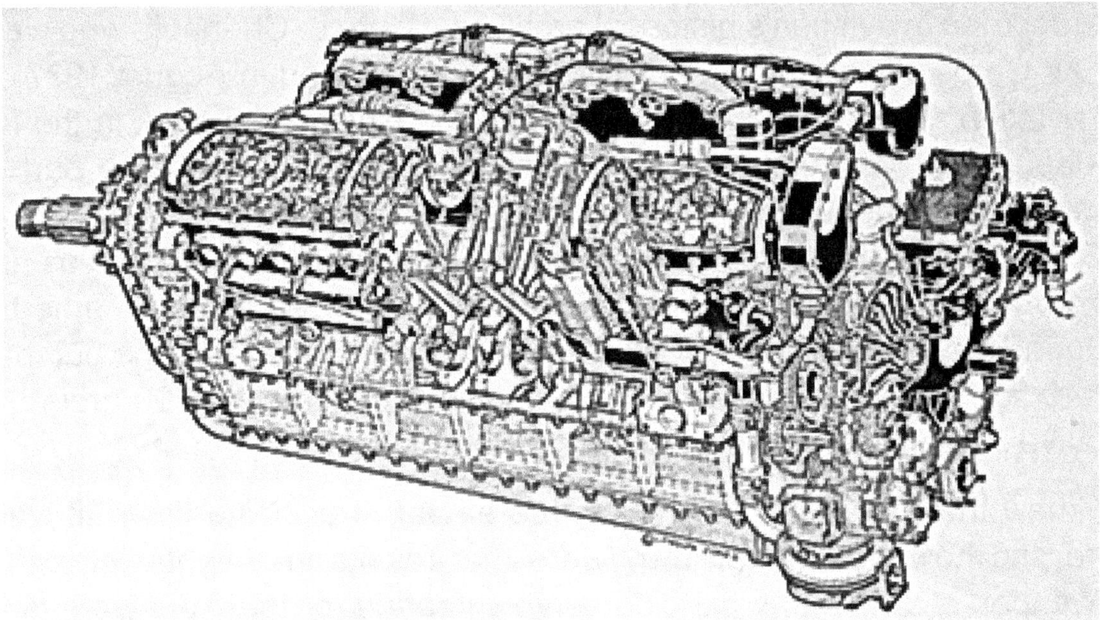

Allison V-1710 Engine, 1938.
(Source: USAF Foreign Technology files

This major modification in the engine configuration and resulting improvement in the propulsive horsepower and aerodynamics of the external configuration of the P-36 Hawk also provided a predicted substantial increase in the maneuvering and top speed of that improved technology. Based on that predicted improvement of the P-36 Hawk, Curtiss began development of a prototype XP-40. The XP-40, with an Allison V-1710 engine installed, had its first flight on 14 October 1938 at the Curtiss Aircraft Company facility in Buffalo, New York. A photograph of the XP-40 on its maiden flight is shown next:

Curtiss XP-40 Hawk on its maiden flight on October 14, 1938.
(Source: USAF photography files)

As is well-recorded in aviation's historical archives, Claire L. Chenault, who was a retired US Army Air Corps officer and who worked in China starting in August 1937, served as a military aviation adviser to China's Generalissimo Chiang Kai-shek in the first part of the Sino-Japanese War. Although not officially recognized as being a part of the US Government, Claire Chenault was authorized "with a wink and a nod" by then-President Franklin D. Roosevelt to form the First American Volunteer Group (AVG) as an advisory organization of the Chinese Air Force. This AVG was given the nickname the "Flying Tigers". During the winter of 1940-1941, Claire Chenault negotiated the purchase of 100 Curtiss P-40 fighters and recruited 100 pilots and approximately 200 ground crew and administrative personnel that would constitute the First AVG.

Chenault knew that it was not only what type aircraft was being flown, it was who the pilots were and how well trained they were that results in a fighter aircraft being an effective weapon of war. Many candidates volunteering for the AVG were doing so out of loyalty to the US and Chinese mutual cause in defending freedom, but many had less than 50 hours of flight time. Any airplane with any level of flight technologies incorporated is only as good as the human at the control stick. While this book traces the genealogy of flight technologies embodied in an aircraft, it cannot emphasize strongly enough the most important technology of all – the pilot. The Curtiss P-40 Warhawk was one of the most talked about in the decade preceding 1941. A photograph of the Curtiss P-40 Warhawk, with the iconic "Flying Tiger" tiger shark "smile" paint scheme is shown next.

Curtiss P-40 Warhawk in the Flying Tigers paint scheme,1940.
(Source: USAF photography files)

As shown in the photograph, the armament of the P-40 Warhawk was initially three .303 caliber guns in each wing (total 6) arranged so as to have the rounds being fired clear the propeller circle. Later versions were equipped with two .50 caliber guns in each wing. Between 1939 and 1945, Curtiss built 13,738 P-40s in various configurations and they flew in every theater of operations during World War II. The Curtiss P-40B was best known as the airplane flown by the American Volunteer Group against the Japanese in their aggression against the Republic of China in 1940, and a Hollywood movie, *Flying Tigers*, starring John Wayne, was made which rather realistically showed how vital a role the Flying Tigers played in the days prior to the start of World War II.

Both the Japanese Mitsubishi Zero and Nakajima Nate fighter aircraft were aircraft-carrier capable, and were the primary fighter aircraft used during the attack on the US fleet at Pearl Harbor on December 7, 1941. In 1939, there was much dispute between the Japanese Navy and Army leaders regarding their air fleet needs, resulting with the Japanese Army causing the development of the Nakajima Ki-43 Hayabusa (peregrine falcon) fighter aircraft, shown in the next photograph.

Nakajima Ki-43 Hayabusa “Oscar”, 1939.
(Source: USAF Foreign Technology files)

Over 5,900 Oscars were built during the period of 1938 to 1945, and being a land-based aircraft (i.e., having no aircraft carrier capability), the Oscars were used by the Japanese mainly in aggressive operations in the invasions of mainland China, Korea, and Manchuria. With its radial engine, the maneuvering performance of the Oscar was not particularly effective when engaged against the US Curtiss P-40 Warhawk of the Flying Tiger squadrons prior to the start of World War II. Much of that disparity in effectiveness was attributed to the much higher piloting skills of highly trained U.S. volunteer pilots, and to the superior performance capability of the P-40 Warhawk with its Allison V-1710 engine. There were no land-based Oscars employed in the Japanese attack of the US naval fleet while at anchor in Pearl Harbor on December 7, 1941.

Continuing with another fighter aircraft design that appealed to the US Army Air Corps (until its disappointing flight performance was later flight demonstrated and gun firing problems were realized) was the Bell P-39 Aira cobra. The P-39 had some unique technology features, internal and external. Some of those, which were external, are visible in the photograph shown next.

Bell P-39 Aira cobra, 1940.
(Source: USAF photography files)

The Bell Aircraft Corporation elected to use the Allison V-1710 engine, which, as on other fighter aircraft applications, resulted in an aerodynamically efficient “pointy” nose of the fuselage. However, to deviate from the armament system designs that featured multiple guns in each wing but mounted outboard from the propeller circle, with the attendant problem of the trajectories of the rounds converging toward a distant point as in a focus on the target, Bell opted to install a much larger single Colt M4 37 mm cannon in the nose on centerline of the engine spinner such that the muzzle was ahead of the plane of the propeller circle and on fuselage centerline. This cannon installation improved the targeting and firing solution at the target being ahead of the nose of the P-39 at any range and not requiring the focusing problem with wing-mounted guns.

Unfortunately, the firing rate of the 37 mm gun was much slower than of the multiple wing mounted guns and the rounds of the large single cannon were aerodynamically less efficient such that the trajectory of the 37 mm cannon rounds required either firing at very close-in range or to fire above the target to account for the earlier drop in trajectory of the slower rounds. Although the probability was higher of hitting the target with a spread of rounds from focused, wing-mounted, smaller guns than hitting the target with a lesser number of larger, aligned, 37 mm cannon rounds from a nose cannon, the probability of destroying the target was greater for the larger, albeit fewer, 37 mm cannon rounds that did hit the target.

A second technology integration problem was that the nose mounted cannon required the single V-1710 engine to be mounted behind the gun in the aircraft nose. This caused the turbo-supercharger of the engine to also be reduced in size and moved from the side of the fuselage and to a location behind the canopy. In "streamlining" (i.e., increasing the fineness ratio of the fuselage length to diameter ratio) the design of the P-39, the canopy height had been reduced. The fuselage was lengthened about a foot, and the wingspan was reduced by about 2 feet. This design resulted in pilots' intensely dislike of having to enter the cockpit through a side door.

The resulting net effect of the external and internal arrangement of the original P-39 slightly decreased its flight performance, but with the unusual location for the engine and the smaller turbo-supercharger, the net center of gravity of the aircraft was much further forward than on conventional designs. This caused the need for an aft landing gear instead of a nose landing gear in order to locate the aircraft center of gravity within the landing gear footprint geometric triangle.

The P-39 also was the first aircraft (military or commercial) to have a tricycle landing gear. Readers should compare the near mid-fuselage wing placement, the location of the landing gear, and the apparent location of the aircraft center of gravity (typically slightly forward of the wing aerodynamic center) of the Bell P-39 Aira cobra with that of the Boeing P-26 Peashooter shown at the start of this chapter. The contrast is startling. The tricycle landing gear technology was continued with incorporation on most subsequent fighter aircraft in most countries developing fighter aircraft. A fallout benefit of the tricycle type landing gear was that the pilot's visibility over the front fuselage was reasonably good when in the takeoff and landing modes. A feature of the Curtiss P-40 Warhawk cockpit which pilots disliked was that the cockpit was far aft on the fuselage, with the forward fuselage blocking the pilot's forward vision. When the P-40 was in the taxi mode, it required "wing-walkers" to walk at a fast pace outboard of each wing (some preferred to be "wing sitters") to steer the pilot and airplane on the taxi strip.

Of possibly greater interest to readers, the United States was planning for, and preparations were being made for, the US to design and manufacture new military weapons of war in the event the United States should be caught up in the increasing military tensions both in Europe and in Asia. To remain within the scope of this book, only the US Army Air Corps fighter aircraft developments for the period 1938 to the end of 1941 will be discussed during the final sections of this chapter. The US Navy had at least an equal number of carrier-capable aircraft in various configurations depending on naval mission needs (e.g., fighters, fighter bombers carrying torpedoes, dive-bombers of enemy shipping, sea reconnaissance and rescue, etc.) and having greater range and combat radius capability.

For the impending war in Asia and Europe, three additional designs of US Army Air Corps fighter aircraft were developed in the late 1930s and early 1940 time period. In order of their approximate development start date in the fighter aircraft inventory of the United States Army Air Corps, these were the iconic Republic P-47 Thunderbolt, the North American P-51 Mustang, and the Lockheed P-38 Lightning.

Design of the Republic P-47 Thunderbolt as a fighter was started in late 1939 as were several others for both the US Army Air Corps and the US Navy and USMC. The US Army Air Corps P-47 was designed around a specified propulsion system – the powerful Pratt & Whitney R-2800 Double Wasp radial engine – which had actually been developed by the US Navy for its two carrier-based fighters, the Grumman F6F Hellcat and the Vought F4U Crusader. Although the P-47 was officially named the Thunderbolt, it was sometimes called the Thud or the Jug (short for juggernaut). Its first flight date was 6 May 1941.

Shown next is a photo of the Republic P-47 Thunderbolt.

Republic P-47 Thunderbolt,1941.
(Source: USAF photography files)

As seen in the photo, the wing planform of the P-47 had a semi-elliptical shape – straight leading edge and elliptical trailing edge. It should also be noted the P-47 also featured a large trailing edge portion of its wing with control surfaces – ailerons and flaps.

The Pratt & Whitney R-2800 Double Wasp was the United States first 18-cylinder radial design. It featured a unique cylinder-head cooling fin configuration in that the fins were machined from the solid metal of the head forging, increasing their cooling efficiency at a lower weight than the more conventional cooling fin configuration. The R-2800 Wasp engine is essential two of the R-1400 9-cylinder banks placed back-to-back on the same crankshaft. The R-2800 generated 2,000 horsepower when it was first flown in 1940 and with various improvements increased to 2,100 horsepower in 1941, and increased further to 2,400 horsepower in 1945. In addition to being flown in the P-47 Thunderbolt, it saw service on 10 other US military and commercial aircraft. A total of 125,334 Pratt & Whitney R-2800 engines of different variations were produced between 1939 and 1960. A photograph of the 18-cylinder Pratt & Whitney R-2800 engine is shown next.

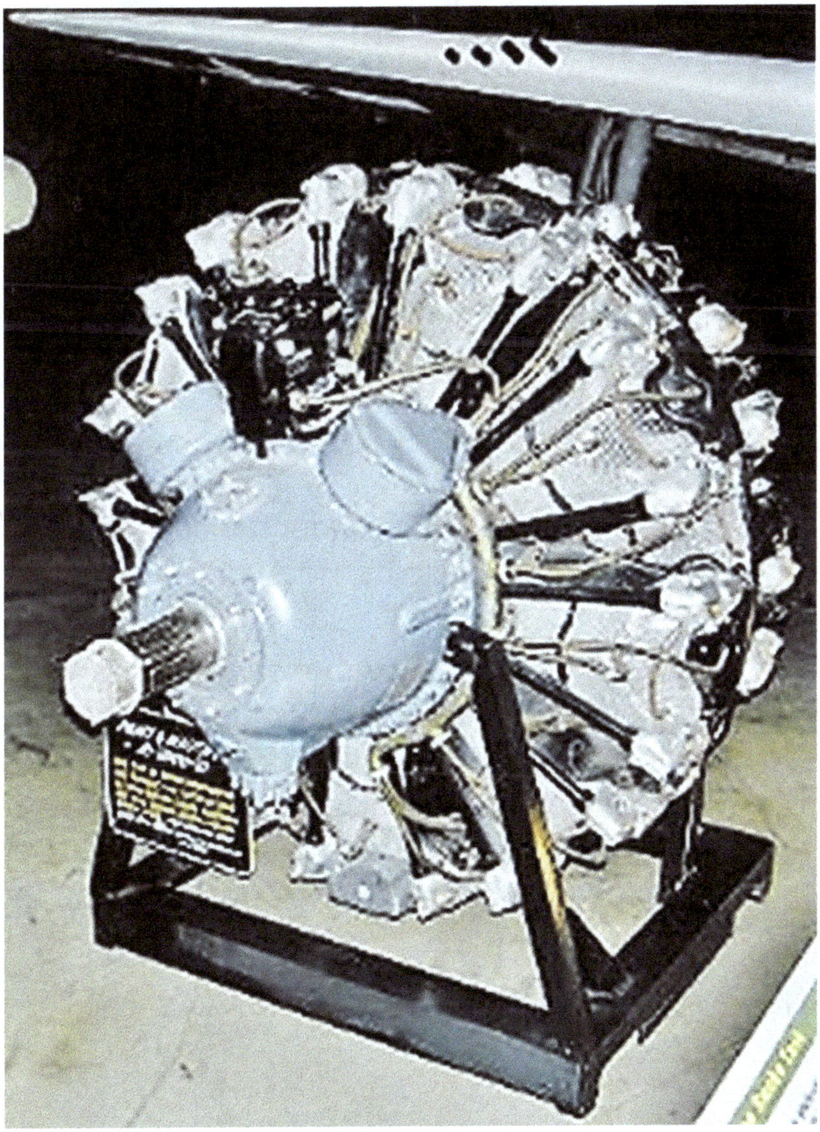

P&W R-2800 18-cylinder radial engine, 1940.
(Source: Photo by the Author)

So as not to mislead Readers that the R-2800 radial engine is as simple as it may appear in the above photograph, a cutaway of that same engine is provided below to reveal its internal complexity.

Cutaway of P&W R-2800 18-cylinder radial engine, 1940.
(Source: Photo by the Author)

At about that same time period (1940), North American Aviation was approached by the British Purchasing Corporation to build Curtiss P-40 fighters under license for the Royal Air Force. However, North American proposed that a newer, better performing fighter configuration be built. North American developed its first prototype for the British Purchasing Corporation and that prototype, the NA-73X was rolled out on 9 September 1940, four months after the contract with the British Purchasing Corporation had been co-signed. The NA-73X first flew on 26 October 1940.

A photograph of the NA-73X, later re-named as the North American P-51 Mustang, is shown next.

North American P-51 Mustang, 1940.
(Source: USAF photography files)

The P-51 originally was designed with the Allison V-1710 engine (shown earlier in this Chapter), but was soon thereafter powered by the more powerful Rolls-Royce Merlin 66 2-stage 2-speed supercharged engine (also shown earlier in this Chapter) which gave it superior high-altitude performance in the last half of World War II. The P-51D version was armed with six .50 caliber M2/ANN Browning machine guns, three guns in the leading edge of both wings and fired rounds clearing the propeller circle.

When equipped with two external fuel tanks carried underwing, the P-51 was used to provide escort protection of the allied bombers at high altitude deep into Germany and to return safely post-mission. A photo of the P-51 Mustang with external wing tanks, capable of being jettisoned before entering an air-to-air engagement, is shown next.

North American P-51 Mustang with 2 external wing tanks, 1943.
(Source: USAF photography files)

Production numbers of the several P-51 variants totaled well over 15,000. Perhaps the most amazing statistic was its low cost of $50,985 in 1945 dollars. A technology feature to especially be noted is in its canopy design which gave the pilot outstanding visibility in the upper hemisphere and especially to his aft (6 o'clock). Its impressive kill record was recorded to have been 4,950 enemy aircraft, most of which were in the European Theater of War. However, it saw service also in North African, Mediterranean, Italian, and Pacific theaters.

In comparing all of the fighters designed in the 1930-1941 time period and described thus far in this chapter, the external configuration of all of those several fighter aircraft had converged to being quite similar. The primary difference was generally in whether a large radial piston propeller engine was installed or an in-line V piston propeller engine.

The one additional piston engine powered design to be described in this Chapter 2 is that of a fighter aircraft having a most unique configuration. While the designation Lockheed P-38 Lightning seems to just be another one of the Army Air Corps attempts at developing a fighter aircraft for countering the German Luftwaffe, it was anything but "just another" fighter aircraft design similar to its contemporaries. The design of the P-38 Lightning is the Lockheed Aircraft Company revolutionary addition to the advanced fighter technologies in the fascinating 1930 to 1941 time period. The Chief Engineer for the Lockheed P-38 was Hall Hibbard and the Configuration Designer was a young Clarence ("Kelly") Johnson. It was the same Kelly Johnson who later designed the U-2 and SR-71.

Instead of describing how the unique design P-38 Lightning came into being, it is much easier to first show a photograph (below) of the unorthodox configuration as an example of a design resulting from "thinking outside the box".

Lockheed P-38 Lightning,1939.
(Source: USAF photography files)

When the P-38 was operationally used in World War II against the German Luftwaffe fighter aircraft, the Luftwaffe pilots referred to the P-38 as *der Gabelschwanz-Teufel* (the "forked tail devil"). It not only looked like one, it performed like one. It had superior altitude, speed, and maneuvering capability than all other fighter aircraft known at the time, and which was attributable to the fact that it had two powerful engines (Allison V-1710 with turbo-superchargers), counter-rotating propellers (to cancel engine gyroscopic torque effects, had a uniquely shaped external twin tail configuration which was both aerodynamically and structurally efficient (producing a low structural weight ratio), and had its armament in the center gondola-like fuselage between the engines. The armament firing accuracy was more accurate than wing mounted guns which were required to focus to a point at the target whereas the on-centerline P-38 guns and cannons were on fuselage centerline line congruent with the flight path and did not require "focusing" of the guns when firing at a distant target. A deficiency in the low amount of ammunition for the gun systems however, and it not having a "spray-pattern" more probable of hitting the target with more than one lucky round, and that detracted slightly from its over-all gun firing probability of kill.

The first XP-38 flew on January 27, 1939, but a carburetor icing accident destroyed the prototype. The USAAC then ordered 13 YP-38s, the first of which flew on September 17, 1940. Soon thereafter, the YP-38 had set transcontinental speed records. A difference in the two engines rotating still counterclockwise but in the opposite direction produced a significant improvement in the gun-firing accuracy, but there remained a too-small an amount of ammunition. As a surprise associated with its unprecedented speed capability, the technology phenomenon of high-speed compressibility was encountered. The compressibility effect was capable of causing structural failure of the wing and gondola combination as well as a controls "lock-up". Kelly Johnson wrote about the problem: "I broke an ulcer over compressibility on the P-38 because we flew into a speed range where no one had ever been before, and we had difficulty convincing people that it wasn't the funny-looking airplane itself, but a fundamental physical problem. We found out when the Lightning shed its tail and we worked during the whole war to get 15 more knots of speed out of the P-38. We saw compressibility as a brick wall for a long time. Then we learned how to get through it."

The ultimate cause and fix for the P-38 compressibility problem is beyond the scope of this book to detail; however, the problem was corrected although not fully until the late phases of World War II. In the interim the max speed of the P-38 was "red-lined" to be below Mach 0.74. Because aircraft had no such instrument as a Machmeter, the information from the altimeter and airspeed indicator in the cockpit was used to consult a graph in the P-38 pilots' handbook to advise the pilot when the pilot was approaching or was at the Mach 0.74 "red-line" speed altitude combination.

[Author's note: As Readers may recall in Chapter 1, the term Mach number or simply Mach, is named after the German physicist Ernst Mach, and represents the ratio of the speed of an object moving through the air relative to the speed of sound through that same air – at the temperature and density of the air.] In the 1880s, when Ernst Mach was developing his theories on supersonic flow and the formation of shock waves, there were of course no supersonic aircraft in existence. However, there were high velocity rifles and cannons developed in Europe for use in ground warfare. As the projectiles from the rifle and cannons varied in flight speed from as low as Mach 0.8 to as high as Mach 1.6, the Schlieren interference type of photography for visualizing those shock waves was a development of Ernst Mach.

From the basic work done by Ernst Mach, the theories and knowledge database of transonic and supersonic aerodynamics were refined by aerodynamicists. Perhaps of interest to readers, Mach had observed that the flash of light from a bolt of lightning was accompanied by either a sharp "crack" sound, if the lightning flash was nearby, or by a "rumbling" sound, if the lightning flash was at a substantially larger distance. Using an advanced stop watch-type instrument, Mach started the timer of the instrument as quickly as he could when the lightning flash was observed and then stopped the timer on the stop watch when the audible sound was heard. The number of seconds very accurately determined to the nearest 1/100th of a second was used in calculating the distance from the location of the lightning strike to the observer's ear based on sound travelling at the speed of sound.

As described hereto in this book, the technologies of aerodynamic external configurations, piston propeller propulsion, and materials and structure for efficient structure having light weight and structural strength as well as cockpit design improvements (especially canopy designs for fighter aircraft) all made significant advances in their respective technology domains during the 1931 to 1941 time period. In the technology of flight control system, the largest advance in flight control system was made by the airplanes developed before and during World War I in Europe. The Wright brothers' first patent for the wing-warping method for aircraft roll axis control, and conventional surface deflections for control of pitch and yaw saw changes which eliminated the wing-warping approach. Newer approaches used conventional aileron and wing flap control surfaces for generating the pilot-commanded motion of the aircraft in the primary three orthogonal axes.

The Author has shown the photographs of several of those aircraft developed by the larger countries as military fighter aircraft, and has attempted to show that those advancements were driven by the ominous geopolitical events looming in Europe and in Southeast Asia. As shown, the altitude, speed, range and payload/armament carrying capabilities of all the aircraft shown and high-lighted in capability improvement that emerged in the decade prior to World War II (to be described in the next chapter). All conducted their developments in a gradually decreasing rate of technology improvement (referred to often as "becoming asymptotic" to their final actual level of efficiency and capability).

In the domain of propulsion technology, there was interest harbored in scientists for centuries as to how sea life (principally octopi, squid, and some shell life) used a form of "jet propulsion" to move their masses through the water. Jet propulsion for moving masses through the air is traced to the Chinese in the 13th century as fireworks that dazzled wondrous eyes of all ages, but which also became used to propel a mass of explosive against an invading enemy force. Very early attempts at using compressed air or later steam pressure were interesting scientific demonstrations of jet propulsion but did not find any useful application in public use. Key to the development of a jet engine was the combining of the technologies of jet propulsion with that of the technology of gas turbines. The patent for a stationary turbine was awarded to John Barber in England in 1791. History of propulsion technology reveals that in 1903, a Norwegian engineer Aegidus Elling developed the first gas turbine to run self-sustaining. Limitations in practical engineering and metallurgy did not allow such engines to be manufacture with acceptable levels of safety and efficiency in sustained operation.

The first Application for Patent for using a gas turbine to power an aircraft was filed in 1921 by Maxime Guillaume of France featuring an axial-flow turbojet, and 5 years later in 1926 an Alan Arnold Griffith published an article "An Aerodynamic Theory of Turbine Design" leading to experiments to be conducted at the British Royal Aeronautical Establishment to explore Gas Turbine Engine applications for aircraft propulsion. In 1928, Frank Whittle, a cadet at the Royal Air Force College at Cranwell, England, formally submitted to his superior officers technologically credible ideas for a turbojet capable of efficiently and safely propelling an aircraft. Frank Whittle was granted a patent protecting his ideas for turbojet propulsion of an aircraft in 1932, but for the next few years there was little if any advancement made in the technology to generate interest within the RAF. In 1937, due to a fuel leak in a turbojet engine demonstration run, the Whittle experiment experienced a "run-away" engine in which the engine continued to run even after the fuel feed system had been shut off. This unfortunate incident cooled the interest the RAF had for adapting jet propulsion to its RAF fighter aircraft development activities, A larger number of effectively-propelled aircraft was more urgently needed for the ongoing Battle of Britain with the German Luftwaffe.

In 1935, work was initiated on turbojet engine technology by Hans von Ohain of Germany with some documentation indicating that Frank Whittle's patented earlier work in Great Britain on that same turbojet engine technology had been brought to the attention of Hans von Ohain. It rarely happens, but does occasionally, that two scientists in two different countries will come up with ideas for a new technology that are quite similar if not identical. When they are identical to the last bolt and nut, it raises the possibility of patent rights having been infringed upon, or worse that Intellectual Property rights had intentionally been ignored.

Aviation historians' credit both Sir Frank Whittle and Dr. Hans von Ohain as being the co-inventors of the centrifugal flow type turbojet engine in the field of propulsion technology. Both inventors are acknowledged as having conceived the same principles for jet propulsion although documentation indicates that Sir Frank Whittle had patented the idea several months before the idea was conceived by Dr. Hans von Ohain. A meeting of Hans von Ohain with the German industrialist Ernst Heinkel in the mid-1930s led to the development of an aircraft built by Ernst Heinkel and powered with the centrifugal flow type turbojet engine developed by Hans von Ohain.

It is likely more interesting to Readers to see graphically what first was envisioned by Frank Whittle who was an officer in the British Royal Air Force (RAF) as a turbojet engine cycle. A cutaway drawing of the Whittle centrifugal flow jet engine cycle is shown next.

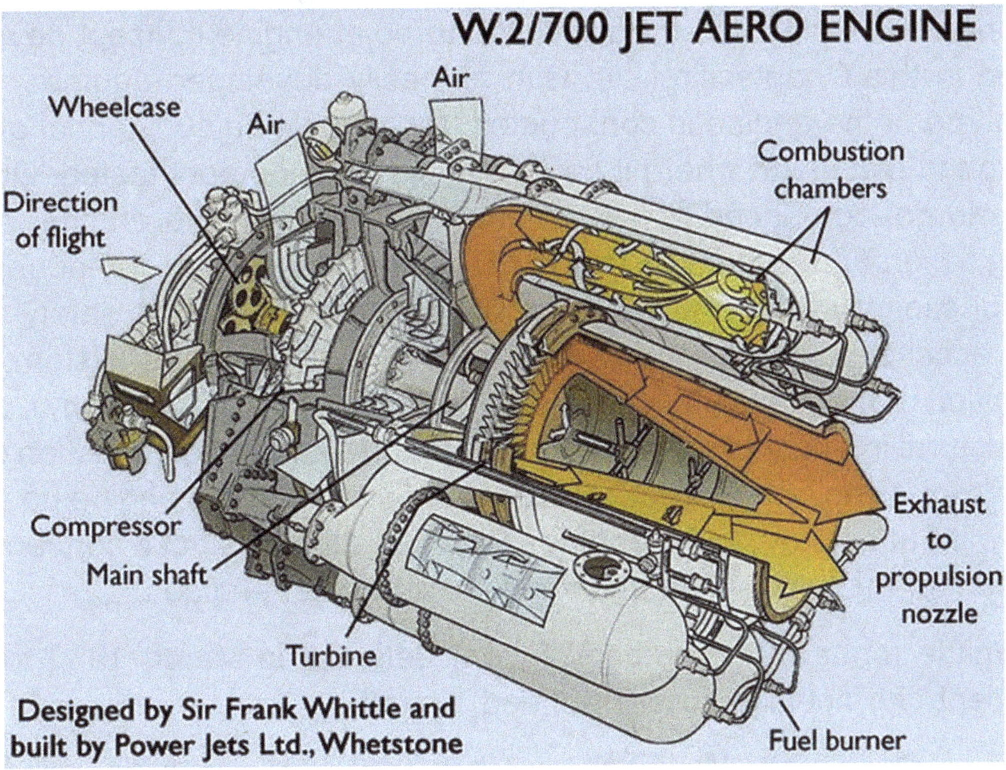

Cutaway of Whittle's centrifugal-flow turbojet engine cycle, circa 1937.
(Source: USAF Foreign Technology files)

For comparison, a cutaway of the Hans von Ohain centrifugal-flow type turbojet engine cycle is shown next.

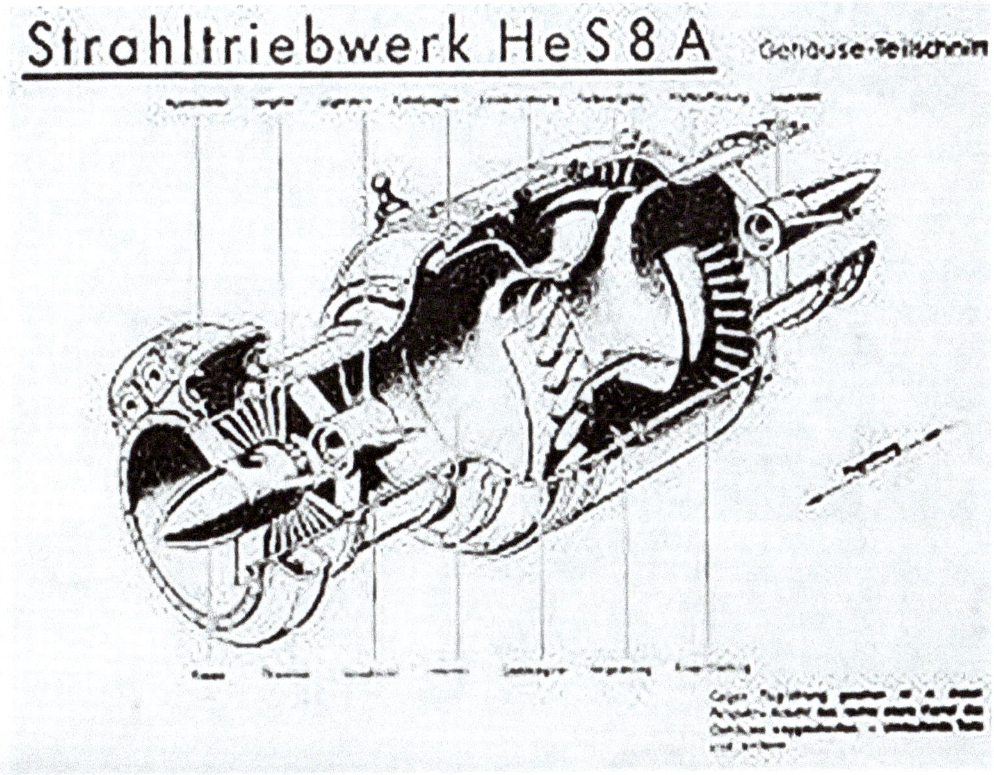

Von Ohain's centrifugal-flow type turbojet engine, He S 8 A, circa 1939
(Source: USAF Foreign Technology files)

As it was conceived earlier, the Frank Whittle turbojet engine concept development will be discussed in this Chapter first. As with all newly developed engines, after a high-quality prototype of the engine is constructed, there must be substantial ground testing done at a ground test stand where it can safely be development tested. On ground test stands there is no high-speed airflow through the inlet into the engine, and the main purpose of initial ground testing is to assure that all of the internal components within the engine are functionally safe, and are capable of demonstrating the safety and reliability of all of the accessories that comprise the engine cycle (cooling system, fuel pumps, electrical ignition of the fuel/air mix in combustors, thrust bearings (if any), turbine rotors integrity, structural integrity of turbine blades (compressor and combustion stages) etc.). [In modern turbojet engines, there is also a requirement that the engine be able to ingest a small bird without destroying the engine.] Bird ingestion was not a concern at the stage of development of the first Whittle and von Ohain turbojet engines.

Sir Frank Whittle formed the company Power Jets Ltd in March 1936 to develop his turbojet concept of propulsion for high-speed aircraft.

Shown next is a photo of the Whittle/Power Jets engine (the multiple cylindrical shaped components are the combustion chambers in the Whittle cycle) as it was undergoing ground testing.

Power Jets/Whittle engine undergoing ground stand run-ups, circa 1937
(Source: USAF Foreign Technology files)

A photograph of the Whittle engine with a human shown next to it to give a frame of reference for its size is shown next. The engine was being prepared for installation in the test bed aircraft. This view is of the exit nozzle.

Power Jets turbojet ready for installation in Gloster Pioneer, 1938
(Source: USAF Foreign Technology files)

Shown next is a drawing of the installation arrangement for the Whittle engine in a Gloster E28/39 Pioneer test bed aircraft. The Gloster Pioneer was constructed with a nose inlet that split the inlet airflow to pass laterally around the pilot's compartment and then be diverted into the engine's face (as shown on the lower half of the diagram).

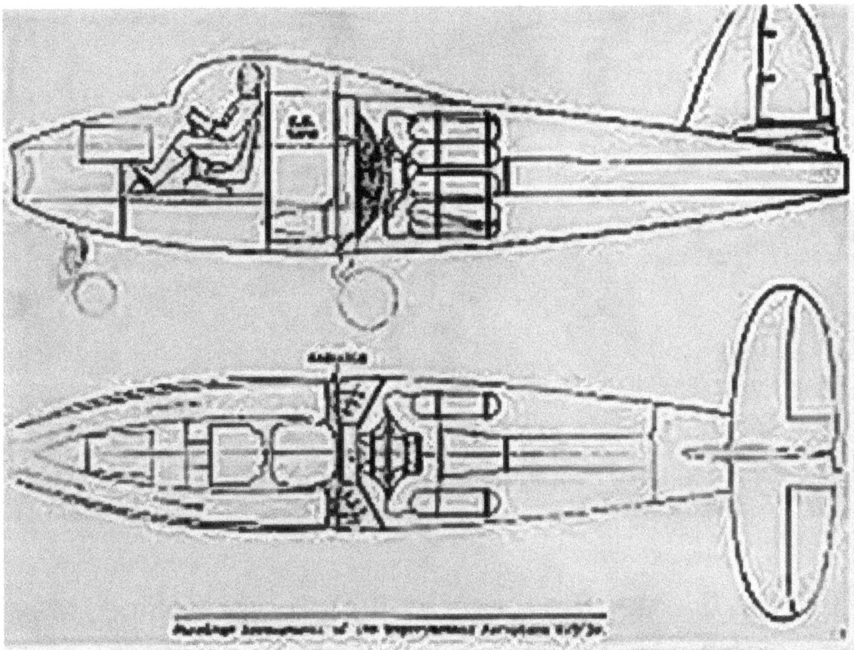

Fuselage Arrangement of the Experimental Aeroplane E28/39

(Source: USAF Foreign Technology files)

The official *first* aircraft to fly with a turbojet engine cycle was the German Heinkel 178 on August 27, 1939. It had a modified Hans von Ohain centrifugal-flow cycle turbojet engine The German jet propulsion turbojet development activity will be discussed later.

The maiden flight of the British Gloster E28/39 Pioneer, flown by Gloster's chief test pilot Philip "Gerry" Sayer, took place on May 15, 1941. This became officially the *second* type aircraft to fly with a turbojet engine cycle, the Whittle WU.2/700 engine.

[Author's note: The honor of being the first turbojet powered fighter aircraft to enter operational use during time of combat goes to the German Messerschmitt Me 262 which, as described next in Chapter 3, first flew on July 18, 1942. This flight was with the Junkers Motorenwerke Jumo 004 axial flow turbojet cycle propulsion system developed by the team headed by Dr. Anselm Franz at Junkers Motorenwerke in Germany. Two Jumo 004 turbojets were each mounted in a through-wing nacelle. The British Meteor F Mk.I aircraft, with modified Whittle turbojet engines also mounted each in a through-wing nacelle – a configuration much like that of the Me 262 - were assigned on operational status on July 27, 1944. They were withheld from air engagement with Luftwaffe aircraft and instead were used to counter the V-1 rockets launched from Germany at London.]

A photograph of the Gloster E28/39 Pioneer which housed a Power Jets Ltd Whittle turbojet is shown next.

First Flight of the Gloster Pioneer Whittle turbojet engine, May 15, 1941.
(Source: USAF Foreign Technology files)

The German Luftwaffe (Air Force) flew their Heinkel178 aircraft with a von Ohain HeS 3 turbojet on August 27, 1939, almost 2 years before the Gloster Pioneer did with the Whittle turbojet. The pilot who flew the Heinkel 178 was Erich Warsitz. Factually, the Heinkel 178 test aircraft is credited as being the world's first aircraft to fly with an all-turbojet propulsion system. Shown below is the Heinkel 178 that was the flight testbed for the von Ohain HeS 3 turbojet:

Heinkel 178 with a von Ohain HeS 3 turbojet engine, 1939.
(Source: USAF Foreign Technology files

The air inlet for the He 178 was also of the through-the-fuselage nose type. Unlike the Gloster Pioneer, which had a circular exit nozzle, the Heinkel exit nozzle was of rectangular shape.

The overall configuration of the Heinkel 178 revealed that it was more of an advanced technology demonstrator aircraft rather than being of a fully operational jet fighter. The cockpit was small, and there was no armament anywhere at the forward fuselage or wing leading edges (and having no propeller to fire through or around certainly eased that problem). The high single wing was of small area and apparently was of continuous structural configuration for its entire span (i.e., was not two half-wings with fuselage attach points). The aft fuselage was little more than a rectangular box which had a flat top to which the vertical tail (resembling a shark fin in shape) was attached. To the flat sides of the rectangular box were attached the horizontal tails (also of shark fin shape). And the shape of the engine nozzle was almost square (a square nozzle being rare in jet aircraft configuration history.

When learning that the Heinkel 178 was undergoing early development tests in 1939, the RAF accelerated its interest in the jet engine which Frank Whittle had developed in the early 1930s and was ready for sale from the Power Jets Ltd. Company, which Frank Whittle had formed to substantiate his full confidence in the turbojet engine technology.

As a result of early flight tests not meeting full expectations, some major redesign was accomplished. The propulsive power with the single Frank Whittle centrifugal flow cycle turbojet engine was found to be disappointing, and the Gloster Aircraft Company elected to design a new aircraft with two of the Whittle engines modified to be more efficient mounted in nacelles on both wings. The fuselage was lengthened, and the horizontal tail was elevated into almost a T-tail configuration to make the variable pitch angle horizontal tail less likely to be "washed-out" by the flow over the top of the lengthened fuselage when at elevated angle of attack. The pilot's cockpit was redesigned and featured a canopy that provided substantially better out-of-cockpit visibility for the pilot, especially to his rear upper hemisphere of view. In addition, two machine guns were mounted on both cheeks of the forward fuselage (total four machine guns).

The resulting Gloster Meteor configuration design changes and attempts to make it more of an operational fighter caused delay of its entry into World War II until 1944. A photograph of the Gloster Meteor will be shown in the next chapter, which covers the World War II and pre-Korean War periods from December 1941 to 1950.

In a similar manner, the Heinkel 178 performance with the Dr. von Ohain HeS 3 engine was not impressive although it met the expectations of Dr. von Ohain in that it proved the turbojet concept was the start of the new turbojet propulsion technology for the world of aviation. Both of the turbojet concepts of Sir Frank Whittle and Dr. Hans von Ohain were technically classed as being centrifugal-flow jet engines. As this chapter contains information for the period ending December 1941, it must identify that in late 1941, after witnessing the developments by Sir Frank Whittle of Great Britain and of Dr. Hans von Ohain of Germany of their centrifugal-flow turbojet engines, Dr. Anselm Franz at the Junkers Motorenwerke in Germany began research into the feasibility of a true axial-flow turbojet engine for use in high-speed aircraft propulsion. The results of Dr. Franz's efforts in the more energy-efficient axial-flow turbojet propulsion cycle did not materialize into hardware form until 1942 and thus will also be described in Chapter 3.

As history does so often to human amazement, there occurs an event which forever changes the future of the World, such as Christopher Columbus sailing across an unchartered Atlantic Ocean and discovering the New World in 1492, or Thomas Alva Edison developing the electric light bulb in 1879, or the Wright brothers (Orville and Wilbur) demonstrating in 1903 the first manned, self-powered aircraft with adequate control of its flight orientation and path, or a foreign government starting a world war, etc. The development of most technologies is either started by, or is advanced by, such type of major events having happened or which are impending.

The whole world was thrown into turmoil when the Japanese Imperial Government elected to perform the surprise attack on the US Naval Pacific Fleet anchored in Pearl Harbor, Hawaii, on Sunday morning December 7, 1941. While the World was still in a state of shock over the surprise attack at Pearl Harbor, four days later on December 11, 1941, Adolf Hitler, the Chancellor of Nazi Germany declared war on the United States and its Allies.

The major participant countries had developed military aircraft as identified in this Chapter 2 that were clearly advancements in aircraft technologies over those for aircraft used in World War I, a war which had raged twenty-three years earlier. Those advancements were also projected to be needed based on intelligence information collected which revealed the technologies being advanced in the aggressor countries for at least 10 years prior to the start of World War II. Therefore, the biggest challenge for the United States and its Allies was not to develop new designs of military aircraft to combat those of the Axis countries; the biggest challenge was producing large quantities of the existing designs of military weapons of war to effectively engage in land, sea, and air combat in two hemispheres of the world – the European theater of war in the Western hemisphere, and the Pacific theater of war in the Eastern hemisphere.

Chapter 3 next describes how the Greatest Generation met that production challenge during the initial years of World War II and how they further advanced military aircraft technologies found lacking in existing aircraft designs and capabilities as the war progressed. Chapter 3 also will describe many of the advancements made in fighter aircraft technologies during that four-year period, December 1941 to September 1945, which changed the world in an unimaginable manner and by incredible proportions. It was ultimately the development of atomic weapons technology that brought World War II to an end; however, it was necessary to use an aircraft developed with the technology advancements needed for carrying the atomic weapons for a long range and with improved technology in the domain of weapons delivery from aircraft. The readers should observe that many technologies developed for one domain (e.g., atomic energy for public utilities) will have profound effect on technology developments in another unrelated domain (e.g., in military aviation operations).

CHAPTER 3
POST-WW II FIGHTER AIRCRAFT TECHNOLOGY DEVELOPMENTS (1941 TO 1950)

The Author chose to begin this chapter than by showing a Japanese Mitsubishi A6M "Zero" taking off from the Japanese aircraft carrier *Akagi* on the flight to attack the US Naval fleet at Pearl Harbor, Hawaii, on Sunday, December 7, 1941. This politically motivated event triggered the entry of the United States and other Allied countries into a full-scale World War.

Japanese Mitsubishi A6M "Zero" taking off from *Akagi*, Dec. 7, 1941.
(Source: US National Archives)

Many books have been written and many motion pictures have been filmed describing almost every perspective of the Pearl Harbor attack, its geopolitical causes, its military objectives and planning (with many flaws), its practicing for and execution, and all the horror stories of the devastation done to warships and crews, aircraft and airfields, and humans on the ground on that day. A detailed description of those historical events cannot be repeated in this book. Military history records reveal that the US War Department (forerunner of the present Department of Defense) had fortified the Hawaiian island of Oahu and had several hundreds of various type aircraft based at the US Army Air Forces, US Navy, and US Marine Corps airfields positioned across the island of Oahu. Due to a series of flawed judgments, the principal one being to disregard the US Army Signal Corps radar detection of there being a flight of a large number of aircraft approaching Oahu from the North, and it being a quiet Sunday morning, almost none of the many military aircraft and their crews were available to mount a defense against the attack.

The Japanese had developed a very detailed and well-planned attack based on much espionage information continuously being relayed to Japanese military intelligence-gathering and analyzing operations for at least a year before the fateful day of the attack. Japanese "tourists" had been filming all naval shipping movement into and out of Pearl Harbor, and had similarly photographically detailed military aircraft locations at all airfields on the island.

In brief, the Japanese plan of attack on Pearl Harbor consisted of three waves of well-coordinated Japanese aircraft launched from six Japanese aircraft carriers (*Akagi, Kaga, Soryu, Hiryu, Shokaku*, and *Zuikaka*) with specialized weapons for use in the attack. However only the first two waves (189 fighters and bombers in wave 1, and 171 fighters and bombers in wave 2) actually conducted the attack. When the Japanese Fleet Commander Admiral Chuichi Nagumo learned that the US carriers were out to sea and not able to be destroyed, he elected to abort launching the third wave of Japanese aircraft.

A graphical version of the Japanese attack by 360 carrier-based aircraft on Pearl Harbor on December 7, 1941, is summarized next.

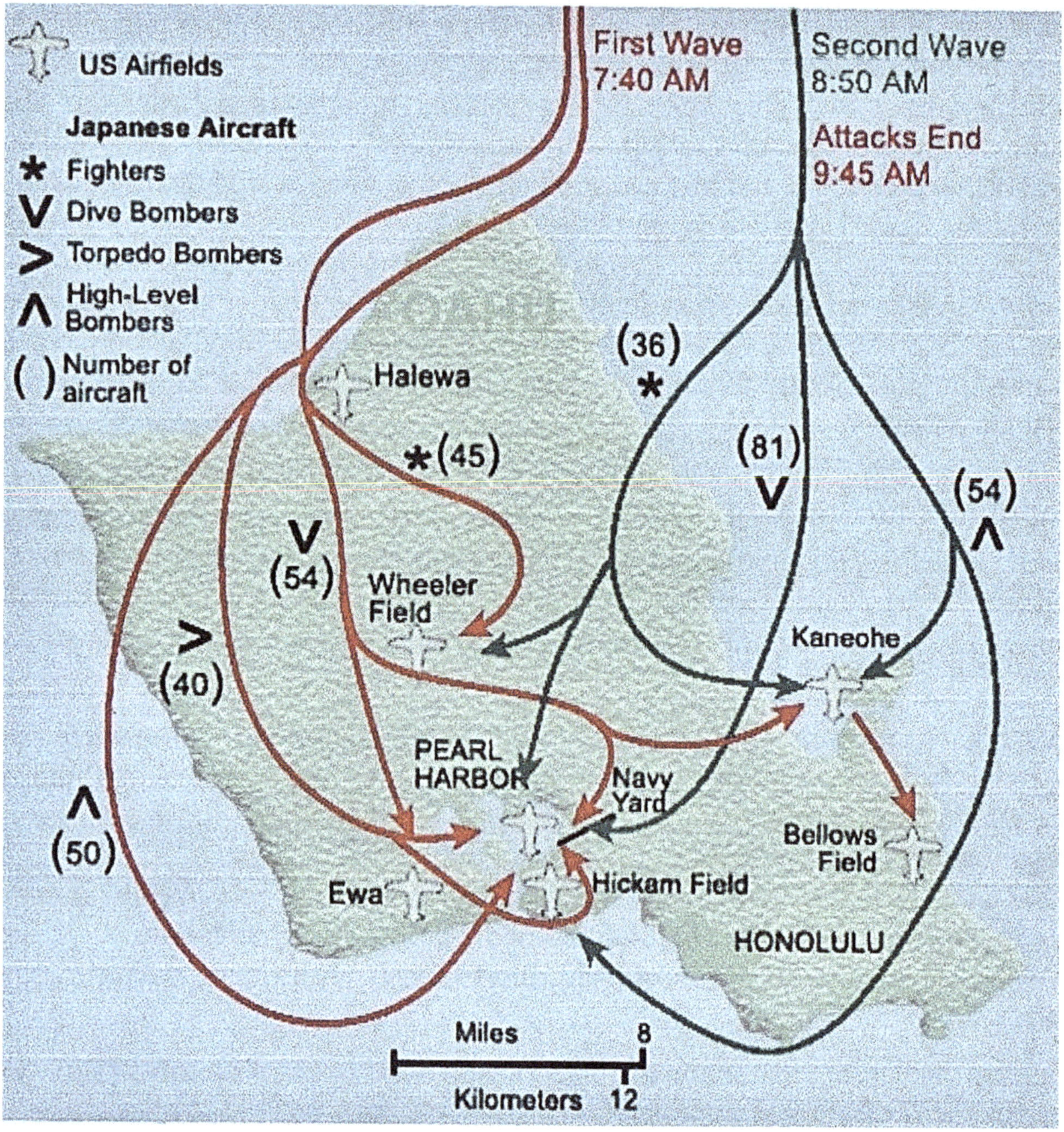

Plan of the Japanese attack on Pearl Harbor, Hawaii on December 7, 1941
(Source: US National Archives)

As shown in the attack plan, the Japanese fighter aircraft attacked six of the seven military airfields, enabling simultaneous strikes by dive bombers and torpedo bombers attacking pre-targeted ships at anchor in the anchorage section of Pearl Harbor. The primary targets were the several battleships lined in a row on the anchorages commonly referred to as "Battleship Row". The objective of having the Japanese fighters first destroying the military aircraft at the six airfields was to prevent a credible air defense from becoming airborne and disrupting the Japanese's plan of attack. A concurrent objective was to then have the dive bombers and torpedo launch aircraft to attack and destroy the priority shipping at will without concern of US fighter aircraft defending the island and its vital harbor of naval shipping.

The total number of US military aircraft (all services and types) on Oahu on December 7, 1941, was 402, of which 188 were destroyed and 159 damaged by the Japanese. The US Army Air Force (renamed from the US Army Air Corps on 20 June 1941) had various types of fighter aircraft based at Hickam, Wheeler, Bellows, Kaneohe, and Ewa airfields – Boeing P-26 Peashooters (fourteen), Curtiss P-36A Mohawks (thirty-nine), Curtiss P-40B Tomahawks (eighty-seven), and Curtiss P-40C Kittyhawks (eleven) on December 7, 1941. Of these, only eight USAAF P-40 aircraft were able to become airborne (from Haleiwa airfield) during the attack, and six of the eight P-40s each shot down at least one Japanese aircraft. The Japanese (a Mitsubishi A6M Zero) shot down one of those eight P-40s. One P-40 was also lost to friendly anti-aircraft gunfire.

Intelligence organizations within the US Army had gleaned much information from the Flying Tigers operating P-40s against Japanese aircraft in China and assessed what were the combat performance characteristics of the various designs of Japanese aircraft. That information was continually fed back to the US Army Air Corps which in turn had issued Request for Proposals (RFPs) to the several aircraft design and manufacturing companies as early as 1930 to create fighter aircraft designs capable of superior performance against the known Japanese aircraft capabilities. However, such information was not of high value if the aircraft embodying the capability cannot be brought into battle for any reason. In later chapters of this book, the highly important term "availability" in weapon systems operations will be described.

[Author's note: On December 8, 1941, the Author (then a 9-year-old boy) was prone on the living room floor of his home listening to the family radio (in 1941 there was no television in the homes) and hearing then-President Franklin Delano Roosevelt give a moving speech to Congress. *"Yesterday, December 7, 1941 – a date which will live in infamy – the United States of America was suddenly and deliberately attacked by naval and air forces of the Empire of Japan."* There were a lot more words in the middle of the speech, but it ended: *"I ask that the Congress declare that since the unprovoked and dastardly attack by Japan on Sunday, December 7, 1941, a state of war has existed between the United States and the Japanese empire."* Those historic words brought tears in the eyes of the Author's parents and are also etched permanently into the Author's memory. It caused the Author to ask "*Why did the Japanese do that?*" The explanation given to the Author by his father was that some countries do not have the space for their people and resources they need in their own country and go to war to take them away from another country which does have those space and resources. The Author interpreted it to be the same as when his older brother took the Author's favorite toy airplane (which the older brother had "borrowed" while the Author was doing his homework) and played with it until it broke, and then gave back the broken balsa wood pieces. The technology of fast-drying and good-smelling balsa wood glue was most responsive to reparability needs of the broken toy aircraft.]

After the two-hour attack on Pearl Harbor, the Japanese Naval Marshal General Isoroku Yamamoto reportedly (although never officially confirmed) made a foreboding statement to his staff on the Japanese carrier *Akagi*: *"I fear all we have done is to awaken a sleeping giant and fill him with a terrible resolve."* The probable cause for the statement was in his knowing that although several US battleships and other large cruisers had been destroyed or damaged beyond repair at Pearl Harbor, the US aircraft carriers (*Lexington* and *Yorktown*) and their aircraft were not in Pearl Harbor on December 7, 1941. To Admiral Yamamoto, that foreboding omen of having *awakened the sleeping giant and filled him with a terrible resolve* meant it was likely that those same aircraft carriers and their hundreds of aircraft would be amid others used against the Japanese aggressors at some later date. That *terrible resolve* would cause the superior technological status of the United States, renowned for their having a superior technological capability of air forces, to retaliate for the Japanese attack on Pearl Harbor.

The historic Jimmy Doolittle Raid on Tokyo on April 18, 1942, using sixteen B-25 Liberator bombers taking off from the US aircraft carrier *Hornet,* proved Admiral Yamamoto's premonition would be realized in less than four months. While the extent of damage done on Tokyo was relatively light, the boost in morale it gave to the American public was immeasurable. The photograph shown below captures the moment when the first (of sixteen) B-25 Liberators took off from the deck of the USN aircraft carrier *Hornet* on a one-way flight path vector toward Tokyo, Japan – retaliating for that attack on Pearl Harbor "with a terrible resolve".

The first B-25 Liberator bomber piloted by Jimmy Doolittle takes off from the Hornet, Apr 18, 1942. (Source: US National Archives)

From a military aviation technology perspective, it marked the first time that a large US Army Air Force (later renamed the US Air Force) bomber would take off from the deck of a US Navy aircraft carrier. War needs have a way of causing a technological capability designed for one purpose to be adapted to a seemingly impossible alternate technological capability. Inasmuch as the B-25s had absolutely no capability for returning to the Hornet and landing, it was planned that as many as possible would fly to landings in friendly China. It was also recognized that many of the 16 B-25s would have to "ditch" when their fuel ran out – wherever that would be. Readers are encouraged to read the book or see the movie, *30 Seconds Over Tokyo,* to get a better appreciation of how those pilots went far above and beyond the call of duty on that assuredly one-way mission.

The fighter aircraft of other countries identified in the previous chapter of this book saw a large amount of operational use in actual military conflicts a few years before the rest of them entered into operational service after Pearl Harbor Day (as December 7, 1941 has come to be known). The British Spitfire and Hurricane fighters flew for two years in opposition against the German Me-109 fighters in the Battle of Britain with heavy losses of aircraft and pilots on both sides of that Battle of Britain. The US P-40 Warhawks also saw combat against Japanese fighter aircraft in China, but it is hastily added that the "Flying Tigers' squadrons were composed of volunteer US pilots (not in the US Army Air Corps) and using P-40s that were officially in the Chinese Air Force inventory, not in the US Army Air Corps inventory. The US had built the Flying Tigers P-40s, and US then-President Franklin Roosevelt gave a "wink and a nod" of approval for ex-General Claire Chenault to use one hundred P-40s along with one hundred US volunteer pilots to fly them (plus two hundred ground crew volunteers to provide maintenance support for their operations) in China as an ally in defense against Japan.

Shown next is a photograph of a P-40 Flying Tigers aircraft being serviced and maintained at a base in Kunming, China, in 1942.

Flying Tigers airfield at Kunming, China ,1942
(Source: USAF photography files)

The preceding scene reflects the resourcefulness needing to be developed by ground crews in servicing and repairing operational fighter aircraft under adverse conditions in harsh environments. Many holes in the metal skin of aircraft from enemy ground gunfire were patched over with flattened beer cans during World War II. Except for when the engine or fuel tanks was hit, the relatively non-complex construction and low amount of "bells and whistles" on World War II fighter aircraft did not require many maintenance hours per flight hour of the operational aircraft. That which was needed was most often done not in a relative well-equipped hangar facility with an abundance of spare parts in supply room and ground personnel enjoying adequate protection from the weather.

The challenge of accelerating the production of weapons of war for World War II was clearly the greatest for the United States because it had to support war operations in the two theaters of war across oceans. The United States had to start with not only a rebuilding of its enormous losses in warships (primarily battleships and cruisers) and combat-ready aircraft (fighters, bombers, and reconnaissance) that were destroyed at Pearl Harbor, but also building additional aircraft carriers. It was in this time period, the first few months after Pearl Harbor, that the United States of America unified into one nation with the singular goal to defeat the Axis forces in all parts of the World. Within the fifty-eight States, Alaska, and Hawaii (Alaska and Hawaii were not yet officially US States but were US Protectorates), the Greatest Generation accepted the challenge and astounded the World by morphing its industrial resources for manufacturing products for peaceful use and manufacturing to almost exclusively manufacturing products for military purposes.

The enormity of the challenge was somewhat lessened by the consideration that the United States did not have to immediately create new technologies or even rapidly make major advancements in its military aircraft technologies that existed on Pearl Harbor Day. All of the US military aircraft identified in the previous chapter were designed during the 1930 to 1941 time period to support the war that seemed to be inevitable in Europe. Then-President Franklin Roosevelt had tried to keep the United States out of the war but the US did provide military materiel logistics support to Great Britain which was being threatened by Germany very soon after German occupied the northern coast of France. Diplomatic correspondences between Japan and the United States also indicated that a war in the Pacific region between Japan and the United States might also erupt – but it was still then unknown when and where.

For Readers who may not be aware, the then two major military services of the United States – the US Army (which included the then Army Air Force), and the US Navy (which includes the US Marine Corps) – had, and still have within their organizational structure, an activity best identified as the Foreign Technology organization. Foreign Technology agents gather as much information as is possible from very clandestine sources as to what technologies potential threat countries are developing. When assessing the information gathered, Foreign Technology agents will assess if that technology the potential threat country is developing poses a military threat, to whom, and possibly in what time period in the near future. Those assessments are provided to the research & development organizations in the US military departments charged with ascertaining if that identified foreign capability warrants upgrading of a current technological capability aboard existing military aircraft, or if the technology is of a threat level requiring development of a new aircraft design with an advanced technology capability for countering the newly identified threat.

In a similar manner, Foreign Technology personnel were uncovering much about the German fighter aircraft capabilities from undercover intelligence operatives – termed as being "the Underground" - in Europe during the Luftwaffe build-up at the start of the rise of the National Socialist Party in Germany at least 8 years before the historic Battle of (possibly better named as "over" rather than "of") Britain between the German Luftwaffe and the British Royal Air Force. It was the National Socialist Party which later morphed to become the dreaded Nazi Party. The Readers are again made aware that the Foreign Technology intelligence information about Luftwaffe military aircraft performance capabilities was provided to analysts within the US Army Air Corps who in turn converted that information into Requirements for fighter aircraft designs contained in requests for proposals (RFPs) sent by the US Army Air Force to the several US military aircraft design and manufacturing companies across the United States.

The urgent need for aircraft which had operational capabilities responsive to the dynamically changing nature of war created activities in the prototyping of new aircraft. Much use of the prototype concept was used during the 1930 to 1941 time period, and prototyping was accelerated during the early World War II years. It became fairly common for major redesign of existing aircraft and development of weapons for those aircraft to be conducted in a prototype approach as a necessity when a war is imminent or ongoing, rather than using the prolonged standard procurement process. The standard process often entails four or five years or more, depending on its technological complexity, of development activity before the procured aircraft is considered developed to "full operational capability" status. In the previous chapter, a brief description was given of the P-38 Lightning aircraft being developed on an accelerated prototyping basis. Additional information on how that P-38 and several other US aircraft applied their respective technological capabilities during World War II are discussed in this chapter.

[Author's note: The main purpose of this book is to describe the prototyping process that was used in development of the F-16 Weapon System, but that process was in itself very much different from the prototyping process used when there are extremely urgent needs for a capability to be responsive to ongoing wartime operational needs. The histories of the technology base leading up to fighter aircraft "state of the art" status were presented in summary form in the first two chapters of this book].

An introductory level description of two of the three technology development "war game-changers", which originated during the period 1920 to 1941, have been provided in Chapter 2 of this book. Those were the initial technological development of radar systems (for detecting moving objects a large distance away), and the technological developments in Germany and Great Britain of the earliest turbojet engines for propelling aircraft at high speed. The third war "game-changer" was that of atomic fission which had its scientific roots in the 1930 to 1941 time period by scientific notables, such as Enrico Fermi and Albert Einstein. It was that revelation of the energy releasing potential of atomic fission that led to development of the atomic bomb in what was known (to those having the highest security clearances) as the Manhattan Project. The dropping of two atomic bombs (the uranium–containing Fat Boy and plutonium-containing Little Man) from US B-29 bombers on the Japanese cities of Hiroshima and Nagasaki in August 1945 was the event that ultimately brought World War II to an end. It also started the atomic age and proliferation of atomic and nuclear weapons which casts an ominous shadow over the world today. It is ironic that a weapon designed to end a war may also be used to start an even more terrible war.

No one alive at the time (August 1945), including the Author, will ever forget the photographs and movie theater motion picture film clips (*Movie tone News*) depicting the ominous mushroom clouds that formed over the two cities that had just been totally obliterated – or images of the thousands of humans below who had suffered radiation burns, or worse had been reduced to skeletons in less than a second. Once again, the foreboding statement of General Marshal of the Navy Isoroku Yamamoto on Pearl Harbor Day of having "awakened the sleeping giant and filled him with a terrible resolve" was to bring unbearable retaliation upon the Empire of Japan. Regrettably, that is the extent of material that can be included within the scope of this book. Readers interested in obtaining further information on atomic weapons technology having been developed that led to the modern-day use of nuclear energy for peaceful (as well as for military) purposes are suggested to reference the myriad of historical accounts of the use of atomic weapons on Japan in the 1945 time period and for the nearly eighty years since then.

Tactical nuclear weapons are now (2017) in the weapon systems (land, sea, and air) inventories of many countries including the USA; but again, this book must constrain its scope to describe only the integration of nonnuclear weapons into the aircraft inventories of the US military services for use in conventional warfare.

As demonstrated by the fighter pilots of World War I, the effectiveness of fighter aircraft technology advancements is only as good as the training and experience of the pilots applying it in air combat. The final pages of Chapter 3 will summarize some lessons learned from a few supercharged aircraft and the pilots who became Aces (an Ace is a pilot having a high number of kills of enemy aircraft) in them.

Redirecting the content of this book at this point then to reflect technology developments specifically for fighter aircraft that emerged during World War II, and especially those that are in the genealogical tree of the F-16 Fighting Falcon weapon system, this chapter will first concentrate on the use of the earliest developments of turbojet propulsion introduced in Chapter 2.

Recognizing that the centrifugal-flow cycle of turbojet propulsion for high-speed military aircraft was proving to not be an efficient form of propulsion for the German Luftwaffe, as demonstrated by flight testing of the von Ohain centrifugal-flow type cycle HeS 3 engine in the Heinkel 178, Dr. Anselm Franz at the Junkers Motorenwerke company in Germany conceived an axial-flow turbojet cycle that was far more promising. It is a general principle in aerodynamics, both external and internal, that the least diversion of the airflow over or through a surface will cause the least expenditure of energy by maintaining linear, streamlined flow of that airflow. Readers may see for themselves that comparing the Dr. Franz axial-flow turbojet is inherently more efficient than the centrifugal cycles of both Dr. Hans von Ohain and of Sir Frank Whittle. The routing of the airflow through combustion chambers causes energy losses to the cycle and which are not present in the Dr. Franz axial -flow cycle. The axial flow of the Dr. Franz turbojet cycle is illustrated in the diagram next.

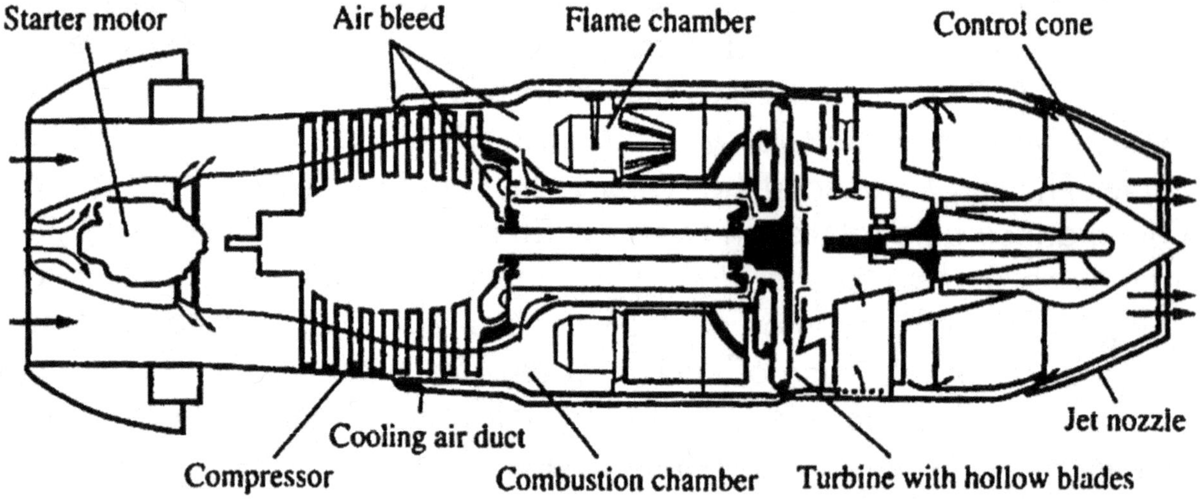

The Dr. Anselm Franz axial-flow turbojet cycle, Jumo 004 Engine, 1942.
(Source: USAF Foreign Technology files)

This axial-flow cycle is the baseline cycle for the development of all turbojet engines developed for high-speed military and commercial aircraft throughout all countries in the world. In aviation history, the technology advance made by the Wright brothers in 1903 for successful controlled flight and the technology advancement made by Dr. Anselm Franz and his team with the axial-flow turbojet cycle are considered to be of equal significance.

For the benefit of those readers who may be propulsion technology challenged, a verbal description of the axial-flow cycle may be of interest. A mass quantity of air, at a density, pressure, temperature, and velocity of the flight vehicle, continuously enters the generally circular-shaped inlet at the front face (left end in the diagram) of the engine. This mass of air is then passed through multiple stages of compression (many small turbine blades attached to a rotating disk fixed on a rotating axle) which greatly raises the pressure and temperature of that mass of gas. The compressed heated gas exits the final stage of compression and enters into the combustion chamber, into which is injected aviation fuel. The compressed, heated air nearly instantaneously ignites that fuel/air mixture. When ignited, the already high pressure, high temperature air is elevated to a higher pressure and temperature and passed over one or more stages of turbine blades mounted on a rotating disk and to which is attached the end of the central shaft. That rotating central shaft proceeds forward through the engine and drives the multi-staged compression disks and turbine blades mounted to that same shaft.

The very high pressure, high temperature mass of air is then passed through and exits from a converging area nozzle which ejects the plume of heated high-pressure air at a substantially higher velocity into the surrounding atmosphere which causes development of a net thrust from the engine. This is a self-sustaining compression, combusting, expansion process which propels the aircraft through the air at a high speed. The pilot will modulate the rate at which fuel is injected and ignited by use of an engine throttle mechanism, and thereby will allow the pilot to have control over the forward speed of the aircraft. It is a process very similar to that of a fuel injection system in a modern automobile engine, but with the major difference being that a torque is generated driving the drive-wheels of the automobile rather than by the combustion process generating a high temperature and pressure column of air producing a thrust to produce the forward motion.

In order to keep the engine as cool as possible when installed inside an airplane, various forms of cooling air are passed over the external combustion chamber surface and walls of the engine covering the high-pressure high temperature turbine blade disk(s). As might be expected, the technology of turbine blade design is an ever improving one especially in the materials used to construct the turbine blades. The blades must not have even the smallest of internal flaws as that flaw will cause fatigue cracks which may lead to turbine blade failure. Many turbines blade failure-caused accidents have happened in the decades since the first turbojet engine, but their rate has been decreasing as the technology database for turbine blade design and construction continues to improve. As with almost every failure and aircraft crash, the lessons learned from the accident are applied to reduce the probability of the cause of the turbine blade failure from recurring.

In modern military turbojet engines, a device is added into the nozzle at the exit from the high pressure/temperature where additional fuel is injected (through multiple stages of rings with fuel injectors) into the already high pressure, high temperature air to produce still higher amount of net thrust by the turbojet engine. In the US, this device is referred to as an afterburner but in some countries, it is referred to as reheat. By whatever terminology, the afterburner or reheat increases the net thrust from the engine very substantially, and permits aircraft to fly at extremely high speeds. A flight Mach number of 3 (three times the speed of sound) or greater is achievable with modern air-breathing turbojet engines. At extreme speeds, it is not uncommon that the materials used in the construction of the airframe to withstand the aerodynamic heating of the external parts of the aircraft from friction of air molecules on the skin of the aircraft becomes the limiting factor on aircraft speed rather than the engine thrust generating capability. For example, aircraft constructed of aluminum are generally limited to a Mach number of approximately 2; whereas aircraft constructed of titanium components are capable of the Mach 3 speeds or greater. [Looking ahead in time, readers should recognize this description of propulsion and airframe integration as being a case of a technology developed to produce a desired capability being impacted by the limits of capability of a totally unrelated technology. It is hastened to be added that the subject of this book being the USAF Lightweight Fighters, it was decided that for cost reasons, the use of aluminum would be extensively used for the manufacture of the USAF Lightweight Fighter aircraft. With that material decision to use aluminum rather than titanium, the associated maximum speed of the Lightweight Fighter aircraft was limited to be approximately Mach 2.]

Returning to the mid-1940s time period, he British converted their turbojet engine cycle developments from that of the Whittle centrifugal flow type to the more efficient axial flow cycle, and developed a two jet engine configuration aircraft named the Gloster Meteor. A diagram of the Gloster Meteor, cutaway to show the installation of the dual turbojet cycles in nacelles which were mounted spanwise in the wings of the Gloster Meteor, is shown next.

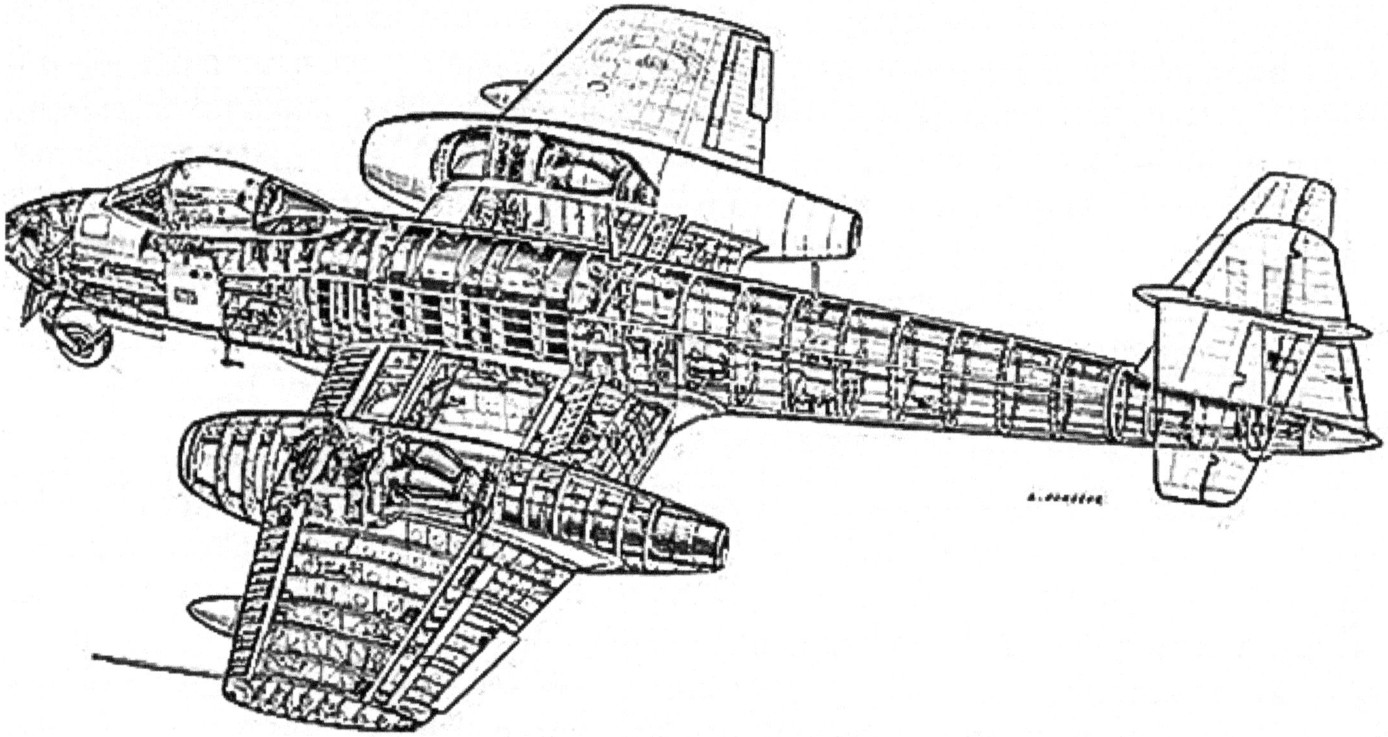

Gloster Meteor F Mk.I with Rolls-Royce Welland engines, circa 1943.
(Source: USAF Foreign Technology files)

The small cutaway diagram of the left engine nacelle illustrates that the cylindrical combustors prominent on the exterior of the original Whittle engine have also been modified to point down toward the engine axial centerline. That would make the internal flow path of air through the Rolls-Royce Welland engine come closer to resembling more of an axial-flow turbojet, and to be more efficient than that of the original Whittle engine. The Rolls-Royce Welland turbojet engines were built under license from the original Whittle design.

The Gloster Meteor F Mk.I had its first flight on March 5, 1943, and first entered the RAF operational inventory during July 1944, near the end of World War II. Shown next is an in-flight photograph of a Gloster Meteor F Mk.III with the Whittle-based Rolls-Royce Welland turbojet engines.

Gloster Meteor F Mk.III with dual axial-flow turbojet engines, early 1945. (Source: USAF Foreign Technology files)

However, the RAF did not want to risk the somewhat unimpressive maneuvering capability of the Gloster Meteor Mk.I in air combat with Luftwaffe aircraft and instead used the Gloster Meteor Mk.I to intercept and destroy incoming V-1 rockets from German launch sites. The first few V-1 encounters failed because the cannons on the Meteor did not fire. The more successful destruction technique was to have the Meteor fly wingtip to wingtip with the V-1 and tip one wing of the V-1 (up to roll the V-1 away from the Meteor) causing the V-1 to roll, thereby confusing its programmed-in guidance, and to crash harmlessly in open countryside outside London. Shown below is a Meteor preparing to tip the V-1 wingtip out of control.

Gloster Meteor preparing to tip German V-1 rocket out of control
(Foreign Technology Photograph)

The RAF did not fly the Gloster Meteor F Mk.III over German occupied territory as the RAF feared disclosing that advanced turbojet technology should a Meteor be downed. As a result, the Gloster Meteor F Mk.III was employed to perform close-air support missions and reconnaissance missions over Holland, which did not fully utilize the high-speed capability of the new jet propulsion technology to its full advantage. The Meteor F Mk.III was limited to 400 mph at altitude below 15,000 feet, or to 450 mph below 8,000 feet in calm air. At 450 mph at 8,000 feet altitude, the straight wing Meteor F Mk.III did not encounter the same adverse effects that the P-38 Lightning had encountered due to its entirely different external configuration and structural design of the Meteor F Mk.III. It did not have the plumb-bob tendency that the P-38 center gondola had when the aerodynamic center of the P-38 wing shifted aft as the flight speed approached transonic speeds. In addition, the P-38 had a horizontal tail mounted on two aft booms of the forked-tailed devil (*Gabelschwanz-Teufel* as the Germans referred to it) or of the "two airplanes, one pilot" (as the Japanese referred to it later in World War II). A flutter-like tendency of the aft structure of the P-38 apparently developed at high speed and was aggravated by compressibility effects also. The Gloster Meteor more conventional configuration was more tolerant to the effects of transonic compressibility.

In the period 1943 to 1955, 3,947 British Gloster Meteor variants were produced for use by three countries besides Great Britain – Australia, Belgium, and Argentina. It had also set several speed records during its life. The first official speed record of a jet powered aircraft was set by a Gloster Meteor F.Mk III on November 7, 1945, when it attained 606 mph. In 1946, a Meteor F.Mk IV reached a speed of 616 mph. Other performance-related records were broken by Gloster Meteors in the categories of flight-time endurance and rate of climb.

A fallout capability of the jet-powered aircraft, which were designed to achieve high speeds, resulted from flight experience with the British Gloster Meteors. That fallout capability was the ability for the turbojet machinery to operate at higher altitudes than did internal-combustion engines on propeller-powered aircraft. Despite the addition of superchargers to the piston engine/propeller-equipped aircraft, which did allow them to operate at high altitudes, the loss of efficiency of the propellers "screwing" through the very low-density air at high altitudes limited their high-altitude capabilities. Perhaps useful in a Game of Trivia, when the first propellers were being designed long before the Wright brothers used them, propellers were often referred to as air screws. When in nautical use to propel boats and ships, propeller blades were also referred to as water screws. So as to be historically correct, Leonardo Da Vinci in the 1600s conceived the principle of the screw through fluids for both water-based and air-based applications.

In essence, turbojet engines use multi-stage compressor turbines to compress the low-density air at high altitude entering through the large inlet(s) (which was the function of conventional superchargers in piston powered internal combustion engines) as that air is entering into the combustion section for combustion at an efficient fuel-air ratio. The high-pressure high temperature air exiting the jet engine nozzle produces thrust without the use of the low efficiency propeller (regardless of being or not being supercharged).

Although not available in the technology applications during World War II of turbojet propulsion, the addition of an afterburner to a basic turbojet engine can greatly augment the thrust of the basic engine. It does so by adding fuel through an array of zoned circular rings which inject the fuel into the very high-temperature high-pressure airflow exiting through the nozzle. The afterburner technology does also require adjustment of the nozzle's exit area to make maximum benefit of having the afterburner. The disadvantage of an afterburner is not that it adds a great deal of weight or complexity to the propulsion system, but it results in fuel being burned in the combustor section and in the afterburner nozzle section at a very high rate. Afterburners are generally used in air combat (to achieve high turn rate and/or acceleration) or for rapid-response takeoffs in smaller-size fighter aircraft having inherently low fuel fractions.

[Author's note: To those readers who are planning to enter into a career of aircraft propulsion, it is again to be noted that when a technology is being developed for some main purpose and application, it is common that other effects will arise, both beneficial and adverse. In the Preface of this book showing the graphical representation of the stages of development of a technology, the final phase in that technology process was shown as being the final flight-testing of a prototype version of that technology. The flight testing of the British Gloster Meteor series of aircraft is a good example of technology fallouts not planned for earlier in its development.]

Returning to the development of turbojets for aircraft propulsion in Germany in the 1930s, the flight test results obtained following the first flight of the Heinkel He 178 on August 27, 1939, the inefficiency of the installed von Ohain type of turbojet engine did not highly impress the Luftwaffe. Unknown to Heinkel, BMW (Bayerischen Motoren Werke) and the Junkers Motorenwerke were developing the alternate axial-flow cycle turbojet form of propulsion. Those companies were finding that axial-flow cycle to have higher efficiency in the three factors of flight; speed, service altitude, and fuel consumption rate. In mid-September 1939, the German Air Ministry ordered German aircraft manufacturers to reduce all exploratory development work and to concentrate all efforts on winning what German officials expected to be a short war in Europe. Luftwaffe Reichmarshall Hermann Goering also proclaimed that what Germany needed to bring a quick end to the war in Europe were more bombers rather than high-speed fighter aircraft. The rate of development of any technology is proportional to the interest and needs of those in command of the Air Force of that country The Air Ministry's order resulted in a twelve months delay in development of the axial-flow turbojet technology.

Following the Japanese raid on Pearl Harbor on December 7, 1941, which brought the United States, a sleeping giant, into the war in both Europe and in the Far East, Messerschmitt and BMW accelerated their effort in further refinement of the design of a single-seat, high-speed fighter concept built around the use of an axial-flow turbojet cycle. On July 18, 1942, the Messerschmitt Me 262 Schwalbe (Swallow), propelled with a Junkers Jumo 004 axial flow turbojet engine, made its first successful flight. It was piloted by test pilot Fritz Wendel.

A photograph of the Me 262 Schwalbe is shown next:

Messerschmitt Me 262 *Schwalbe* with Jumo 004 turbojet engines,1944.
(Source: USAF Foreign Technology files)

The paint scheme on most of the Messerschmitt Me 262s was such that when seen from below (most often, as the Me 262 had a higher altitude capability than any other aircraft in World War II), the light coloration was difficult to discern against white clouds in a blue sky. When seen from above, the dark green paint scheme on upper portions blended better against ground foliage coloration. The mottled color appearance at the junction between light and dark painted regions was said to make the Me 262 look like a shark in the ocean. Regardless of the paint scheme, the smooth aerodynamics of the external configuration invoked the old credo of experienced aerodynamicists: "If it looks good, it probably flies good." The Me 262 did just that and far better than the Swallow (*Schwalbe*) designated as its official name.

Soon after its initial flight, a design change was made to the Me 262 landing gear from being a conventional landing gear (two main landing gear under the wing and a tail wheel [under the tail, which was found to interfere with aircraft rotation at takeoff speed]) to being a retractable tricycle landing gear. The Me 262 was declared to be in operational status after its first operational flight on April 19, 1944, and it officially became the first jet fighter in any country to enter operational service. Further hampered by the lack of engines (due to the earlier Air Ministry proclamation), serial production of the Me 262 did not begin until June 1944, and the delivery rates to operational organizations were slow at best (twenty-eight Me 262s in June, fifty-nine in July, and only twenty in August).

The wing of the Me 262 was redesigned from having a straight wing leading edge to having 18.5 degrees of sweep to its leading edge The small sweepback angle was introduced primarily because the construction weight of the wings was much heavier than that which designer had planned, and sweeping the leading edge back 18.5 degrees was done primarily to redistribute the wing's weight effect on the aircraft's center of gravity. This low amount of sweepback was not significantly effective on reducing compressibility effects on the wing, however. It did little to minimize the effect of the shift of the aerodynamic center of the wing aft at high speeds and degrading the longitudinal stability (static margin distance between aerodynamic center and weight center of gravity) of the aircraft.

The speed and altitude capability were less than what the Luftwaffe performance goals had been asked for. Once airborne, the Me 262 could accelerate to speeds over 530 mph (reportedly red-lined at 540 mph), over 90 mph faster than any other Allied propeller-driven fighter operational in the European Theater. The Me 262 also had a better altitude capability than the P-51 Mustang with supercharging, making it very effective against the P-51s escorting Allied bombers into raids on manufacturing facilities well inside Germany late in World War II. The inertial effect of the two large engines, mounted in pods under the wing and quite a distance outboard from the fuselage centerline, was a significant retardation of roll onset rate and thus the maneuvering capability of the Me 262 was not impressive. The Me 262 was most effective in slicing–type tactics on both the P-51 escorts and the bombers being escorted by the P-51s.

Gunners of Allied bombers found that their electrically powered gun turrets had problems tracking the Me 262s when in a high-speed slicing attack. Target acquisition was very difficult because the Me 262s closed into firing range very quickly, remained in firing position only for a very short time (less than 2 seconds), and then broke off the attack before they could be acquired in a target acquisition solution by gunners.

An excerpt from documents at the US National Archives well describes the eventual counter tactics to the Me 262s: *"Allied pilots soon found that the only reliable way to destroy the Luftwaffe Me 262 aircraft was to attack them on the ground or during takeoff or landing. Luftwaffe airfields identified as jet bases were frequently bombed by medium bombers, and Allied fighters patrolled over the fields to attack jets (Me 262s) trying to land. The Luftwaffe countered by installing flak alleys of anti-aircraft guns along the approach lines to protect the Me 262s from the ground – and by providing top cover during the Me 262s takeoff and landing with the most advanced Luftwaffe single-engine (propeller driven) fighters, the Focke-Wulf Fw 190D and (just becoming available in 1945) Focke-Wulf Ta 152H. Nevertheless, in March-April 1945, Allied fighter patrol patterns over Me 262 airfields resulted in numerous Me 262 losses."* Another account (using a familiar name known to anyone interested in US military aircraft history) referenced *"Lt. Chuck Yeager of the 357th Fighter Group was one of the first American pilots to shoot down an Me 262, which he caught during its landing approach."*

Operations of the Me 262 (as well as all other war machines in German armed forces) with the Luftwaffe were also curtailed somewhat in the late 1944 period after the D-Day invasion by Allied Forces on the Normandy coast when POL (petroleum, oil, and lubricants) became in short supply. The destruction ratio statistics of the Me 262s were rather good considering their short history in air combat and the lack of support from dwindling ground resources (especially POL and repair parts) as the air war decreased over Europe. Me 262 pilots reported destroying 542 Allied airplanes (primarily bombers and fighter escorts), with the Allies reporting having destroyed about 50 Me 262s in the air and another 150 on the ground.

Several losses of Me 262s were also due to the in-flight failure of its turbojet engines. German engine manufacturers had difficulty finding materials and alloys of sufficient strength at high temperatures and stress loading for the last stage of the high temperature turbine blades in their turbojet engines. Allied bomb raids on strategic manufacturing facilities were responsible for destroying vital manufacturing of ball bearings, airplane and tank manufacturing facilities, hydroelectric facilities, and POL refineries and storage facilities. In the last months of the war, the Me 262s had a failure rate of about one engine per fifty flight hours, and there were no spare engines or parts to repair or replace those that failed.

At the end of World War II in Europe (April 1945), a total of 1,430 Me 262s had been manufactured. The Czechoslovakian Air Force continued to operate the Me 262 in its air force inventory until 1951.

The United States clearly from all historical documentation was not unaware of the probability that a second World War would erupt, and the US Army Air Corps, designated as a major organization of the US Army in 1926, was therefore instrumental in, and deserving of highest credit for, the major advances made in all of the technologies developed for land-based military aircraft from 1930 through June 1941 preparing for the inevitable war. In the early years of World War II in the European and Pacific Theaters, it was that military aviation base of pre-war technology developed 1930 to 1941 by the Allied countries preparing for air war in Europe that bought time by thwarting Axis countries from using their then-superior air power to perform air cover and support for the Axis ground forces overrunning neighboring countries and capturing their mineral and POL resources.

On June 20, 1941, the US Army Air Corps was re-designated as the US Army Air Force. The US Army Air Force was delegated responsibility for all US Army (i.e., not including the US Navy) development of all land-based military tactical aircraft from June 1941 (6 months prior to the Japanese attack on Pearl Harbor), through the entire duration of World War II (December 1941 and up to September 17, 1945). On September 18, 1947, the United States Air Force was designated as the third United States military department – the three departments being the Department of the Army (formed in 1784), the Department of the Navy (formed in 1775), and the Department of the Air Force (formed in 1947, two years after the end of World War II).

In the last year of World War II, as the German Third Reich was collapsing, the three World powers, the United States, Great Britain, and Russia, all had military and political sights on Berlin, the capital of Germany and Adolf Hitler's bunker beneath the Chancellery.

Five-star General Dwight D. "Ike" Eisenhower, who was the Supreme Commander of the Allied forces in Europe, made many monumental decisions during World War II that steered the course of the Allies toward victory in Europe and Africa. The decision as to what day and how there would be the invasion of the Normandy Coast of France on D-Day, June 6, 1944, was solely his to make – and was not challenged by the other Allied Forces Commanders (most notably General Sir Bernard Montgomery of Great Britain). It was another decision Ike made that would haunt World history and that was to reward the USSR (Union of Soviet Socialist Republics) for their having suffered the loss of over 23.4 million of its soldiers and civilians during World War II. (World War II had cost 72 million lives in all countries). Ike's decision was to not advance the British and American Forces into Berlin until it had been taken possession by the Soviet Union military forces (on May 2, 1944). On May 7, 1944, the ranking German Army officer, General Alfred

Jodl accepted the terms of unconditional surrender and World War II in Europe was ended (see photo below.)

Gen. Alfred Jodl signs Germany's unconditional surrender, ending WWII in Europe. (Source: US National Archives)

In August 1945, the USAAF dropped two atomic bombs on Hiroshima and Nagasaki (both air bursts, so that more correctly the bombs exploded *over* Japan). In historical irony, the day that the first atomic bomb was dropped on Hiroshima (August 4, 1945), the newspaper headlines of that first atomic bomb drop overshadowed the tragic news that the USAAF top ace, Major Richard I. Bong, who had forty "kills" in his P-38 Lightning of Japanese aircraft in the Pacific Theater of Operations, had been killed in a crash (in California) during an acceptance flight of a newly-manufactured Lockheed P-38. On August 15, the Emperor of Japan announced to the World that the Japanese Empire would cease all war activities to end the World War. The photograph shown below of the Japanese surrender in Tokyo Bay aboard the USS Battleship *Missouri* (Big Mo) was taken on September 2, 1945, as more than 220 military aircraft from Allied countries did a flyover above the formal surrender ceremonies. It fittingly displays the "resolve" premonition that was foreseen by Japanese Fleet Admiral Yamamoto only four years earlier.

The surrender of Japan in Tokyo Bay to end World War II, September 2, 1945.
(Source: US National Archives)

Various estimates have been made by military aviation historians as to the number of aircraft that participated in the flyover of the US battleship Missouri on September 2, 1945, but the average of the estimates was that the flyover fleet of aircraft numbered 210 aircraft from each of the Allied countries of World War II.

[Author's note: Not to discount the influence of the awesome power of atomic weapons causing the emperor of Japan to unconditionally surrender and end World War II, another influence, less well known but equally as influential, existed. The Author was too young to serve in the active military services during World War II, but felt it to be his duty to become a Junior Military Cadet in 1944 at the age of 12 and a half. Shown next is a photograph of Cadet Herb Hutchinson on duty in front of his residence in Maspeth, Queens, New York.

Cadet Herb Hutchinson, Queens, New York, 1944.
(Author's family photo)

It is believed (but not by many) that those same Japanese "tourists" who had passed photographic intelligence back to Japanese military planners during their planning for attacking Pearl Harbor had also passed a copy of this photo back to the Emperor of Japan which caused him to unconditionally surrender rather than risk encountering such a fierce American warrior on some island battlefield.]

After the war in Europe, the United States and Soviet Union military forces quickly acted to capture the total technological resources of Germany in terms of scientific research facilities and scientific and engineering brainpower. As an example, the United States captured, dismantled, and shipped back to the United States the world's two largest air compressors in Germany, which had been used by the Germans doing research in transonic and supersonic aerodynamics flow using the large wind tunnels. Those two compressors are still in operation at the Arnold Engineering Development Center at Tullahoma, Tennessee, and used to operate the AEDC 16T (T designating "transonic") and AEDC 16S (S designating "supersonic") wind tunnels, both having 16-feet-diameter test sections. They have been used since the end of World War II to ground test all military (and a few commercial) large-scale wind tunnel models of aircraft that have entered the US military aviation inventory.

There were also over 1,600 scientists and engineers who were brought (voluntarily, rather than falling into Soviet captivity) to the United States in Operation Paperclip in the years from 1945 to 1952. The USAF sponsored the largest number of Paperclip scientists, importing 260 men. 86 aeronautical engineers were transferred to Wright Field, near Dayton, Ohio, where the United States had Luftwaffe aircraft and equipment captured under Operation Lusty (*Lu*ftwaffe *S*ecret *T*echnolog*y*). Thirty-six returned to Germany and one (Walter Schreiber) re-emigrated to Argentina. One of the most notable among those who were transferred to Wright Field (now Wright-Patterson AFB) was Dr. Hans von Ohain who shared development of the first gas turbine (turbojet engine) cycle with Frank Whittle of Great Britain.

[Author's note: Dr. von Ohain was assigned to the Propulsion Laboratory at W-PAFB and was instrumental in the early development of United States gas turbine engines used in US Air Force military aircraft starting with the F-90 Lockheed Shooting Star (the USAF first turbojet-powered fighter interceptor aircraft shortly after World War II ended) and at the start of the design of the North American F-86 Sabrejet. This Author is honored to have known Dr. von Ohain personally and professionally at Wright Patterson Air Force Base where the Author was employed as a civilian Aeronautical Systems Engineer at the USAF Aeronautical Systems Division starting in 1961. Dr. Hans von Ohain was a resident of the community of Kettering, Ohio, five miles from the W-PAFB complex and where he and his wife Hanny attended the same Lutheran Church in Kettering as the Author. Hanny had a beautiful singing voice and sang in the church choir. Dr. von Ohain died in 1998 at age eighty-six (in retirement in Melbourne, Florida) and is buried in a cemetery in Dayton, Ohio next to his beloved wife Hanny. Dr. and Frau Ohain are survived by four children.]

A second notable scientist who came to the USA after World War II was Werner von Braun who had worked at Peenemunde, Germany in developing the V-1 and V-2 rockets used against England. Dr. von Braun went to the Army Redstone Arsenal (now the NASA Huntsville Space Center) at Huntsville, Alabama, and was instrumental in developing the liquid fueled rocket propulsion engines for the first US man-in-space programs.

Before continuing with the advancement of airframe and propulsion technologies that were achieved for/during World War II, the Author wishes to remind the readers that technology advancements do not happen by themselves; there is always human involvement that "bends the twig" leading to those advancements of existing and newly emerging technologies. In addition, the measure of how effective a technology is after it has been developed, with much analysis and laboratory testing, lies with those who employ the technology and use it in its final intended environment. Those who are most successful in advancing the technologies of the weapons of war are those who have close experience with how weapons of war actually perform in the hazardous, physically challenging environment of war and must use the sometimes-agonizing measure of effectiveness of kills.

Readers are also again reminded that the objective of every combatant in a war is to defeat the enemy's capability to wage war. A pilot in a fighter aircraft engages in air combat with an enemy pilot who is operating his/her aircraft with the same objective. Whichever pilot in a one-on-one engagement is able to maneuver his/her aircraft and employ the air-to-air weaponry built into his/her aircraft so as to destroy the enemy aircraft is credited with an air victory, sadly referred to as a kill. It is of little victory to kill an opposing pilot; whereas, it is of great pride to a pilot to have removed a weapon of war that the enemy could no longer use. Those who remove five or more enemy aircraft from the air are dubbed as being aces. Both sides (Axis and Allied) of World War II produced many hundreds of aces in their military air force ranks, as shown in the following lengthy tabulation by country and in descending order of the number of aces in that country during 1941 to 1945.

Military "Aces" of World War II - Allied and Axis (European and Pacific Theaters of Operation (ETO & PTO)					
Country (Name of Air Force)	Affiliation / ETO / PTO	# Aces	# Air Victories	Avg Victories per Ace	Top Ace in Country and Number of Air Victories
United States (US Army AF, USN, USMC)	Allied / ETO & PTO	871	6711	8	Richard I. Bong **40** (in PTO) Thomas B. McGuire **38** (in ETO)
Germany (Luftwaffe)	Axis / ETO	435	33929	78	Erich Hartmann **352** (Top Axis Ace) Kurt Welter **25** (1st Ace in jet Me262)
Finland (Finnish AF)	Allied / ETO	53	1426	27	Ilmari Juutilainen **94** (Top Ace)
Romania (Romanian AF)	Allied / ETO	43	586	14	Constantin Cantacuzino **69** (Top Ace)
United Kingdom (Royal AF) (Great Britain, Ireland)	Allied / ETO	38	673	18	James E. Johnson **38** (Top RAF Ace) "Paddy" Finucane **28** (Top Irish Ace)
Poland (Polish Air Force)	Allied / ETO	35	310	9	Stanislaw Skalski **18.9** (Top Ace)
Italy (Regia Aeronautica)	Axis / ETO	31	341	11	Teresio Vittorio Martinoli **22** (Top Ace)
Australia (Royal Australian AF)	Allied / PTO	30	362	12	Clive Caldwell **28.5** (Top Ace)
Japan (Imperial Army AF & Imperial Navy AF)	Axis / PTO	21	726	35	Tetsuzo Iwamoto **80** (Top Ace)
Soviet Union (Soviet AF)	Allied / ETO	21	806	38	Ivan Kozhedub **66** (Top Allied Ace)
Croatia (Croatian Air Force)	Allied / ETO	12	207	17	Mato Dukovac **44** (Top Ace)
Canada (Royal Canadian AF)	Allied / ETO	10	187	19	George F. Beurling **31.5** (Top Ace)
New Zealand (Royal NZ AF)	Allied / PTO	9	161	18	Alan Christopher Deere **22** (Top Ace)
Slovakia (Slovak AF)	Allied / ETO	9	168	19	Jan Reznak **32** (Top Ace)
China (Chinese National AF, Flying Tigers)	Allied / PTO	5	47	9	Lui Chui-Kong **7** (Top Ace) Robert Neale **13** US Volunteer (Top Ace)
France (Free French AF)	Allied / ETO	5	111	22	Pierre Clostermann **33** (Top Ace)
Czechoslovakia (Czech AF)	Allied / ETO	4	62	15	Karel Kullewascher **18** (Top Ace)
South Africa (Ryl S African AF)	Allied / ETO	4	94	24	John "Jack" Frost **15** (Top Ace)
Greece (Greek Air Force)	Allied / ETO	3	20	7	Steve Pisanos **10** (Top Ace)
Hungary (Royal Hungarian AF)	Allied / ETO	2	57	28	Dezso Szenigyorgi **30.5** (Top Ace)
Belgium (Royal Belgian AF)	Allied / ETO	1	6	6	Remy Van Lierde **6** (Top Ace)
Bulgaria (Ryl Bulgarian AF)	Allied / ETO	1	15	15	Stoyan Stoyanov **15** (Top Ace)
Iceland (Royal Iceland AF)	Allied / ETO	1	5	5	Tony Jonsson **5** (Top Ace)
Netherlands (w/RAF & RCAF)	Allied / ETO	1	7	7	Hugh C. Godefroy **7** (Top Ace)
Norway (Norwegian AF)	Allied / ETO	1	14	14	Svein Heglund **14** (Top Ace)
Rhodesia (Ryl Rhodesian AF)	Allied / ETO	1	16	16	John Plagis **16** (Top Ace)
Sweden (w/ US Army AF)	Allied / ETO	1	7	7	William Y. Anderson **7** (Top Ace)
Axis Countries Total	Both TO	487	34,996	72	
Allied Countries Total	Both TO	1,161	12,058	10	
World War II Total	Both TO	1,648	47,054	29	

List of World War II Aces (Allied and Axis)
(Source: USAF Military Records)

As a somewhat astounding big-picture statistic, the total number of aces from all countries during World War II years (1941-1945) is seen to be that there were 487 aces accounting for 34,996 air victories in the Axis countries (the German Luftwaffe statistics dominating those numbers), and there were 1,161 aces accounting for 12,058 air victories in the Allied countries. The statistics are not sufficiently detailed as presented in the original list (fifty-three pages) to distinguish what was the victory rate in the European theater of operations (ETO) or the Pacific theater of operations (PTO) as the United States aces achieved victories in both theaters of war. However, the statistics for the Axis countries are presented with the location of the Axis country identifying its theater of operations (i.e., German Luftwaffe data are exclusively in the ETO, and Japanese Imperial Air Force (Army and/or Navy) are exclusively in the PTO). Missing also from the list is any indication during what part of the war - beginning, middle, or end – did the individual ace achieve his air victories.

The list of aces presented in USA military records does have side notes with notations showing whether the Luftwaffe aces gained that status in the first operational jet fighter of World War II, the German Luftwaffe Me 262 Schwalbe. That does permit some observations to be drawn about the propulsion technology of a turbojet-powered (two engines in 18- degrees sweptback wing) fighter aircraft in operational use against piston-powered opponents. Accounts from Luftwaffe Me-262 aces clearly show that the vast difference in speed (about 91 mph faster than the USAAF P-51D Mustangs) provided the tactical advantage to the Me 262 by allowing it to make slicing descent attacks down from higher altitudes, and with the sun at its back when possible. The list of aces above also supports the observation that a country will possess a massive air superiority advantage by virtue of having produced large numbers of technologically advanced fighter aircraft over smaller countries having much fewer numbers of near-equally technologically advanced fighters in their inventories. (Later in this book, the fact that those countries having pilots with greatly superior training in air combat tactics will eventually dominate airspace against those countries having aircraft of generally equal technology and numbers of aircraft in its inventory but with less quality of pilot training.)

During the first few years after Victory in Europe (VE) Day, geopolitical tension increased between Joseph Stalin of the Soviet Union and both Harry Truman of the United States and Winston Churchill of Great Britain. In essence, the Communist Party's ideology varied significantly from that of the economic and political US democratic way of life and the British Imperial way of life. This saw the erection of the infamous Berlin Wall, denoting the eastern sector (Soviet Union) of Berlin from the western sector (USA and Great Britain) of Berlin and divided German families who had survived the devastation of Berlin in World War II. Germany had hoped to rebuild its economy and living standards, not see them divided. The end of the shooting war in Europe was replaced by a "Cold War" (coined by Winston Churchill to describe a non-shooting war, but a war nevertheless between major countries). The end of World War II also saw the beginning of a nuclear arms race between primarily the Soviet Union and the Western World countries (primarily the United States and Great Britain) with all of its terrible potential for world destruction. The use for nuclear energy harnessed as a power-producing technology had not yet emerged.

On the other side of the World in Southeast Asia, the end of World War II with Japan also created some unexpected events and major geopolitical concerns which faced the victorious Allied countries of World War II. Japan had occupied the country of Korea from 1910 to 1945. As World War II was in its final months, the Soviet Union Army had invaded from Siberia, through Manchuria and through the Northern half of Korea at about the 38th

latitudinal parallel. In August 1945, forces of the United States had invaded the Southern half of Korea up to the 38th latitudinal parallel. The occupation of Korea by Japan therefore ended in September 1945 when Japan unconditionally surrendered in Tokyo Bay to officially end World War II in the Pacific Theater. One result of the Japanese surrender was that Korea became a divided nation, the Northern half being under Soviet military control, and the Southern half under United States military control. In effect, the 38th parallel was a geopolitical dividing line between Communist-controlled North Korea and US democratic-controlled South Korea.

[Author's note: As this book is structured to trace the genealogy of technological advances in military aviation, the ramifications of the surrender of both Japan and Germany to end World War II causes this book to continue with the constraint of describing only those factors resulting from World War II that had a significant impact on the development of military aviation technological activities. For the readers interested in the overall history of the wars in which United States military forces participated and suffered losses, there are countless numbers of references on the internet and public libraries that are suggested for reading.]

There were several key activities that were underway in the genealogy of turbojet powered fighter aircraft during and after World War II and before the Korean War had begun (in June 1950). The first dates back to the developments and the entry into the jet aircraft inventory of the British Gloster Meteor and the Messerschmitt Me 262 late in World War II. Through foreign intelligence feedback from agents in Europe to the United States, the United States began developing a jet-powered aircraft of their own design that met with the jet propulsion technology goals and specific performance requirements that were superior to those known threats emerging in the jet-powered aircraft inventory of Germany. Not only did the technology advances represented in the Me 262 demonstrate it to be highly effective in the last year of World War II, but it became obvious that new piloting tactics would need to be revised to take advantage of the high-speed difference of the jet propelled Me 262 against even the fastest US fighter piston powered aircraft - the Lockheed P-38 Lightning and North American P-51 Mustangs.

In parallel with those efforts in the United States but with a slight lag in start time, the Soviet Union also observed the potential of jet-propelled fighter aircraft in the earlier efforts by Great Britain and Germany. Using a reverse-engineered Jumo-004 engine (from Germany's Me 262 propulsion technology) to replace the piston-propeller engine of its Yakovlev Yak-3 fighter developed earlier in World War II, the Yak-15 started into development by the Yakovlev OKB manufacturer for the Soviet Air Force in April 1945. A photograph of the Yak-15 is shown next.

Soviet Air Force Yakovlev Yak-15, 1945.
(Source: USAF Foreign Technology files)

The Soviet Yak-15 was in the configuration category of a "pod and boom", with the Jumo-004 engine (from the German Messerschmitt Me 262 engine technology) occupying the forward bottom of the fuselage and with the wings and remainder of the airframe being attached to the propulsion pod. [The Author would have been inclined to describe the configuration more appropriately as being a large winged tadpole.] The main straight-through spar of the wing had to be bent in order to detour over the top of the engine, and the original air intakes in the wing root for oil-cooling of the piston engine were eliminated. Before the first installed jet engine tests, a steel heatshield was added under the boom, above the exit nozzle, to eliminate heat softening of the aluminum skin on the underside of the boom. The original rubber tailwheel, which got burned off by the first ignition of the engine, was replaced by steel leaf springs capable of surviving the jet exhaust as well as the contact at runway impact when landing and abrasion when taxiing. In addition, the steel heatshield had to be extended aft to the end of the boom when taxi tests showed that the jet exhaust heat was softening/burning the aluminum skin of the boom aft of the shorter original steel heatshield. The Yak-15 featured no armament. The first flight of a Yak-15 prototype was on April 24, 1946.

Using the competitive prototype approach (which was later so successfully applied in the USAF Lightweight Fighter Prototype Program evaluating the General Dynamics YF-16 and the Northrop YF-17, to be described in later chapters of this book), in February 1945 the Soviet Council of People's Commissars ordered the Mikoyan-Gurevich (MiG) OKB manufacturing company to develop a single-seat jet fighter to be equipped with two German BMW 003 engines. It was also required to have a single 57 mm (2.2 inches) gun or a 37 mm (1.5 inches) cannons, plus two 23 mm (0.9 inch) guns.

A photograph of the resulting Soviet MiG-9 is shown next.

Soviet Air Force Mikoyan-Gurevich MiG-9, 1945.
(Source: USAF Foreign Technology files)

The Council of People's Commissars ordered that the MiG-9 be equipped with a 57 mm cannon and two 23 mm guns. Due to poor system engineering, the large 57 mm NL-57 cannon was mounted in the center of the engine intake bulkhead (much like the aquatic narwhal), and the two 23 mm guns were mounted on the lower lip of the nose inlet. When the cannon and two guns were fired later in flight, the muzzle gasses emitted from all three (cannon and guns) were introduced into the inlet flow to the two engines and caused a high incidence rate of engine stall (one or both) at high altitudes. That was most likely due to a combination of the gun gas chemical composition (not highly combustible) as well as the severe inlet distortion (turbulent air with a random pressure variation in its circular disk distribution). The MiG-15 did feature a tricycle landing gear that eliminated the Yak-15 problem of burning off the rubber tailwheel or having steel leaf springs affixed instead of a wheeled tail-gear. Similar to the Yak-15, the first MiG-9 also had the steel heatshield to protect the underside of the tail "boom" from the jet engine hot high pressure exhaust gasses. However, the difference in coefficient of expansion between the steel heat shield and the aluminum construction of the boom above the heatshield initially caused the tail structure to bend downward slightly and acerbate the tail heating problem in the jet exhaust. Redesign of the heatshield attachment to the aft boom support structure eliminated the heat-induced boom-bending problem.

At a fly-off competition on April 24, 1946, both the prototype versions of the Yak-15 and the MiG-9 were flown and both had severe problems in their structural design and propulsion performance (engine stalls and engine failures). Eventually, the Council of People's Commissars caused modifications (to fix all major flight problems) of the Soviet Air Force's Yak-15, and it later became the Soviet Air Force Yak-17. The Soviet Union produced 280 of the Yak-15s and Yak-17s but to be used for flight training purposes only. Neither the YAK-15 nor the Yak-17 ever was equipped with any functional armament.

The MiG-9 had a similar series of failures and performance deficiencies, but the MiG-9 had been configured at the beginning of the fly-off with the Yak-15 to be more operationally configured with armament, had better range, better stability and control, and modified to use larger jet engines. The MiG-9 eventually was produced in a total quantity of 598, which included twelve prototypes. The MiG-9s in the Soviet Air Force inventories saw operations during the Cold War years of 1947-1950 in East Germany and in the Kaliningrad region of the Soviet Union. Approximately two hundred of the MiG-9s were transferred to the Soviets' Communist ally China in late 1950 for use with the People's Liberation Army Air Force.

It is most common for countries having an Air Force in its inventory of weapons of war to maintain an awareness of what other countries are developing as advanced technologies for their weapons of war. Whenever a war that was fought between two or more countries comes to one country unconditionally surrendering to another victorious country, the defeated surrendering country was required to surrender all of its weapons of war and all of its associated technologies advanced during development of those weapons. Historians often refer to that as "to the victor goes the spoils of war". When the Japanese Empire surrendered unconditionally in 1945, the Allied countries received very little spoils of war that were of any technological benefit to the victorious Allies. Conversely, when Germany surrendered unconditionally in 1945 an incredibly valuable treasure of aerodynamics, propulsion (turbojet and rocket), alloyed materials, and nuclear physics were the spoils of war that were acquired by the United States, Great Britain, and the Soviet Union. The German Government and almost all German people regarded the United States and Great Britain to have fought an honorable war but that the Soviet Union had not fought the war as honorably as the USA and Great Britain, especially when the Soviet Union entered into Berlin and went on a frenzy of rape and pillage. It therefore became a race between the USA and Great Britain to capture research centers, laboratories, and test facilities which the Germans had developed to dominate the World with the Third Reich.

Two technology areas that fell into the hands of all three Allied leaders (United States, Great Britain, and Soviet Union) were that of aircraft designs for high-speed (i.e., swept wing design, and compressibility effects when an aircraft is flown into transonic environments) and that of increasingly more powerful and efficient turbojet machinery (especially the works of Dr. Hans von Ohain, and of Dr. Anselm Franz) and rocket propulsion (especially that of

Dr. Werner von Braun). In the Soviet Union, a major research effort was begun at Mikoyan Gorevich OKB in late 1945 to explore the application of swept wings to their advanced fighter aircraft configuration work. The Council of People's Commissars saw the more urgent technology advancement need to be in the area of turbojet propulsion, and that is what caused ordering of the Mikoyan-Gorevich efforts to focus on the MiG-9 development first. However, research on swept wing configurations continued in earnest in parallel with the propulsion technology advancements. It was the swept-wing research coupled with the technology from Great Britain for the Rolls-Royce Nene engine that generated what was to become the MiG-15. The Rolls-Royce Nene had a weight of 1,600 pounds and produced 5,000 pounds of thrust, making it the World's most powerful and highest Thrust/Weight engine in propulsion technology in 1945. A photograph of the Rolls-Royce Nene on display at an RAAF Base in Australia is shown next.

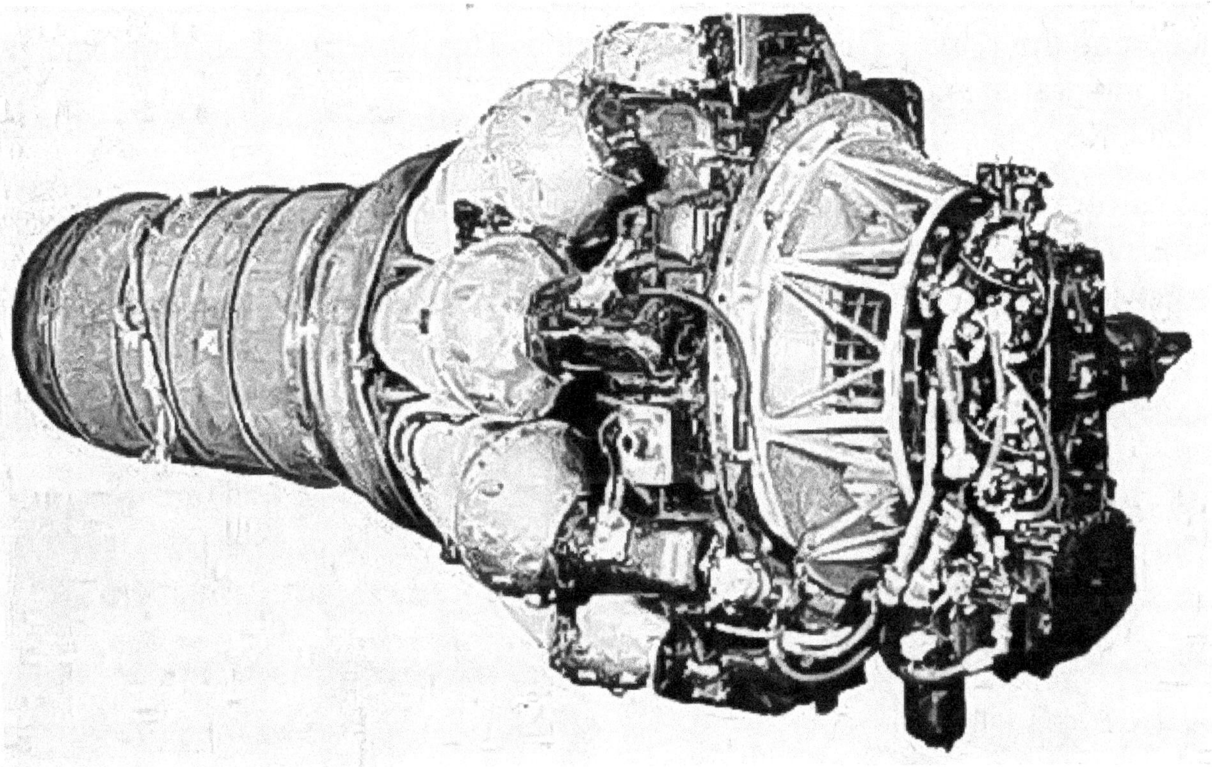

British Rolls-Royce Nene turbojet engine for the Soviet MiG-15, 1945.
(Source: USAF Foreign Technology files)

An amusing event happened in May 1945 (just after the Germans had surrendered to end the war in Europe) which portrays how war sometimes makes strange bedfellows. The Soviet aviation minister Mikhail Khrunichev and aircraft designer A. S. Yakovlev had suggested to Premier Josef Stalin that the Soviet Union buy three Rolls-Royce Nene engines from the British as the Nene was a much better engine than anything they could develop in the Soviet Union, inasmuch as the state of turbojet propulsion technology was years in advance in Great Britain than in the Soviet Union. Frank Whittle had developed the first turbojet engine cycle in 1935 and the Rolls-Royce derivative of Whittle's technology was flying very well in the British Meteor. Khrunichev and Yakovlev stated they could reverse engineer the Rolls-Royce Nene and produce an equally fine jet engine for the proposed new design replacing the MiG-9 (MiG-15). Stalin is said to have commented: *"What fool will sell us his secrets?"*

As Foreign Technology files describes, with Stalin's skeptical approval, aircraft designer Mikoyan, the Mikoyan-Gurevich OKB engine designer Vladimir Klimov, and a few other Soviets travelled to Great Britain in early June 1945 to request the Rolls-Royce Nene turbojet engines. To Stalin's amazement, the British Labour government and its Minister of Trade were perfectly willing to sell the Rolls-Royce Nene engines with supporting technical information and blueprints to the Soviet Union (the vital Allies with Great Britain to bring about the end of war in Europe) and with a license agreement for the Soviets to manufacture the Rolls-Royce Nene. Following evaluation and adaption to the Soviet conditions and needs, the windfall, Rolls-Royce Nene technology, was tooled for mass-production as the Klimov RD-45 turbojet engine. The Klimov RD-45 propulsion system was then integrated into the Soviet Air Force MiG-15 under development by the Mikoyan-Gurevich OKB.

With the swept wing technology captured from German technological facilities and with the Klimov RD-45 engine, the Mikoyan Gurevich OKB proceeded to develop the MiG-15 for the Soviet Air Force. The next photo shows the modular fuselage design used in the MiG-15 with the Klimov RD-45 turbojet engine installed.

Mikoyan Gurevich MiG-15 with Klimov RD-45 engine installed, 1947.
(Source: USAF Foreign Technology files)

The first flight of the MiG-15 was on December 30, 1947, and it entered operational status in 1949, primarily in East Germany, as a symbolic air defender of Soviet territory during the "Cold War" period (as described by British Prime Minister Sir Winston Churchill). The Author wonders what Sir Winston Churchill's retrospective thoughts were about the licensing agreement that Rolls-Royce Engine Company gave to the Mikoyan Gurevich OKB to fly the Rolls-Royce Nene propulsion technology in the Soviet MiG-15 in such an adversarial geopolitical atmosphere.

Of note in the preceding photograph are some design technology details apparent in the MiG-15 external configuration. The first is that the wing on the MiG-15 is of the swept-back technology obtained from captured German high-speed aerodynamics technology. The wing planform has a low taper ratio, moderately high aspect ratio and with a leading-edge sweepback angle of 35 degrees. From later knowledge gained in the United States about wing sweepback angle and its interaction with high-speed compressible flow, a 35 degrees wing sweepback is most appropriate for a fighter aircraft in the near transonic flight regime. The wing is mounted approximately at the middle of the fuselage. The horizontal tail is mounted about two-thirds up on the vertical tail and also features a sweepback angle of 35 degrees (same as the wing) to provide tail effectiveness by not being submersed in the turbulent flow from the forward portions of the fuselage and wing when at high angles of attack during maneuvering.

There were also a series of "fences" on the top surface of the wings which are basically small (height variation from nil to about three inches above the wing surface) metallic strips that serve to reduce the spanwise flow of air on the top of the wing in high-speed flight that would produce an adverse stall locally at the wing root and wingtip regions. The fences are alternatively called "flow-straighteners".

[Author's note: In a later chapter, the Author will describe flight testing conducted by USAF engineers on two MiG-15s that were clandestinely in the possession of the USAF and were flight tested at the now-declassified location at Groom Lake, California – Area 51. The two MiG-15s bore the code names HAVE DRILL and HAVE FERRY, and this Author was a member of the Foreign Technology flight test team at Groom Lake for six months in 1968. Later to be described will be the testing done to evaluate the effectiveness of the wing fences on the HAVE DRILL MiG-15.]

Another strange-looking feature that appear in the photograph on the previous page were the larger thicker wing "fences" on the upper surfaces of the wings near the wing root junction with the fuselage. These were learned later to not perform the flow-straightening function of the further outboard fences intended to straighten the flow over the top surface of the wing. The manufacturing and assembly of the MiG-15 airframe was not to the highest of quality control as exists in the manufacturing assembly lines producing aircraft in the United States. The earliest MiGs had a tendency to roll either to the left or right due to manufacturing variances. When Mikoyan Gurevich OKB identified and corrected the quality control during manufacturing problem, over one hundred of the MiG-15s had been built. The fix for those MiG-15s already built was to add "aerodynamic trimmers" (called *nozhi* (knives) by the Soviets) which were adjusted by ground crews on the flight line until the MiG-15 flew correctly. For those MiGs which were manufactured and assembled correctly, the *nozhi* aerodynamic trimming feature was unnecessary and was eliminated. [Author's note: There was another quality control feature on the MiG-15s identified by the Author during the HAVE DRILL and HAVE FERRY MiG-15s aerodynamics inspection/analysis and flight testing at Area 51 which involved the polishing of skin fasteners in some regions of the MiG-15s fuselage but unpolished fasteners in other regions. These will be discussed in a later chapter of this book.]

Not highly apparent in the photo on the previous page is the manner in which the armament on the MiG-15 (a 57 mm cannon and two 23 mm guns, specified by the Council of People's Commissars to be on the MiG-9 and MiG-15) was integrated into the design; it is immediately under the nose inlet. A photograph of the cannon/gun's installation is shown next.

Armament tray on a MiG-15 undergoing USAF technology evaluation at Area 51,1968.
(Source: USAF Foreign Technology files)

In the photo of the MiG-15 armament tray subassembly, the tray with the armament installed is shown in its lowered position and making the cannon, guns and ammunition storage containers accessible to ground personnel for servicing (re-arming with munitions) and maintaining. The tray had 4 cables located near each corner of the tray for lowering/raising the armament subassembly. This accessibility significantly contributed to the MiG-15 having an impressively low Turn Around Time (TAT) between sorties. The tray arrangement was a part of the MiG-15 being configured to avoid the multitude of problems the MiG-9 had earlier. By relocating the cannon/guns installation further aft from the nose inlet (with the bifurcated inlet splitter shown in the photo), Mikoyan-Gurevich OKB avoided the adverse effects of engine stalls during cannon/guns firing. The non-combustible gun gasses and adverse pressure distortion generated when firing the cannon/guns at higher altitudes caused poor quality air entering into the nose inlet of the MiG-9. It should also be noted that the spacing between the 57 mm cannon and the visible 23 mm gun was such that there was no interference with the nose landing gear mechanism.

Diverting Readers' attention back to 1941, after learning of the British development of the Frank Whittle centrifugal flow turbojet engine for the Gloster Pioneer, General Henry "Hap" Arnold of the US Army Air Force caused a secret activity at the General Electric Company to be initiated. That activity was to develop a centrifugal-flow turbojet engine of similar construction and function as that of the Whittle engine. General Electric began construction in November 1941 and the first engine ran in March 1942, generating 1,650 pounds of thrust, almost 200 pounds more thrust than that of the British Whittle engine that had generated 1,460 pounds of thrust. This first jet engine to be built in the United States was designated as the GE I-16 Whittle Jet Engine (it did not receive a US Patent because the Whittle Engine had already been patented in Great Britain. [Author's note: The Author was visiting at the Aerospace Museum of California at the McClellan Air Base in McClellan, California near Sacramento in June 2017, and was totally impressed by the collection of aircraft engines and aircraft on display at this Museum. Among the engines on display was seen a GE I-16/J31 engine which had been graciously loaned to the Aerospace Museum in California from the National Museum of the USAF located at Wright-Patterson Air Force Base in Dayton, Ohio. The Aerospace Museum in California commissioned Daniel D. Whitney to prepare a booklet entitled simply *"Engines"* containing information, specifications, and photographs of the aircraft engines on display including the GE I-16/J31. Readers who have a special interest in the history of aviation propulsion should make a serious effort to visit the Aerospace Museum of California to see a display of interesting aircraft and their engines that were constructed during their lifetime].

A copy of a photograph taken by this Author at the Aerospace Museum of California at McLellan Air Base near Sacramento of a Whittle Jet Engine is shown next.

GE I-16/J31 Whittle jet engine
(Photo by the Author)

As explained in the "*Engines*" booklet by Daniel Whitney, the US version of the Whittle turbojet engine concept produced approximately 200 pounds more thrust than did the original British Whittle turbojet engine reproduced by GE. The GE engine featured a better matching of the nozzles to direct hot gas generated within the multiple combustion cylinders onto the turbine wheel. First flight of the GE J-31 Whittle type jet engine occurred on October 1, 1942, in the first US jet propelled aircraft, the Bell XP-59.

As stated in the "*Engines*" booklet: *"The J-31 is a significant engine because it revolutionized aviation in America by introducing the jet engine. It showed that a jet engine was practical and overnight brought out new concepts in propulsion and aircraft design. Development of the GE I-16/J31 soon resulted in the GE I-40/J33 4,000 pounds thrust, Whittle type, jet engine. Tens of thousands (GE J-33 turbojet engine) eventually were built, most by the Allison Engine Company after World War II ended and demand for the Allison V-1710 piston engine in combat aircraft had been greatly reduced. The GE J-33 turbojet engine was installed in the Lockheed P-80 Shooting Star and T-33 airplanes. Some GE & Allison J33 turbojet engines are still flying in T-33 trainer airplanes today."*

The engine on display (shown in the preceding photo) is a GE I-16-6/J-31) built for the Ryan FR-1 airplane. The Ryan FR-1 airplane was the second US jet airplane. Designed for the US Navy in 1943 it first flew in June 1944. Six hundred FR-1 aircraft were ordered built and the first Squadron VF-66 was prepared to deploy (with the GE J-31 jet engine) to the Pacific Fleet Operations when World War II ended in August 1945. Only sixty-six FR-1 aircraft were built.

On September 5, 1941, the US Army Air Corps asked the Bell Aircraft Corporation to build an aircraft to utilize the General Electric J-31 turbojet engine. On September 12, 1942, the first (of three) XP-59A prototype aircraft was trucked to Muroc Army Air Field (today, Edwards Air Force Base) in the California desert near Rosamond, California. As the project was highly classified, a dummy propeller had been affixed to the nose of the XP-59A so that public observers of the passing truck and aircraft on its bed would be misled to believe it was just another propeller driven aircraft on its way to be flight tested. The first XP-59A inadvertently became airborne after a high-speed taxi test on October 1, 1942. The first official flight of the XP-59A, with the GE J-31 turbojet engine, was conducted successfully in a take-off on October 2, 1942. Over the following three months of flight testing, the three XP-59As revealed a multitude of problems including poor engine response and reliability (common problems with all early turbojets), insufficient lateral stability (especially in the roll axis), and general performance that was far below expectations. The original request from the US Army Air Corps for the XP-59A prototypes contained only desired performance and handling goals, no firm requirements specified. Chuck Yeager flew the aircraft and was dissatisfied with its maximum speed, but was impressed by its smooth handling characteristics.

A design modification program was begun to convert the XP-59A prototypes into the YP-59A configuration (which also included being armed with one 37 mm M4 cannon and three .50 caliber machine guns), and the USAAC ordered 80 production versions of the P-59A Aira comet even before the YP-59A had entered into the development flight test phase of the program. [Author's Note: This was the first government program using the "buy before fly" procurement concept, due to the urgency of the needs in early World War II]. A photo of the Bell YP-59A in flight is shown next, and below that photo is a nose-view photograph showing the cannon and 3 guns installed on the YP-59A.

Bell YP-59A Airacomet,1942.
(Source: USAF Photographic files)

YP-59A armament installed, 1 x 37 mm cannon and 3 x .50 Cal guns.
(Source: USAF Photographic files)

While attempting to correct all of the engine reliability, aerodynamic anomalies, and flight control/handling deficiencies of the P-59 Aira comets, Bell Aircraft Corporation manufactured twenty P-59As and thirty P-59Bs. None of these fifty P-59s were deemed to ever be operationally ready and never saw combat. The P-59Bs were relegated to familiarizing pilots with the handling and performance of aircraft having jet propulsion in preparation for the more advanced types of jet powered aircraft that would shortly become available with operational capability.

The impetus for development of an operationally capable turbojet powered fighter aircraft came in spring 1943 when allied foreign technology experts became aware of the German Me 262 *Schwalbe* jet aircraft development. The information General Henry "Hap" Arnold had been given about the British turbojet-powered British Meteor aircraft in 1942 also pointed to the need for an improved high-speed, well-armed, highly maneuverable fighter aircraft. The USAAF Materiel Command at Wright Field (later Wright-Patterson AFB) tasked its Research and Development Division to work with Lockheed to develop the new aircraft in as short a time as possible. In mid-June 1943, Lockheed Aircraft Corporation submitted a design proposal of a prototype fighter aircraft which Kelly Johnson had created and promised that the new prototype would be ready for testing in 180 days. The Lockheed team, beginning on June 26, 1943, produced the prototype known as XP-80 in 143 days (37 days ahead of schedule) and delivered it to the Muroc Army Airfield (now Edwards Air Force Base) on November 16, 1943. The first flight of the XP-80 took place on January 8, 1944.

After the first British-built engine (a Halford H1 axial-flow turbojet) was mated to the XP-80 airframe, foreign object damage (FOD) occurred during the first run-up and destroyed the engine. With a second Halford H1 engine expeditiously delivered from Great Britain, the first XP-80 prototype had its first flight on January 8, 1944. With plans for powering the XP-80 with an American-made turbojet engine, the second prototype designated the XP-80A, was fitted with a larger General Electric I-40 engine (an improved GE J31, later produced by Allison Engine Company as the GE J33).

The P-80 testing program proved very dangerous. Lockheed Chief Test Pilot Milo Burcham was killed on October 20, 1944, while flying the third YP-80A (44-83025). Newly-promoted as Chief Test Pilot at Lockheed to replace Milo Burcham, pilot Tony LeVier, bailed out of the XP-80A Gray Ghost on March 20 1945 when one of the engine's turbine blades failed, causing structural failure in the aircraft's tail. LeVier landed hard and broke his back, but returned to the test program after six months of recovery. The top-scoring World War II USAAF ace, Major Richard Bong (in Lockheed P-38 Lightning aircraft), was also killed during an acceptance flight of a production Lockheed P-80 Shooting Star on August 6, 1945, in the United States. Both Milo Burcham and Richard Bong crashed as a result of a main fuel pump failure. Milo Burcham's death was the result of a failure to brief him on a newly installed emergency fuel pump backup (redundancy) system. The investigation of Major Richard Bong's crash found that

Richard Bong had apparently forgotten to switch on this pump, which would have likely prevented the accident. Richard Bong bailed out when the aircraft rolled inverted but was too close to the ground for his emergency parachute to deploy.

A photograph of the XP-80A "Gray Ghost" in flight over Muroc Army Field is shown next.

Lockheed XP-80A Gray Ghost in flight, Muroc Air Base,1944.
(Source: USAF Photographic files)

Because of delays in delivery of production P-80 Shooting Stars, the P-80 saw no actual air combat during World War II. It did see limited operational service in Italy with the USAAF on reconnaissance sorties during February and March 1945. A total of eighty-three P-80s had been delivered by Lockheed by the end of July 1945. Production continued after World War II ended, with 1,714 single-seat F-80A, F-80B, F-80C, and RF-80s manufactured by the end of production in 1950. The two-seat TF-80C was first flown on March 22, 1948, and became the basis for the T-33 trainer aircraft. 6,557 Lockheed T-33 trainers were produced by the end of 1950. In the next chapter, operation of the F-80C and RF-80 photo-reconnaissance variants during the Korean War, which started in June 1950, will be described.

When World War II in Europe ended (in August 1945, when the German Government unconditionally surrendered), a vast treasure of technological information was captured by United States, British, and Soviet military personnel. Among the technology database captured by the United States were most of the Me 262 *Schwalbe* wind tunnel test data and a large percent of the research and testing done on high-speed aircraft with swept wings to better perform in air compressibility effects at high speed. The Me 262 had an 18 degrees swept-back wing, but this was due to having to adjust the center of gravity position of the total aircraft when the Me 262 wing (in an unwept variant) had been manufactured and found to weigh much more than originally had been calculated.

Extending the database on compressibility effects obtained from flight testing with the Lockheed P-38 Lightning (which held high speed flight records for a long time), and coupled with the captured German data on swept-back wings in mid-1945, the US Army Air Force (and its successor in 1947, the United States Air Force) embarked on the design of a new configuration fighter aircraft having swept wings. In late 1944, North American Aviation proposed a jet-powered fighter to the US Navy designated as the FJ-1 Fury. It was an unexceptional transitional jet fighter that had a straight wing derived from the outstanding piston engine powered with supercharging North American P-51 Mustang which was one of the best fighter aircraft during World War II. In mid-1944, the USAAF drafted initial proposals for a medium-range, high-altitude, jet-powered, day escort fighter/fighter-bomber. In early 1945, North American submitted four designs meeting the several requirements each to a varied degree of conformance to the USAAF RFP. The USAAF selected one design from the four responding proposed designs which best met the set of RFP design requirements, and gave North American Aviation a contract to build three examples of an XP-86 (experimental Pursuit) aircraft of the selected single design. The XP-86 design, however, featured a straight wing similar to the highly successful P-51 Mustang wing.

A crucial weakness in the straight-wing XP-86 design was that of not being able to meet the required maximum high speed of 600 mph specified by the USAAF. Development of the XP-86 had reached an advanced stage and changing from a straight wing design to a swept wing design (based on analysis of the newly captured German database on swept wing technology) was met with resistance from some senior North American engineering and manufacturing staff members. Despite the stiff opposition, wind tunnel tests of the XP-86 with a 35-degrees swept-back wing were most promising, and the swept wing design change was made rapidly. Testing of the XP-86 swept wing wind tunnel model in NACA's (later to become NASA) high-speed wind tunnels at the Langley Research Center near Hampton, Virginia, confirmed that a maximum speed of 600 mph was totally feasible without adverse compressibility effects on the airframe. The XP-86 prototype did not roll out until August 8, 1947, far too late to be entered into any World War II operations.

The XP-86 prototype, powered by a General Electric J35-C-3 turbojet engine had its first flight on October 1, 1947, with pilot George Welch at the controls, at the Muroc Army Field Dry Lake (now Edwards AFB near Rosamond, California). [Perhaps as an interesting item of trivia to readers, the name Muroc is derived from a backward spelling of the name Corum. The Corum family owned and donated the land on which the

Government constructed the Muroc airfield on the dry lake bed, but it was considered inappropriate at that time to name government property with a family name, and therefore the name Muroc was devised to resolve the concern.] The XP-86 prototype had an unusually successful flight test program and was given the designation by the USAAF as the North American F-86A Sabre in its production configuration. An F-86A powered by a General Electric J47-GE-7 engine was reported to have set the first official world speed record of 671 mph on September 15, 1948 at Muroc Army Field piloted by Major Richard L. Johnson.

In late 1943, a classified project was begun by the Miles Aircraft Company of Great Britain to develop the world's first aircraft capable of supersonic speed in level, steady speed flight. The British Air Ministry had issued a top-secret Air Ministry Specification E.24/43 calling for "an aero plane capable of flying over 1,000 mph in level flight, along with the ability to climb to 36,000 feet in 1.5 minutes". A speed of 1,000 mph equates to about Mach 1.3. The project resulted in the development of a prototype Miles M.52 turbojet-powered aircraft, which was designed to reach the required 1,000 miles per hour (a M 1.3 requirement) – over twice the existing official speed record – and to climb to an altitude of 36,000 feet in one minute thirty seconds. A sketch of the resulting design entitled the Miles M.52 turbojet is shown next.

British Miles M.52 turbojet supersonic (M 1.3) prototype, 1943.
(Source: USAF Foreign Technology files)

A 3-view drawing, shown next, of the British Miles M.52 turbojet better shows the external features of the M.52 and what the British considered important in the advancement of supersonic aerodynamics technology to be embodied in an aircraft designed for low supersonic speeds.

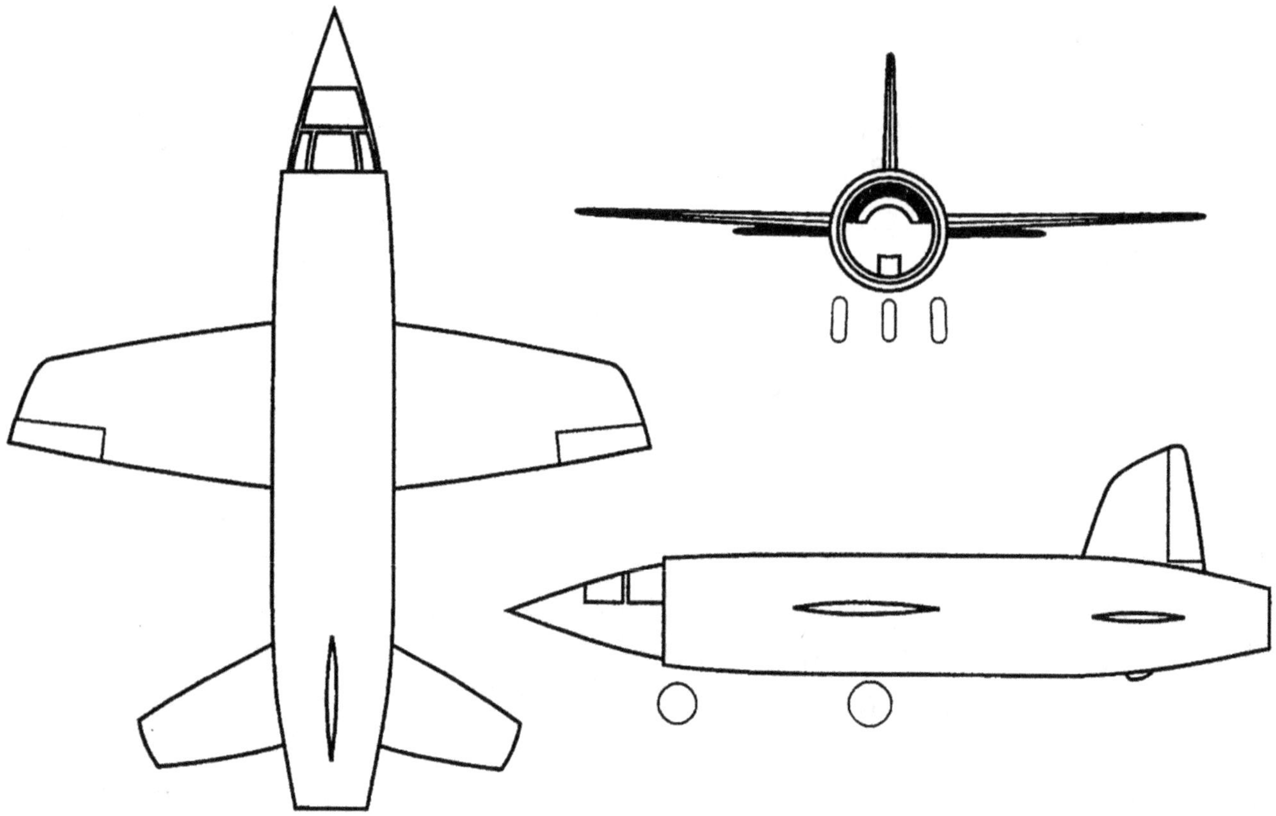

A 3-view drawing of the British Miles M.52 supersonic (M1.3) turbojet, 1943
(Source: USAF Foreign Technology files)

The designer of the Miles M.52 was a Fred Miles, who used technology knowledge from the shape of gun-fired projectiles at supersonic speeds and adapted that as a template for designing an aircraft that resembled an artillery shell round fitted with wings and empennage tail surfaces to gain stability in flight rather than being spin-stabilized (as artillery rounds are designed). Intuitively, Miles shaped the un-swept wings to be of a nearly-elliptical shaped planform (derived from the elliptical lift distribution on the British Supermarine Spitfire wings of the late 1930s and early 1940s), and with a very sharp leading edge and very thin wing thickness ratio. Because Miles had photographically seen shock waves from artillery shells trailing from the nose of the artillery rounds and rearward in a conical shape, Miles clipped the tips of the elliptical wings to keep them fully within the flow field of that conical shaped bow wave. The weakness of the M.52 wing design was that it did not have any sweepback, later found to be essential in achieving level steady high-speed flight in the high transonic regime and exhibiting compressibility effects on un-swept aerodynamic surfaces.

A most important feature of the M.52 was the design of its horizontal tail surfaces. Miles designed the horizontal stabilators to have a sweepback angle and to be an all-moving tail (sometimes referred to as a “flying tail”) which were power operated and connected mechanically to the pilot’s control column (“stick”). Based on earlier Spitfire aerodynamics tests (wind tunnel and flight), Miles saw that conventional control surfaces became ineffective at high subsonic speeds such as that being achieved by high-speed fighters in dives, due to the formation of shock waves at the hinge lines of the control surfaces. Coupled with the rearward movement of the center of pressure on surfaces in the transonic flow speed regime, those coupled factors caused control "reversal", a phenomenon which caused the pilot's stick inputs to have the opposite effect on the control surface deflection at high speeds. This compressibility and shock-induced control surface reversal phenomenon was to be the cause of many accidents and near-accidents when pilots entered a high-speed dive and approached transonic speeds. Despite there being no such physical condition as a barrier being encountered, the phenomenon was labeled as being the "sonic barrier". An all-flying tail is now considered to be a minimum design feature enabling high speed aircraft, especially fighter and interceptor aircraft, to pass through the transonic regime without loss of pilot flight control. The M.52 was the first design to provide this technological solution of an all-flying tail and this design feature is universally applied on all high-speed aircraft.

The M.52 design also showed the effectiveness of having a conical spike in the nose's circular-shaped inlet. The pointed spike creates a shock wave that allows inlet air to flow in an annular manner into the face of the turbojet engine and improves the air intake performance of such nose inlets at supersonic speeds. The field of propulsion technology has applied this spiked-inlet in various applications on high-speed aircraft and further developed the technology of an axially translating spike to cause a better "match" of the inlet design with the engine variable airflow needs.

From a pilot's perspective, having the cockpit integrated into the nose inlet spike was most disturbing for two primary reasons: first was that the pilot would have no ability to "check his 6" while in a high-speed air-combat engagement. [The phrase "check 6" is commonly used by fighter pilots to make a visual check to the rear of the aircraft – in the 6 o'clock direction – to gain situation awareness of the potential presence of a threat aircraft approaching the rear of the friendly aircraft.] More importantly, pilot adverse comments were in the vein of "If the airplane were to run into an obstruction during take-off or landing, the nose spike and human occupant would be pushed backward and into the "meat-grinder" turbine engine. (One pilot reportedly chirped that he would be "too overdone for a good hamburger".)

In 1944, the British Air Ministry and the United States War Department agreed to exchange high-speed research and data. The Lockheed Aircraft Company (Kelly Johnson and his P-38 Lightning design team) and the Bell Aircraft Company (with the design team of the P-39 Aira comet) were given access to the drawings and research on the M.52; however, the Lockheed Company (Kelly Johnson) partially defaulted on the agreement and little data about high-speed compressibility effects was actually exchanged. [The Author had professionally known Kelly Johnson for a long time and he characteristically did not share data about problems encountered with his designs - such as those the P-38 Lightning encountered in its development phase.] Fred Miles and the

Bell Aira comet design team did exchange some information, and it was used, to limited extent, in the design of the Bell X-1 high speed research aircraft... The Bell X-1 was also shaped much like a large artillery round with wings and empennage, but was powered by rockets rather than an air-breathing turbojet propulsion system.

Ironically, the M.52 design and project were discontinued in 1945 when World War II ended and the British Labour government then headed by Sir Clement Atlee cancelled the entire Miles M.52 project for budgetary reasons. That is entirely understandable as the British Government in 1945 was faced with the higher priority and enormous costs of re-building Great Britain which had been so devastated by the German bombings during World War II.

Another of the "firsts" associated with military aviation with jet powered aircraft resulting from World War II was the development of the McDonnell FH Phantom as the first operational jet providing the vital fleet defense capability for the US Navy and US Marine Corps. A photograph of the McDonnell FH Phantom is shown next.

McDonnell FH Phantom, 1946.
(Source: USN Photographic files)

The photograph (admittedly of less than sharp photographic quality) shows the first operational jet landing of a McDonnell FH Phantom on the deck of a US Navy aircraft carrier. The date was July 21 1946, and the ship was the *Franklin D. Roosevelt*, named after the former President of the United States who led the United States out of the Great Depression of the 1930s and through World War II. The US Navy's practice of naming aircraft carriers after Presidents of the USA, and after battles in which US military ground forces that included US Marines, was and is a fitting tribute to those who gave their lives for the defense of their home country.

The McDonnell FH Phantom featured inlets for the two jet engines in the wing roots and air ducting of the air into the two engines located in the middle and aft sections of the fuselage. A successor to the McDonnell FH Phantom was the USAF McDonnell F2H Banshee, another two-turbojet engine fleet-defense fighter, which along with the USAF F-86 Saber jet, were to be operational in the Korean War. The FH Phantom had less-than-impressive performance capabilities because of a less than optimum jet engine cycle efficiency and exacerbated by the size of the inlets being adequate for the smaller turbojet cycle needs with the present engine but not being large enough for the larger improved turbine engine cycles that gave the F2H Banshee far more impressive performance capability.

In the next Chapter, the performance of the MiG-15s in the Korean War, which broke out in June 1950, will be discussed in a comparative manner with that of the fighter aircraft sent by the air forces of the "friendly" countries under command of the United Nations Security Council. This Chapter 3 will continue describing those friendly aircraft that were in operation during the Korean air war, those being primarily of the United States Air Force and United States Navy air arm.

The US Air Force (formed in 1947 after the US Army Air Force was separated into a separate air arm of the military department) announced that the XF-86 had exceeded Mach 1 in a dive on April 26, 1948. However, in 1996, North American aerodynamicist Ed Horkey claimed that pilot George "Wheaties" Welch had exceeded Mach 1 much earlier than April 1948 (possibly as early as October 1, 1947). The legendary pilot Chuck Yeager broke the sound barrier on October 14, 1947 in the rocket-propelled Bell X-1, during a level flight, making it the first true supersonic aircraft (albeit it with rocket engine power). The X-1 had an un-swept wing, and had a conventional hinged stabilator, which probably would have shown the adverse compressibility effects causing "control reversal" had pilot Chuck Yeager maneuvered the X-1 from a straight and level flight path on his "breaking the sound barrier" historical flight.

Unaware of the high-speed compressibility work being done by the Miles Company (which was classified top secret at the time), the North American Company in California, proceeded with their design work on the configuration of what was to become the legendary F-86A Saber jet. A photograph of an F-86A Sabre in flight is shown next.

North American F-86A Sabre, 1948.
(Source: USAF Photographic files)

What should be noted is that the horizontal tail on the F-86A indicates that there is a hinged aft portion of the stabilator, and had the Miles all-flying tail technology been made known to North American, there is little doubt that the F-86 design would have incorporated that all-flying tail concept to avoid the control surface reversal problems which other contemporary high speed fighter aircraft were having when pulling out from steep dives.

The North American F-86A as a fighter-interceptor was powered by the General Electric J47-GE-7 engine, which generated a thrust of 5,200 pounds. The F-86A entered operational service with USAF in 1949. A photograph of a General Electric J47-GE-7 is shown next.

General Electric J47-GE-7 axial flow turbojet engine, 1948.
(Source: Photo by the Author)

The armament on the F-86A Sabre was six 0.50-inch (12.7 mm) M3 Browning machine guns, with electrically-boosted feed, in the nose of the aircraft. Later versions of the F-86 Sabre (i.e., F-86D/E/F/H) developed after 1950 will be described in the next chapter (1950-to-1960-time span).

It should be noted by Readers that the United States capitalized on the technology of British and German technologies in fighter aircraft propulsion, with the advent of the turbojet engine cycle, and capitalizing on the German technological advancements in aerodynamics at high speed, from which the swept wing technology evolved. The period 1941 through 1950 was the most pivotal decade in military aviation technologies. The change from piston-engine powered to turbojet-powered aircraft is considered to be as significant to aviation history as was the Wright brothers' accomplishment in flight control in 1903.

Also, to be noted were the technological advances the Soviet Union made near the end of World War II and during a Cold War environment between the Communist-controlled Soviet Union and its allies and the democratic nation-members of the United Nations General Assembly established in October 1945. When the Korean War erupted in 1950, Soviet jet-powered aircraft in the inventory of Communist-controlled countries became the adversaries rather than the friendlies of Allied countries operating under United Nations international peacekeeping auspices.

The North Atlantic Treaty Organization (NATO) was established in April 1949 - and indirectly affected countries in Europe - to assure that none could ever cause a repeat of a war such as that which was started by Germany and its allies. The defensive war machinery for those countries in NATO banding together with "defense in unity" relying on the aviation technologies and war machinery, including military aircraft, that were made available to smaller NATO members by larger NATO members. The NATO effect on technologies was not as major a factor as that of the United Nations as the NATO scope was limited to the North Atlantic region, not World-wide like that of the United Nations.

In parallel with the United States Air Force developing the Century Series Fighters and the B-58A Hustler Strategic Supersonic Bomber, the USN developing the F-4 Phantom II and other carrier-capable supersonic fighter aircraft as described in Chapter 4, the Soviet Union developed long-range bombers (capable of delivering nuclear weapons at long ranges) and three supersonic fighter aircraft as escorts for those long range bombers and for air combat with the Century Series fighters and the USN/USMC F-4H-1 Phantom II fighter. The Cold War tensions created the leapfrogging of aircraft with technologies providing a capability which the adversary would need to overcome.

Based on the loss rate of the North Korean MiG-15 against the USAF F-86 Saber jet in the Korean War, the Soviet Union expeditiously began design activities, which consisted of a number of modifications to update the MiG-15 design such that its deficiencies against the F-86 Saber would bring more parity in the capability difference between the new MiG-17 and the F-86. These were accomplished on an expedited basis, but the MiG-17, which had its first flight in January 1950, was not introduced into operation until October 1952, which was too late to have any significant impact on the air war over Korea. A photograph of the underside of a Soviet MiG-17 is shown next and can be seen to not have a large number of external upgrades (other than the ability to weapons on its external wing hard-point stations) from that of the MiG-15.

Soviet MiG-17 "Fresco" with Egyptian AF symbols on its wings, 1952.
(Source: USAF Foreign Technology files)

The MiG-15 air combat capability was hampered by the lack of a ranging-radar and optical gunsight, but in 1951, Soviet engineers obtained a captured F-86 Saber during the Korean War and copied (reverse engineered) the F-86 Ranging Radar and Optical Gunsight. The copied ASP-4N gunsight and SRC-3 radar were integrated into the MiG-17, and this upgrade would prove to later be extremely important during the Vietnam air war.

In this chapter, the Author could only include a small sample of what technologies were introduced and/or advanced during that most important decade in fighter aircraft technology history. Included have been those which have had a direct influence on military flight since the early 1940 time period. By necessity (scope), there has been no in-depth identification or description of those fighter aircraft developed by the US Navy for USN and US Marine Corps military aviation. Operating naval aircraft on carrier ships introduces a host of other technological challenges and design requirements for their military aircraft, not the least of which is the requirement for folding of the outer portion of the wings to increase the "spotting factor" of aircraft on the top deck. There is much transfer of technology between the USAF and the USN at aviation research and development facilities when it is for the benefit of the United States as a country. It is coupled with an integrated military force structure to dominate the airspace, over land, over open-ocean waters, and under the water surface with the appropriate weapons systems. Technology development is much like a child's game of leapfrog. The technological edge is achieved largely by developing a better technology sooner than the adversary can. Chapter 4 relates what technologies gave the edge to one country over its adversary.

CHAPTER 4
POST-KOREAN WAR FIGHTER AIRCRAFT TECHNOLOGY DEVELOPMENTS (1950 TO 1961)

On June 25, 1950, the Korean People's Army (KPA) invaded South Korea. Similar to that of the Japanese attack on Pearl Harbor in 1941 to start World War II, there was little or no indication in monitored diplomatic communications channels that the KPA attack was that imminent. The North Korean Government claimed that the South Koreans had attacked North Korea first but this was never substantiated. The more plausible reason was that North Korea had long been wanting to oust and execute "the bandit traitor Syngman Rhee" and to unify the two Koreas with the entire Korean Peninsula being under Communist control. There was of course much diplomatic correspondence surrounding the building "Cold War" tension existing in Communist-controlled countries such as the Soviet Union, the People's Republic of China, and several of their communist-supporting countries, opposing the capitalistic, democratic countries of the Western Hemisphere.

To support the KPA ground troops, the Korean People's Air Force (KPAF) employed 210 Yakovlev Yak-7 and Ilyushin IL-10 piston-engine- powered aircraft provided by the Soviet Union to North Korea in the three years following the end of World War II as part of the Cold War spread of Communism in Asia. At the time of the KPA attack on June 25, 1950, the United States Army had approximately 275 military troop personnel on the Korean Peninsula serving mainly with the Korean Military Assistance Group (KMAG) supporting the South Korean Army. The US Air Force had 1,172 aircraft of various types in the Pacific Region including hundreds of Lockheed F-80 Shooting Stars (jet fighter interceptors) and numerous North American F-82 Twin Mustangs (two engine piston powered). based at air bases in Japan. A photograph of the uniquely configured "two pilots, one wing" F-82 Twin Mustangs (formed by the attachment of two North American P-51 Mustangs from World War II to resemble a P-38 like configuration) is shown next.

North American F-82 Twin Mustangs, 1950.
(Source: USAF Photographic files)

There were also at least 2 USN Aircraft Carriers in Japanese waters equipped with McDonnell F2 Banshees, Douglas F3 Sky knight (night fighters), Chance Vought F4U Corsairs, Grumman F7 Tigercat (night fighters), and Grumman F9F Panther jets. Unfortunately, there was a delay in authorization of the use of USN/USMC military fighter aircraft and it was left to the USAF F-80 Shooting Stars and F-82 Twin Mustangs to engage the KPAF Yak-7 and IL-10s.

The airfields in Korea had all been given a K- designation rather than a confusing Korean name. The K-13 airfield was located at a location near the town of Suwon, about 50 miles southwest from the capital city of Seoul in South Korea, and was about 75 miles south of the 38th latitudinal parallel.

A number of F-80C Shooting Stars used against the KPAF during the early days of the Korean War are shown on the flight line at K-13 (Suwon) airfield in the photo next.

Lockheed F-80C Shooting Stars at K-13, Suwon, S. Korea, 1950.
(Source: USAF Photographic files)

By late August 1950, the 95,000 South Korean and 275 US military forces occupying South Korea after World War II ended, were driven to the port city of Pusan, at the southern end of the Korean Peninsula where they stood ground with their backs to the sea.

The United Nations Security Council condemned the act of war by North Korea, On June 27, 1950, the United Nations Security Council authorized the formation and dispatch of UN forces to repel what was recognized as a North Korean invasion. Twenty-one countries of the United Nations eventually contributed to the UN force, with US personnel comprising 88% of the UN force. The UN military force, including mostly US military personnel and equipment, and augmented by British and Australian ground forces, was sent to Korea in September 1950 to retake South Korea from North Korean communist control. The UN force counteroffensive was launched at the port city of Inchon (about 50 miles SW from Seoul, the capital of South Korea).

In very short time, the USAF F-82 Twin Mustangs and F-80 Shooting Stars had downed almost all of the NKPAF Yak-7 and IL-10 piston engine aircraft, and the attacking NKPA ground forces lost their air cover. With the loss of air cover and confronted by a large, well-equipped UN Security Council military team (mostly American, British, and Australian), the Communistic North Korean military was forced north and back across the 38th parallel.

In another surprise event that elevated the seriousness of the North Korean invasion of South Korea, the Communist Chinese People's Army crossed the Yalu River border between China and North Korea in October 1950, merged with the Korean People's Army forces, and drove the UN forces south, back again to the 38th Parallel. The Chinese People's Army however also brought Soviet-made MiG-15 jet aircraft into the Korean War at this stage of the war to support the NKPA forces. Temporarily, the North Koreans regained air superiority over the US military aircraft that already were in range of North Korea when World War II ended and Cold War tensions increased.

The USN and USMC jet-powered aircraft inventory, constituting a major portion of the UN Command forces sent to Korea, brought carrier-based Grumman F-9F Panthers into Korean airspace alongside the Lockheed F-80 Shooting Stars to combat the MiG-15 aircraft. A photo next is of the Grumman F-9F Panther operating from a USN carrier off the Korean coast:

A USN Grumman F9F-2 Panther above Korea, 1950.
(Source: USN Photographic files)

Development of the Soviet MiG-15 fighter aircraft technology was described earlier in Chapter 3. Readers should recall that the dominant external features of the MiG-15 configuration were that it had a 35-degrees swept-back wing (that wing design technology being acquired by the Soviets at the end of World War II from captured German records at German aircraft design facilities). It also had a single engine with a bifurcated inlet in the aircraft nose that flowed inlet air rearward on either side of the pilot to the engine face near the mid-way location inside the fuselage. Another prominent external design feature of the MiG-15 was that it had a comparatively large-area vertical tail and large rudder with the horizontal tail mounted approximately midway on the vertical tail. A less obvious external design feature of the MiG-15 was that it featured a span-wise pair of vertical "fences" on each wing which the Soviets considered necessary to reduce the spanwise component of airflow that tended to flow outward on top of the swept-back wings. That spanwise flow component resulted in a loss of aerodynamic efficiency of the wings.

To provide Readers with a visual reminder of the MiG-15 external configuration, a photograph of a Soviet MiG-15 is shown next.

Soviet MiG-15 "Fagot" of 1950 vintage at the Proud Bird Museum, Los Angeles
(Photo by the Author)

[The Author photographed the MiG-15 above on static display, along with several other US and foreign military aircraft from World War II and Korean War time period, on display at the rear of the Proud Bird Restaurant in Los Angeles. The Proud Bird Restaurant is located very close to the end of the runway at the Los Angeles International Airport (LAX), and passengers aboard commercial airline flights arriving and departing LAX on Runway 24L may catch a fleeting glimpse of this MiG-15 resident at the rear of the display lot at the Proud Bird Restaurant. The MiG is seen to bear the roundel (red star symbol) of the Soviet Air Force. Those MiG-15s were in the inventory of the KPAAF bore the  used in the Korean War that roundel symbol, - an encircled Red Star reflecting the symbol of Communism.]

Readers should note that both the F-80 Shooting Star and the F9F-2 Panther were straight-wing aircraft; whereas, the MiG-15 had a 35-degrees swept-wing design. All other factors being equal, a jet fighter with a swept-back leading-edge wing should perform better at high subsonic Mach Numbers than does a fighter aircraft having a non-swept back leading-edge wing. The aerodynamic flow improvement of the wing fences on the MiG-15 wings are not large but are of significance as an extra edge in wing design technology. [In the next chapter, the Author will describe having served in an aerodynamics flight test assignment for the USAF at the Groom Lake Test Facility – now openly identified as Area 51 – in the Nevada desert. This was to exploit the aerodynamics of the wing fences on a clandestinely "borrowed" MiG-15 given the code name HAVE DRILL in the 1968 time period.]

The first all jet-versus-jet fighter air combat recorded in aviation history occurred on November 8, 1950, and was between a USAF F-80C Shooting Star (piloted by USAF First Lieutenant Russel Brown) and a KPAAF MiG-15. The engagement started with a flight of USN F9F Panther jets attacking the Sinuiju bridges near the mouth of the Yalu River, and with a flight of USAF F-80 Shooting Stars providing air cover for the F9F Panthers. Other pilots in Lt. Russel's flight confirmed Lt. Russel's claims that he had shot down the MiG-15, but, although all confirmed seeing the smoke-trailing MiG-15 going down through the cloud layer, none could verify seeing the MiG-15 actually crashing into the terrain below. Several decades later when the Soviet Union collapsed, and the Soviet military records became known, those records indicated that the MiG-15 damaged from Lt. Russel's F-80 Shooting Star gunfire had made a safe emergency landing at a KPAAF base north of the Sinuiju bridge region. The MiG pilot's name was not released as it could have been either a pilot of the Korean People's Air Force, a pilot of the Chinese People's Air Force, or a pilot of the Soviet Air Force.

Given that the air victory (kill) of a MiG-15 by Lieutenant Brown in an F-80C Shooting Star as claimed on November 8, 1950, was later shown to not be a confirmed kill, the honor of the first confirmed kill in a jet-versus-jet fighter air combat engagement reverted to USN Lieutenant Commander William T. Amen. On November 9, 1950, in a Grumman F9F Panther from the USN aircraft carrier *Philippine Sea* (CVA-47), Lt. Cdr. Amen followed an evading MiG-15 into a very steep dive and delivered an unknown number of 20mm rounds into the MiG-15. Lt. Cdr. Amen fought the adverse effects of rudder reversal (similar to that which the pilots of P-38s in World War II had encountered at high subsonic speeds due to compressibility effects) and pulled out of the dive with only 200 feet of ground clearance as estimated by Lt. Cdr. Amen's wingman, USN Lt. George Holloman. The damaged MiG-15 did not pull out of the dive (for unknown reasons, possibly from Lt. Cdr. Amen's gunfire hits or again because of high-speed compressibility effects) and crashed into the terrain below the engagement. Lt. Cdr. Amen did not see the crash, but it was verified by his wingman, Lt. Holloman. The Soviet military records, obtained after the Soviet Union collapsed decades later, indicate that the pilot of the downed MiG-15 was a Soviet Air Force Capt. Mikhail Grachev. (To avoid sanctions of the United Nations Security Council, the Soviet Union did not acknowledge during the Korean War that Soviet pilots were in fact flying in the MiG-15s that were being used by the KPAF and CPAF during the entire Korean War from 1950 to 1953.)

The USAF F-80 Shooting Stars were not as highly effective as had been hoped by the USAF commanders, and their kill ratio with the MiG-15s was at its highest, about 5 to 1, in the late 1950 to mid-1951 time period. Although the two jets were of approximately equal technology and performance, the kill ratio was accepted to be more dependent on the training of the combatant pilots in their respective F-80s and MiG-15s. The KPAF MiG-15 pilots were poorly trained, the Chinese People's Air Force pilots were somewhat better trained than the KPAF pilots, but the Soviet pilots were trained quite well, compared to that of the USAF pilots in the F-80 Shooting Stars. A difference in cockpit design technology also had bearing on the outcome of many F-80 vs MiG-15 engagements. This important difference was attributed to the F-80 pilots wearing anti-G suits; whereas, the MiG-15 pilots (of any nationality) did not have any anti-G protection whatsoever.

In December 1950, three squadrons of swept-wing North American F-86A Sabre jets were shipped to the Korean Peninsula and based at K-13 (Suwon) to replace the outclassed straight-wing F-80 Shooting Stars. Shown next are two photographs of F-86A Sabres of the 51st Fighter Interceptor Wing Checker tails newly arrived at K-13 (Suwon) and in flight on Combat Air Patrol (CAP) over Korea.

F-86A Sabre Jets of the 51st Fighter Wing at K-13 (Suwon), South Korea, 1950
(Source: USAF Photographic files)

The F-86A Sabre jets carried six 0.50 caliber machine guns with 300 rounds per gun as its armament. The lower photo shows three of the muzzle ports on the sides of the nose inlet and there were three muzzle ports on the opposite side (not shown in the photo) of the nose inlet.

The muzzle ports are seen to not be aerodynamically closed, and the openings did have a minor adverse effect on the aerodynamics of the forward fuselage to the aft of the muzzle ports. The original XP-86A configuration (1948) did feature a set of sliding door panels to close off the open muzzle ports, but that mechanism was not incorporated in the operational versions to avoid maintenance problems (as well as complexity and cost). In the photo next is shown with the guns access panel removed on the side of the F-86 for replenishing the 300 rounds per gun and routine maintenance of the guns.

F-86A Sabre with guns access panel removed, 1949.
(Source: USAF Photographic files)

Because of the external similarity of the F-86's external configuration to that of the Soviet MiG-15, the characteristic high-visibility yellow bands were painted on the fuselage (just aft of the cockpit), on the vertical tail, and on the outer wing tips of the early F-86As to avoid the loss of F-86s to air gunfire from friendly aircraft. However, when the F-86As of the 51st Fighter Wing were delivered to Korea, the placement of the yellow high- visibility markings were modified to that as shown in the earlier photograph of the 51st Fighter Wing Checker tails which were based at the K-13 Air Base in Suwon, South Korea.

Also shown in the photograph is a General Electric J47 engine used to power the F-86A Sabre jets. The General Electric J47 engine is an improved version of the GE J35 engine which powered the F-80 Shooting Star. The J47 was produced by General Electric in at least seventeen different series before its production ended in 1956, and was the source of propulsive power for several USAF aircraft during the 1949 to 1956 time period. Most notable applications of the GE J47 were in the F-86, XF-91, B-36, B-45, B-47, and XB-51 aircraft.

A cutaway diagram of the GE J47 axial-flow turbojet is shown next, revealing the state-of-the- art in turbojet engine technology developed for the previously identified applications and installed in the F-86 Sabre.

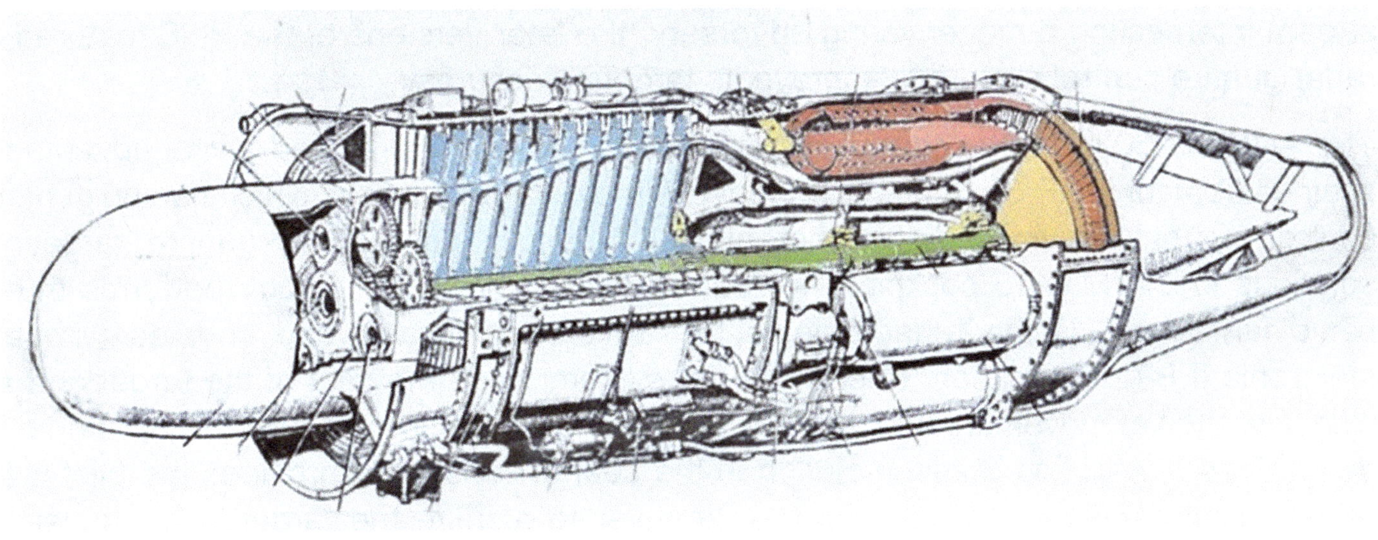

Cutaway Diagram of GE J47 engine used in the F-86 Sabre, 1951.
(Source: USAF Engineering files)

As a side note, when the GE J47 engine was installed in the early F-86 Sabres, it was internal to the fuselage and directly below the yellow stripe painted on the exterior of the fuselage. While the yellow fuselage strip was intended to avoid misidentification of the F-86 by other F-86 pilots, it being vey similarly externally configured as its MiG-15 adversary, the yellow strip on the fuselage gave a very visible aim-point for the MiG-15 pilots to fire their autocannons (two 23 mm and one 37 mm) at the F-86. Based on the lessons learned from World War II with the German Me-262 jet aircraft and with Battle Damage Assessments (BDAs) of the jet aircraft doing combat in the Korean War, gunfire rounds or missile fragments damaging a critical functional part(s) of the jet engine in any jet fighter is likely to destroy the aircraft. [It will be described in later chapters that the technology area of Survivability and Vulnerability – to minimize the probability of loss of the aircraft to enemy gunfire - was given major emphasis in the early configuration phase of the USAF Lightweight Fighters.]

The MiG-15 was originally designed to fire at bomber and reconnaissance aircraft and not to engage in dogfights with other jet aircraft. Accordingly, they were equipped with just the three larger caliber cannons but with a limited number of rounds per cannon. The 37 mm cannon had only forty rounds and the 23 mm had eighty rounds per gun. In addition, the trajectory paths of the rounds fired from the 37 mm cannon had a lower trajectory than the rounds fired from the 23 mm cannons such that gun-firing solutions were nearly impossible against high maneuvering targets. The six .50 caliber guns on the F-86 Sabre had three hundred rounds each and were designed with the same trajectories of the gunfire solutions for increased probability that one or two .50 Cal rounds would hit a critical part the jet engine. To increase the probability of the gunfire stream intersecting a maneuvering adversary, the later versions of the F-86 featured a radar gunfire control system to improve its targeting accuracy.

The AN/APG-30 radar/fire control system in the F-86 Sabre provided a vital advantage in air combat for the F-86 Sabre jet engaged with the MiG-15 Fagot, which had no gunfire control system other than the pilot himself mentally envisioning gunfire targeting solutions in the MiG-15 cockpit. The radar gunfire control technology performs three basic functions: (1) the radar detects the adversary aircraft, and coupled with an electronic IFF (Identification, Friend or Foe) system alerts the pilot of the target and at what direction of movement and range relative to the aircraft; (2) the fire control computer determines a targeting gunfire solution in the coarse sense and provides the pilot with targeting information, allowing him to maneuver to acquire the target in his gunsight aperture; (3) the fire control computer assists the pilot to acquire the target in his gunsight "pipper". The fire control computer computes the gun-firing solution that maximizes the probability of hit of the gunfire stream of rounds on the target. A sketch of the radar/fire control system on the F-86F is shown next:

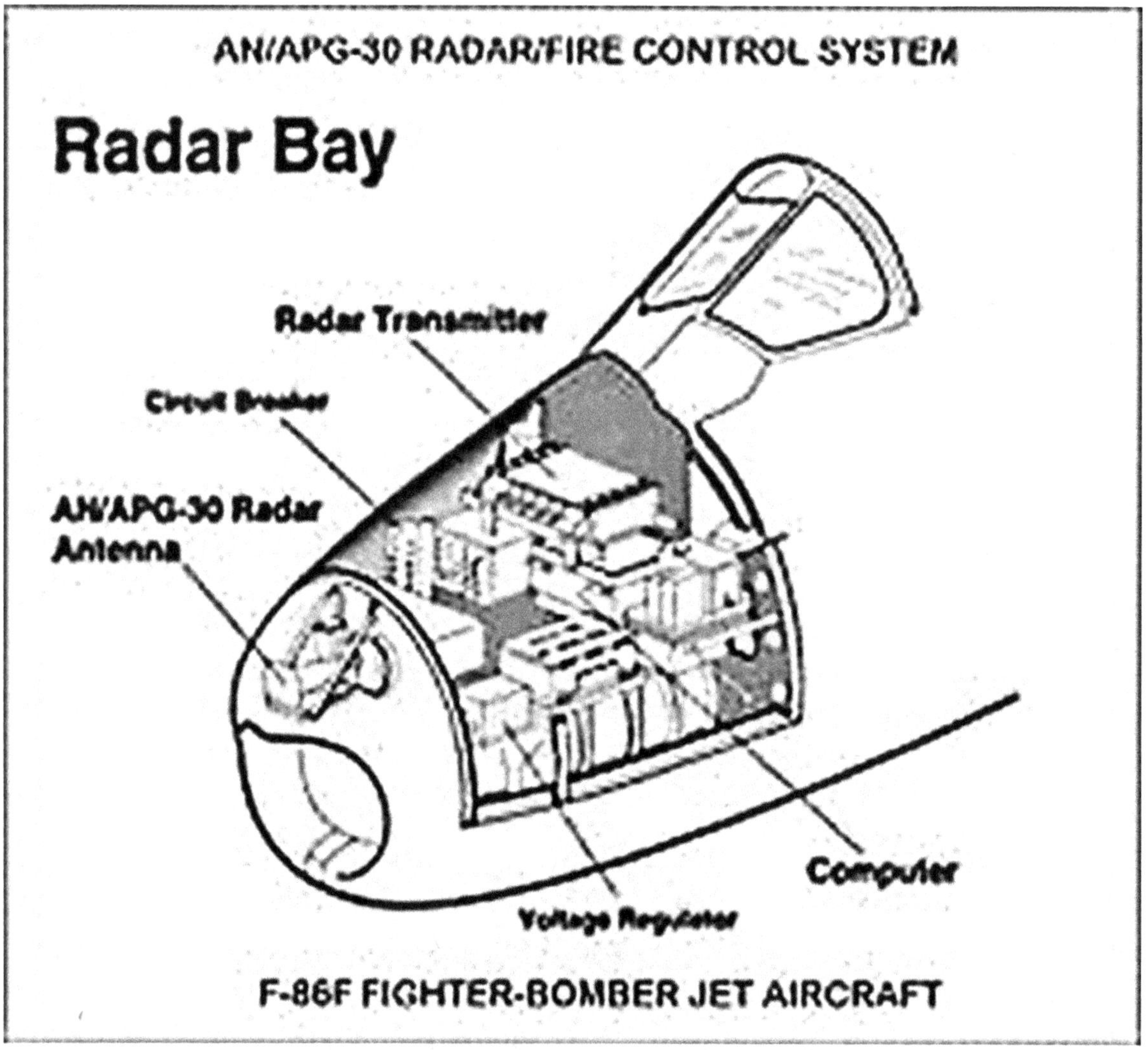

AN/APG-30 radar/gunfire control system in the F-86F.
(Source: USAF Engineering Files)

Upon arrival of the F-86 Sabre jets in the Korean theater of conflict, the rate of air victories of F-86 Sabre jets over the Soviet MiG-15 jets saw an immediate improvement. This improvement varied in magnitude depending most on whether the MiG-15 encountered and engaged in the combat was piloted by a KPAF pilot, a CPAF pilot, or a Soviet Air Force pilot. While there were widely varying claims as to air victories by both sides of the air war over Korea, military archives data after the Korean war indicate that of the 224 F-86 Sabre jets lost during the Korean War, only about 100 of them were due to losses in air-to-air engagements. From Soviet archival records obtained later when the Soviet Union collapsed indicates that 566 MiG-15s had been destroyed by UN aircraft (all but one being downed by US F-86 Sabre jets). On an overall basis then, the air victory ratio of F-86 vs MiG-15 averages out to be about 5.6:1.

The nature of the air war in Korea was somewhat clouded by the fact that the United Nations Security Council had dictated that there was to be no entry of UN Command air forces across the Yalu River into Communist China. Air warfare was therefore restricted to be conducted only over the Korean Peninsula against the MiG-15 aircraft regardless of who was actually piloting the MiG-15. The typical tactic of the MiG-15 aircraft was to penetrate across the Yalu River, do an incursion raid against UN Command forces south of the 38th Parallel, and flee back across the Yalu River into Communist China at a sanctuary base. That flight corridor came to be known as MiG Alley. As the negotiations were proceeding between UN Command representatives and the North Korean representatives, both sides agreed that there would be no air-to-air engagements north of what was to become the Demilitarized Zone (DMZ) which separated North Korea from South Korea at approximately the same 38th latitudinal parallel that existed when World War II had ended. F-86 pilots who violated that constraint and flew north of the DMZ were reprimanded severely; whereas, those who were known to have flown further north and across the Yalu River after MiG-15 adversaries were court-martialed and returned to CONUS (Continental United States) for jeopardizing the negotiations underway at Panmunjom. But there were a few pilots who did penetrate the Yalu River restriction and attacked MiG-15s operating over the base nearest the Yalu River in China. By not reporting that they had done so, they could not claim any air victories obtained while disobeying official UN Command orders.

It is at this point in this book that the single person most responsible for the development of the USAF Lightweight Fighters and their successor, the F-16 Fighting Falcon weapon system, must be identified as having been one of those daring F-86 Sabre pilots who unofficially attacked MiG-15s on the north side of the Yalu River. That person was John Boyd, then a First Lieutenant in the USAF. Lt. Boyd reported in during March 1953 to the air base in South Korea designated as K-13. The K-13 Air Base is adjacent to the town of Suwon in northwestern South Korea some 50 miles southwest from Seoul, the capital city of South Korea. He was rotated back to CONUS in early 1954, and served the next 6 years (until 1960) at Nellis AFB, in North Las Vegas, Nevada. It was during that assignment at Nellis AFB that John Boyd gained the reputation for having been perhaps the greatest air combat tactician ever to have served in the US Army Air Force or USAF.

[The Author did not know of John Boyd for his air combat episodes (official or unofficial) in Korea, but rather for the work he did in developing the tactics for fighter aircraft while he was in Korea and which later were refined at Nellis AFB and Eglin AFB. The first meeting the Author had with John Boyd was in 1969 at a highly classified meeting at the Foreign Technology Division at Wright-Patterson AFB in Dayton, Ohio, to discuss the results of exploitation flight testing of a "borrowed" MiG-15 (Code Name HAVE DRILL) that was done at the Groom Lake facility now known as Area 51. For continuity of this book, that activity of assessing the aerodynamics capability of the MiG-15 in a comparative sense with the USAF F-4 Phantom II and the USAF Lightweight Fighter is deferred until a later chapter in this book.]

UN-sponsored negotiations between North Korea and the United States took place for over 2 years at the town of Panmunjom straddling the 38th Parallel. On July 15, 1953, a cease fire was negotiated by the UN Peace Commission between North Korea and the UN Forces led by the United States in Korea. Readers should note that a cease fire is not to be interpreted as the end to the Korean War. Neither side capitulated, and the best outcome that the negotiations at Panmunjom could produce was an agreement to a cease fire in Korea. Technically, a state of war still exists (in 2017) between North Korea and South Korea.

[Author's note: The Author pursued Army ROTC courses for the 4 years of his attendance at Georgia Tech (September 1950 to June 1953). While a sophomore, in July 1950, the Korean War erupted. Upon graduating with a Bachelor of Aeronautical Engineering degree from Georgia Tech in Atlanta, Georgia, on June 15, 1953, the Author received a US Army ROTC commission as a Second Lieutenant in the Army Corps of Engineers the next day on June 16, 1953. Accompanying that commission was a set of orders from US Third Army Headquarters directing that Lt. Hutchinson was to expeditiously report for active duty and undergo combat training at Fort Bragg, North Carolina (in preparation for reassignment after completion of combat training). The orders continued that he then was to travel to the Personnel Replacement Depot in Inchon, South Korea, for duty with the US Eighth Army (under command of the UN Korean Command Headquarters in Seoul). Noting that the cease fire in the Korean War had been signed on July 15, 1953, the Author (with tongue in cheek) can report that the North Korean (Communist) negotiators had obviously become aware that Lt. Hutchinson was on his way to Korea and had hurriedly agreed to a cease fire before Lt. Hutchinson arrived in Inchon for duty in Korea. (That duty assignment was at a US Army camp not far from the southern portion of the demilitarized zone (DMZ) at the 38th parallel dividing North Korea and South Korea.) The sight of F-86 Sabre jets passing overhead in South Korea will never be forgotten, even if the Korean War has been.]

The number of aces who achieved that honor during the Korean War was far less than those who did so during World War II. This was not unexpected considering the scalar magnitudes of the scope of World War II and that in Korea. The aces of the UN Command force were those who flew in the North American F-86 Sabre against the Soviet-made MiG-15s piloted by North Korean, Communist Chinese, and Soviet Air Force pilots. The Soviet Union credited the top aces of the Korean War to be four of their Soviet Air Force pilots and they are listed first:

Soviet Union Top Aces in Korea in MiG-15 (67 aces with 514.5 air victories claimed)

1. Yevgeny Pepelyaev 22.5 air victories
2. Nikolay V. Sutyagin 22
3. Alexandr P. Smortzkow 15
4. Lev K Schukin 15
5. Dimitri P. Oskin 14

USAF Top Aces in Korea in F-86 Sabre (41 aces with 315.5 air victories claimed)

1. Joseph C. McConnell 16 air victories
2. James Jabara 15
3. Manuel J. Fernandez 14.5
4. George A. Davis 14
5. Royal N. Baker 13

China Top Aces in Korea in MiG-15 (7 aces with 54 air victories claimed)

1. Zhao Baotong 9 air victories
2. Wang Hai 9
3. Li han 8
4. Lu Min 8
5. Fan Wanzhang 8

North Korea Top Aces in Korea in MiG-15 (2 aces with 14 air victories claimed)

1. Kam Den Dek 8 air victories
2. Kim Di San 6

(Source: USAF Military records)

The observation should be made that the sum of all of the three Communist countries (Soviet Union, China, and North Korea) pilot aces claiming to have scored a total of

582.5 air victories in the Korean War, casts serious doubt that those claims are all valid. The claims of USAF aces for having downed 315.5 MiG-15s and coupled with the total of 566 MiG-15s having been claimed to have been destroyed by UN Command pilots during the total Korean War period is perhaps more plausible. No matter the validity of all the claims data, the Korean War proved the effectiveness of jet-powered, first-generation fighters. The era of piston-engine, propeller -powered aircraft in air-to-air combat operations had become a matter of aviation history after the Korean War.

The end of the Korean War (with the disappointing technicality that only a cease fire had been agreed to in order to end the battlefield hostilities on the Korean Peninsula) was regarded as a United Nations victory as an international peace-keeping activity from 1953 until the present (2017). Many regard the Korean hostilities as the Forgotten War. [The Author cannot forget the Korean War and for two primary reasons: (1) because Soviet Air Force pilots in the North Korean MiG-15s were used in direct air combat with USAF and USN/USMC pilots; and (2) because the Author served in the US Army under the UN Command, and saw first-hand what the North Korean People's military forces did to destroy the lives and residences of many hundreds of thousands of South Koreans under the guise of unifying the Korean Peninsula under the Communistic form of government. Those comrades of several different nations who sacrificed their lives to protect those of others are true heroes and cannot and must not ever be forgotten.]

If anything, the tensions existing between Communist controlled countries and Democratic governed countries during the Cold War were intensified during the Korean hostilities of 1950 to 1953. Adding to the Cold War tensions was the troublesome activities of the superpower nations (United States and the Soviet Union especially) building up an arsenal of nuclear weapons capable of the destruction that was presaged by the atomic bombings of Hiroshima and Nagasaki in Japan in 1945. The expression "nuclear arms race" saturated the news media and cast ominous shadows on the World.

Very correctly, the delivery of those nuclear weapons could be by either Inter-Continental Ballistic Missiles (ICBMs) or could be carried and delivered on targets having mass population density by bombers with high-altitude capability of either Soviet manufacture or of United States manufacture. As stated earlier, there is no victor when a war is concluded; both sides of the conflict will suffer irreplaceably in human lives. Should a nuclear war ever occur in which both sides employ nuclear-based weapons of mass destruction against each other, not only will both sides suffer unimaginable losses, the entire World will have lost all hope for a peaceful environment in which to enjoy the natural beauty on all of the World's continents. Given that as being an unacceptable environment, the United Nations Security Council took immediate steps to assure that the Cold War would not become a nuclear war, but as history proved once again there would be another set of more conventional wars that would occur among countries in the years following the Korean War. This book's scope covers only the effect war has on the technological advancements made in the military aircraft developed by the US Air

Force (and supported by the aircraft capabilities other military services) to bring conventional wars to as rapid a conclusion as is possible.

The United States Department of Defense saw the need for the air forces of the US military services to defend the United States from nuclear attack by any other nation (with the Soviet Union being considered the primary threat) and approved the development of various aircraft development programs in parallel with nuclear weapons programs developed as being a deterrent or as a retaliatory capability. Constraining the contents of this chapter to programs to be identified as having significant influence on what would later be the technology genealogy of the F-16 Fighting Falcon weapon system, the development by the USAF of the multiple aircraft production programs which became known as the Century Series fighter interceptors (to intercept in-coming Soviet aircraft perhaps carrying nuclear weapons) and fighter bombers to be used in any ensuing conventional war that might occur in the near future (which, as history shows, was the Vietnam War). In a tabulated manner shown below, the six Century Series aircraft which went forward into production and operational use are identified and in their order of first flight date (of the initial YF prototype to reflect the technology baseline of each. The propulsion system for each is shown in the right portion of the tabulation):

North American F-100 Super Sabre	May 1953	P&W J57-P-21A
Lockheed F-104 Starfighter	February 1954	GE J79
McDonnell F-101 Voodoo	September 1954	P&W J57-P-13
Convair F-102 Delta Dagger	December 1954	P&W J57/Curtiss Wright J67
Republic F-105 Thunderchief	October 1955 (F-105A)	P&W J75
" " "	May 1956 (F-105B)	P&W J75
Convair F-106 Delta Dart	December 1957	P&W J75-17

(Source: USAF Engineering files)

The F-103, F-107, and F-108 configurations, missing from the list, were initially built-in prototype form, but were later eliminated from further development by the USAF because of a lack of mission need or because of the technologies attempted to be applied not being of sufficient maturity to risk further development.

[The Author was an aerodynamicist at the Republic Aviation Corporation in Farmingdale, New York, from early 1955 to late 1956 and conducted aerodynamic drag reduction analyses of both the F-105A and F-105B. In the prototype stage was an XF-103 high-speed interceptor, and that concept was a very needle-nosed forward fuselage, no external lines discontinuities from a cockpit, had an inlet with a two-position horizontal ramp, and featured a dual-propulsion system. At the top in the aft fuselage was a turbojet engine, which was used for takeoff, climb to altitude, and landing. At the bottom was a ramjet engine, used to accelerate the XF-103 to maximum Mach number (approximately 3.0). Normal position of the inlet ramp was down to ingest and deflect the inlet air into the turbojet mode. When the ramjet mode was to be initiated, the inlet ramp was transitioned from the down position to the up position and the turbojet power decreased toward shutoff, the ramjet was ignited and generated full ramjet thrust. The USAF considered the technologies of the XF-103 configuration to not be sufficiently matured for further consideration, despite the advantage of the XF-103 having the potential for a M3 speed and quicker intercept of incoming Soviet aircraft than the other Century Series fighter interceptors being considered. As a result, the XF-103 activity was canceled. The Fairchild Company purchased the Republic Aviation Corporation shortly thereafter and the company name became Fairchild Republic.]

The XF-107 and XF-108 configurations were also truncated for reasons similar to those of the XF-103.

In the photograph next is shown a grouping of five of the six Century Series fighters on static display at the present Edwards AFB Visitors Registration area: (The F-106 is missing from the group photo but is essentially only a larger-size version of the F-102 delta wing configuration included in the Century Series group photo).

Clockwise from bottom: F-104 Starfighter, F-100 Super Sabre F-102 Delta Dagger, F-101 Voodoo, F-105A Thunderchief (Source: USAF Photographic files)

Readers who have a special interest in the detailed accounting of the design development and testing (ground and flight) of each of the six tabulated Century Series fighter–interceptors or fighter-bombers will find a comprehensive and very factual detailed accounting of each configuration in a report (actually a Research Memorandum (RM) created for, but not sanctioned by, the USAF) easily found on the internet. The identification for the Research Memorandum is "US Air Force Project RAND Research Memorandum RM-2549, *"The Century Series Fighters: A Study In Research and*

Development", issued May 20, 1960, by Leland L. Johnson of the Rand Corporation offices in Santa Monica, California.

As indicated in the RAND RM-2549 (105 pages), the USAF evaluated all of the tabulated Century's configurations very carefully and comprehensively, and came to the overall general conclusion that the North American F-100 Super Sabre was the best overall aircraft to perform the highly maneuvering air combat missions. The Lockheed F-104 Starfighter was assessed as the best aircraft to perform the intercept missions. The other four Century Series aircraft configurations were assessed as being generally of equally high capability for performing the fighter-bomber or fighter-interceptor missions. The assessment that the F-100 Super Sabre was the best for the air combat missions was attributed to the experience and vital database obtained by the North American Company from its prior model, the F-86 Sabre, and the outstanding combat performance of the F-86 Sabre against the Soviet MiG-15 Faggot in the Korean War.

It is beyond the scope of this book to do an in-depth independent analysis of the set of technologies advanced by each of the 6 Century Series aircraft that eventually went into production and operational use for the USAF. It must be stated though that the wind tunnel testing done on each of the 6 configurations at Dr. Richard Whitcomb's transonic test facility at the NASA Langley Research Center (adjacent to Langley Air Force Base) benefitted immensely from Dr. Whitcomb's talents in applied aerodynamics. [The Author was honored to present many different aircraft design models to the Dick Whitcomb test facility and which resulted in the greatly improved aerodynamics improvements suggested by Dick Whitcomb. The Author and Dick Whitcomb became quite good friends, and it was he who taught the Author to "think like a molecule", and to consider "how would you act if you were that molecule being disturbed by an aircraft and being disrupted from a placid stationary point in the atmosphere?". In a later chapter, the contributions Dick Whitcomb made to the Lightweight Fighter configurations (General Dynamics YF-16 and Northrop YF-17) will be discussed later in greater detail.]

The area rule (sometimes referred to as the "coke bottle" or "wasp waist" shape) configuration concept developed by Dick Whitcomb converted many barely-transonic capable aircraft into supersonic aircraft. The original cylindrical-shaped fuselage on the F-102 Delta Dagger limited the maximum speed of the F-102 to be about Mach 0.93 due to wave drag (associated with the supersonic flow over the configuration external components) being quite a bit higher than Convair had originally estimated. With additional wind tunnel testing at the NASA Langley Research Center Transonic Wind Tunnel, Dick Whitcomb contoured the fuselage shape to feature a much-improved cross-sectional area distribution along the horizontal axis to minimize the disturbance to the mass of air molecules through which the pilot would command the flight path of the F-102. Shown next is a top view drawing of the F-102 after its fuselage above the wing had been contoured by Whitcomb:

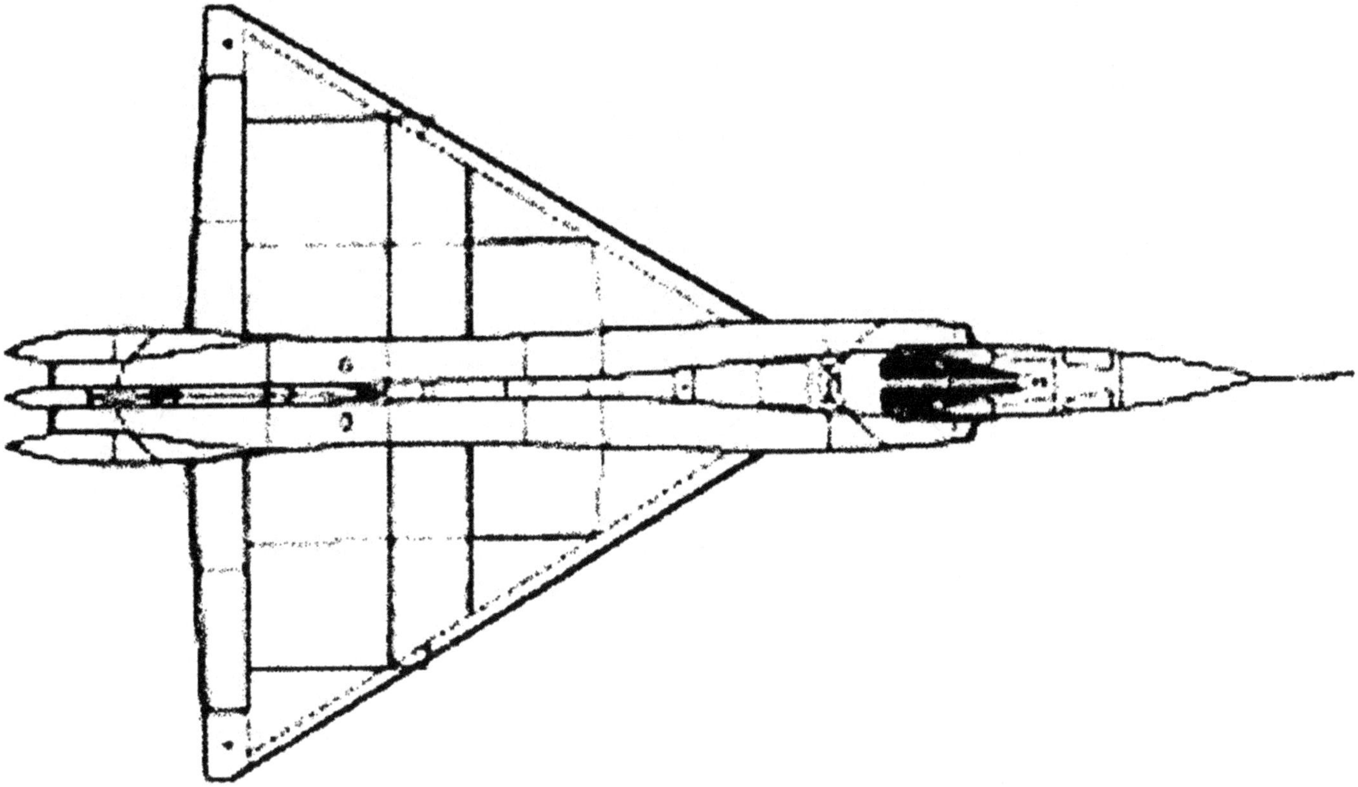

F-102 Delta Dagger with area-ruled fuselage by NASA/Whitcomb
(Source: USAF Engineering files)

The F-102 configuration is seen to feature a delta-wing planform and the conventional horizontal tail is seen also to not exist in the F-102 configuration. The Configuration Design Department of Convair in Fort Worth, Texas, was headed by a very talented Bob Widmer who had an even more advanced concept thinking associate, Harry J. Hillaker. Harry Hillaker would go on to later be the Chief Designer of the Convair B-58 Huster Supersonic Strategic Bomber, the General Dynamics F-111 Aardvark swing-wing fighter bomber, and of the family of fighter aircraft configurations which were developed in an Advanced Day Fighter Study for the USAF, sponsored by none other than the Col. John Boyd of energy-maneuverability fame) that converged to become the external configuration of the F-16 Fighting Falcon. [The Author has dedicated this book to two persons who had the most influence on the USAF Lightweight Fighter configurations, Col. John Boyd and Harry J. Hillaker. As this book continues, much more will be revealed about the ingenuity and creativeness of John Boyd, Harry Hillaker, and the General Dynamics team in development of the YF-16 Lightweight Fighter prototype configuration.

Perhaps the most advances in the technologies for the Century Series supersonic fighter aircraft came in the propulsion area. The Pratt & Whitney J57 engine had been developed and was used in the F-100 Sabre, the F-101 Voodoo, and the F-102 Delta Dagger developments. A photo showing the cutaway of a P&W J57 engine on display at the National Aerospace Museum is provided next.

Cutaway of Pratt & Whitney J57 Engine (1954)
(Source: Photo by the Author)

The P&W J57-P-7 version of the J57 generated 14,800 pounds of thrust (without afterburner). The P&W J57-P-21A version of the J57 incorporated an afterburner and generated 16,000 pounds of thrust. The non-afterburning version of the J57 was used on the F-100A, the two-engine F-101 Voodoo, and the F-102 Delta Dagger. The afterburning version of the P&W J57 was used on the initial F-104A Starfighter (but was changed to the GE J-79 engine for the later F-104 variants providing them with the rapid acceleration capability desired for an interceptor aircraft).

In parallel development with the P&W J57 engine was the General Electric J75 engine to be used on the F-105A (and later F-105B). A cutaway diagram of the GE J75 engine is shown next.

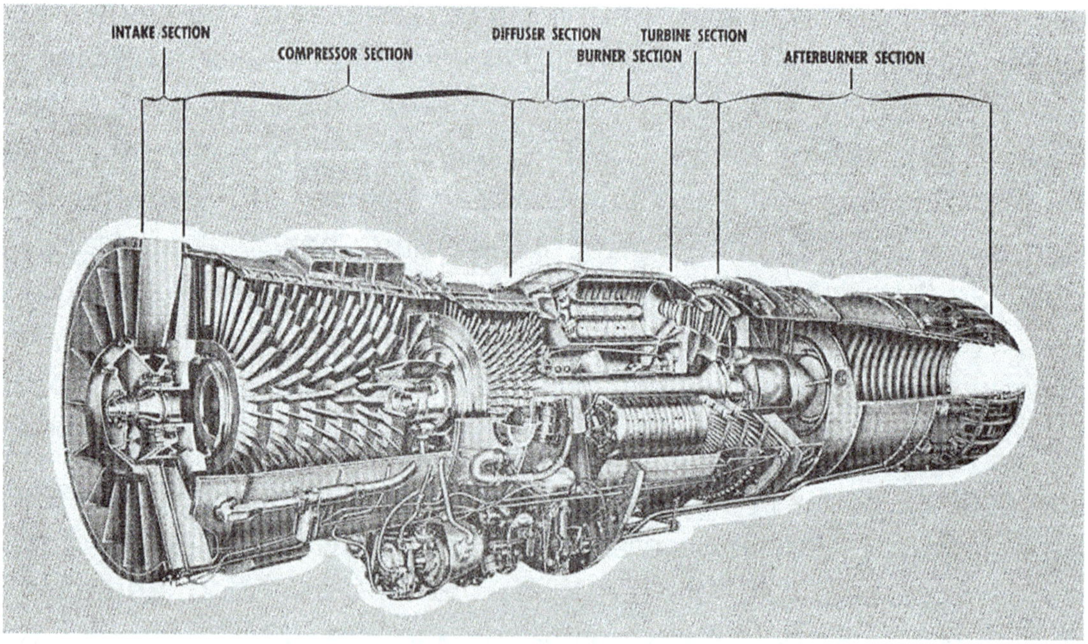

Pratt & Whitney J75 turbojet engine, 1955.
(Source: USAF Engineering files)

The F-105 configuration was the largest and heaviest of the Century Series fighter-bombers at over 50,000 pounds empty weight. (The F-105A and F-105B were about the same size as the B-17 Flying Fortress strategic bomber of World War II). The J75 was used also on the F-106 Delta Dart, which was in essence a larger and heavier version of the F-102 Delta Dagger and also required the afterburning thrust level for accomplishing its primary role as a fighter-interceptor. The P&W J75-17 engine with afterburner generated 24,500 pounds of thrust.

At the same time, the USAF was developing the Century Series fighter interceptors and fighter bombers having the capability to intercept and destroy Soviet bomber perhaps carrying nuclear weapons, that the United Nations Security Council postulated could be a very real aspect of the Cold War and threatening creation of a nuclear arms race. The retaliatory strike capability aspects of the Cold War gave rise to a Supersonic Strategic Bomber concept for rapid retaliatory delivery of a nuclear weapon on a target(s) within the Soviet Union. In February 1949, the Air Research and Development Command at Wright-Patterson AFB issued a request to the US aircraft industry for the conduct of a Generalized Bomber Study (GEBO II). Responding to the request were (in alphabetical order) Boeing, Convair, Curtiss, Douglas, Martin, and North American Aviation. After evaluating the responses to the USAF request, Boeing and Convair were given study contracts for a Phase 1 study for the first supersonic bomber program. Early in the study, the USAF also recognized the need for a new engine development and caused General Electric to start development of their GE 79 series of engines in the 16,000 pounds of thrust class. The carriage of nuclear weapons on manned Strategic Air Command bomber aircraft theoretically gave some degree of added control and ability to abort the delivery of a nuclear weapon before it erroneously created an international crisis, and was favored by many rather than accepting that an ICBM unmanned missile had been launched autonomously and would strike its intended target without the option for recall.

Jumping over many interesting results of phase 1 of the GEBO II study, in February 1953 the USAF issued a contract to Convair/Fort Worth for development of the Convair four-engine, delta- wing, supersonic strategic bomber having a nuclear weapon carriage capability. Much of the development data generated by Convair for its F-102 Delta Dagger and F-106 Delta Dart programs, especially the area rule aerodynamic configuration to minimize wave drag at transonic and supersonic speeds, was directly transferable to that of the Convair strategic supersonic-bomber configuration chosen by the USAF.

With again the invaluable advice of Dr. Richard Whitcomb at the NASA Transonic Wind Tunnel at Langley AFB, Virginia, and the brilliant aerodynamic configuration talents of Convair Design Engineer Harry J. Hillaker, the fuselage of the proposed new configuration was even more area ruled such that the fuselage was extremely and smoothly contoured on the top and bottom above and below the wing-fuselage juncture. The bottom contour of the fuselage was such that the nuclear weapon carried below the fuselage actually fit snugly into that fuselage contour without adverse effect on the total cross-sectional area of the configuration at that location of the center fuselage. The four engines were attached (#1 and #4 on wing hard points and #2 and #3 on wing faired pylons) and again spaced fore and aft so as to achieve a gradual increasing and then decreasing total cross-sectional area distribution to minimize supersonic wave drag of the aircraft. A photograph of the Convair B-58A Hustler configuration is shown next to illustrate the ingenuous configuration for that supersonic strategic bomber.

Convair B-58A Hustler, 1959.
(Photo USAF Photographic files)

The four engines installed on the B-58A were the General Electric J79 turbojet engines with afterburners. A cutaway diagram of a J79 engine on display is shown next (to enable Readers to better visualize the state-of-the-art in the technologies of an aircraft with a superior aerodynamic configuration and having a large, high-technology, propulsion system integrated into that configuration).

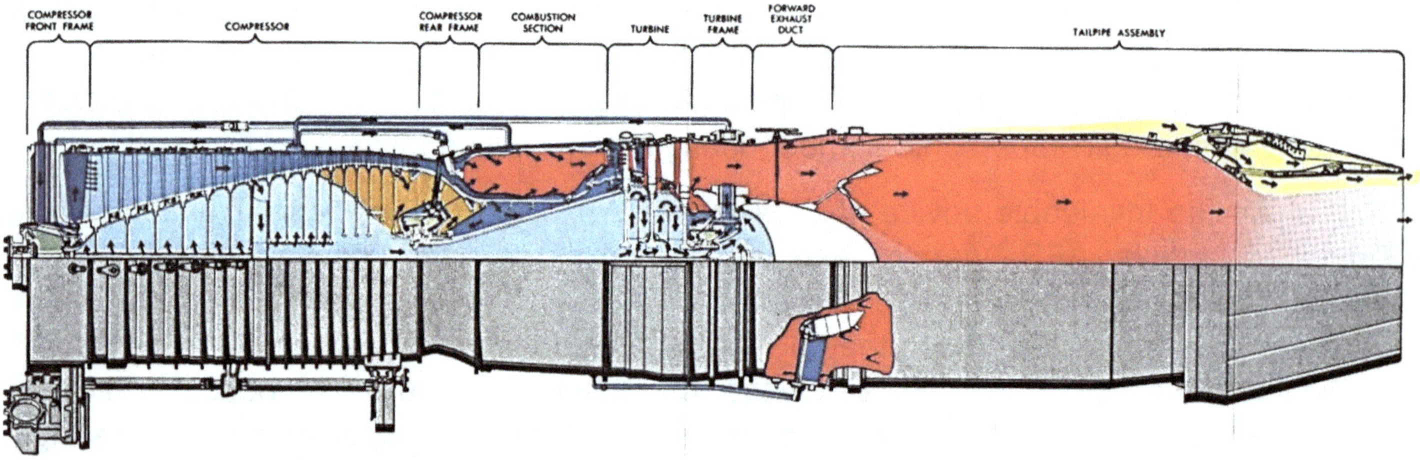

Cutaway diagram of General Electric J79 engine, 1959.
(Source: USAF Engineering files)

The General Electric J79 engine was an axial-flow afterburning turbojet capable of generating 17,835 pounds of thrust in afterburner mode (GE-J79-17 version). The GE J79 was designed specifically for powering the Convair B-58A Hustler strategic supersonic bomber but was the propulsive system used for the Lockheed F-104 Starfighter, the McDonnell Douglas F-4 Phantom II, and the North American A-5 Vigilante. Over 17,000 J79 engines were built in the period from 1954 (first ground static run) until the late 1960s, when the technology advancement of turbojet machinery was increased to afterburning turbofan engines that powered the General Dynamics/USAF F-111 Aardvark and the Grumman/USN&USMC F-14 Tomcat fighter-bombers. The J79 engine had a weight of 3,850 pounds which gave it a thrust- to-weight ratio of 4.6, making it one of the most efficient turbojet engines in propulsion system technology history.

Also, to be observed is that the B-58A configuration featured the large delta wing (with 60-degrees leading edge sweep angle) and no horizontal tail, absent also on the Convair F-102 and F-106 fighter configurations. Readers may question how then did the B-58 Hustler achieve longitudinal control as the aircraft transitioned through the transonic regime when accelerating from subsonic to supersonic speeds without a horizontal tail to accommodate the shift of the wing center of pressure aft from approximately 25 percent of the MAC (Mean Aerodynamic Chord) at subsonic speeds to approximately 50 percent of the MAC at supersonic speeds. The solution to that control challenge was chosen to be a fuel transfer system internal to the B-58A, with which a mass of fuel was transferred aft in the internal tanks before accelerating from subsonic through transonic speeds to supersonic flight. When decelerating from supersonic speeds through the transonic speed regime to subsonic speeds, a large mass of fuel had to be transferred forward to maintain the static margin (distance between aircraft center of gravity and center of pressure) needed for adequate longitudinal stability.

A humorous viewpoint expressed by B-58A test pilots was that #3, the third man in the crew - the B-58 had a flight crew of three seated in a fore and aft arrangement - being the Electronics and Weapons System Operator (EWSO), needed having something useful (such as fuel- transfer management) to do during the long flight inbound toward the target and not being occupied with delivering the nuke weapons. Although this fuel-transferring activity did cause a need for high reliability of components for managing the transfer rates and amounts as well as complexity, which always costs more, the advantage was that it eliminated the weight of the horizontal tail assembly and more importantly avoided the drag associated with horizontal tail deflection to achieve a positive longitudinal static margin.

At this point in this book and with Reader's indulgence, the Author wishes to point out that in any aircraft development program, it is not only the choice of technologies selected for that particular design, but it is the combined talents of the entire development team which determines the success (or less than success as is sometimes the unfortunate case) of the end product of the development activity. All participants each have a unique career path and experience base. For those Readers planning on entering a career path in aircraft development, the Author has been injecting "Author's notes" at various places in this book to indicate the type of events or technical problems that design team members encounter during the development of an aircraft that require adaptation of the approach being planned. Often, a revision of the testing program is needed to thoroughly validate the use of one or more emerging technologies in a flight system.

In the vein of providing to newcomers on an aircraft design and test team an example of what might be expected to occur, the Author will present next a small segment of his career path as an actual example of the factors that can, and will, alter the path of a career path envisioned by a newcomer to a technology advancement development team. The Author used the Korean War educational benefits to obtain a Master of Science in Aeronautical Engineering at Georgia Tech in April 1959. Former classmates in undergraduate AE courses at Georgia Tech had accepted employment at Convair in Fort Worth, Texas, to get in on the ground floor (the aspect repeated often throughout this book as stated in the metaphor "As the twig is bent, so shall the tree grow.") in design development of the B-58A Hustler. The Author received a job offer from Convair upper management (Bob Widmer, Harry Hillaker, George Kahler, etc.) to come and join the B-58A Hustler development team in the Flight Test Engineering Department at Convair. The Author quickly moved his family to Fort Worth and started employment as a Flight Test Engineer in late April 1959. As a piece of trivia, the photograph of the B-58A Hustler shown earlier was taken by an aerial flight photographer who, when not off flying, occupied a desk next to the Author in B-58A Flight Test Engineering at Convair. That sequence of events occurred most uniquely for the Author, but such events should not be expected to repeat exactly the same during the career of anyone else. It requires a person to have the flexibility built in his/her career plan to take advantage of opportunities as they can emerge very suddenly and very unexpectedly.

Most Readers are probably aware that the aerodynamic characteristics of any new flight system/aircraft configuration are always established in a series of wind tunnel tests long before the new flight system will actually be flown for the first time. For those not familiar with how the drag of an airplane is obtained by flight testing, there is a process of flying speed-power combinations of altitude, Mach number, and engine power setting at which the net thrust output of the propulsion system is equal and opposite to the drag of the aircraft at stabilized (non-maneuvering) flight conditions. The purpose of these flight tests is to validate the flight envelope for the aircraft which had been estimated based on the aerodynamics obtained in wind tunnel testing (and using propulsive thrust obtained from ground stand runs conducted by the engine manufacturer.) As there is no measurement sensor/meter on the airplane to measure aerodynamic drag directly and sufficiently accurately, drag is derived by numerically determining the net thrust of the propulsion system at those stabilized speed power flight conditions. Since the thrust vector is equal and opposite to the drag vector, establishing the thrust vector is the means by which the drag vector can be defined. At the same speed-power condition, the lift vector is opposite and equal to the weight vector. Establishing the weight of the aircraft at the time of the speed-power flight event is the means by which the lift vector can be defined.

In the case of the B-58A, that meant determining the thrust of each of the four J79 engines for a wide range of conditions in which engine ground test stand runs had mapped out the engine net thrust output depending on the difference between the pressure and temperature of the air entering into the engine face and the pressure and temperature of the air exiting the nozzle and creating the resulting reaction thrust force. Critical to the speed power flight testing then is the ability of flight test engineers to measure or otherwise determine those inlet and nozzle exit temperatures and pressures sufficiently accurately for determining the aerodynamic drag of the flight vehicle being tested.

In the early months of the B-58A flight test program during the conduct of the B-58A speed-power tests, the meteorological data (i.e., balloons with temperature sensors) for the intended flight path of the B-58A was measured and made available to the flight test coordinator approximately one hour before the planned flight test. Because the length of the flight path was many miles, the meteorological measurements were made at three balloon stations bracketing the flight path and the temperature at the aircraft location within that triangle of temperatures to be interpolated for temperature at the test aircraft. The flight test speed-power data was found to not correlate well at all with the wind tunnel and engine ground run data because of the inaccuracy in determining the temperature of the air entering the face of the engines. The Author was tasked by the General Dynamics Flight Test Director to analyze the problem (inaccuracy of obtaining temperature by triangulation of balloon temperatures) and to identify what type of on-board instrumentation and temperature sensors were needed to improve the flight test drag data. After a few days of analysis, the Author prepared a report "*Instrumentation Accuracy Requirements for the B-58A Flight Test Program*" in 1959. This led to an activity at/with the Rosemount Instrument Company to develop a total temperature probe that would be mounted on the nose of the B-58A and sense/record accurately the temperature of the air through which the test article was being flown, totally eliminating the need for the archaic manner of triangulation between balloon-based meteorological measurements.

That all went well, but early tests then with the Rosemount Temperature Probe still showed very erratic temperature measurements that reflected poorly on the J79 thrust data being derived. Analysis of the Rosemount System (which featured a coarse and a fine temperature using the Wheatstone Bridge principle of electronics) did not explain the erratic behavior of the measure temperature data. Inspection of the actual installation of the Rosemount Temperature Probe revealed that inadvertently (due to the lack of detailed installation instructions describing how to properly install the Rosemount Probe and associated electrical wiring), the voltage from one Buss bar (power supply) had been connected into the coarse Wheatstone Bridge circuit but the voltage from a different Buss bar had been connected into the fine Wheatstone Bridge circuit. Apparently, the voltage differences between the two different Buss bars were the root cause for the erratic temperature sensed by the Rosemount Probe. Subsequently, both Wheatstone Bridge circuits (coarse and fine) were connected to the same Buss bar and thereafter accurate in-flight temperature data was obtained during the speed power testing. The inferred aerodynamic drag data obtained with the Rosemount Temperature Probe matched almost perfectly between the flight test data and the earlier wind tunnel data. The greatest personal reward came to the Author in the form of a hand-shake and verbal "Atta boy" from Harry Hillaker.

The foregoing example of a simple error in properly installing instrumentation for measurement of a key element of data is intended to serve as a caution to newcomers (and old-timers as well) that stuff happens for a reason, and that lessons learned from problems and/or accidents need to be well documented and disseminated throughout the test team rapidly, clearly, and comprehensively. The experience of the Author has been that more often than not, when a functional component in the aircraft design does not perform as designed or when a component fails to perform for unknown reason, the problem is not within the design of the component itself but with one of the interfaces (mechanical, electrical, hydraulic, or environmental) into or out of that component. The Author has been chided many times for being an engineer who if asked 'What time is it?', the details for how to build a clock would be forthcoming in response. The reputation for being a good clock-maker soon was replaced by being regarded as being a good system engineer for having enabled the test team to resolve an unprecedented problem. All of which points to the situation when faced with a challenging problem, new (and old) engineers should use the human gift of ability to *think*. Knowing which button to depress on a computer keyboard to call for help with a "design your own airplane" app is no substitute for plain, old-school, thinking.

While the USAF was developing the Century Series fighters, the US Navy initiated development of a carrier-based air superiority fighter designated as the McDonnell Douglas F-4 Phantom II in an unsolicited proposal on September 19, 1953. The proposal entitled "Super Demon" was a much-improved version of the USN F3H Demon naval carrier-based fighter and was to be powered by two J79 engines. The USN felt that the upcoming Grumman XF9F-9 and Vought XF8U-1 already satisfied the USN/USMC needs for a supersonic fighter. The McDonnell Douglas Company reworked its Super Demon concept into an all-weather fighter bomber with eleven external hard-points for external weapons carriage. On October 18, 1954, the McDonnell Douglas Company received an LOI (Letter Of Intent) from the US Navy and was encouraged to design two YAH-1 prototypes. On May 26, 1955, four USN officers arrived at the McDonnell Douglas facilities in St. Louis, Missouri, and presented McDonnell Douglas with an entirely new set of capability requirements for an all-weather, fleet-defense interceptor. Following an expedited design study by McDonnell Douglas, the USN on July 25, 1955, ordered two XF4H-1 test aircraft and five YF4H-1 pre-production prototypes. There were various candidates for naming the F4H aircraft (candidate names such as Satan and Mithras) however the less controversial name Phantom II was selected after the earlier McDonnell jet fighter, the FH-1 Phantom.

The first flight of the McDonnell F-4 Phantom II was on May 27, 1958. A photo of the F-4 Phantom configuration is shown next.

McDonnell Douglas F-4 Phantom II,1960.
(Source: USAF Photographic files)

In a demonstration on December 6, 1959, (entitled *Operation Top Flight*) of the flight capabilities of the F-4 Phantom II, USN Commander Lawrence E. Flint, Jr., in the second XF4H-1 produced did an acceleration at 47,000 feet and attained a speed of Mach 2.5, following which he climbed the aircraft to 90,000 feet at a 45-degrees flight path angle. Cdr. Flint then shut down both J79 engines and performed a zoom-climb to a world record 98,557 feet altitude. An F4H-1 on September 5, 1960, averaged a speed of 1,216.78 mph (Mach 1.84) over a 500 km (311 miles) closed-course. Twenty days later (September 25, 1960), an F4H-1F averaged 1, 390.24 mph (Mach 2.11) over a 100 km (62.1 miles) closed-circuit course.

Based on the loss rate of the North Korean MiG-15s against the USAF F-86 Sabre jets in the Korean War, the Soviet Union began design activities which consisted of a number of modifications to update the MiG-15 design such that its deficiencies against the F-86 Sabre would bring more parity in the capability difference between the new MiG-17 and the F-86. These design modifications were accomplished on an expedited basis but the MiG-17, which had its first flight in January 1950, was not introduced into operation until October 1952, too late to have any significant impact on the air war over Korea.

The MiG-15 air combat capability was hampered by the lack of a ranging radar and optical gunsight, but in 1951, Soviet engineers obtained a captured F-86 Sabre from North Korea and reverse engineered the F-86 Ranging Radar and Optical Gunsight. The copied ASP-

4N gunsight and SRC-3 radar were integrated into the MiG-17 and its successor the MiG-19. This upgrade would later prove to be extremely important during the Vietnam air war.

A major change for the MiG-19 was the addition of an afterburner to provide the Soviet Air Force with its first fighter aircraft having a supersonic capability. The resulting supersonic design (embodying the modifications and upgrades made for the MiG-15 and MiG-17) was designated as the MiG-19. A second major change in the MiG-19 was in the wing design to be a swept-back wing with a smaller taper ratio (length of the tip chord divided by length of the root chord). A small taper ratio wing is favorable to aerodynamics at high-speed flow over fighter aircraft wing planforms. The MiG-19 had its first flight on May 27, 1952, but was not introduced into operational status with the Soviet Air Force until March 1955.

A photograph of the supersonic MiG-19 is shown next.

Soviet MiG-19 "Farmer",1956.
(Source: USAF Foreign Technology files)

It can be seen that the horizontal tail underwent three primary changes: (1) it was moved downward from the vertical tail and onto the aft fuselage, (2) its area had been enlarged, and (3) had a higher sweepback angle. Noticeable also is that the Soviet penchant for having fences (a.k.a. flow- straighteners) on its upper wing surfaces is apparent with the pair of large fences on the MiG-19 upper wing surfaces. Except for having the wing fences, the MiG-19 was a very similar external configuration to the USAF F-86 Sabre.

For additional visual description of the MiG-19, a small 3-view drawing is shown next.

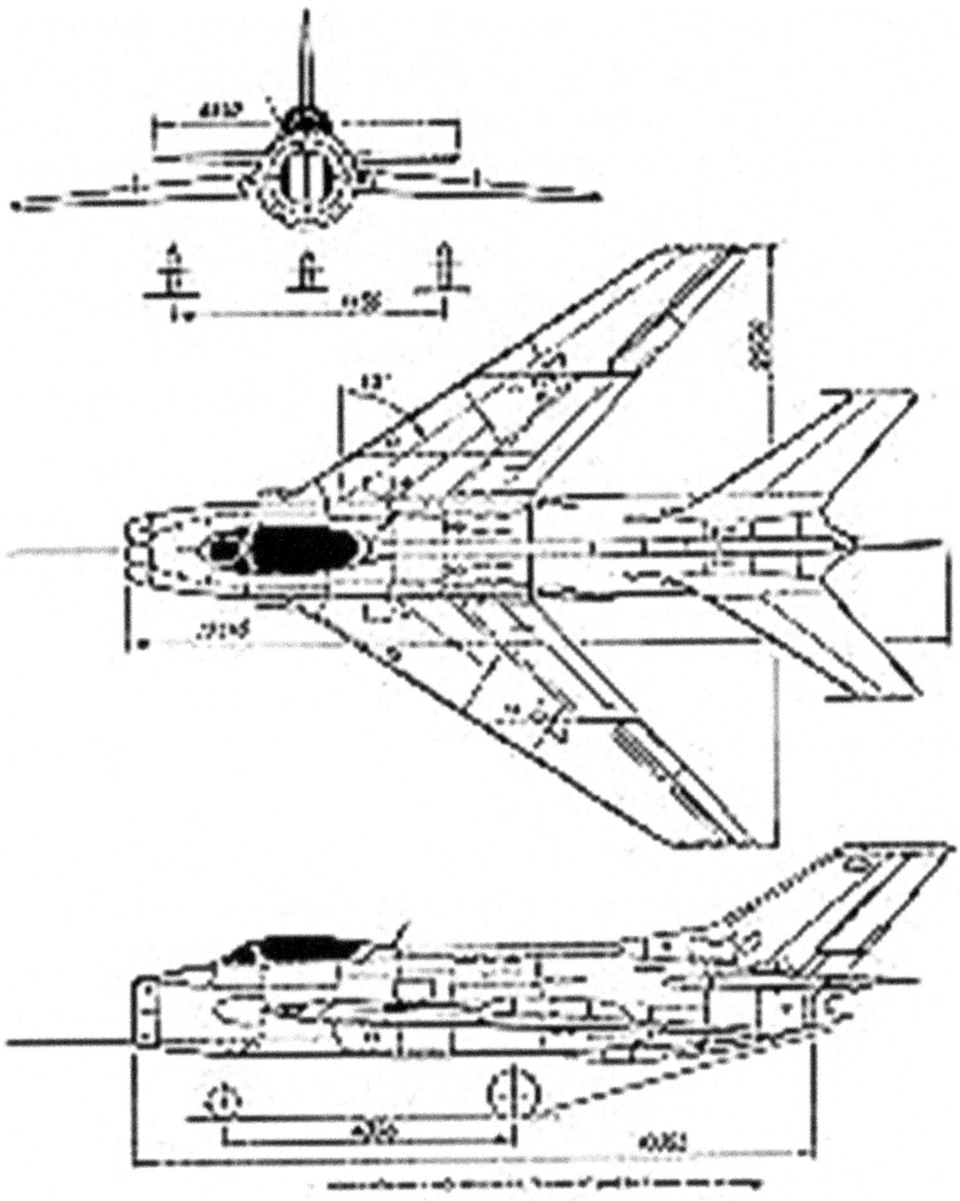

A 3-View drawing of the Soviet MiG-19, 1954.
(Source: USAF Foreign Technology files)

It is seen however, that the horizontal tail of the MiG-19 still featured a hinged stabilator, which had been found by the NASA and the British aerodynamics engineers to be a contributor to, if not the main cause for, the adverse “control reversal” phenomenon (due to compressibility effects) associated with flight during transonic and supersonic speeds.

When the Korean War ended (in July 1953), the Soviet Union placed maximum design effort toward a Mach 2 supersonic fighter-interceptor that would be a technology equal (or superior) to the supersonic fighters of the United States (USAF and USN). Virtually all of the short-comings and lessons learned, (in all technology disciplines) in the MiG-15 (from the Korean War), in the MiG-17, and in the MiG-19s were avoided in the design of the new Mach 2 supersonic fighter MiG-21 for the Soviet Air Force. In addition, the Soviet engineers had also copied many technology features, especially in the avionics technology, of the F-86 Sabre (a few which had fallen into Soviet hands during the Korean War). Further, the Soviet Union had a Foreign Technology organization that was known to be as effective as was that of the United States Foreign Technology Division at Wright-Patterson AFB. The Soviet Union also embodied some technology advances from emerging technology advancements being made in the Research Laboratories at the Air Research and Development Center laboratories at Wright-Patterson AFB in Dayton, Ohio, that were being developed for USAF advanced concepts in fighter aircraft design. One of those emerging developments in the propulsion technology area was the translating cone in the inlet of turbojets at supersonic speeds to improve inlet air intake efficiency.

The first prototype of the Soviet Union MiG-21 concept bore the designation Ye-1, but was shortly after designed with an improved engine cycle and re-designated as the Ye-2. The first flight of the Ye-2 was on February 14, 1956, and it made its first public appearance in July 1956 at the Tushino airfield near Moscow as part of the annual Soviet Aviation Display. The re-designated MiG-21F (and incorporating some upgrades based on development flight tests of the Ye-2 prototype) was introduced into the Soviet Air Force inventory in 1959.

Shown next is an in-flight photograph of two MiG-21 "Fish bed's illustrating the side profile and the underside weapons-carrying configuration of the MiG-21".

Mikoyan-Gurevich MiG-21 "Fishbed"s, 1959.
(Source: USAF Foreign Technology files)

A small portion of the translating conical spike in the circular-shaped nose inlet of the MiG-21 is seen protruding ahead of the inlet. When extended to its foremost position, the shock cone from the tip of the spike provided more efficient inlet efficiency at supersonic speeds than could be achieved without the spike. The nose boom (seen just above and extending forward of the inlet) provided high quality (non-interfered) air data (pressure and temperature) to the pilot for conversion to an accurate Mach number.

In an effort to obtain a unified fighter for all branches of the military, in the early 1960s then-Secretary of Defense Robert McNamara caused the USAF to receive versions of the USN F4H-1 Phantom IIs. With modifications to alleviate some of the extra weight attributed to carrier-operations, the USAF F4H-1 Phantom II was re-designated as being the USAF F-4C Phantom II and with the USN/USMC version being designated as the USN F-4B. The first USAF F-4C flew on May 27, 1963, exceeding Mach 2 on its maiden flight. The avionics and weapons carried on the USAF F-4C were tailored to the USAF mission as a fighter-interceptor.

Chapter 4 concludes with the observation that in the eleven years covered in the chapter, the advancement of the number and sophistication level of the multitude of technologies directly applicable to the wide gamut of supersonic fighter aircraft (air-superiority, interceptor, bomber, and reconnaissance) developed for operational use by the USAF and USN against its Cold War adversaries was remarkable and will go down in aviation history as possibly the most influential force that eventually brought the Cold War threat to a satisfactory conclusion. Not to go unnoticed, however, is that the Soviet Union also developed its military air force capabilities with the same leap in advancement of its technological capabilities, and continued to supply the Communist allies of the Soviet Union with military capabilities that continue to pose a formidable challenge to Free-World countries.

Chapter 5, spanning the period 1962 to 1968, will also cover the influence the air war over Vietnam between USAF and Soviet fighter aircraft had on fighter aircraft design. Chapter 5 will not only discuss how many of the fighter aircraft developed in this 1950-1961 time period performed in operational use in the Vietnam War, but will shift emphasis toward those attributes needed in the design of an affordable, light weight, low-cost fighter aircraft. John Boyd saw the urgent need to change the trend of fighter aircraft from becoming large, heavy, complex, and costly such that an insufficient number of them will be ready when/if called upon in the ever-changing nature of future wars. The efforts of John Boyd and the formation of the Lightweight Fighter Mafia form the center of topics in Chapter 5 to improve that ominous trend in fighter aircraft growth.

A major portion of Chapter 5 summarizes how that complex, convoluted acquisition process is organized and integrates all of the elements needed to put a weapon system into operational use at operational bases. To provide some guidance to those newcomers unfamiliar with US Government military research, design, test, and procurement activities, the procurement strategies of whether to prototype or not, whether to sole-source or conduct a competitive design phase with multiple contractors, and whether or not to conduct concurrent production while development testing is still not completed, are addressed.

CHAPTER 5
TFX (F-111), BOYD ENERGY-MANEUVERABILITY, POST-VIETNAM WAR (1961 TO 1972)

When the Korean War ended in 1953, there was considerable unrest associated with the Cold War in the French Indochina region, especially in the countries of Cambodia, Laos, and Vietnam. The United States provided military support, mostly in military advisory form during the early years of those hostilities, to the beleaguered French troops in French Indochina. In the battle for the village of Dien Bien Phu in South Vietnam in March-May 1954, the French ground forces were nearly totally massacred by the Viet Minh, and the French Government abandoned its participation in the Vietnam War. That left the US ground forces to continue the war against the Communist (Soviet-backed) forces of North Vietnam, and necessitated the rapid build-up of US ground and air force capabilities to do combat against the North Vietnamese Army using weapons provided by the Soviet Union. The scenario in French Indonesia was much like that of the Korean War time period just a few years earlier. Much has been written and many motion picture films produced by Hollywood about the ground war in Vietnam so that aspect will not be addressed in this book. It is the lessons learned from the USAF flight operations in the air war over Vietnam which will be described later in this chapter.

There were also two activities vital to the field of fighter aircraft technology development that unfolded in parallel with the course of the Vietnam War during the period 1960 to 1968, and those will also be described in this chapter. One activity was driven by the ongoing Cold War environment, that being the activity of the TFX study which led to the development of the General Dynamics F-111 Aardvark fighter-bomber program featuring variable-sweep wing technology. Another activity was not directly caused by the influence of the ongoing Cold War, but which resulted from the creativity and scientific analytical mind of one man, John Boyd, which he entitled as the Energy-Maneuverability (E-M) concept for improved fighter aircraft effectiveness in air combat.

The Author wished to avoid writing this chapter with all three of those aforementioned activities intertwined like a rattlesnake-ball (for those Readers familiar with the breeding habits of rattlesnakes seen in Texas desert areas during their mating season), as it is difficult to spotlight the important factors derived from the separate but intertwined activities. With Readers indulgence, the three activities will be described separately and in the order in which the title of this chapter identifies them.

It is promised that at the end of this chapter, Readers will have a better understanding of how many seemingly unrelated activities coalesced to indicate the need for development of the prototype of a lightweight, low-cost (achieved by low complexity), but effective multimode fighter aircraft (such as the General Dynamics YF-16 advanced technology demonstrator prototype).

TFX / USAF F-111A Aardvark

A number of incidents and events happened in the early 1960s which were perceived as being fuel for the geopolitical Cold War between the Soviet Union and the United States, and which also had significant influence on the technologies that future US military aircraft developments would need to have in order to stay ahead of Soviet Union military aviation developments (i.e., the leapfrog game in technology developments).

In May 1960, the Soviet Union shot down a US CIA reconnaissance Lockheed U-2 aircraft (piloted by Francis Gary Powers) on a mission at 60,000 feet over the Soviet Union. This incident revealed that the Soviet Union had developed a surface-to-air missile (SAM) that could intercept US aircraft at altitudes of 60,000 feet or more. This cast a serious shadow on the B-58A Hustler Supersonic Strategic Bomber program which was designed to penetrate the Soviet radar defense network at high altitude to deliver retaliatory nuclear weapons. By virtue of the B-58A Hustler also requiring a number of in-flight refuels on a typical nuclear delivery strike mission (its range at supersonic speeds was not impressive), it caused B-58A flight path attack mission-planning to include at least three in-flight refuels between the eastern United States and the European half of the Soviet Union as far inland as Moscow. Foreign technology assessments showed that by the Soviet Union eliminating the last refueling aircraft over the eastern Atlantic Ocean, the B-58A Hustler would easily be destroyed by simply running out of fuel over the ocean before making landfall over Europe. Accordingly, the USAF Strategic Air Command turned its technology thrusts toward the development of a low-level penetration aircraft capable of flying under the Soviet radar defense network. Also, the low-level penetration aircraft would be based at a NATO base in Western Europe and eliminate the need for the transoceanic refueling waypoints.

In 1961, the USAF and the USN/USMC were both seeking a low-level interdiction aircraft but having an air superiority capability as well for fleet defense and for low-level penetration into far inland targets. When newly appointed (in January 1961) US Secretary of Defense Robert McNamara was appraised of the situation, he endorsed the development of a jointly procured (the commonality approach in acquisition) new interdiction fighter-bomber and directed the US aircraft industry to proceed with the study, design, and development of what was termed the Tactical Fighter Experimental

(TFX) Program. Readers should be informed that members of the United States aircraft design and manufacturing industry are granted funding from US military organizations to conduct Independent Research and Development (IRAD) programs within their design organizations to continually be aware of and to explore the potential of applying emerging technologies to conceptual designs of aircraft without being burdened by a formal contract and set of "cast in concrete" requirements being levied upon contractors' IRAD in-house activities. By the early 1960 time period, several aircraft manufacturers had already begun, with and without IRAD funding from the USAF, in-house studies and designs of a low-level interdiction aircraft having a new terrain-following technology. The goal of IRAD funded activities (referred to often as "spending seed money") was to allow aircraft manufacturers to explore those benefits and risks of emerging technologies in a complete aircraft design that might be of sufficient interest to those military research and technology organizations (such as the Research and Development Laboratories at Wright-Patterson AFB) wishing to further pursue and to conduct more in-depth design and test activities to advance that technology domain.

NASA had also been developing (starting in the late 1950s) the aerodynamic technology of an aircraft with a variable sweep wing as an option in the design of low-level interdiction aircraft that were to have Mach number capability from low subsonic speeds and up to supersonic Mach number of 2.0. The aerothermodynamic heating of aircraft having exposed skin constructed from aluminum materials limits the aircraft to sustained speeds no higher than about Mach 2.2 for structural safety reasons. [Aircraft such as the Lockheed SR-71 Blackbird sustaining Mach 3+ flight for extended periods have titanium skin for structural safety reasons in their unique high aerothermodynamic heating environment.] While the variable sweep technology being wind tunnel tested at the NASA Langley Research Center demonstrated very promising attributes by that testing, there persisted accompanying risks of applying that technology to a complete aircraft design and integrating the new technology with other emerging promising new technologies. Varying the sweep angle of the wing of an aircraft in flight presents many mechanical, hydraulic, and flight control challenges not addressed in the NASA conceptual aerodynamic design wind tunnel tests.

USAF Headquarters requirements planners' interest was heightened to consider the NASA emerging variable sweep technology being applied in the design of a low-level penetration interdiction type aircraft. With variable sweep, the aircraft wings could be swept forward for operation (such as at takeoff, during cruise at subsonic speeds, and on landing) where high aspect ratio wings are aerodynamically efficient. For flight at supersonic speeds (Mach number higher than Mach 1, i.e., above the speed of sound), the wing leading edges then could be swept back so as to form a delta-shaped wing planform having a low aspect ratio which is aerodynamically efficient for flight above Mach 1. The variable swept-wing technology would necessarily be integrated with other technologies, such as terrain-following radar, and piloting considerations, such as pilot ride quality in turbulent flight near the ground having less than smooth terrain.

In addition, the propulsion system needed for this type of operation across a wide speed and Mach number spectrum spurred the engine manufacturers into developing a turbofan version of the afterburning turbojet engines available. Here again, IRAD funding from US military organizations were provided to the two primary manufacturers of large jet engines in the US – General Electric and Pratt & Whitney – to develop the technology of advanced jet propulsion for supersonic aircraft. This included the overall cycle design, the accessories that are incumbent in the engine cycle design, and the materials needed for turbine blades and compressor blades ("buckets") capable of withstanding the higher temperatures and pressures that stress the turbine blades. The smallest flaw going undetected during the fabrication of a turbine blade will grow as the cyclic stress loading on that blade is applied in a constantly varying manner (as the power level and rpm of the engine varies continually in response to pilot commands).

For safety reasons, the turbine blades on jet-powered aircraft (both military and commercial) are inspected during frequent scheduled maintenance using a borescope instrument. (From a "chicken or egg coming first" perspective, it is not known if the borescope instrument (inserted through small access ports in the engine outer shroud at several locations between the stages of the turbine blades) was developed before the instruments medical proctologists use during colonoscopies or vice versa. Whichever was developed first, the principle of frequent visual inspection of vital components in the complex host machine is the same.)

[The Author will inject another example here of the type of decisions that will need to be made in the course of a career to Readers who are contemplating making the aircraft design/testing/manufacturing the career path they wish to follow. In late 1960, the USAF was clearly planning to "terminate for Government convenience" the B-58A Hustler supersonic strategic bomber program in favor of pursuing the low-level penetration attributes of a TFX variable wing sweep type aircraft design. Those who are working on a program that is known to be on the brink of termination will soon find themselves seeking an employment opportunity elsewhere. It is inherent when taking employment with an employer deriving its income from contractual efforts with the US Government that when work on the Government contract is completed (or otherwise terminated), there will often be a Reduction In Force (RIF) activity at that employer's facility. The decision of an employer for who stays and who goes is typically a most painful decision to make, but it must be made for economic reasons. There will often be an income from IRAD funding to permit reassignment of an employee, usually a specialist in some technology domain, on an IRAD project to retain the employee for the purposes of advancing the technology in that technology domain in which he/she has acquired experience and demonstrated a valuable talent. Unavoidably, there will be some employees (usually junior, less experienced workers) given notice of Termination Of

Employment (a.k.a. the big TOE) – usually with sixty days' notice ahead of the actual date of employment termination (and coinciding with the last day of a pay cycle). Most considerate employers will offer assistance in locating reemployment elsewhere.

Such were the circumstances the Author faced in his career decision-steering process to take advantage of unexpected adverse events (such as termination of the B-58A Program at the convenience of the Government). The Advanced Concepts Division (organization symbol ASD/ASRSSV) of the USAF (in the Aeronautical Systems Division within the Air Force Systems Command at Wright-Patterson Air Force Base near Dayton, Ohio), had an employment vacancy in that organization for an Aeronautical Engineer (as a Federal Government Civil Servant at grade GS-13) with experience in the design and testing of aircraft concepts and preferably also having an advanced degree to indicate a capability for analyzing emerging technologies in the aircraft design domain.

In early 1961, a small team of hiring interviewers from the Aeronautical Systems Division arrived in Fort Worth and asked Convair management if there were any of the Convair experienced engineers employed on the B-58A Hustler Program who might be losing their employment at Convair and were willing to consider transferring their employment from Convair to Wright-Patterson AFB. At the time, the Author was in B-58A Flight Test Engineering at Convair and had just completed writing the software program (in Fortran IV coding language) for extracting the J79 jet engine in-flight performance parameters (thrust, specific fuel consumption, etc.) for use in analyzing the test data being collected in the B-58 flight test program. The Author was made aware of the opportunity for interview with the USAF ASD head-hunting team, and had a lengthy interview with Mr. Hank Stachowski from the USAF Advanced Concepts Division. Two weeks after the interview, the Author was made a job offer to join the Advanced Concepts Division at Wright-Patterson AFB and at Federal Civil Service Grade GS-13. In itself that was an unusual opportunity because most job vacancies in Government organizations are filled by merit promotion of someone already on the Civil Service payroll. The Author had disclosed having the experience and capability for programming the emerging digital computer technology to replace the manual process in place for analyzing the performance of advanced flight concepts at the USAF Aeronautical Systems Division.

Skipping over all the inconveniences of moving a young family from Texas to Ohio, the Author began in April 1961 what was to be a totally rewarding career for twenty years as a civilian Aeronautical Engineer with the USAF. That single career change decision was the most important one the Author had to make during his career, and the decision made vindicated him, when, in 1970, the Author was assigned to be the Chief System Engineer for the USAF Aeronautical System Division in the development of the USAF Lightweight Fighter Program, and which later became the F-16 Air Combat Fighter Weapon System Program in 1974.

This accounting of a career change decision is offered to newcomers to the engineering profession, not to sound as a boast, but rather to indicate that opportunities for betterment of a career path will arise at any point in a career and may be pursued or not, depending on the assessment of many factors, not the least of which is to make a major job change entailing moving a long distance from friends and relatives. From experience, the Author advises "Go to where the job opportunities appear best. They are not likely to come to you".]

After a round of USAF issued RFPs and modified RFPs for the TFX Program, the TFX competition was narrowed down to two designs, one by the Boeing Company, and the other by General Dynamics (formerly Convair). In November 1962, the USAF selected the General Dynamics proposal and aircraft design due to its greater commonality in integrating the unique Air Force and unique Navy operational requirements into "common" requirements. Also, the propulsion system to be developed specifically for the two engines in the TFX was selected to be the Pratt & Whitney TF30-P-1 afterburning turbofan engine. The USAF/USN contract with General Dynamics for the TFX program was signed in December 1962. The Author was assigned to be an advisor to the TFX RFP proposal evaluation team at the Aeronautical Systems Division, and with a brilliant colleague and fellow aeronautical engineer, Douglas N. Beatty, helped evaluate the supersonic wave drag aspects of the TFX designs submitted by Boeing and General Dynamics.

A photograph of Doug Beatty and the Author (Doug is the one holding a six-inch slide rule, which was the computational instrument used before the advent of digital computers to automatically analyze DIY (design it yourself) aircraft designs at the push of a button, is shown next.

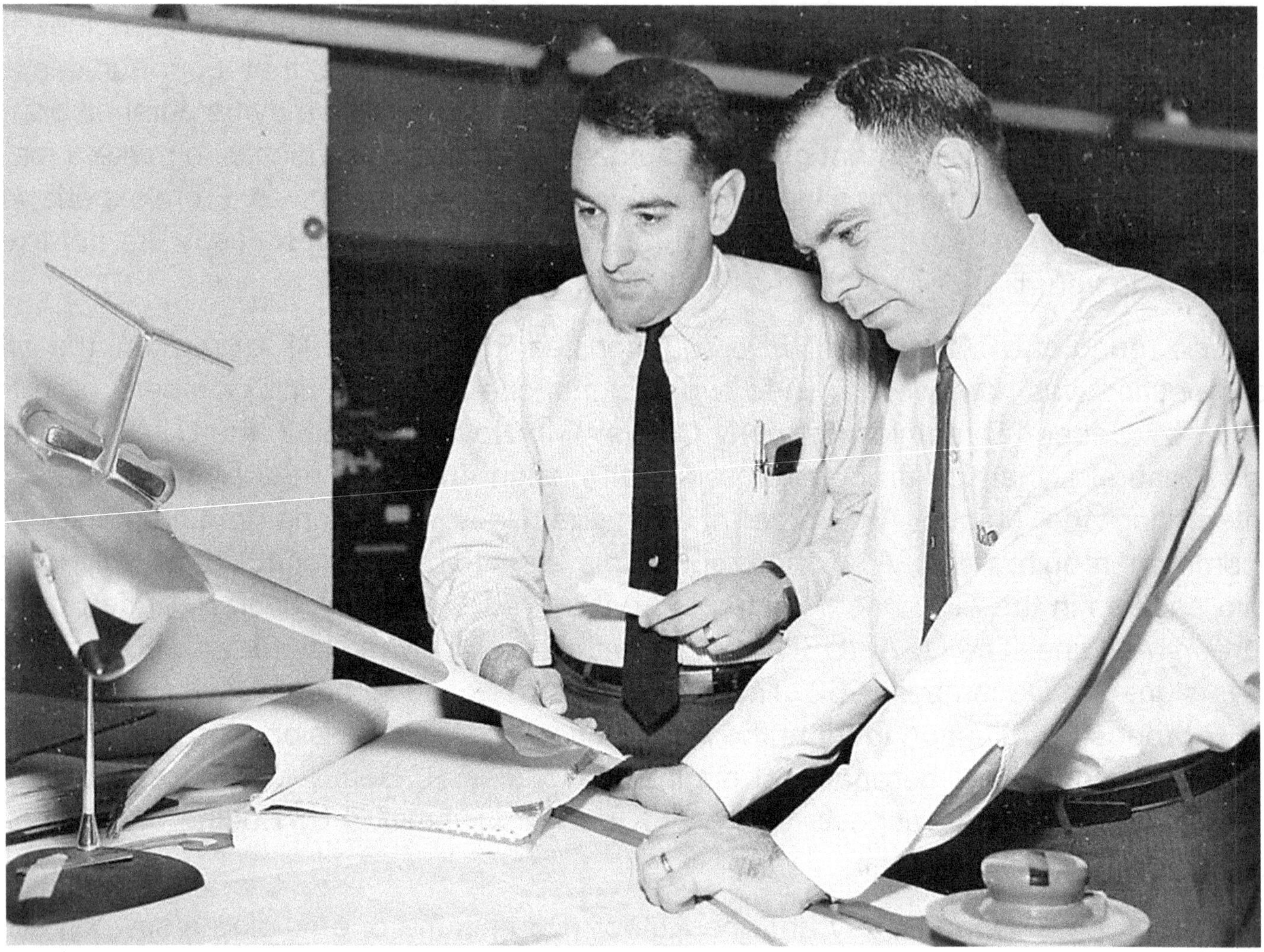

Doug Beatty and the Author, analyzing a variable-sweep wing model at ASD, W-PAFB, 1962. (Photo from Author's personal file)

The slide rule – held by Doug Beatty - was admittedly an archaic tool for use in analyzing aircraft designs; however, the slide rule required the user to determine where to place the decimal point in a three-digit calculated answer. This caused users to apply a test of reasonableness to their calculations with the slide rule to determine where to place the decimal point. Was the reasonable answer 395, 39.5, or 3.95? Using digital computers admittedly eliminates the need for a test of reasonableness and leads to the statement by too many newcomers to the engineering profession of "The computer says . . .. ". If a numeric input is mistakenly keyed in as an input, the computer will produce a wrong answer at electron speed. The Author advises that no one should ever tell a supervisor that "*The computer says . . .*". It is what *you* say that earns *you* a paycheck, not the computer.

Digital computations are best achieved by computations that involve a complex database of test inputs and when comparing a set of small variations in one feature on the calculated performance of the aircraft design when on a multi-segment mission profile. The Author and his engineer colleagues, including Doug Beatty, developed a computerized process (entitled NSEG) for analyzing a new aircraft concept (knowing its general aerodynamics and propulsion characteristics) by "flying" it on the NSEG computer program simulating the segments of an arbitrary design mission, from takeoff to landing. The need for double- precision (i.e., 32 bytes) computational accuracy possessed with large digital computers may be overkill of precision, but it is more often useful to compare the answers in a per cent change in range type of analysis rather than on an absolute aerodynamic drag-count level basis.

Returning attention to the TFX activities, the Author considers it to have been an honor to be chosen to evaluate the response of the General Dynamics technical team for the TFX proposal as that proposal response was headed by none other than Harry J. Hillaker (who had designed the B-58A Hustler at Convair and for which the Author had verified by flight test the excellence of the aerodynamics of the B-58A while employed by Convair from 1959 to 1961 in Fort Worth, Texas). [Author's note: That is the same Harry Hillaker who will be described later in this book as having created the basic external configuration of the General Dynamics Model 401, which was wind tunnel and flight tested as the YF-16 Lightweight Fighter prototype, and was ultimately selected for development as the F-16 Multinational Weapon System. It was awesome to realize that Harry Hillaker's engineering talents were on the same technology path through military aviation as pursued by the Author.]

The designation USAF F-111A was given to the USAF version of the General Dynamics TFX and the designation F-111B was given to the USN/USMC version of the General Dynamics TFX. The first test F-111A was rolled out of Plant 4 (same plant that had produced the B-58A Hustler) on October 15, 1964 at Fort Worth, Texas, and first flew on December 21, 1964.

A photograph of an early USAF F-111A Aardvark in flight is shown next:

USAF F-111A Aardvark, wing swept forward in subsonic mode, 1965.
(Source: USAF Photographic files)

The USAF F-111A had an elongated nose to carry the AN/APQ-113 attack radar and the AN/APQ-110 terrain-following radar and also had air-to-ground armament. The USN F-111B had an 8.5 feet shorter nose length so as to fit on existing carrier elevator decks. It also had a wing span 3.5 feet longer to increase its aspect ratio that results in longer on-station endurance. The USN F-111B was designed to carry an AN/AWG-9 Pulse-Doppler Radar and AIM-54 Phoenix missiles. The nose radome is noted to be of the black fiberglass technology that causes an acceptable level of transmissibility of radar electromagnetic waves through the radome.

It should be realized that the aerodynamics (especially the stability parameters) of the F-111A with its wings in the swept forward position, and with the center of pressure at about 25 percent of the Mean Aerodynamic Chord (MAC) at subsonic speeds, will be substantially different than those when the aircraft wings are fully swept back to form more of a delta wing planform and with the center of pressure of that delta wing being translated aft to be at about 50 percent of the MAC at supersonic speeds. The flight control system was designed by the General Dynamics Flight Control engineers to accommodate that variation in static margin (difference between the aircraft center of gravity and the aircraft center of lift) as the flight speed varied during a mission.

Shown next is a four-scene photograph of an F-111A in flight with the wings in different position of wing leading edge sweep angle from full forward (16 degrees) to full aft (72.5 degrees):

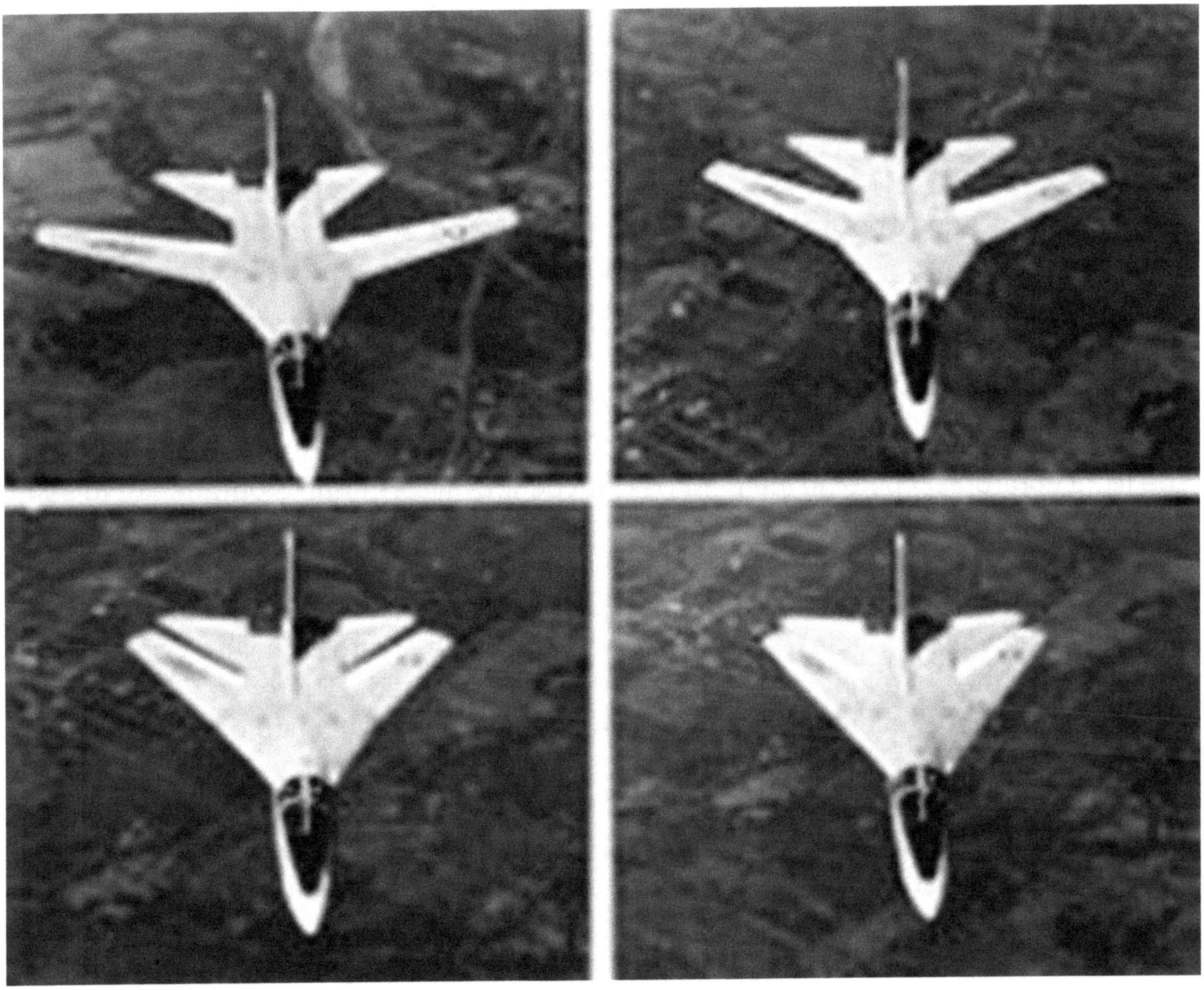

Four angles of wing sweepback of the USAF F-111A, 1964.
(Source: USAF Photographic files)

At quasi-steady sweepback angles, however, the actual dynamics of the sweepback angle changing as well as the flight Mach number (and perhaps altitude) introduces dynamic and inertial stability and control terms into the flight control system which were a severe challenge to the flight control system design team at General Dynamics. To avoid the pilot being distracted while dynamically changing flight altitude and speed, a flight control computer was integrated into the F-111(all variants) flight control system to make the fine tuning of trim control surfaces deflections to be fully automated, with only pilot manual override capability if battle damage or functional failure were to be experienced in the automated system. The pilot did manually control the angle of wing sweepback with a pistol handle type grip on a short column next to the engine throttle.

Heated discussions were held to determine if the wing sweep angle controller handle should be moved forward to indicate if the movable outer panel of the wing was being commanded to a smaller sweep angle (the smaller sweep angle being commanded only when in low speed flight), or should the wing sweep angle controller be moved toward the aft - in the same direction as the engine throttle – when the pilot is configuring the wing sweep angle to be decreased in preparation for low speed operation? The engine throttle control handle is moved aft on all jet-powered aircraft (standardized for flight safety) when reducing speed (requiring a decrease in engine thrust) of the aircraft such as for landing. The majority of pilots preferred the consistency of moving both the engine throttle and wing sweep back control handle in the same direction as speed being commanded to increase or decrease to avoid the possibility of pilot error under high adrenalin speed variation situations.

The flight control computer in the F-111A flight control system was programmed to sense the position of the wing sweep angle controller as well as the power-setting pilot command to the engine. The wing sweep angle controller had an angular degree's scale (in gradation from 16 degrees at the most swept-aft position and to 72 degrees at the most swept-forward position) mounted at the base of the controller column with an index tab showing the sweepback angle degrees commanded. There was no Accident Investigation Board record found to indicate an improper positioning of the wing sweep angle controller (pilot error) as being the cause of any of the F-111A aircraft lost. (A voice warning alarm, by Screaming Sally, was installed in the F-111A cockpit to alert the pilot of an improper Wing Sweep Angle Controller command that was not consistent with safe flight speed and altitude conditions. [Screaming Sally's voice in the F-16 cockpit voice alarm system would later be referred to as being that of Bitching Betty, but the alarm system provided the identical flight safety warning function regardless of her name.]

The selection of the P&W TF30-P-1 turbofan jet engine with afterburner for the USAF F-111A was a large advance in the engine cycle technology for supersonic aircraft. The design feature of an added fan section at the front face of the engine (at the interface with the inlet duct) was to act in much the same manner as the superchargers of fighter aircraft (P-51 and P-38) in World War II aircraft that escorted bombers at high altitudes. In the turbofan application, the movement of air by-passing that which would otherwise enter the first multiple stage compressor section was measured in terms of a BPR (by-pass ratio) and was used for cooling the engine shroud in the section of the combustor where high-pressure, high-temperature turbine gasses were being expelled from the combustor section of the engine.

Shown next is a cutaway diagram of the Pratt & Whitney TF30-P-1 afterburning turbofan engine (a pair of which powered the two-engine USAF F-111A Aardvark).

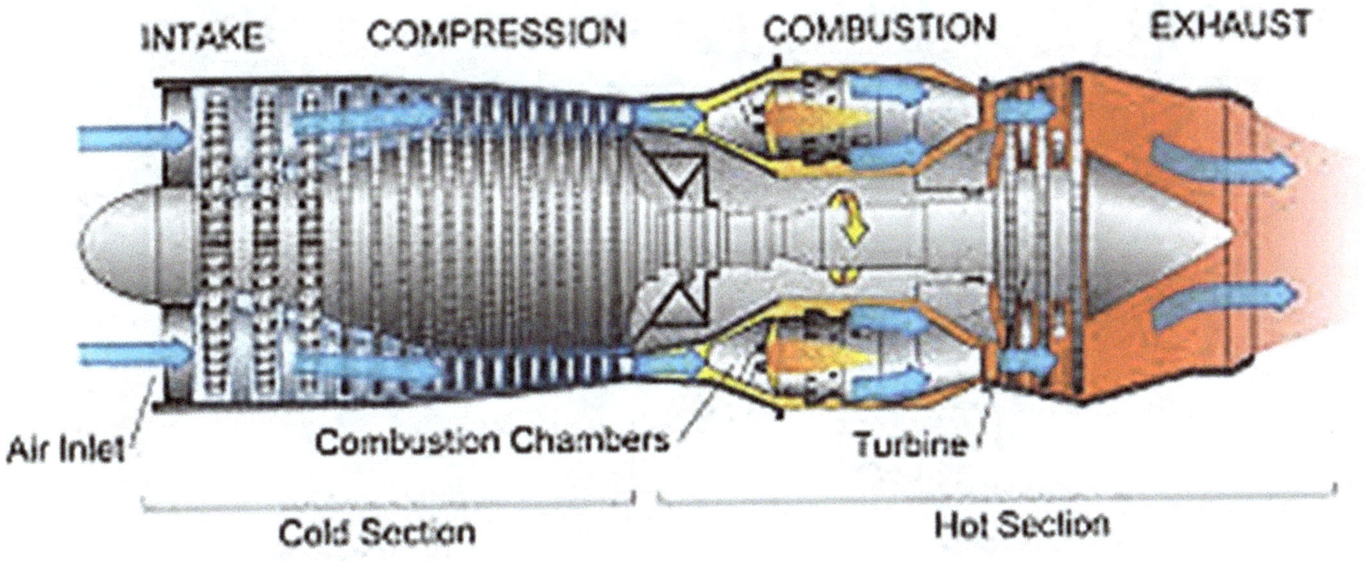

Pratt & Whitney TF30-P-1 afterburning turbofan engine, 1964.
(Source: USAF Engineering files)

The TF30 engine represented the state of the art in afterburning turbofan jet engines in the late 1960s. The compressor section was a 2-spool axial flow with three fan stages, six low- pressure stages, and seven high-pressure stages. The turbine had one stage high pressure and 3 stages low pressure. The bypass ratio was 0.878:1, and the overall pressure ratio (ratio of pressure P6 at the nozzle exit to pressure P1 at the face of the engine) of the engine was 19.8. The totally assembled TF-30 engine weighed 3,985 pounds, generated 14,560 pounds of thrust in dry power, and generated 25,100 pounds of thrust with afterburning. The thrust-to- weight ratio of the engine was then an impressive state-of-the-art at 6.0.

The engine air inlets for the F-111 were of the quarter round with a spike underwing "shoulder" location ("armpit" being more descriptive) type. The quarter-round shaped inlet had a splitter plate to divert boundary layer air from the elongated nose forward of the inlet from entering into the inlet. There was a spike (segment of a tapered cone) which was translatable to adjust the quantity of air into the inlet to match the engine intake air requirements at variable Mach number, altitude, and power levels commanded by the pilot. There were a large number (about 22) of small vortex generators inside the curved outer inlet contour to maintain attached flow of the air as it entered into the inlet duct.

A photograph of the left (port) engine air inlet on the USAF F-111A is shown next:

Inlet (port side) of the F-111A Aardvark, 1964.
(Source: USAF Engineering files)

While seemingly quite well designed, the F-111 inlets did not perform as efficiently as hoped, and required several reconfigurations to aerodynamically tune the inlet to better match the engine inlet airflow requirements as they varied significantly across the F-111A flight envelope (altitude and Mach number). A particularly troublesome and hazardous characteristic of the F-111A propulsion system was that rapid changes in power levels commanded by the pilot moving the engine power lever too rapidly or not smoothly would cause compressor stall in either or both engines. Because of the restrictions imposed on power level rate of change, the F-111A had rather overall poor maneuvering capability. However, the F-111A was designed to be a low-level interdiction fighter- bomber and did not require the frequent, aggressive maneuvering inherent in air combat engagements.

The largest problem the F-111A design had was in its weight – from empty weight and at every intermediate weight condition up to maximum takeoff weight (MTOW). At MTOW, the F-111A weighed 98,000 pounds. The comment made by a General Officer at HQ USAF was classic, *"I am not going to accept any fighter aircraft weighing 98,000 pounds"*.

The weight of aircraft designed for USN carrier landing operations is inherently high due to the structural strength needed to endure those high stress landings. The very fact that the USN prefers aircraft used in carrier-based fleet-defense operations to have two engines for safety and to have some means of folding-up the outer wing sections to improve its "spotting factor" aboard the carrier and below deck storage adds weight. The added complexity and weight of wings that could be changed in angle of sweepback, coupled with all the other incremental weights in order to achieve commonality of use by the USAF and USN/USMC caused the USN in 1968 to cancel further development of the F-111B intended for USN/USMC fleet defense usage. Shortly following their withdrawal from the F-111 Program, the USN proceeded with development of the Grumman F-14 Tomcat which featured use of the outstanding P&W TF30 engine and had the variable sweep wing technology (which provided slower, safer speeds over the fantail on landing, better than any other USN supersonic fighter design).

The USAF F-111A entered operational service with the USAF in July 1967. Its geometry, physical, and performance specifications are tabulated next:

USAF F-111A Specifications

Wingspan (16 degrees sweep angle)	63 feet
Wingspan (72.5 degrees sweep angle)	32 feet
Wing area (16 degrees sweep angle)	657.1 sq. feet
Wing area (72. degrees sweep angle	525 sq. feet

Length	73 feet 6 inches
Height	17 feet 1 inch

Empty weight	47,172 pounds
Max TO weight	98,850 pounds

Propulsion	P&W TF30-P-1 Afterburning Turbofans (2)

Max speed at altitude	1,455 MPH / 1,265 Knots (M2.2)
Service ceiling	58,000 feet
Range, internal fuel	2,315 nautical miles

(Wingspan of the USN F-111B with wings swept 16 degrees was 70 feet; and its length was 65 feet.)

Twenty-four of the F-111C variants were ordered by Australia and their first F-111C was delivered to the Australian Air Force in September 1968. The delivery of the remaining 23 F-111Cs to the Australian Air Force were delayed several months, however, to correct some structural deficiencies unique to the F-111C design. Forty-two F-111As were converted in 1972 into EF-111As for use as electronic warfare aircraft.

Updates of the F-111 to the EF-111 are seen in the photograph of the EF-111 next

Gen. Dynamics EF-111A electronics warfare updates, 1972.
(Source: USAF Photographic files

A nose radome of grey fiberglass and a16 feet long canoe-shaped fiberglass radome was added on the underside of the fuselage to house high-powered transmitter antennas. A fin-tip pod was added on the vertical stabilizer to house receiving antennas, a processor to detect hostile radar emissions, and other electronics equipment.

Rarely does a new technology have only beneficial characteristics and no adverse characteristics. For instance, an early TFX design considered having two wing stations under each wing for the carriage of external fuel or for other types of weapons (air-to-air or air-to-ground). The two-inboard pair of wing stations were to be under the fixed "glove" section of the wing and the other pair under the variable sweep angle wing panel. When the external stores (fuel tanks or weapons) are installed on those four wing station pylons, the wings are generally in a most forward sweepback angle position (i.e. the external stores will be parallel to the fuselage centerline with the wings swept fully forward for subsonic take-off.) If the wings were swept back for supersonic speeds, the pylons and stores below the wing panel that is being swept back would no longer be in a streamlined aerodynamic orientation and create drag problems as well as structural mounting problems.

The additional drag with those non-aligned stores could likely jeopardize completion of the mission. General Dynamics cleverly developed articulating pylons that caused the wing pylons and external stores to remain streamlined no matter what changes were made in the wing sweepback angle. However, that caused additional mechanisms and hydraulic and electrical services to be installed for maintaining the pylons and external stores in a streamlined orientation. There surely was an increment in weight of the mechanisms and hydraulic tubing/electrical cabling for the services to operate the streamlining mechanism. There also was a change in reliability and safety if the mechanism was to fail, and with its increase in complexity the cost of the entire system would be increased. Accordingly, the pair of stores stations on the movable panel of the wing were eliminated altogether. [The Author repeats the earlier stated admonition to anyone developing a new technology to consider all factors that will change in the total flight system when applying one isolated technology into that system.]

A unique feature of the USAF F-111A was that the two-man crew sat side by side in the cockpit of the F-111 which itself was in a crew escape capsule. If/when ejection was commanded, the entire capsule was separated from the aircraft instead of ejecting the two crew members each in a seat escape system. A photograph of the F-111A crew escape capsule is shown next.

USAF F-111A Aardvark crew escape capsule, 1964.
(Source: USAF Photographic files)

The crew escape capsule was installed in the forward fuselage and with a detonating cord (Prim Acord obtained from military demolitions) around its peripheral surfaces such that when the pilot commanded an ejection, the detonating cord exploded, cutting the capsule into the shape as shown in the photograph, and separating the entire crew escape capsule from the fuselage. The capsule was configured such that it had a rudimentary aerodynamic shape with a glide path ratio somewhat better than that of a large rock that caused it to glide safely until a parachute opened and took the capsule to a ground recovery. The F-111A pilot community did not rave about the crew escape module and most indicated they would have preferred having individual seat escape systems. Pilots also expressed that in a side-by-side seating arrangement, it reduced the pilot's visibility to the right outside the cockpit. More than occasional comments expressed the need to see better "over the rail" views both left and right when in the terrain- following mode where the terrain might be very different on one side of the aircraft than on the other side.

As indication that Foreign Technology organizations exist in the Soviet Union, as well as in the United States and in most other countries having a military aircraft design and manufacturing capability, an interesting comparison is shown next of the General Dynamics F-111A configuration in the USAF inventory and the Soviet Union Sukhoi Su-24 configuration in the Soviet Air Force inventory. Both are seen to have their variable sweepback angle wings in the full aft (supersonic Mach regime) position.

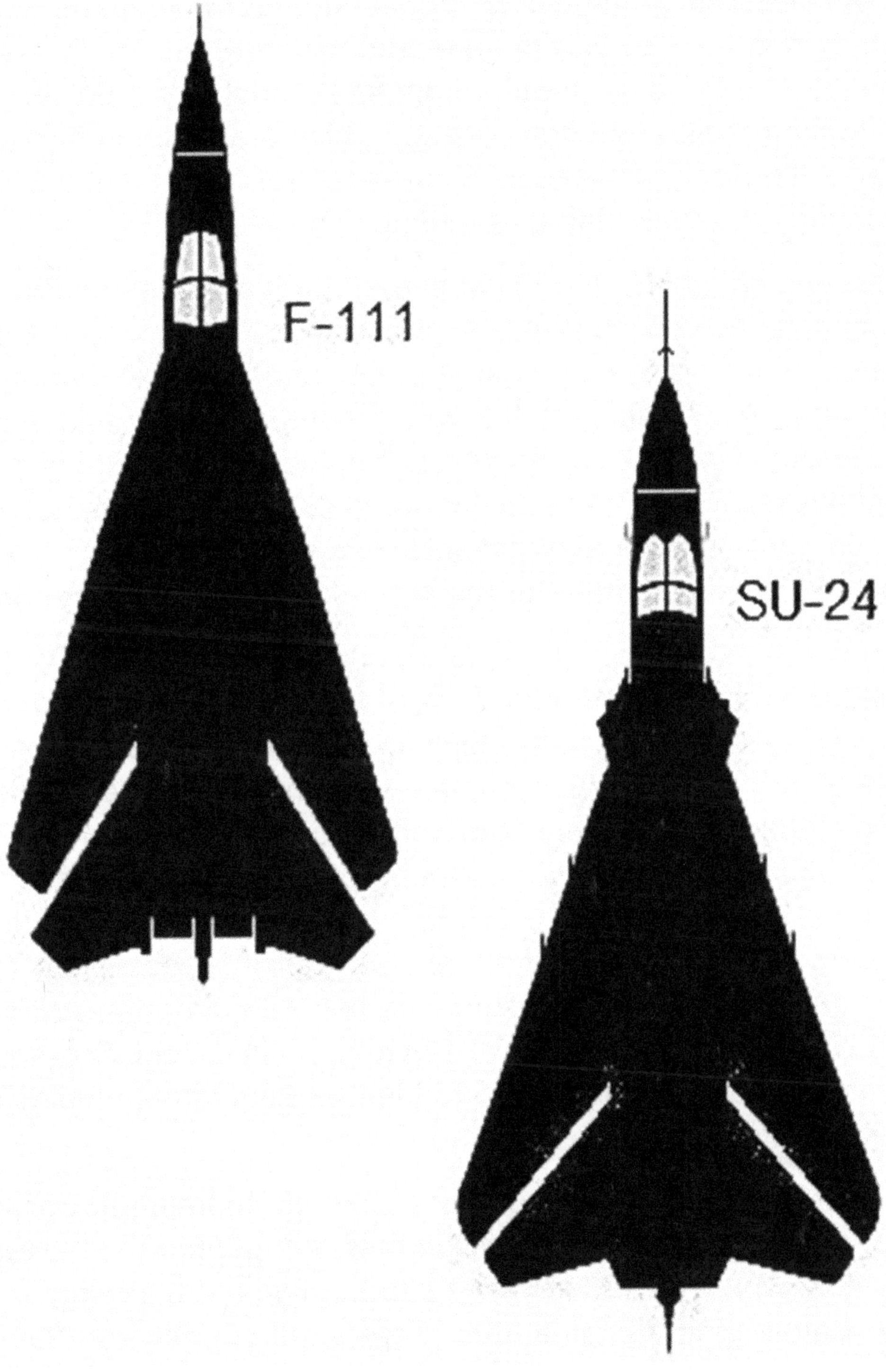

Comparison of USAF F-111 and Soviet Su-24 planforms
(Source: USAF Foreign Technology files).

The Soviet Su-24 had its first flight on July 2, 1967. On November 24, 2015, a Turkish F-16 shot down a Soviet Su-24, making it the first shoot-down of a Soviet-made aircraft by a US- made aircraft since US F-4 Phantom IIs shot down Soviet MiG-15s in the Korean War. Although protested by the Soviet Union, recorded voice warnings given by Turkish pilots and ground tracking station staff members to the Soviet pilot that he was headed towards Turkish airspace, the Soviet pilot penetrated Turkish air space by at least 2.8 nautical miles. Both the Soviet pilot and WSO (Weapon System Operator) ejected safely from the doomed Su-24. The pilot was rescued by Turkish civilians and returned expeditiously into Soviet military hands. Tragically, the Soviet WSO was shot and killed by Turkish rebels while descending in his parachute, and that evoked a UN declaration of violation of the Geneva Convention of War Conduct prohibiting any country from shooting at any human descending in a parachute.

While the official name of the F-111 was the Aardvark, an unofficial (and of low complimentary nature) name was given by ground personnel to the Aardvark (or Vark) as "Pig" because of its elongated nose (snout) and terrain-following (in the mud) capability. The operational history of the F-111 fighter-bomber and that of any of its variants are not included in this book as it had no influence on the technology genealogy of the USAF Lightweight Fighter to be discussed in detail later in this book. Nonetheless, there were three technology developments made during the F-111 development program that added lessons learned to the technology base of fighter aircraft ensuing the F-111:

1) The longitudinal stability characteristics of the variable swept wing design varied with the wing sweepback angle, causing the center of pressure to have large magnitude of excursions in static margin in a dynamic manner. Development of an automated flight control system (i.e., with a flight control computer) that actively positioned aerodynamic surfaces to their optimum settings that otherwise would have had to be commanded manually by the pilot was a highly important advancement to supersonic fighter aircraft design capabilities. (This database being accumulated by, and experience gained with dynamic stability and control, would later be vital to the General Dynamics flight control organization in the design of the General Dynamics YF-16 Lightweight Fighter all-electric, fly-by-wire, flight control system.)

2) Development of the first dual-spool axial-flow afterburning turbofan engine cycle by Pratt & Whitney for the F-111 was later refined and formed the technology base for the Pratt & Whitney F100-PW-100 afterburning turbofan engine developed for the F-15 Eagle (and became the propulsion system for the USAF YF-16 Lightweight Fighter design).

3) The requirement that the design of a new aircraft with the commonality" approach, to satisfy the performance capability for operating safely from a naval ocean-based

aircraft carrier or from a land-based operational base, causes the structural design of that common aircraft to be heavier than that of the design of two separate aircraft each designed for the different type of operational basing (ocean versus land). This same commonality approach causes the design of the aircraft optimized for ocean basing to feature two engines for operational safety reasons; whereas, the design of an aircraft optimized for land-based operations may be better satisfied with a large single-engine propulsion system. (This commonality approach was shown to not be acceptable on the F-4 Phantom II aircraft development program and on the F-111 Aardvark aircraft development program. The commonality approach was purposefully not applied for development of the USAF F-15 Eagle program or the USAF F-16 Air Combat Fighter program to avoid compromising the weight and performance of both designs (F-15 and F-16) by having the added weight associated with operating from an ocean-based platform. The USN/USMC development of the Grumman F-14 Tomcat and the McDonnell Douglas F/A-18 Hornet and Super Hornet programs were also not designed for commonality and did not incur any weight penalties for incorporating any design or performance capabilities needed for USAF operations.)

John Boyd – "Aerial Attack Study" and the Energy-Maneuverability Theory

This book has been dedicated to 2 principal men without whom the USAF Lightweight Fighter program and its successor the F-16 Fighting Falcon Program would not have become the great success which it did and has been for over 40 years. John Boyd is singled out for having conceived and championed the development of a lightweight fighter embodying advanced flight technologies but with the caveat that those technologies integrated into the lightweight fighter design do not detract (for any reasons, such as high cost, high weight, high complexity and low reliability, or intolerably high risk) from its effectiveness in air combat maneuvering against an aggressor or cause a hazard to pilot safety in the harsh operational environment.

John Boyd had much liked a Doctor Jekyll and Mr. Hyde personality – he could be most congenial if you supported his ideas about fighter aircraft needing to be of simple, low-cost design. Conversely, he was a most abrasive and loud critic of your capabilities to understand how to chew gum if you dared to not support his ideas. You would quickly know if you were on his team if he allowed you to call him "Boyd" or "John" – but if he told you to call him "Colonel" or "Sir" that would be the signal where you stood in his acceptance rating for you. [The Author was allowed to call him Boyd for reasons to be explained later in this book.] The objective of this book however is not to describe what Boyd was like to work with – an excellent book about Boyd (entitled just that, *Boyd – the fighter pilot who changed the art of war*) was written by Robert Coram in May 2004 seven years after John Boyd had passed away (in March 1997).

A photograph of USAF 1st Lieutenant John R. Boyd in an F-86 Sabre cockpit is shown next:

USAF 1st Lt. John R. Boyd, Suwon, S Korea,1953.
(Source: USAF Photography files)

Like many persons whose lives were dedicated to making life better for others, their achievements are not made known until after their passing away. While alive, they just went ahead and did what they felt was right to do without expectation of any reward. Of the two men to whom this book is dedicated, John Boyd and Harry Hillaker (Chief Designer at General Dynamics), both are known to have passed away several years ago. The Author has included a copy of their obituaries in this book as Appendix 1, and Readers are encouraged to read those obituaries, as well as to read the Robert Coram book on *Boyd*, which captures the essence of their professional lives. The Author was honored to call them both as friends, and it was an indescribable experience to have worked closely with them (and so many others) to create the USAF YF-16 Lightweight Fighter and the undeniably best of the fourth-generation fighter aircraft of the world inventory of fighter aircraft – the F-16 Fighting Falcon.

Earlier in this book, USAF 1st Lt. John Boyd was identified as being an F-86 Sabre pilot in the Korean War in the 1952 to 1953 time period. When the Korean War ended (July 1953), Boyd was assigned to Nellis AFB near Las Vegas, Nevada, to instruct newbies in the art of air combat tactics. While there, he was promoted to Captain and significantly improved the curriculum then in use at the USAF Fighter Weapons School (analogous to the USN/USMC counterpart "Top Gun" School at the Miramar Naval Station near San Diego, California). He earned the reputation as being Forty Seconds Boyd as he boasted he could defeat any other pilot daring to challenge him in an air combat engagement in the Nevada skies. Almost one hundred fighter pilots tested the challenge and Boyd lived up to his 40 second's names. He defeated every "aggressor" in 40 seconds or less before a kill was declared in his favor. [In an informal conversation in 1972, Boyd admitted to the Author that the engagements were sort of stacked in his favor because he knew what was the best altitude and Mach number within the flight envelope of his F-86 Sabre into which to draw the aggressor. Boyd had analyzed the flight capabilities of Soviet aircraft (obtained by Foreign Technology activities) and had determined where there was a less-than- optimum altitude and Mach region for that other type fighter aircraft. It was to that performance capability weakness that Boyd would force the engagement. Furthermore, in training flights, Boyd always flew in a clean (no external stores) configuration with a minimal amount of internal fuel to keep the Sabre to its minimum safe weight during the simulated air combat engagement. He also admitted to having occasionally maneuvered his Sabre to exceed its safe structural load limits during the heat of maneuvering, sometimes causing his Maintenance Crew Chief to make structural repairs without bothering to (or daring to) enter that maintenance action on official USAF maintenance records.]

Boyd was promoted to Captain in 1957 while at Nellis AFB. In 1960, the last year of his assignment at Nellis flying in the F-100 Super Sabre as an instructor at the Fighter Weapon School, Boyd was given permission to unofficially create, provided it was done during his off-duty hours and without expending USAF funds, an air combat tactics study paper entitled *Aerial Attack Study.* The goal of Boyd's extra-curricular activity was to record on paper the analytical genius and air combat experience of Boyd who was/is regarded as being one of the best fighter pilots and flight instructors ever having served in the USAF.

The Author found it futile to write a concise synopsis of John Boyd's *Aerial Attack Study*, just as it would be the same challenge to write a synopsis describing the classic works of art such as the Michelangelo painting on the ceiling in the dome of the Sistine Chapel at the Vatican in Rome. It has to be seen in the big picture to be fully appreciated. As a substitute, to show the scope and details of the contents within Boyd's *Aerial Attack Study*, the Author has extracted the three-pages Table of Contents and which are presented on the next three pages.

Extracted from *Aerial Attack Study*
Capt. John R. Boyd, Nellis AFB, March 1960. (Approved by USAF for Public Release)

TABLE OF CONTENTS

Extracted from *Aerial Attack Study*

Capt. John R. Boyd, Nellis AFB, March 1960. (Approved by USAF for Public Release)

Part II – Fighter vs Fighter

Extracted from *Aerial Attack Study* Capt. John R. Boyd, Nellis AFB, March 1960.
(Approved by USAF for Public Release)

For each section describing a particular type of attack maneuver which Boyd described should prove to be victorious against an aggressor, Boyd presented a narrative explanation for the maneuver, presented a graphic representation of the maneuvering flight path, presented a step-by-step procedure for conducting the attack maneuver on the aggressor, and in the event it was the friendly evading an attack by an aggressor,

Boyd presented a step-by-step procedure for conducting the maneuvers that would cause a successful evasion (and perhaps conversion to a kill. A sample of those Maneuvering Procedure and High-Speed YoYo maneuver descriptions are presented next.

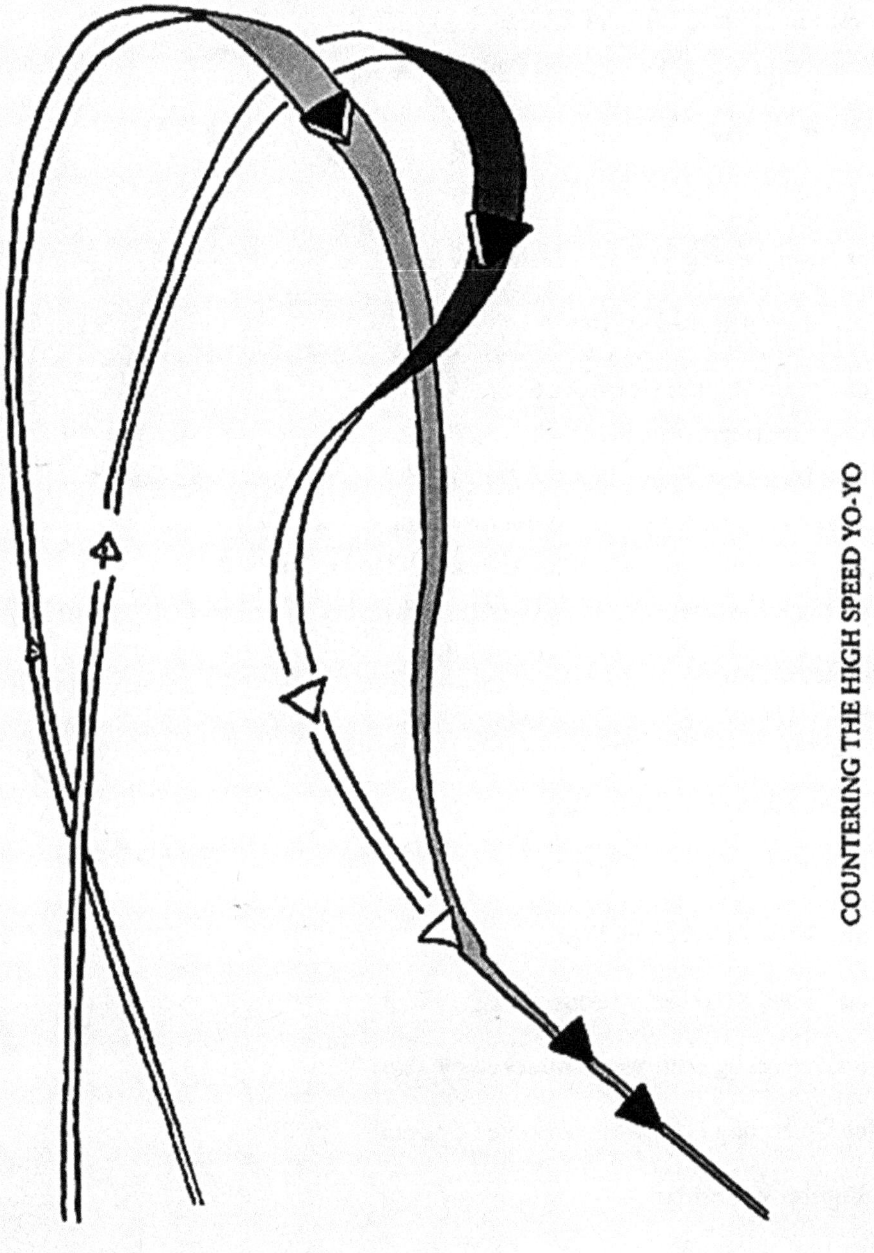

Figure 25

69

Extracted from *Aerial Attack Study*
Capt. John R. Boyd, Nellis AFB, March 1960. (Approved by USAF for Public Release)

Procedures for the High-Speed Yo-Yo

1. Attempt to reduce angle-off and stay on the inside of the turn as range diminishes.

2. Maintain back-pressure, employ top rudder and roll away from the turn, up through the vertical plane, when you can no longer stay inside your opponent's turn radius. As the pipper slides behind the target – you are no longer matching your opponent's turn – you must zoom off to prevent an overshoot. In other words, any time it becomes impossible to maintain parallel-fuselage with your opponent, your angle-off will be too great and you must zoom into a high-speed yo-yo.

3. Play the zoom so as to maintain nose-tail separation and yet retain a low yo-yo apex. If proper stick and rudder technique is employed, this is easily accomplished.

4. Roll, or slide down to your opponent's six-o'clock position. Employ the roll, if you have nose-tail separation, to acquire a six-o'clock advantage.

5. Use extreme caution when dropping down to your opponent's six-o'clock position. Do not acquire a steep nose-down attitude, otherwise your opponent may reverse up into you. This is a difficult situation to recover from. However, if it does occur, employ the following procedures.

6. Roll one-quarter turn away from your opponent's line of flight, the instant your opponent begins a pull-up, and you recognize a steep nose-low attitude as you approach tracking range. This maneuver will provide nose-tail separation and lateral separation in the vertical plane.

7. Initiate a smooth pull-up – after the quarter-roll – without overshooting your opponent's original altitude, continue the pull-up, turn and zoom toward his six-o'clock position. If your opponent attempts to dive away, cut off in the horizontal plane and slide toward his six-o'clock position.

Procedures for Countering the High-Speed Yo-Yo

1. Play the attack in an attempt to force your opponent to overshoot. This will force him to counter with a yo-yo maneuver, in order to maintain his offensive advantage.

2. Determine whether or not the attacker is going high and to the rear. If he has an extremely high rate of closure, he may be forced into extreme altitude separation. If so, employ the dive-away for separation and place the attacker in an overhead attack with a negative delta Mach. If the attacker maintains nose-tail separation, and a low yo-yo apex, employ the procedures outline below:

3. Maintain angle of bank and relax G, as your opponent slides high. This will place you in a slight descent and will allow you to maintain airspeed for future maneuvering potential. At the same time, it will force your opponent to commit himself to a nose-low attack. Continue to maneuver until he does so.

4. Perform a hard-rolling reversal – with rudder and back pressure – up into the attack. This will cause your opponent to overshoot below and forward of your line of flight, or force him to employ the quarter roll-away. If he overshoots below, simply roll or "S" down toward his six-o'clock position. If he employs the quarter roll-away, employ the procedures below.

5. Continue to zoom, and turn through the vertical plane down toward your opponent as he crosses your flight path and begins his zoom toward your angular velocity cone.

6. Relax G, light afterburner and dive for separation. This forces the attacker to perform a 180° turn through the vertical plane to get back into your six-o'clock position. By the time this is accomplished, you will have placed him out of 20mm cannon range and in an overhead attack with a negative delta Mach.

Extracted from *Aerial Attack Study*
Capt. John R. Boyd, Nellis AFB, March 1960. (Approved by USAF for Public Release)

Barrel-Roll Attack

The high-speed yo-yo appears to be an excellent offensive maneuver to use any time an attacker has a rate of closure and cannot match an opponent's defensive turn. Generally, this is true; however, in a situation in which an attacker approaches a defender at high angle-off (40° or more) and long range (10,000 feet or more) the high speed yo-yo has questionable value. Under these circumstances an attacker would have to yo-yo to an extremely high apex to maintain nose-tail separation and stay inside his opponent's turn radius. Naturally, if this occurs, we can expect the defender to immediately dive for separation and airspeed. The attacker is placed in an overhead attack with a negative delta Mach. This means that he must initiate a new attack. Noting this, the defender turns into the attack and generates a high angle-off at long range, and forces the attacker into another yo-yo with high vertical displacement. Once again, the attacker performs a new attack and the process repeats itself. The result is obvious: if the defender counters each attack successfully, the attacker gains little additional advantage and a stalemate exists. The purpose of the barrel-roll attack is to prevent this stalemate from developing.

Geometric examination of this situation – high angle-off, long-range attack – reveals that the attacker would not prefer to execute the yo-yo in this manner. Instead, he would appreciate the opportunity to reduce his angle-off and slide into the defender's angular velocity cone, without the defender being able to gain longitudinal separation. In other words, the attacker would like to reduce his velocity, cut off, and turn inside his opponent, then regain his velocity after he diminishes his angle-off, and slide toward his opponent's six-o'clock position. Is this possible? Yes, if we know how to correctly apply the three-dimensional concept of turn and velocity. As previously stated in the introduction to fighter maneuvers, it is possible to reduce turn and velocity by two methods: (1) Through both the vertical and horizontal planes by employing a two-dimensional maneuver (yo-yo) or (2) Maneuver through both the vertical and horizontal planes by employing a tree-dimensional maneuver (barrel-roll). Thus far, our emphasis has been on the first method. Now, we will employ both methods to solve our present dilemma – to reduce airspeed and angle-off, then regain airspeed to prevent longitudinal separation.

After its completion in March 1960, that "Aerial Attack Study" created by Captain John R. Boyd was recognized by the senior officers at the USAF Fighter Weapons School at

Nellis AFB as being “essential” for use at the USAF Fighter Weapons School. A Statement of Review extracted from Boyd’s “Aerial Attack Study” is presented next.

STATEMENT OF REVIEW

1. In accordance with AFT 5-5 and in compliance with AFR 5-5/Wing Supplement-1, Nellis Air Force Base, dated 21 March 1960, this publication has been thoroughly reviewed by the undersigned officers, and:

   a. Is considered essential for the following reasons:

      (1) It satisfies a Tactical Air Command need for GAR-8 employment information.

      (2) It provides a complete and authoritative analysis of Aerial Studies for use by all fighter units with TAC.

      (3) It will be used as a reference work in support of Course No. 1115053, and by units undergoing training in the Academic Section of the USAF Fighter Weapons School.

   b. It has been determined to be an original work which does not duplicate any similar publication.

   c. It does not contain any matter which might be considered objectionable or subject to ridicule.

   d. No further coordination is deemed necessary.

Norman C Gaddis, Major

Norman C Gaddis - Director of Operations and Training.
Fighter Weapons School

Billie R. Cothern, Major

Billie R Cothern - Director of Maintenance. Fighter Weapons School

Allen S. Nelson, Major

Allen S Nelson - Director of TR&D. Fighter Weapons School

Extracted from *Aerial Attack Study*
Capt. John R. Boyd, Nellis AFB, March 1960. (Approved by USAF for Public Release)

Because the 1960 version of the *Aerial Attack Study* by Capt. John Boyd (then the Academic Supervisor at the USAF Fighter Weapons School) contained sensitive classified material, the USAF designated the original 1960 version of the *Aerial Attack Study* as being Secret security classification. In 1964, USAF removed the sensitive classified material and de-classified the *Aerial Attack Study* for Release to the Public. All student pilots and instructor pilots at the USAF Fighter Weapons School after 1960 have used the Boyd *Aerial Attack Study* in their BFM and ACM pilot training. Captain Boyd was awarded the USAF Legion of Merit for his inspired efforts and unwavering persistence in excellence. His final OER (Officer Efficiency Report) at Nellis AFB also recommended that Captain Boyd be promoted to Major ahead of his contemporaries.

In early 1960, Boyd requested as his next assignment that he be permitted to attend an engineering college to study a field of engineering which would further educate him in the laws of physics and in mathematics so that he could translate his "seat of the pants" flying skills into technically honed flying skills. Captain Boyd was offered the opportunity of going to AFIT (the Air Force Institute of Technology) which agreed to accept him provided he studied electrical engineering to fill the needs of the USAF in applying new electronics-oriented technologies on fighter aircraft. Boyd declined the caveated AFIT acceptance offer (he did not want to be an electrical engineer) and instead applied to and was accepted at the Georgia Institute of Technology (Georgia Tech) in Atlanta, Georgia) to start in 1960 in the School of Industrial Engineering. Boyd's choice of Georgia Tech was also partly based on its proximity to Warm Springs, Georgia, where his first son, who had contracted polio at a very young age, could continue to receive therapy for his debilitated condition.

[The Author attended Georgia Tech from 1949 to 1953 in the Daniel Guggenheim School of Aeronautical Engineering at the undergraduate level where John Boyd later would receive his desired Industrial Engineering degree, and can attest to the excellence of all of the engineering courses taught at Georgia Tech. It was also to Georgia Tech where the Author elected to return in 1956 to attend Graduate School in the Department of Aeronautical Engineering and received a Master of Science in Aeronautical Engineering in April 1959. John Boyd's path and that of the Author had missed being engineering classmates at Georgia Tech by a few years, but they enjoyed later conversations about the same engineering courses and professors they had during their respective years at Georgia Tech.]

Not surprisingly, the Mathematics and Physics curricula at Georgia Tech taught John Boyd how to convert his "seat of the pants" piloting talents into the more scientific concept of what Boyd entitled energy-maneuverability.

For Readers who may not be aware of the principles of flight of an aircraft, the four primary forces acting upon an aircraft to cause its dynamic motion in flight - lift, drag, thrust, and weight - are illustrated on the photograph of an airplane shown next.

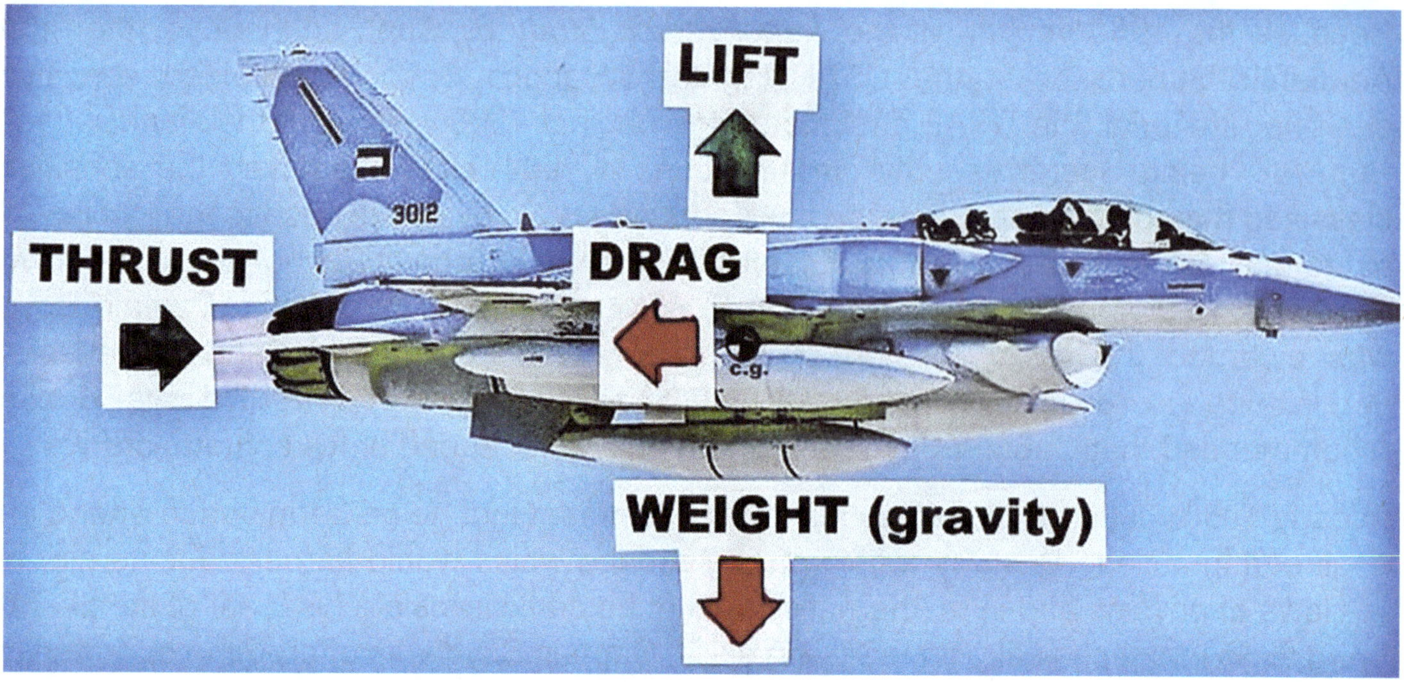

Four primary forces acting on an aircraft
(Source: USAF Engineering files)

[Author's note: In engineering analyses of a system, the technique of using a free-body diagram such as that above showing all the forces acting upon the free-body in a vector form of illustration is a common way to analyze the motion the free body will be caused to take as a result of those applied forces. In the illustration, the drag and weight vectors are colored red as these are the forces opposing the aircraft to have forward motion and to climb above the ground. The lift and thrust vectors are beneficial forces, with the thrust force vector overcoming the drag force vector to accelerate the forward motion of the aircraft. The lift force vector must overcome the weight force vector to cause the aircraft to ascend in altitude above the ground. In a free-body analysis, the shape and size of the body are ignored and all forces are assumed to be acting through the center of gravity (c.g.) of the body as if it were a point mass. The c.g. of the aircraft above is approximately located where noted by the c.g. callout.]

For an aircraft flying at steady-state conditions (i.e., no climb/descent and no acceleration or deceleration of speed), the lift necessary to be generated to maintain constant altitude will be equal and opposite to the weight of the aircraft. The Author was taught the principles of aerodynamics and flight at Georgia Tech from the textbook *Principles of Aerodynamics* by James Dwinnell published by McGraw Hill in 1949. By adjusting two of the four forces acting on the aircraft, lift and thrust - drag and weight forces are not controlled by pilot command, the aircraft can be made to accelerate or decelerate in speed, and/or climb or descend in altitude. Within that *Principles of*

Aerodynamics textbook, Dwinnell includes a formula (not copyrighted) for estimating the rate of climb potential for an aircraft:

Rate of climb (R/C) = k x velocity x (thrust – drag) / weight

(The factor k being a units conversion constant to result in rate of climb in feet per minute)

John Boyd had not elected to take the AE 101 course at Georgia Tech as part of his Industrial Engineering curriculum, and so he was not aware of this simple equation affecting the path of motion an aircraft can take, in this case its rate of climb. It would be two years later that when taking a formal course in Thermodynamics that the concept of entropy, a measure of the energy state of a mass, would lead John Boyd to better understand how the principles of physics (applicable to thermodynamics as well as aerodynamics) would apply in development of the analysis tool which Boyd would derive and title as his Energy-Maneuverability theory.

Also taught in Physics 101 at Georgia Tech in early undergraduate school was the Law of Conservation of Energy. This law identifies that energy can neither be created nor negated, it can only be transformed from one state to another. As applied to the motion of a mass such as an aircraft moving from one set of altitude and speed conditions to a different set of altitude and speed conditions, the potential energy and kinetic energy possessed by the aircraft in its state of motion remain as a constant sum of total energy.

The physicist Sir Isaac Newton in the 1600s had developed heuristically what is now accepted as Newton's Laws of Motion:

1) A body at rest tends to remain at rest; and
2) A body in motion tends to remain in motion.

Mathematicians would later convert Sir Newton's observations into physical stated formulation and mathematical relationships.

A physical mass at some height above the ground possesses potential energy as described by the relationship:

Potential energy (P.E.) = weight of mass x height above ground (in foot pounds)

In order for that mass to be at some height above the ground, there is necessarily a force opposing the force of gravitational acceleration to elevate that mass through that distance to that height above the ground.

A physical mass in motion through the air will also possess Kinetic Energy, as described by the mathematical formula representing Newton's 2nd Law of Motion:

Kinetic energy (K.E.) = ½ mass x velocity^2 (in foot pounds) and since mass is weight / g (g being the gravitational acceleration which is equal to approximately 32.17 ft/sec^2 in Earth's near-atmosphere), kinetic energy (K.E.) may alternatively be expressed as:

$$\text{Kinetic energy (K.E.)} = \tfrac{1}{2}\,(w / g) \times \text{velocity}^2 \qquad \text{(in foot pounds)}$$

Substituting the aircraft for the arbitrary mass and recognizing that the height above the ground may be the altitude H at which the aircraft is flying

For an aircraft, the weight force, caused by the acceleration of gravity on the mass, is always in the direction toward the center of the earth regardless of how the aircraft longitudinal axis is pointed. The lift force, which as its name implies, lifts the aircraft away from being at rest on the ground. The heavier the weight that has to be lifted, the greater the lift force must be generated by the aircraft. The propulsion system on the aircraft produces a thrust force to cause the mass to move forward on its flight path. Because there is friction between the air molecules and the skin of the aircraft and all of its external components (wing, vertical tail, horizontal tail, and the fuselage) the higher the speed of the aircraft through the air the higher will be that friction (minimum) drag force. Because the lift force vector is inclined when at an angle of attack of the aircraft lifting surfaces (with respect to the fuselage centerline relative to the flight path), another component of the drag force opposing forward motion is introduced. That component of drag is referred to as induced drag or drag-due-to-lift. Further, if the speed of the aircraft is approaching the speed of sound (which varies with the air temperature at that altitude), there will be the formation of shock waves in front of the forward edges of the aircraft's external components. These local shock waves all add up to produce a total wave drag that is added to friction drag, pressure drag, and drag-due-to-lift.

To give the Readers some indication as to what this all means in the practical sense, the expression is used that a pilot "flies by the seat of his pants". It is inferred that by maneuvering the aircraft, the resultant of all the four forces will cause a force to be developed between the pilot's body mass and the bottom of his anatomy being pressed into or pulled away from the seat in which he /she is seated. When any person sits in a chair which is not moving, the condition of one g exists on the person's body. The person's weight will be sensed as a force against the seat (usually cushioned, as sitting on a hard surface for a length of time can be most uncomfortable, even at one g conditions). When a pilot maneuvers a fighter aircraft at nine g's, his/her weight causes the weight force into the seat to be sensed as being nine times his/her weight. A more hazardous condition for pilots is the pilot being required to wear a protective flight helmet which weighs approximately eight or nine pounds when it is sitting on a shelf. When on the pilot's head and in an aircraft maneuver at nine g's, that helmet induces a force of about 80 pounds of force that must be reacted in the pilot's neck muscles and cervical spine. That very real hazard to fighter pilots' cervical and lumbar spine has been and is currently being studied for its medical implications and long-term effects on fighter pilots' health and safety.

To adhere to the Law of Conservation of Energy (i.e. energy can neither be created or destroyed, it can only be transformed from one state to another), the sum of potential energy and kinetic energy as applied to an aircraft moving through the air from point 1 to point 2 requires that the total energy possessed by the aircraft in motion (having a weight w, at a speed v, and at an altitude h) at point 1 be exactly the same as the total energy possessed by the same aircraft moving to point 2 in its flight path (the gravitational acceleration g remaining constant):

$$(w \times h + \tfrac{1}{2} (w / g \times v^2)_1 = (w \times h + \tfrac{1}{2} (w / g \times v^2)_2$$

For those Readers not familiar with the equation, that mathematical expression simply reflects that the aircraft pilot can alter (i.e., maneuver by commanding wing angle-of-attack change and/or command higher or lower thrust from the propulsion system), the altitude h and forward speed v of the aircraft at point 1 to move the aircraft to a desired different altitude and speed at point 2. However, the total energy at every point in the flight path of the aircraft between point 1 and point 2 must remain exactly the same.

So as not to incur the criticism of physicists, the total energy includes three terms:

Total energy = potential energy + kinetic energy + stored chemical energy.

Again, a simplified explanation for Readers is that the fuel combusted in the aircraft's propulsion system is a conversion of the stored chemical energy in the fuel to heat energy which creates the thrust at the engine exit nozzle needed to move the aircraft from point 1 to point 2. The change in the aircraft weight due to the combustion of fuel (converting it from liquid state to gaseous state) must also be accounted for in the computations of the aircraft's maneuvering path of motion. This is accounted for by not assuming the w to be a constant numerical quantity at both points 1 and point 2, but rather to be w_1 at point 1 and w_2 (a few pounds less than w_1) at point 2. [Flight performance engineers in space astronautics must account for the rapid change in weight between point 1 and point 2 due to the extremely high burn rate of the rocket propellant fuel on rocket propelled aircraft or during the launch of vehicles such as NASA's Space Shuttle launched on large, liquid-fueled booster-rockets. During the few seconds actually engaged in the maneuvering of an air-breathing, conventional fuel combustion-propelled aircraft against another, the change in weight of both the friendly and the aggressor aircraft is virtually negligible as the fuel burn rate will also be approximately the same rate for both.]

All of this mathematical gymnastics is not of concern to the pilot (even to John Boyd) when engaged in the maneuvering of the fighter aircraft against an aggressor in an attempt to achieve a kill, either with gunfire or with an air-to-air missile. Pilots do have to monitor their fuel remaining at all times, especially when engaged in air-to- air maneuvering at high fuel flow rates, and when "bingo fuel" state is reached (probably with an audible warning from Bitching Betty that *bingo fuel* status is present) and dictates the pilot to return to base (RTB). If the pilot is actively engaged in air-to-air combat, it will be necessary for him/her to disengage expeditiously and safely, and "head for the barn" (or toward a nearby refueling tanker).

Again, so as not to incur the criticism of performance engineers calculating the maneuvering performance of fighter aircraft, it must be recognized also that maneuvering at constant altitude and near constant speed, there inevitably will be pilot commands to "pull gs" and engage in a high-g turn rate to achieve a gun firing solution or missile launch condition orientation toward the aggressor. Because drag is directly related to lift at an angle of attack, the pilot will necessarily add power (i.e., increase thrust in non-afterburning mode of the engine or by selecting a zone of afterburning) to overcome the additional drag and maintain the large thrust excess (i.e., thrust – drag) during the high-g turning maneuver.

Pilots flight train in Basic Flight Maneuvering (BFM) so as to develop effective tactics in maneuvering by engine power lever movement, and movement of (or pressure on) the flight control stick to change angle of attack (and/or move control surfaces much like a "one-man-band at a carnival") to achieve a kill. It is left to the fighter aircraft cockpit designers to minimize the actions the pilot must manually perform to manage all the systems aboard the aircraft at the same time while in maneuvering flight mode. [Later in this book, the concept of HOTAS will be described. HOTAS refers to the pilot keeping his/her Hands on Throttle and Stick while concentrating on flying the aircraft and interpreting the several numerical flight symbols and geometric images presented to him/her on the cockpit Head Up Display (HUD)].

After graduation from Georgia Tech in 1962, Boyd completed development of his Energy-Maneuverability concept

[The equation in the Dwinnell textbook *Principles of Aerodynamics* was shown earlier to be:

$$\text{Rate of climb} = k \times V \times (T - D) / W \qquad \text{(in feet per minute)}$$

where k is a conversion constant (dependent on whether V is input as feet per second, statute miles per hour, or knots – nautical miles per hour) to convert the final units of rate of climb to be in conventional feet per minute nomenclature]. The difference is that John Boyd envisioned that instead of applying the excess thrust (i.e., T – D) only to achieve rate of climb, the excess thrust could be used to maneuver the aircraft to trade potential energy for kinetic energy in some combination of change of altitude H, speed V, turn radius, and at elevated g other than 1g in level steady-state flight. As a result of his derivation and combining of the equations of Physics and Laws of Motion, John Boyd's expression for what he termed as Specific Excess Power in his Energy-Maneuverability concept was written as:

Specific Excess Power, $P_S = k \times V \times (T - D) / W$ (in feet per minute)

In both relationships, the key to performing the rate of climb or using P_S to maneuver the aircraft at elevated g is the maximizing of the (T – D) / W term. The means for maximizing (T – D) / W can be to maximize T, to minimize D, and to minimize W, or to achieve a combination of all three variables at the speed V at which the pilot maneuvers the aircraft in air-air combat appropriate to the conditions of the engagement. That suggested maneuvering for maximizing the chance for a kill or to avoid being killed was precisely what John Boyd had described in his *Aerial Attack Study* described earlier in this chapter.

Examining the (T – D) / W factor closer, Readers should note that the factor can be used when establishing the requirements for any new fighter aircraft designed to engage in air-to-air combat. By requiring a fighter aircraft to achieve a high T / W at start of combat, a designer of the aircraft can select a propulsion system which produces high thrust at a low weight of the propulsion system and lowest weight W of the aircraft at start of combat. As fighter design technology has evolved, it has been learned that a fighter with a T/ W greater than 1.1, much higher than that of his opponent at the start of combat, and at the hands of a well-trained pilot, the pilot/aircraft with the higher T / W at the start of combat has a higher potential to achieve a kill. At the same time, the D / W term must be held to a minimum also. By designing the fighter configuration with an aerodynamically efficient external configuration and also keeping the weight of the aircraft at a minimum at the start of combat, the entire P_S factor in the Boyd Energy - Maneuverability concept can be maximized.

After Major John Boyd received his engineering degree from Georgia Tech in early 1963, he was given his choice of next assignment to be either to the USAF Requirements Planning Office at the Pentagon, or a base of his choosing. John detested the strict discipline environment at any Headquarters where it is more important to have sparkling shiny shoes than to be conducting an analysis of anything of true military importance. For a different reason, John asked for assignment to Eglin AFB, Florida, which is the home of the USAF Armament and Weapons System Research and Development Center. It was a combination of Boyd feeling that it was at Eglin AFB that he stood the best chance for completing development of his Energy-Maneuverability theory and applying it to newly developed missiles and fire control systems. Of equal importance to Boyd was that Eglin AFB was close to Warm Springs, Georgia, where his first son had been receiving physical therapy for the polio disease the lad had contracted while very young. The Warm Springs Physical Therapy Center is where President Franklin Roosevelt had received therapy for the polio that had stricken Franklin D. Roosevelt as a young man – and where FDR died in April 1945 in his fourth term as President of the United States.

Upon reporting for duty at Eglin AFB, Major Boyd was assigned, as one of the worst personnel mal-assignments of any officer's career in various do-nothing jobs that were offensive to his brilliance and experience, to be a standardization officer, a training officer on the operations staff, a briefing officer to distinguished visitors to Eglin AFB, and as the Base Housing Officer. After a few months, Boyd then was assigned as a Maintenance Officer which experienced pilots consider as being only a minor promotion above sweeping floors. At night, Boyd would go home and fill yellow writing pads with equations and short analyses using his technical education at Georgia Tech in development of his Energy-Maneuverability Theory.

At Eglin AFB, Boyd then crossed paths with Tom Christie, a high-grade civilian engineer and supervisor of the technical analysis organization at Eglin AFB which did the performance computations for missiles having various methods of flight path guidance and control, and bombs (ballistic, and then later, guided) being developed at Eglin AFB. Tom had received a Master of Arts in Applied Mathematics from New York University in 1962. It was a most fateful chance meeting but one which was to become one of aviation history's most important stories with a fairy tale-like ending "and they lived happily ever after". Much like Boyd had done his magnificent *Aerial Attack Study* at Nellis AFB, Tom Christie enabled Boyd to refine his Energy-Maneuverability theory with mathematical precision and elegance, and all on weekends and after duty hours with no supporting funds from any government source. (Well, *almost* no funding support. Tom Christie did boot-leg some computer time to process Boyd's Energy-Maneuverability charts for various US and Soviet fighter aircraft. With these charts in color and on transparent overlays, the Boyd E-M theory showed regions of the flight envelope where one aircraft possessed higher maneuvering capability over the other. These overlays were of course highly classified and restricted Tom Christie and John Boyd from widely and officially advertising their monumental work accomplishments.

One day, their extra-curricular work was exposed to the Eglin AFB Deputy Base Commander who came close to dismissing both Tom Christie and John Boyd from Government employment for having potentially violated security rules governing protection of classified information and compounded the potential violation by the fact they misappropriated computer time on a government computer to do it. Letters of Reprimand in both of their personnel files were issued, but when the Eglin Base Commander saw what Boyd and Christie had accomplished, the Base Commander ordered dismissal of the charges of breach of security and misappropriation of expensive computer time. Both Boyd and Christie were admonished to "Go forth and sin no more", but John was not noted for terminating an engagement with a clear kill in sight. Christie continued supporting Boyd with computations on the Eglin base computer, but did so under an approved research and development "slush fund" permitting such computations.

Shown next is a copy of an Energy-Maneuverability summary analysis chart which Major Boyd developed, with Tom Christie's invaluable mathematical assistance, and with Harry Hillaker's invaluable fighter aircraft technology and design expertise). John Boyd, and his "Lightweight Fighter Mafia" associates throughout the Government, used this form of chart extensively during their crusade at the Pentagon for development of a light-weight, highly-maneuverable, low-cost, multimode fighter aircraft.

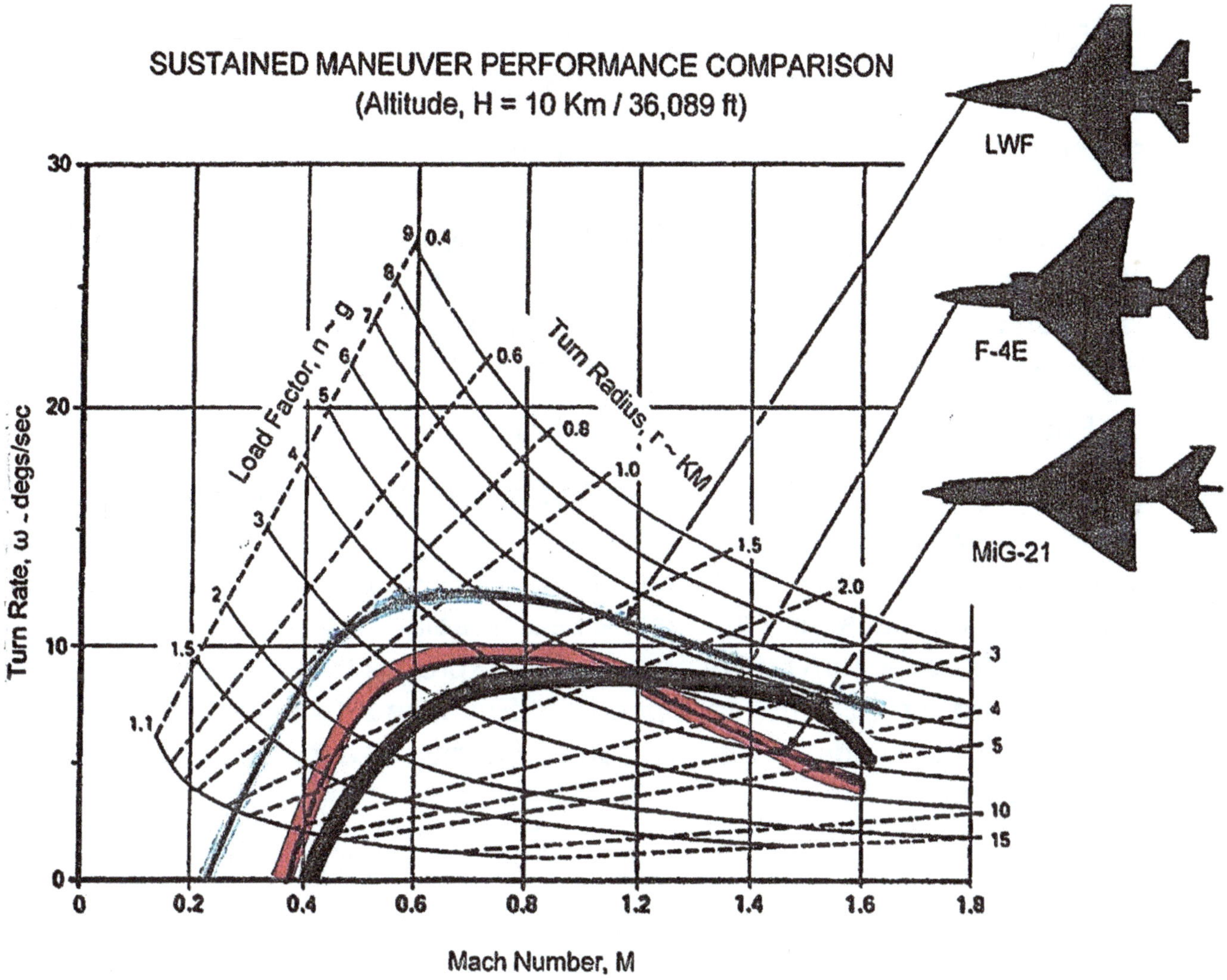

Boyd / Christie energy-maneuverability aircraft comparisons
(Source: USAF Engineering files)

[For Readers who may be interested in how such a chart was created, the mathematical relationships, found in any college-level Physics textbook, used in developing the "warped mesh" visible on the chart will be shown next. The overlay information for a specific aircraft design does require numerical definition of the lift, thrust, drag, and weight of the aircraft at the speed and altitude flight conditions for each aircraft being compared. Those numerical details are beyond the scope of this book; however, all four of those essential inputs were provided by the manufacturer of the aircraft (lift, drag and weight) and by the engine manufacturer (installed thrust). For the foreign aircraft used in the comparative analysis, those four essential inputs, lift, drag, weight, and installed engine thrust, were obtained from Foreign Technology exploitation files.]

The chart is seen to use the variable for speed of the aircraft to be Mach number rather than velocity in units of miles per hour or knots. Mach number M was created by the physicist Ernst Mach (a founder of the compressibility effects and shock waves in supersonic flow) and is the ratio of the speed of the aircraft to the speed of sound at that temperature and density of the surrounding air. That is written as

M = V of mass / V of sound (a ratio in decimal form, no units)

The basic chart (i.e., without any colored overlays for specific aircraft designs) is constructed from the equations that were developed by mathematicians a few centuries ago and have been long-used by physicists and engineers when analyzing the dynamic motion of a mass in space. The first equation used is that which relates the tangential velocity of a particle or mass rotating at an angular rate ω about a point at the center of the arc (e.g., the path of the aircraft) having a radius R.

Equation 1 $V = R \times \omega$ (in feet per second)

where V is the tangential velocity (in feet per second), R is the radius of curvature of the path of motion (in feet), and ω is the angular rate of change of the turn angle ($d\theta/dt$ in radians per second). There are 2π radians in a 360-degrees circle, as established by early Greek mathematicians, and that defines each radian to subtend an angle of 57.3 degrees. Therefore, as the turn rate of an aircraft is expressed conventionally in degrees per second, the Equation 1 can be rearranged into Equation 2 (with velocity V expressed in feet per second, turn radius R in feet, and turn rate ω expressed in degrees per second) as shown next:

Equation 2 $\omega = 57.3 \times V / R$ (in degrees per second)

Equation 2 can also be rearranged to the form in Equation 3 below in order to estimate the turn radius of the aircraft if its velocity V and turn rate ω are known from flight test measurements:

Equation 3 $R = 57.3 \times V / \omega$ (in feet)

Equation 3 is of use when ω and V have been determined from the aircraft Mach number at altitude H during flight testing, and when R is unable to be measured. R is a vital parameter used in developing fighter aircraft maneuvering tactics.

Other equations from Physics are used to augment Equations 1, 2, and 3 above in creating the warped mesh in the basic chart to express the normal acceleration a_n of that point/mass on a curvilinear path knowing the tangential velocity V and the radius R of the curvilinear path:

Equation 4 $a_n = V^2 / R$ (in feet per sec^2)

Equation 4 can be rearranged to the form:

Equation 5 $$R = V^2 / a_n$$ (in feet)

Equation 5 is of use when directly measuring a_n and V (derived from velocity Mach at altitude H) during flight testing and when R instantaneous is unable to be measured. The two estimates of R determined with Equation 3 and Equation 5 should be numerically very close during the analysis of flight test data for the same point in the path of maneuvering flight.

In analyzing the flight path motion of a maneuvering aircraft, it is most common to use the number of gs being commanded by the pilot (more commonly referred to as "pulling gs") as a multiplier of the acceleration of gravity (acceleration of gravity = 32.17 feet / sec^2 in Earth's atmospheric region). Instead of expressing the velocity of a fighter at high speed in terms of V feet per second, the Mach number M of the aircraft is used.

To construct the Boyd Energy-Maneuverability basic chart (the "warped mesh" portion), arbitrary values are used for the independent variables (i.e., not requiring knowledge of the aerodynamics, propulsive forces and the weight of the aircraft) and dependent variables in the flight envelope of the maneuvering fighter aircraft. The chart is typically done for a nominal altitude and held constant in the chart. The speed of sound is calculated for that altitude and is a constant for each chart when converting Mach Number into velocity in the units of feet per second to be consistent with the dimensional units in the equations relating turn rate, and turn radius. Care is necessary to use common units of scalar magnitude during all computations. The computational process was developed by Tom Christie and his associates and each basic "warped mesh" chart is able to be printed out in a matter of very few seconds.

With the aerodynamics and propulsive forces known from tests (ground and possibly flight) of the configuration and the engine, coupled with the weight estimated by the aircraft design team, the four force terms (thrust, drag, lift, and weight) for the aircraft at that Mach number, altitude, and maximum power setting are used to calculate what number of gs and the turn rate that aircraft can perform in its max maneuver. Again, by varying the Mach number at the altitude for the chart, the resulting maneuvering performance can be plotted in an overlay fashion on top of the "warped mesh" basic chart. In that manner, as shown on the example chart, one fighter aircraft can be compared to another in its ability to pull gs, turn at a rate in multiple degrees per minute, and with a turn radius that would give that aircraft a favorable maneuvering advantage over an adversary aircraft.

Using the information shown in the Energy-Maneuverability comparison chart, Readers should be able to conclude that the USAF F-4E was potentially able to outmaneuver the Soviet MiG-21at all Mach numbers above M 1.2 at 36,069 feet altitude, but that the MiG-21 was able to outmaneuver the USAF F-4E at all Mach numbers below M1.2 at 36,089 feet. When comparative maneuverability analyses are done by professional analysts, charts similar to this one, for altitude of 36,089 feet, are created for other altitudes, usually from 10,000 feet to 50,000 feet altitude in 10,000 feet increments. That is done to account for the effect changes in altitude and Mach number have upon the engine performance and the aerodynamics of the different aircraft being comparatively analyzed. Another important factor Readers must always consider that, when reviewing graphical presentations of the maximum maneuvering capability an aircraft design can achieve, the capability shown on the chart is only a potential. The potential capability is only as useful and as effective as the training of the pilot commanding that maneuverability. John Boyd well recognized that fact, and generated his *Aerial Attack Study* as a means of properly training fighter pilots in the tactics to make maximum use of the capability of the aircraft and its weapons.

Whether a kill is achieved also depends very importantly on the accuracy, effectiveness, and reliability of the air-to-air weapons employed to make the kill. The early chapters of this book have described how the technologies involved in the complex interaction of the pilot with the aircraft capabilities and with the air-to-air weapons installed on the aircraft have been improved during the first 50 years of military aviation history.

For a statistical portrayal of how the F-4s did in killing MiGs, the Table shown next illustrates the number of unofficially verified kills estimated to have been obtained by three variants of the F-4 aircraft engaged in air combat with three variants of Soviet MiG aircraft designs during the air war over Vietnam:

Aircraft	MiG-17	MiG-19	MiG-21	Total
F-4C	21	0	21	42
F-4D	12.5	4	28	44.5
F-4E	0	4	17	21
Total	33.5	8	66	107.5

Estimated F-4 kills of MiGs in the Vietnam air war
(Source: USAF military records)

The three different variants of the F-4 Phantom II entered into the Vietnam war in sequence of their alphabetical variant designator, and the MiGs entered in their numerical sequence. Analysis of the kill records (not contained herein) showed that all three variants of the F-4s achieved a higher percentage of the total kills with the use of missiles rather than by 20 mm gunfire. The original USN F-4A had been designed with no gun at all, which was recognized early in the Vietnam War as being a poor decision not to include a gun. For the USAF F-4C and F-4D, that deficiency of no gun was rectified somewhat by mounting a 20 mm gun in a pod mounted on a wing station. Gun firing caused the pod to vibrate significantly and the actual trajectory of the fired rounds deviated very erratically compared to their intended trajectory. Mounting the 20 mm gun internally on the F-4E eliminated the gun vibration induced oscillation, and gun kills in the F-4E were much improved over that of the F-4 C and D variants with the gun in a pod.

As a result, missiles achieved a better kill probability over that of the kills by cannon fire. In the early Vietnam War, the AIM-9 Sidewinder missile on the USAF F-4C had the highest kill efficiency on the F-4C against the MiG-17 and was also a good weapon against the MiG-21 later in the Vietnam War. The AIM-7 Sparrow missile was found to be very effective on all three F-4C/D/E variants against the MiG-21 during the later stages of the air war in Vietnam.

However, there was massive damage done to all USAF (and USN/USMC) aircraft in the Vietnam War from enemy ground fire (small arms/automatic weapons, AAA artillery, SAMs, etc.). The extent and location of the battle damage on the F-4 aircraft determined if the aircraft would be able to RTB (Return to Base). The battle damage ranged from one small caliber bullet hole in the skin of the aircraft in an area having little effect on flight safety, and up to massive damage done by a Surface to Air Missile (SAM) to critical flight safety functions. Shown next is a photograph of a USAF F-4 Phantom II that despite having suffered extensive damage to its empennage from a near-burst of an AAA shell over Vietnam was able to return to its base safely:

Extensively battle-damaged F-4 Phantom II in Vietnam War
(USAF Photographic files)

The pilots in the badly-damaged F-4 deserve much credit for being able to compensate for the loss of so much area of flight control surfaces and being able to return the aircraft and themselves back to a safe landing. Although badly damaged, the F-4 was able to be repaired and returned to operational flight status.

In late1963, USAF planners had been planning the conduct of a study to develop a new fighter aircraft that would replace the Century Series fighters and the F-4 Phantom II that was not performing as well as desired in the Vietnam War against especially the Soviet MiG-21 Fish bed fighter being flown by pilots of the North Vietnamese Air Force. (That poor performance will be discussed at the last part of this chapter). There were other lessons learned about the losses of F-4 Phantom IIs due to enemy ground gunfire bringing down the very visible smoke-stream producing F-4 Phantom II (derided with the title Smokey Bear) at an unacceptable loss rate. (Lessons learned from the poor survivability/vulnerability aspects of the F-4 Phantom II in the Vietnam air war exposure will be described further in Chapter 6).

In 1965, feedback from the Vietnam War and his being involved in deriving the performance requirements being generated for the FX design, caused John Boyd to construct a chart entitled The Growth of the Fighter. The Growth of the Fighter chart, presented next, shows the exponential weight growth of all fighter aircraft with time from World War II and includes data points for the Century Series fighters, the F-4D/E variants, and the projected F-15 Air Superiority Fighter for the USAF. Both USAF aircraft and Soviet MiG aircraft are shown in the Growth of the Fighter chart in the gray shadowed band. The chart does indicate the F-15 having a position at the lower edge of the weight growth trend, but it was still most disturbing to John Boyd that so many "bells and whistles" associated with emerging technologies in the rapidly-growing domain of avionics were being added into the F-15 design. The USAF F-111A, which was a commonality design, is seen to be located on the 70,000 pounds TOGW upper boundary of the Growth of the Fighter chart, as shown next:

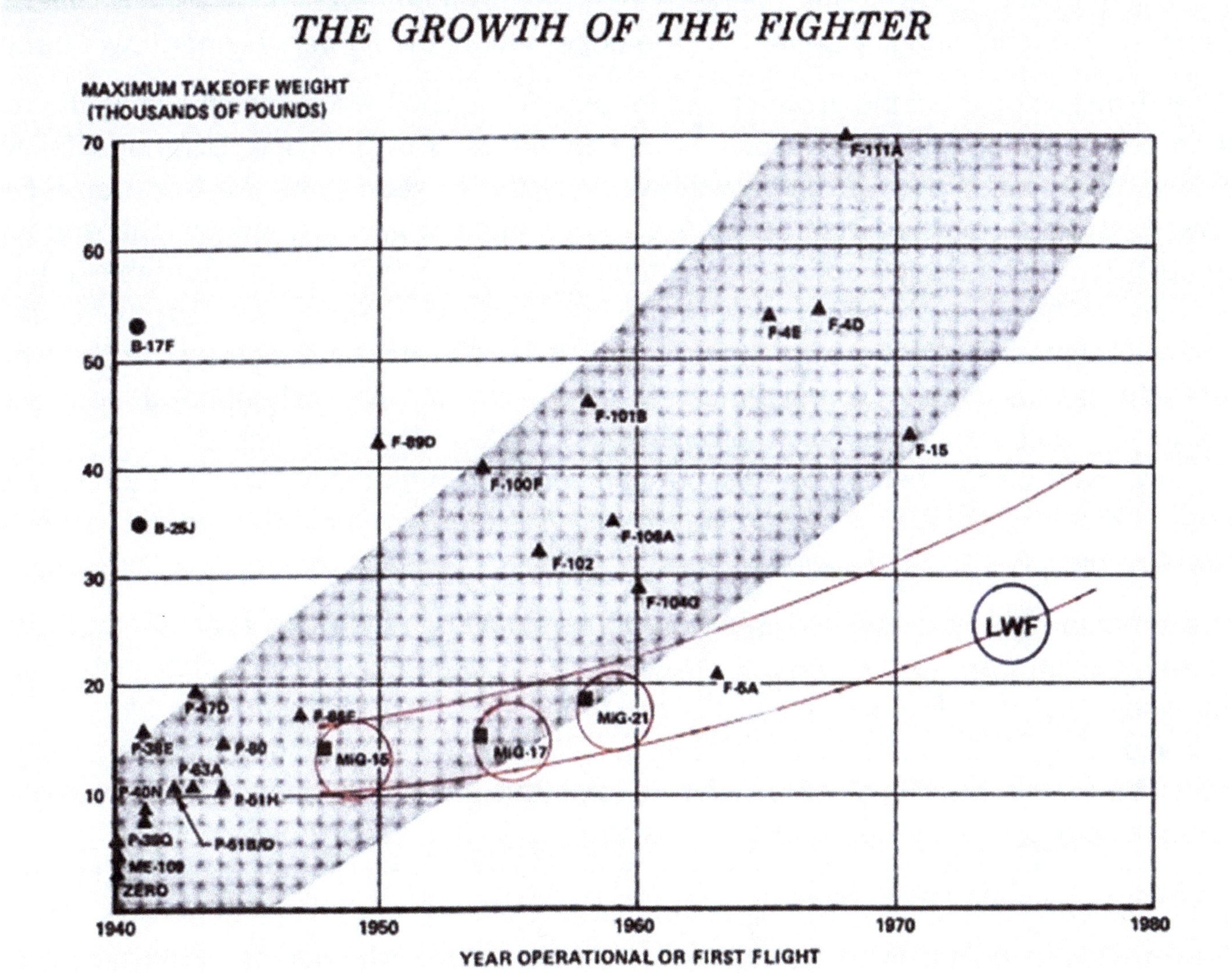

Growth of Fighters 1940-1970 chart created by John Boyd for an Advanced Day Fighter study (Source: USAF Engineering files)

[The Author suggested to John Boyd that instead of showing the growth trend as a concave upward exponential it be shown as a concave downward exponential but it is apparent by the chart who won that debate. The red lines and circles on the chart represent an interpretation by the Author in 1970 of the weight growths of the Soviet Fighter aircraft and with the suggested weight goal for a lightweight fighter having an estimated weight goal of 25,000 pounds.]

The message Boyd conveyed in the weight growth trend chart was that aircraft designed for carrier-based operations are inherently higher than land-based aircraft due to the high loadings into the carrier-based aircraft being hooked and brought to a short distance stop. On the chart above, there is not shown a symbol for the USN Grumman F-14 Tomcat, which had variable sweep wings, as it is above the upper weight boundary of the chart. Those aircraft designed with commonality between USAF land-based and USN carrier-based capabilities, especially that of the original USN/USMC F-4A Phantom II leading to the USAF F-4B counterpart, the maximum take-off weight of aircraft designed with commonality appears in the upper half of the fighter weight growth trend.

When Boyd singled out the growth trend for only Soviet-built MiGs (as shown by the red lines defining the growth trend of only the Soviet aircraft) the only US-made fighter design falling within that Soviet growth trend band was seen to be the Northrop F-5A. Boyd had praise for the Northrop F-5A because of its small size, low weight, and low complexity. Those all add up to achievement of a good fighter that is affordable in high quantity for the inventories of smaller NATO and UN Coalition countries. Boyd was often heard to comment *What counts most is having a lot of noses of fighters on the flight ready line* as opposed to having a much fewer number of high-cost hangar queens with the costly maintenance and failure problems associated with high complexity. Boyd used the argument for the benefit of low cost and resulting affordability with the analogy of what is most hazardous to human life, *to be attacked by a single wolverine, or to be attacked by a swarm of killer bees*?

The largest single fault at which fingers could be pointed in the Boyd approach to fighter aircraft design was to the austere "daylight, clear weather" design approach being demanded by John Boyd as being unrealistic based on experiences with all previous operational fighter aircraft designs. Others said that Boyd's approach with austere weight reductions by leaving out some safety equipment having redundancy compromised safety of the pilot and survivability of the aircraft in the harsh operational air combat environment. Few objected (but not extremely passionately) to the need to hold down costs and weight on all aircraft. Older Readers may recall the absurdity of "The Golden Fleece" Award being given by Senator William Proxmire to defense contractors for overcharging for equipment items on government acquisitions. Who can forget the "Golden Fleece" award for the $2,000 toilet seat or the $600 screw driver? John Boyd certainly never could or did. Bureaucracy was his arch enemy.

As Boyd parted from participation (in total disgust over the weight growth of the aircraft) with the FX and F-15 development activities, he acquired the support (moral rather than economic) of others within the Defense industry and Department of Defense and resumed his crusade for the need for a Light Weight Fighter (LWF) – one which has a weight goal as shown to be on the lower bound of the Soviet-built fighter aircraft growth trend (labeled LWF). There were five persons on the Boyd support list (which soon became known as the "Lightweight Fighter Mafia"); Major John Boyd and Tom Christie from Eglin AFB, Col. Everett "Rich" Riccioni, an ex-pilot then assigned in the Operational Requirements Office at HQ USAF, Pierre Sprey, who was a Defense Department operations analyst, and Charles "Chuck" Meyers who was Deputy for Acquisition in the Department of Defense and ex-pilot.

At the General Dynamics facility in Fort Worth, Texas, Chief Designer Harry Hillaker (who had designed the B-58A Hustler supersonic strategic bomber for the Strategic Air Command, and later the F-111 Aardvark supersonic fighter bomber for the USAF/USN commonality replacement of the F-4 Phantom II) was closely associated with the advanced technologies for the Lightweight Fighter being championed by the Lightweight Fighter Mafia. Harry Hillaker could not officially be a member of the Major Boyd "Lightweight Fighter Mafia"; otherwise, there would be a Conflict-of-Interest situation arise which could kill the whole thrust of the Lightweight Fighter crusade. Nonetheless, Hillaker did provide the technology reality to validate what the Lightweight Fighter Mafia had set as its goal. [The story has to be told that when John Boyd and Harry Hillaker had a first unplanned meeting, at the Officers Club at Eglin AFB in late 1962, Boyd went into his attack mode and told Hillaker *I hear you designed the F-111?* Hillaker responded that, yes, he had led the team at General Dynamics that designed the F-111. In his characteristic abrasive manner, Boyd told Hillaker; *That thing is a piece of (expletive). It's too big to be a fighter and that goddamn little wing it's got, it must take two states to turn the thing around. I'll tell you something else. The pilot can't see behind and he can't see out the right window. He has to depend on his copilot to tell him what's out there.* From there, the conversation between Boyd and Hillaker had only one way to go – up. Boyd described his E-M theory and Hillaker ate up every word of it. What those two accomplished later as a team with the Lightweight Fighter and F-16 design has become legendary in US aviation history.] As this book unfolds, the role each of the Lightweight Fighter Mafia "soldiers" played in the development of the USAF Lightweight Fighter will be described in more detail.

The rest of this chapter will relate the pertinent highlights from the air war over Vietnam during the period 1963 to 1968 which provided very valuable lessons learned and pointed to technology improvements needed for future USAF fighter aircraft designs.

The Air War Over Vietnam

In the big picture, the USAF flew 5.25 million sorties over South Vietnam, North Vietnam, Laos, and Cambodia, losing 2,251 aircraft. Of those 2,251 losses, 2,197 were fixed-wing aircraft, and 54 were rotary-wing aircraft (the majority being the CH/HH-3 Jolly Green Giants). In another statistical category, 1,737 USAF aircraft were lost to hostile action and 514 were lost due to accidents. Of the 1,737 USAF aircraft lost to hostile action, over 55 percent (861) of the F-4 Phantom IIs and F-105 Thunderchief's were destroyed in air-to-air combat (by Soviet-made MiGs) and by enemy ground gunfire or SAMs. 382 USAF F-4 Phantom IIs were destroyed by enemy ground fire.

The records kept for the number of air-to-air kills of aircraft (both US Military and UN coalition air forces records - and North Vietnamese Air Force claims) were compiled after the end of the Vietnam War and are polluted with many unconfirmed kills by the pilot community. As best as can be determined, 119 F-4 Phantom IIs were destroyed by MiGs in air-to-air combat, and 108 MiGs were destroyed by F-4 Phantom IIs in air-to-air combat.

In the next table shown, the score-sheet displays the statistics for what type MiG was killed by which variant of the F-4 Phantom II and what type armament/missile was used to achieve the kill (where as best could be determined from later inspection of USAF and USN records, and North Vietnamese claims).

Aircraft	Weapons/Tactics	MiG-17	MiG-19	MiG-21	Total
F-4C	AIM-7 Sparrow	4	0	10	14
	AIM-9 Sidewinder	12	0	10	22
	20 mm gun pod	3	0	1	4
	Maneuvering tactics	2	0	0	2
F-4D	AIM-4 Falcon	4	0	1	5
	AIM-7 Sparrow	4	2	20	26
	AIM-9 Sidewinder	0	2	3	5
	20 mm gun pod	4.5	0	2	6.5
	Maneuvering tactics	0	0	2	2
F-4E	AIM-7 Sparrow	0	2	8	10
	AIM-9 Sidewinder	0	0	4	4
	AIM-9+20 mm gun	0	0	1	1
	20 mm gun	0	1	4	5
	Maneuvering tactics	0	1	0	1
	Total	33.5	8	66	107.5

Weapons credited for MiG Kills by F-4 C/D/E
(Source: USAF Military records)

Reversing the roles, shown in the next table were the number of F-4s (of all variants and of three US military services (USAF, USN, and USMC)) destroyed during air-to-air engagements with MiGs (MiG-17, MiG-19, and MiG-21) operated by the North Vietnamese Air Force. The table is presented in chronological order for the years 1965-1972. [The Author recommends that any Reader wishing to know more about the air war over Vietnam for the entire period 1965-72 should read the classic account given in factual detail and outstanding description by Marshall L. Michel, III in the book *Clashes – Air Combat over North Vietnam 1965-1972* published in 1997.]

There were no official records found or valid claims expressed that any of the MiG-21s used in the Vietnam air war were flown by Soviet pilots. The North Vietnamese claims are that all kills of US aircraft achieved by Soviet-made MiGs were made by North Vietnamese pilots. However, there was every probability that Soviet pilots were in North Vietnam as advisers, instructor pilots, and air combat maneuvering training instructors.

Year	Number of F-4s Downed	By MiG-17 (Cannon)	By MiG-19 (Cannon)	By MiG-21 (Atoll Missile)
1965	6	6	0	0
1966	13	7	0	6
1967	30	20	0	10
1968	10	3	0	6
1969	0	0	0	0
1970	1	0	0	1
1971	3	0	0	3
1972	56	2	7	47
Total	119	38	7	73

MiGs' weapons used for F-4 C/D/E kills in Vietnam
(Source: USAF Military records)

If only the totals shown in the two foregoing tables of kills during the entire Vietnam War are considered (MiGs killed by F-4s and F-4s killed by MiGs), it could be concluded that the kill ratio derived from 107.5 F-4s killing MiGs (as in the first table) and 118 MiGs killing F-4s (as in the second table) is close to 1:1 for both the F-4s and the MiGs. That is hardly to be considered as a good kill ratio for the overall US military aircraft in the Vietnam air war. The second table also shows that as the air war over Vietnam was in its later stages of the war, the MiG-21s in 1972 were cleaning the clock of the F-4s. Noting from the first table that F-4 kills of the MiGs were best achieved by Sparrow and AIM-9 missiles is an indicator that maneuvering of the F-4 to get a cannon shot at the MiG was not often achieved and that missile kills were necessary to overcome the deficiencies in the F-4-gun system. With the Boyd energy–maneuverability comparison chart shown earlier, it was shown that in the region between Mach 0.8 and 1.2, the maneuverability potential of the F-4 and the MiG-21 was comparatively equal, and this too adds to the explanation of there being a virtual stand-off in achieving a kill of the MiG-21 by cannon fire of the F-4.

Unlike World War II, which produced so many air aces (having 5 or more victories in air-to-air engagements) as shown earlier, and even fewer air aces in the jet aircraft of the Korean War, the air war over Vietnam produced very few aces. Shown next is a listing of the low number of confirmed aces produced during the air war over Vietnam in the period 1965 – 1973.

United States

USAF Capt. Charles B. "Chuck" DeBellevue (WSO)	6 victories
USN Lt. Randall "Duke" Cunningham	5 victories
USN Lt j.g. William P. Driscoll (RIO)	5 victories
USAF Capt. Steve Ritchie	5 victories
USAF Capt. Jeff Feinstein (WSO)	5 victories

North Vietnam

Nguyen Van Coc	9 victories
Nguyen Van Bay	7 victories
Nguyen Duc Soat	5 victories
Vu Ngoc Dinh	5 victories

(Source of Vietnam aces: USAF Military records)

By coincidence, the aces listed above scored 26 total victories for the United States and an equal 26 total victories for North Vietnam. The North Vietnamese Air Force claims having twelve more aces than those four shown, but those additional twelve aces have not been confirmed from official records. USAF Col. Robin Olds was a triple ace who flew in the Vietnam War, but only four of his sixteen kills were from Vietnam, the other twelve were from World War II and Korea.

The more important statistic about the F-4 Phantom II operational performance during the Vietnam War was its high losses due to enemy ground gunfire and to SAMs. There were 445 F-4s and RF-4Cs lost in the Southeast Asia conflicts (primarily the Vietnam War) during the period 1965 to 1972. 382 of these were F-4s of the C, D, E, and J variants and 76 were RF-4Cs attributed to being lost in combat (air-to-air and air-to-ground operations). Shortly after the F-4s entered into the air war over Vietnam and were sustaining the high loss rate to several different kill mechanisms (small arms/automatic weapons, anti-aircraft guns, surface to air missiles, and to the MiG aircraft use of cannons and air-to-air missiles), the USAF activated an organization at Wright-Patterson AFB to assess what were the features in the F-4 design which made it so vulnerable to the different threats and kill mechanisms.

The Survivability/Vulnerability Branch of the Vehicle Equipment Division in the AFSC Flight Dynamics Laboratory at Wright-Patterson AFB was originated and chartered with conducting Battle Damage Assessments of the USAF F-4s in Southeast Asia shortly after the USAF observed the high rate at which the F-4s were being lost. A technical report AFFDL-TR-77-115 *A Comparative Analysis of USAF Fixed-Wing Aircraft Losses in Southeast Asia Combat*, generated by the Survivability/Vulnerability Branch of the AFFDL Vehicle Equipment Division, was released in December 1977 - 5 years after the Vietnam War ended. The original version of AFFDL-TR-77-115 was classified as a secret document when first issued in 1977 - understandably, as it revealed the deficiencies and weaknesses in a USAF weapon system. However, AFFDL-TR-77-115 was subsequently declassified (some information was removed from the report) and Approved for Public Release with unlimited distribution under the Freedom of Information Act in June 2002. This full report is available on the internet as ADC016682.

In Table 7 shown next and Table 8 (shown below Table 7), both extracted from AFFDL-TR-77-115, are shown the number and type of kills experienced by USAF F-4 aircraft, and what damage mechanism caused the resulting loss of the F-4 aircraft. (Security classifications blacked out and downgraded to unclassified have been removed to avoid question about the classification status of the information and if it had been approved for public release).

The following abbreviations were used throughout the AFFDL-TR-77-115 Report for each of the several USAF aircraft assessed for the Battle Damage Assessment and serve to clarify the symbols listed in the following Table 7 and Table 8.

1) To designate the source of the projectile/fragment causing the damage:
   AAA – AntiAircraft Artillery
   MIG – MiG Aircraft
   SA/AW – Small Arms / Automatic Weapons
   SAM – Surface-to-Air Missile
   UGF – Unspecified Ground Fire

2) To designate the severity of the kill damage:
"A" – the aircraft flew 5-50 nautical miles after being hit
"B" – the aircraft flew more than 50 nautical miles after being hit
"K" – the aircraft flew less than 5 nautical miles after being hit

TABLE 7

F-4, THREAT CLASS VERSUS KILL SEVERITY

	"K"	"A"	"B"	UNKNOWN	TOTAL	PERCENT
SA/AW	45	22	4	11	82	22.9
AAA	82	64	37	29	212	59.2
SAM	13	10	2	3	28	7.8
MIG	17	12	5	2	36	10.1
TOTAL	157	108	48	45	358	
PERCENT	43.9	30.2	13.4	12.5		

Table 7 from AFFDL-TR-77-115
(Source: USAF Military records)

TABLE 8

F-4, THREAT CLASS VERSUS REASON FOR CRASH, 1971-1973 (GROUND FIRE ONLY)

	LOSS OF CONTROL	CREW/ CONTROL	CREW	LOSS OF PROPULSION	ENGINE FIRE	FIRE/ EXPLOSION	MISC.	INSUFFICIENT DATA	TOTAL	%
SA/AW	0	0	1	1	2	3	2	1	10	15.4
UGF	1	1	0	0	0	2	2	16	22	33.8
AAA	6	1	0	7	3	8	2	6	33	50.8
TOTAL	7	2	1	8	5	13	6	23	65	

Since more than one lethal event may occur in a single aircraft loss, the numbers shown in this table are not necessarily mutually exclusive.

Table 8 from AFFDL-TR-77-115
(Source: USAF Military records)

A statement of the consolidated experience of the F/RF-4 is extracted also from AFFDL-TR-77-115 to relate the significance of the Battle Damage Assessment made by the BDA assessment team of the Survivability/Vulnerability Branch of the Flight Dynamics Laboratory is shown below (direct extraction from AFFDL-TR-77-115):

"3. F / RF-4 CONSOLIDATED EXPERIENCE

The threat spectrum encountered in all areas of Southeast Asia changed considerably during the 1971–1973-time frame as compared with that of prior years. In North Vietnam, SAMs and MiGs accounted for only 17.8% of the F/RF-4 losses in the earlier years but during the 1971–1973-time frame, 68.5% of the F/RF-4 lost were downed by SAMs and MiGs. At the same time, small arms/automatic weapons activity decreased markedly and more emphasis was placed on a coordinated AAA-SAM-MiG defense. In South Vietnam, AAA activity increased in 1971-1973 and the introduction of the SA-7 accounted for 15.3% of the F/RF-4 losses in this time frame. There was no appreciable change in the defenses in Laos. In summary, the North Vietnam air war started as an AAA war with sporadic use of SAMs and MiGs and evolved into an integrated air defense, coordinating all phases for optimum effectiveness. In Laos, the threat spectrum remained fairly stable (mainly ground fire) with some emphasis on AAA weapons. In Cambodia the defenses consisted almost exclusively of small arms/automatic weapons. Early in the war in South Vietnam, small arms/automatic weapons were the statistical threat, but during the 1971–1973-time frame, a significantly high number of aircraft were lost to 23-37mm AAA and SA-7 missiles.

Aircrew members survived in 59.9% of the F/RF-4 losses. The highest survival rate was noted in North Vietnam (54.1%), the lowest in Cambodia (50%). Where kill severity could be determined in F/RF-4 losses due to SAMs and MiGs, 87.9% of the aircraft flew 50 nautical miles or less. Over 48% were "K" kills and over 39% were "A" kills. Under these conditions, crewmembers experienced an extremely high survival rate, 85.9% for "K" kills 82.7% for "A" kills.

Where the reason for crash could be determined for F/RF-4s downed by ground fire, 38% were lost due to fire/explosion, 23.5% due to engine damage/fire, and 13-14% due to flight control damage.

These results compare favorably with those generated in Reference 1*."

*[Reference 1 was AFFDL-TR-72-15, "Analysis of USAF Fixed-Wing Aircraft Losses, Aircrew Casualties, and F-105 Damages in SE Asia Combat", P.C. Hewett et al, July 1971).

Another important Reference which provided important information that was used in the Lightweight Fighter Program – in development of a Survivable Flight Control System – was Reference 4.

Reference 4 was AFFDL-TR-75-12, "An Assessment of USAF BDART Data for A-1 and F-4 Series Aircraft", P.C. Hewett, March 1975)].

[Author's note: The Author received a Merit Promotion to Civil Service Grade GS-14 to transfer from the Advanced Concepts Branch in the Aeronautical Systems Division to the Aeromechanics Branch of the Flight Mechanics Division in the AFSC Flight Dynamics Laboratory (AFFDL) at Wright-Patterson AFB in August 1966. Of the several technology areas the Author was assigned in his new assignment in the Aeromechanics Branch, two were to (1) support the Battle Damage Assessment activities of the Survivability/Vulnerability Branch in the Vehicle Equipment Division of AFFDL; and (2) support the development of a Survivable Flight Control System (SFCS) technology effort underway in the Flight Control Division of AFFDL. The SFCS support activity led to the adaptation of a fly-by-wire Flight Control System to be demonstrated on an F-4 Phantom II, modified to test the selectable FBW FCS with the hydro-mechanical flight control system existing on the F-4 FBW prototype aircraft.]

The Chief Scientist at the Air Force Flight Dynamics Laboratory at Wright-Patterson AFB was William E. Lamar, Civil Service Grade PL-313. It was Bill Lamar's prime responsibility to coordinate the development of advanced technologies for the USAF flight vehicles (aircraft and missiles) that were being analyzed, tested, and readied for application in the Divisions (Flight Mechanics, Flight Control, Vehicle Equipment, Mechanical Subsystems, etc.) of the Flight Dynamics Laboratory. Bill Lamar was also the primary interface between the USAF Foreign Technology Division (located at Wright-Patterson AFB) and the Air Force Flight Dynamics Laboratory to assure that the Foreign Technology engineers and analysts were aware of the technology advancements being accomplished by AFFDL engineers and scientists, and conversely to assure that AFFDL engineers and scientists were aware of the findings of the Foreign Technology engineers and scientists on the developments by other countries posing a military threat to the USA.

The AFSC Research Laboratories at Wright-Patterson in the 1960s were comprised of

1) the Air Force Flight Dynamics Laboratory (AFFDL),

2) the Air Force Propulsion Laboratory (AFPL),

3) the Air Force Avionics Laboratory (AFAL),

4) the Air Force Materials Laboratory (AFML), and

5) the Aeromedical Research Laboratory (AMRL).

Another prime responsibility of the six Chief Scientists (one of which was Bill Lamar) in the AFSC Research Laboratories was to assure that their technology advancement efforts were relevant and responsive to the projected needs of the USAF Planning and Requirements Development organizations at the Pentagon, and also to the Program

Offices formed within the Aeronautical System at Wright-Patterson AFB for applying those emerging technologies to Weapons Systems Programs being produced under contract with the US military contractors.

With regard to the Foreign Technology Division (FTD), in early January 1968 FTD contacted Bill Lamar and requested that an engineer having a solid background in aerodynamics, systems integration, and flight testing of flight vehicles provide support to an unspecified activity at an unspecified location, and unspecified as to what the engineer was expected to do. Having earned the trust of Bill Lamar to do a professional job at any project assigned, the Author was requested by Bill Lamar (on a Friday morning) to report to that unspecified location on the following Monday morning and to plan to remain there at least 120 days in support of the FTD activity. Plans had already been made for travel on a special Mil-Air flight departing from Wright-Patterson AFB at noon on Sunday to Nellis AFB, Nevada, where there was a room reserved for overnight lodging at the Nellis AFB VOQ. The Author was to report to Nellis Base Ops at 0700 the next morning for another Mil-Air flight to the unspecified location.

Skipping the intriguing logistics, the Author arrived at Site Alpha near a dry lake bed somewhere in a rattlesnake and Gila monster heaven region and was met by a most gruff Security Police agent and taken to a room for a special security briefing about the test location and about what was being tested. Immediately after the briefing the group of 8 of us were taken to a hangar on the flight line and given another briefing as to why we had been called to be there. Inside the hangar were three aircraft which certainly looked strange – they looked like Soviet MiGs but they had the USAF Star and Bar emblem on their nose and aft fuselage. In the briefing a few minutes later, the three aircraft were identified as two Soviet-made MiG-15 FRESCOs (given the code names HAVE DRILL and HAVE FERRY) and a MiG-21 FISHBED E with code name HAVE DOUGHNUT). In the briefing, it was explained that these three aircraft had been "borrowed", that FTD planned to examine and flight test all three of the HAVEs, and that there was a lot of time to do whatever was decided had to be done – so long as it was completed within the next 120 days. Shown next is a photograph of the MiG-21 HAVE DOUGHNUT being examined and readied for a test out on the apron in front of the hangar.

MiG-21 FISHBED (HAVE DOUGHNUT) Area 51,1968.
(Source: USAF Foreign Technology files)

Readers probably are asking themselves “Didn’t the Soviets know we had one –or all three - of their MiGs?” There were precautions taken to minimize the probability that the Soviets would know that the USA had “borrowed” their aircraft and were testing it. In the first briefing we were told to listen to the sound of a Claxon horn which if it was a steady blare meant that a Soviet spy satellite would be passing over in about 20 minutes – and we had to get the MiGs (all 3 of them) back into their hangar within the next 10 minutes. The Soviet satellite was known to have a look-ahead capability of its sensors (optical, IR, etc.) of 8-9 minutes. But we then had to drive trucks to the spot on the apron where the MiG-21 had been parked. Readers should note in the photograph of HAVE DOUGHNUT on the previous page that the Sun overhead causes a very distinct shadow below the DOUGHNUT. That shadow is several degrees cooler than the surrounding openly-exposed concrete, and the IR sensors on the approaching Soviet satellite would certainly be capable of detecting the shadow and its shape as being exactly that of a MiG-21. Accordingly, the trucks to be driven to the spot of the shadow were to be parked on top of the shadow so as to break up its outline and look to be just a small group of trucks parked there – a matter of little military Foreign Technology value.

The Author had the specific assignment while at Groom Lake to evaluate the aerodynamics of the HAVE FERRY MiG-15 Fresco located there and being tested in parallel with the HAVE DRILL MiG-15 and the HAVE DOUGHNUT MiG-21 Fish bed. A photo of the HAVE FERRY MiG-15, "borrowed" for the flight test program, is shown next.

Soviet MiG-15 HAVE FERRY, Area 51, January 1968.
(Source: USAF Foreign Technology files)

As seen in the photograph, the Soviet aircraft wing designers applied wing fences (three fences seen above the top surface of the right wing in the photograph) to most of their fighter aircraft with swept-back wings. The purpose of the wing fences was to decrease the span-wise component of airflow over the top of the wing and which caused the flow to separate and create drag. The series of MiGs from the MiG-15 through the MiG-23 feature those wing fences but in different height, spacing and spanwise location for each wing design. To evaluate the effectiveness of the wing fences on the HAVE FERRY MiG-15, the Author opted to employ the flow visualization technique of tufts attached to the surface of the vehicle being tested. (Tufts are short segments of a flexible non-metallic material which follow the air flow in the boundary layer next to the aircraft external skin surface. Tufts used in low subsonic wind tunnel tests are commonly made of wool yarn, but a stronger material such as white nylon parachute-riser line is used in high- subsonic wind tunnel and some flight tests.)

To provide a dark background for photographic contrast, the Author painted the top of the left wing of the HAVE FERRY MiG-15 with black water-based paint. A requirement of the Foreign Technology exploitation team leadership was that since the "borrowed" MiGs had to be returned to the "Lender", nothing could be mechanically attached to the MiGs (externally or internally) or painted on leaving any paint residue that would reveal to the Soviet government that the aircraft had been flight tested. Hence, the water-based black paint that could easily be hosed-off after completion of the fence tuft tests.

A photographic summary of the tufts used is shown next.

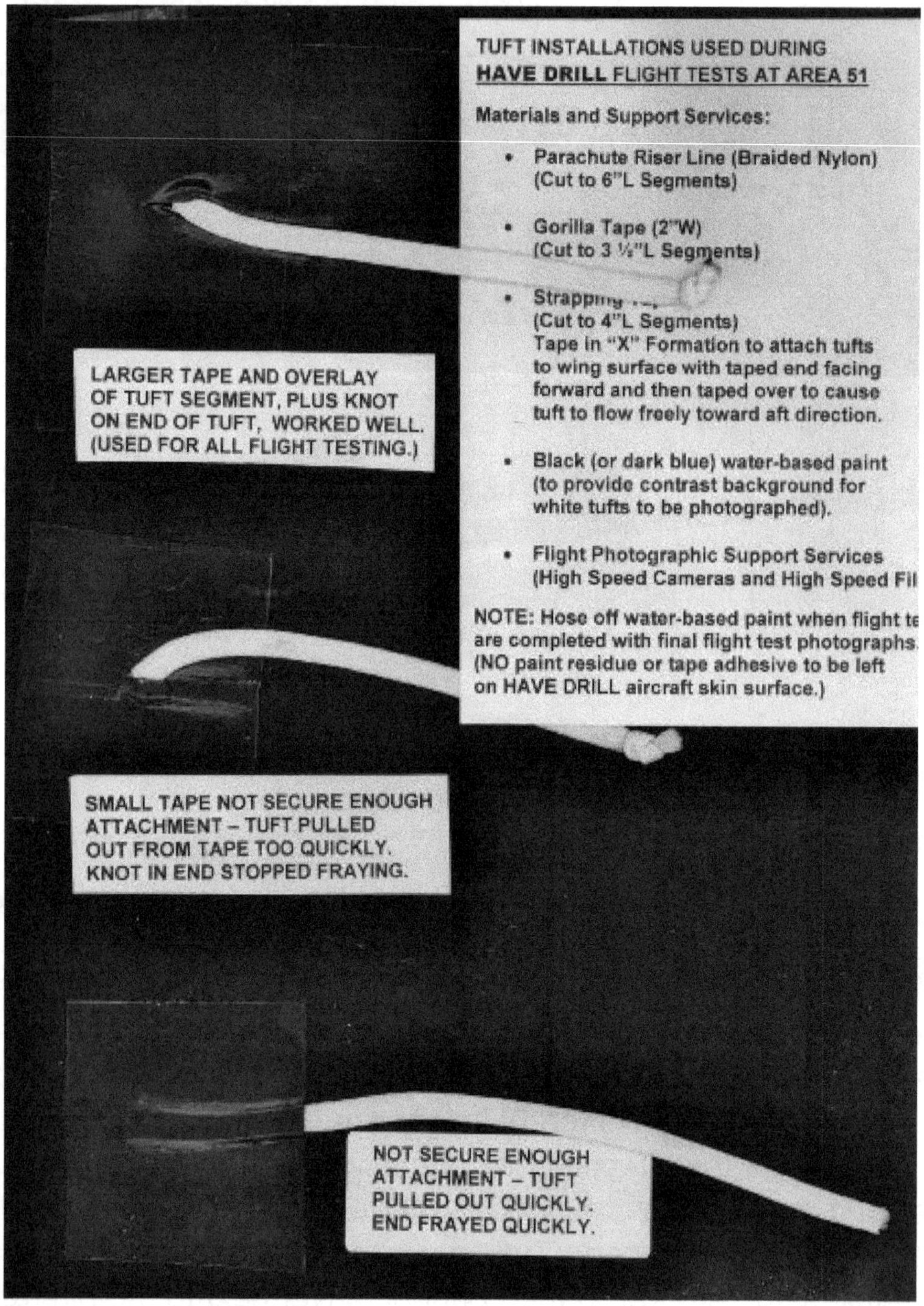

Flow visualization tufts used in Area 51 flight tests
(Created and applied by the Author)

The tufts were arranged in a pattern alongside both sides of each of the three fences and spaced so as to not get entangled when dancing. The test aircraft was then taken to a prearranged Mach number and altitude in combinations from Mach 0.5 to Mach 0.9 (in 0.1 Mach increments) and at altitudes from 20,000 feet to 40,000 feet (in 5,000 feet increments). This covered the Mach and altitude envelope within which air engagements with the MiGs were most likely to occur. At each stabilized flight condition, an Aerial Photographer in the rear seat of an F-4, hanging in his cockpit straps while the F-4 was inverted to give him a clear line of sight, took very high-speed photography of the left wing of the MiG instrumented with the dancing tufts. Analysis of the tufts showed that the wing fences did have a positive effect on improving the flow over the wings of the MiGs throughout the air engagement envelope, but the added friction drag of the fences and the weight of the fences were of approximately equal and opposite influence as their improvements on total MiG aircraft maneuverability capability.

When not engaged in aerodynamic exploitation tests on the HAVE DRILL and HAVE FERRY MiG-15s, the Author served as a "donkey" assisting the electronics engineers who were making radar cross section measurements on HAVE DOUGHNUT. The RCS measurement equipment was loaded on top of what is best described as a wooden "Trojan Horse" platform, and the platform was positioned at a large number of lateral aspect angles with respect to the HAVE DOUGHNUT MiG-21. Propelling the Trojan Horse were two able-bodied otherwise gainfully employed 1 manpower "engines" – one being the Author - positioning the Trojan Horse in its orbital travels around the HAVE DOUGHNUT vehicle being measured for RCS. There were many opportunities for applying resourcefulness at the Groom Lake test facility as, at that time, there was not much of anything available at the isolated Groom Lake facility for conducting instrumented flight testing or RCS measurements. The motto was "If you need something special for a test, go find it somewhere else and bring it in – with the utmost of security protection".

The next two charts are extracted from a Foreign Technology Division briefing now unclassified and released to the public under the Federal Freedom Of Information Act. The first chart below identifies the highlights of the HAVE DOUGHNUT exploitation project and the second chart after that identifies some of the MiG-21 shortcomings identified during the exploitation examination and testing.

Project Background

- The U.S. borrowed a Soviet-built MiG-21F-13 FISHBED E from 23 January 1968 to 8 April 1968
  - The Foreign Technology Division of AFSC led the exploitation utilizing expertise from AFFTC, ASD, TAC, NWC
  - The purpose was to substantiate and supplement existing threat data
  - Included ground and flight testing
  - 102 flights (77 flying hours) in 40 days of flying
  - The U.S. gave the jet back when they were done with it

(Source: Declassified Foreigh Technology Report - USAF Approved for Public Release FOIA)

What we learned

G-load factor	8gs without stores, 6 with stores
Max indicated airspeed	595 kt below 15,000 ft, 640 kt above 15,000 ft
Maximum indicated Mach	2.05 without stores, 1.6 with stores
Strike radius	370 NM with external fuel
Poor forward and rearward visibility	F-4 acquired at 3-5 miles range
Low Q limit	Below 15,000 limited to .98 Mach or 595kt – severe buffet
Afterburner puff	Above 15,000 FISHBED E produces a puff in/out of AB
Engine response	Extremely slow
Cockpit noise	Extremely low
Gunsight capabilities	3.7NM, missile mode; 1.6NM, gun. Gun/missile target tracking impossible over 3gs
Slow speed	The MiG-21 could maneuver at 115 KIAS
Easy to kill	Non-sealing tanks, unprotected engine, light metal structure, high pressure O2 bottles – 85% kill probability

(Source: Declassified Foreign Technology Report - USAF Approved for Public Release FOIA)

The feature of the MiG-21 (HAVE DOUGHNUT) that the Author found technologically most intriguing was the very long probe located on the nose of the aircraft under the inlet and extending far forward from the inlet spike as shown next.

MiG-21 "FISHBED" with long nose probe (1968)
(Source: USAF Foreign Technology files)

While its function for sensing undisturbed air data and angles of attack and yaw of the aircraft, it must have accounted for numerous injuries of ground crew personnel being stabbed when accidentally walking into the tip of the probe. The Author had some difficulty seeing the USAF Star and Bars symbol on the sides of the two MiG-15 Frescos (HAVE DRILL and HAVE FERRY) and on the MiG-21 Fish bed (HAVE DOUGHNUT). Although commercial airlines flights were forbidden over the Groom Lake test facility (now identified as "Area 51"), there were a few instances where airline Captains passing near the restricted area having seen the MiGs flying in the distance during flight testing inquiring on their Comm system why MiGs are in USA airspace? The pilot of the F-4 chase aircraft advised the airline Captain "Forget what you think you have just seen."

An incident occurred one day, which Readers might enjoy reading, when the HAVE DRILL MiG-15 was engaged in simulated air combat with an F-4 Phantom II (piloted by an unnamed USN Top Gun pilot). In an evasive maneuver, the USAF HAVE DRILL pilot took the MiG-15 down to just a few feet above the dry lakebed (Groom Lake) and slowed down to the incredibly low minimum airspeed of the MiG-15. The USN F-4 pilot followed the MiG-15 down to the desert floor and tried for a "six-shot" up the MiG-15's nozzle, but failed to observe that his speed had dropped below minimum airspeed for the F-4. The F-4 stalled out and slid to a stop on the dry lakebed. The MiG-15 pilot flew back over the F-4 and after the mandatory victory wing-rock of the MiG-15 wings, landed and taxied back into its hangar. A flatbed truck and crane went out and lifted the F-4 onto the "lowboy" flat bed for return to the maintenance hangar where the F-4 was again lifted, its landing gear lowered, and the minor bottom fuselage skin damage repaired. No injuries were involved, other than to the USN Top Gun pilot's ego. The following morning, the HAVE DRILL MiG-15 was rolled out of the hanagar and on the side of the fuselage just below the pilot's cockpit entry was painted a twice-normal size F-4 kill silhouette. There are no symbols on the Author's keyboard that adequately capture the comments of the Top Gun victim who left in his repaired F-4 for Miramar NAS (home of the Top Gun facility) without saying a fond farewell to those on the flight line apron snickering behind his back.

The Area 51 housing facilities have recently been greatly upgraded for those who are assigned there on the inspection, measurements, and exploitation flight test team. At the time of the MiG HAVEs (DRILL, FERRY, and DOUGHNUT) in early 1968, the Author was assigned to a room in a Quonset-Hut-type structure. It was difficult getting any sleep, however, because the four walls were papered with Playboy Centerfolds courtesy of the previous occupant (name unknown, but can most likely be found on the Registry List of Sex Offenders for the State of Nevada). There was a parking lot light that illuminated the room for the entire hours of darkness, and it was impossible to fall asleep while being stared at by all those lovely faces – and all their external weapons. Despite those distractions, there was much learned about the MiG-21 Fish bed design during that short four-month period in early 1968 – virtually all of it applicable to the Lightweight Fighter Program performance goals (as part of the Know Thy Enemy database).

This Chapter 5 has now been completed with the information for, and lessons learned from, the TFX (F-111) Program, the John Boyd *Aerial Attack Study* and derivation of his Energy-Maneuverability Theory, the operational experiences of the F-4 Phantom IIs in the air war over Vietnam, and from testing of the Soviet MiG-21 Fish bed HAVE DOUGHNUT at the Groom Lake (Area 51) facilities. Chapter 6 will include the Author's description of a few other significant events that took place during the period 1963 to 1971 which were all entered into the development funnel of the USAF Lightweight Fighter Program.

CHAPTER 6
AX (A-10) STUDY, FX (F-15) STUDY, ADF STUDY, USAF PROTOTYPES STUDY. (1963 TO 1971)

The previous chapters of this book have described how the primary technologies of fighter aircraft designs – the aerodynamics configurations, the propulsion systems, the performance capabilities in terms of speed and altitude, and the weapons used in air combat engagements – are developed and advanced in a "leap frog" manner to stay ahead of the technologies being developed for the same purpose by the adversary countries. The names of the countries participating in wars change over the course of history, but the very nature of war changes little with time. There is much carnage and destruction done in countries as the geopolitical aims of misguided leaders of some countries caused wars to erupt as the fastest means for those misguided leaders to achieve their aims. Early wars had primarily either or both of two purposes – to acquire the land of neighboring countries and/or to acquire the natural resources of other countries needed by the aggressor country Such was the case of World Wars I and II.

The Cold War that emerged after World War II was more of a war of ideals and cultures between Communist-controlled countries and those having more democratic ideals and governments. The capitalistic goals of many democratically controlled governments are in direct opposition to those governments following the Communist doctrine which opposes capitalism and requires more dictatorial control over the population of the Communist countries. The mother countries for Communism were the Union of Soviet Socialist Republics (USSR), led by what was earlier known as Russia, and was joined by the several Asian countries neighboring Russia.

In essence the Korean War (1950 to 1953) and the Vietnam War (1963 to 1973) were wars to remove the control of democratic-controlled nations (the United States, Great Britain, and France) from Communist-controlled regions of the World. The military forces of North Korea and North Vietnam were based on build-up of massive manpower for the invasion of their targeted neighbors of South Korea and South Vietnam, but neither North Korea or North Vietnam had progressed technologically in their industrial capabilities to produce the advanced technologies of the weapons of war to support the ground forces.

The Soviet Union provided that missing element to North Korea and to North Vietnam in the form of Soviet-made aircraft, tanks, artillery pieces, automated small arms weapons. As a neighboring country to both North Korea and Vietnam, Communist China also provided military manpower and some weapons of war to their war efforts.

There is little doubt that there were both Soviet and Chinese pilots better trained than were North Korean and North Vietnamese pilots in the Soviet-made aircraft in the Korean War and Vietnam War. Chapter 4 and Chapter 5 have shown how the technologies of Soviet-made fighter aircraft and those made in the United States have leapfrogged during the period 1950 to 1973. This chapter is intended to describe additional activities that took place within the US Department of Defense for the generation of aircraft used in the Vietnam War and leading to the concept of a low-cost, low-complexity, highly-maneuverable fighter aircraft capable of defeating the primary threat aircraft – the Soviet MiG-21.

As in most aircraft development programs, there is a mythical funnel into which all of the pertinent events and events that have occurred prior to and in parallel with the new development are accumulated, digested and integrated into the new design. In previous chapters, readers have been introduced to the General Dynamics F-111A Aardvark and the McDonnell F-4 Phantom II programs and their origins. Both had been originated with the design and acquisition approach of commonality – using the same basic design of an aircraft to satisfy the operational needs of both the USAF (land-based) and the USN/USMC (aircraft carrier-based) inventories of aircraft. The several major differences in operations (the USN/USMC carrier-based designs necessarily being more complex and with heavier structure. The additional structural strength and weight enables carrier-based aircraft to endure the high peak forces associated with the accelerations induced into the aircraft structure and landing gear subsystem by catapults (for take-off from the short carrier deck), and the decelerations into the aircraft aft structure (through the trap of the arresting hook engaging a deck arresting cable).

On USN aircraft carriers, real estate is far more limited than is the real estate at USAF land bases of operations. Another complexity and weight producer is the need for the outer portions of wings of carrier-based aircraft to be capable of folding (upward on a powered hinge) to decrease the wing span and improve the spotting factor when on the carrier launch deck and decks below for storage, maintenance, and servicing of the aircraft aboard the carrier.

Accordingly, the same basic design for the aircraft is achieved by producing that design in variants (e.g., the USAF F-111A and the USN/USMC F-111B). While the USAF A and the USN/USMC B variant may be equipped differently, the landing gear subsystem and the aft fuselage structure is essentially the same for both variants, resulting in the USAF A variant carrying some additional structural weight and complexity as a compromise due to "commonality". As stated in Chapter 5, the USN/USMC B variant of the Joint Services F-111 Program was cancelled due to the size and weight of the USN/USMC F-111B.

Both the USN/USMC variant and the USAF variant achieving only a 1:1 exchange ratio in air combat engagements. Coupled with the much lower than impressive performance of the commonality designed McDonnell F-4 Phantom IIs, the USN proceeded down a different design path for a new fighter not compromised by commonality. That led to the acquisition of the Grumman F-14 Tomcat for the USN fleet defense requirement and the USMC requirement for an overland air-to-air and air-to-ground capability in support of US ground forces (USMC and Army).

The USN divorce from the commonality approach, and the recognition by HQ USAF planners and requirements organizations that the USAF needs for an air superiority fighter and a fighter- attack aircraft to support ground forces, HQ USAF proceeded with the design and acquisition of an FX air superiority fighter and an AX ground support aircraft

AX (A-10) Attack Aircraft and Survivability/Vulnerability Features

As a result of the damage done to US aircraft, especially the F-4, due to the large extent of enemy ground fire in the Vietnam War, the USAF began studies of an attack aircraft (AX) in the early 1960s with competitive design contracts issued to the aircraft industry. The scope of this book does not allow detailed coverage of what transpired for the AX aircraft, which saw a fly-off competition of two prototypes – the Northrop YA-9 and the Fairchild Republic YA-10 – and which the USAF ultimately selected the Fairchild Republic YA-10 as the better (i.e., having better-designed survivability/vulnerability features) of the two designs. A major reason for theYA-10 being down-selected for further development was in the survivability and vulnerability features of its configuration. The external configuration of the Northrop YA-9 shown in the photograph next reflects a deficiency in the vulnerability of the YA-9 configuration.

Northrop YA-9 attack aircraft prototype, 1965.
(Source: USAF Photographic files)

As can be seen, having the engines mounted in the "armpits below the wing and having large inlets acting much like large funnels ingesting debris from operation very close to ground targets, the vulnerability of the YA-9 to ground fire (small arms/automatic weapons, AAA artillery, and SAMs) was much higher than it was for the Fairchild Republic YA-10.

A photograph showing the Fairchild Republic YA-10 external configuration is presented next.

Fairchild Republic A-10 Thunderbolt II, 1967.
(Source: USAF Photographic files)

To especially be noted is the placement of the engines in the aft region of the fuselage and with the inlets above the plane of the wing. This unusual placement of the engines permits the wing to act as a shield protecting the inlets from ingesting much of the ground fire projectiles or fragments.

Only a few features designed into the USAF A-10 Thunderbolt II, more popularly referred to as the Warthog, will be described here. Readers are certainly encouraged to read more about the A-10 as there are hundreds of accounts as to how well it did in the ground wars that have taken place in the past thirty years in the Middle East.

The A-10 was built around the carriage of the 30mm GAU-8 Gatling cannon which gave it a place in aviation history for its unprecedented ability to destroy enemy tanks and armored vehicles. The large gun muzzle is seen in the above photograph projecting forward from the lower nose of the fuselage. To give relative size of the GAU-8 Avenger cannon, a photograph next shows its size compared to a Volkswagen Beetle.

GAU-8 Cannon size comparison with Volkswagen Beetle (Source: USAF Photographic files)

Integrating the large GAU-8 cannon was no small design challenge. Shown next is a cutaway diagram of the forward fuselage on an A-10 with the GAU-8 installed:

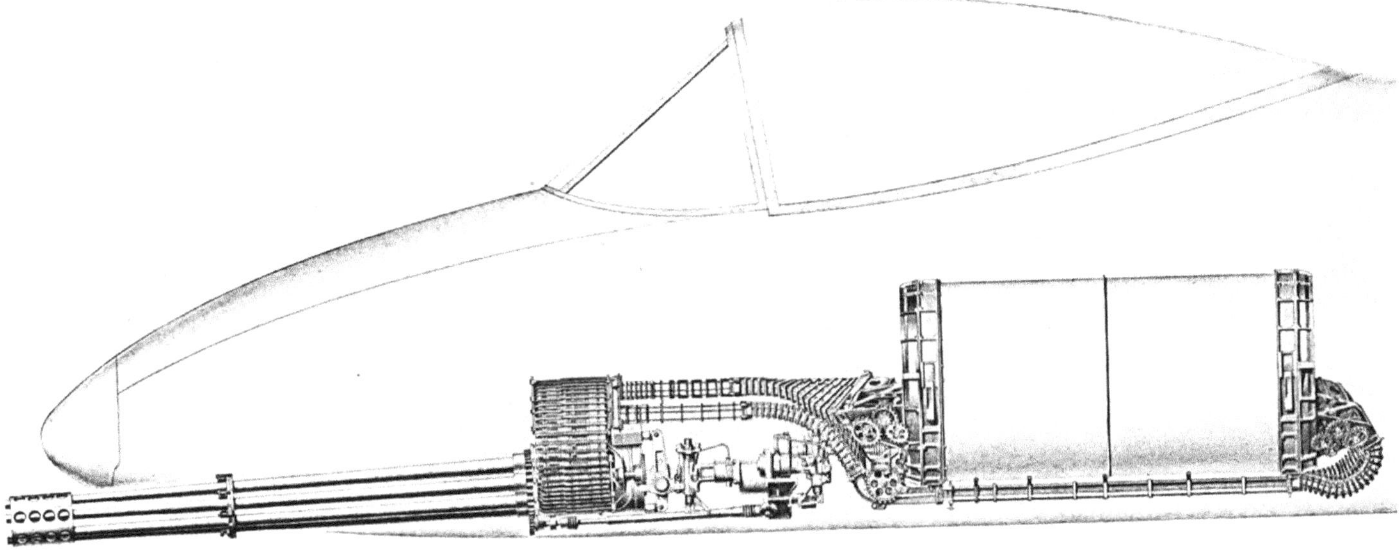

GAU-8 Cannon integrated into the A-10 fuselage (Source: USAF Engineering files)

An Inboard Profile of the A-10 is shown next to indicate how critical components were arranged within the A-10 to afford some protection from expected ground gunfire when in the ground attack mode of tanks, armored vehicles, and static emplacements.

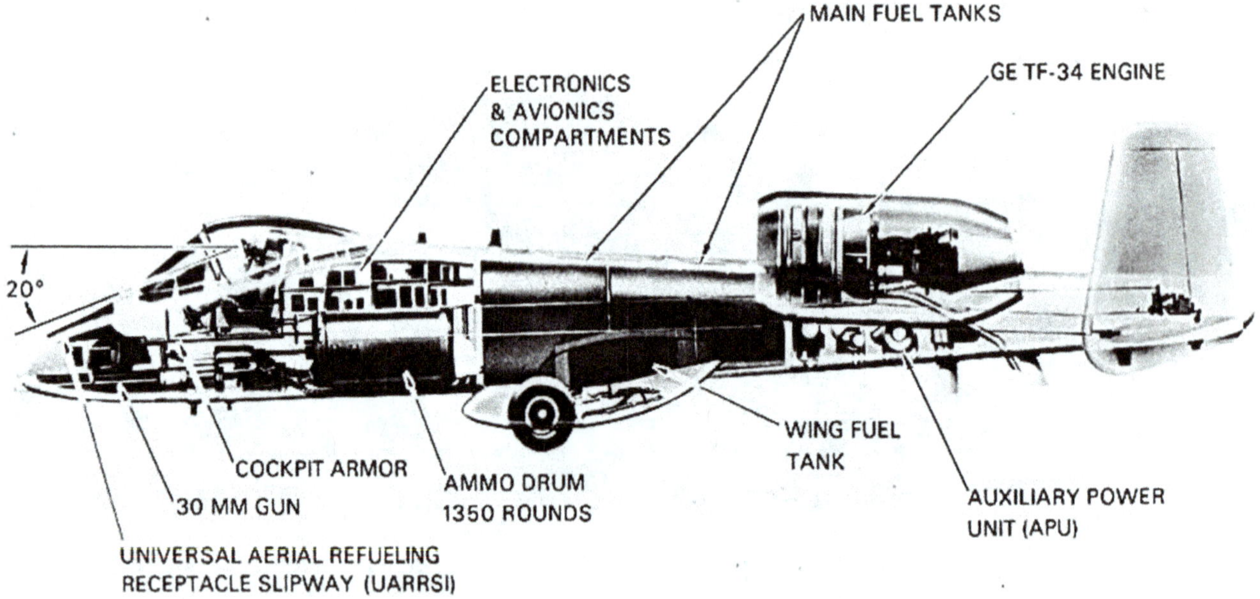

Internal arrangement of vulnerability critical components
(Source: USAF Engineering files)

The pilot sits in a titanium bathtub-like enclosure and which also protects the critical area where control cables of its hydro-mechanical flight control system are attached to the bottom of the control column. All fuel tanks are self-sealed in the event a round of gunfire or a nearby missile burst sends a fragment into a fuel tank.

Shown next is a frontal view of the gun muzzle (7 revolving barrels, Gatling type) and providing a comparison of its size relative to a ground crewmember holding a 30 mm round fired electrically (i.e., no mechanical percussion hammer and firing pin.)

Relative Size of GAU-8 30 mm Round and 7 Barrel Muzzle
(Source: USAF Photographic Files)

The size of the GAU-8 round is thirty millimeters. Each round has a depleted uranium core surrounded by an aluminum jacket. This round is a wondrous step in airborne cannon technology not merely because of its large size and high firing rate, but the round when impacting the armor of a tank will create enormous heat which melts the aluminum jacket and acts as a lubricant for the depleted uranium core to penetrate the thick steel armor. The round has a time-delayed fuse to explode after it has penetrated into the tank crew compartment. Surprisingly, when the GAU-8 is in its firing mode (at a rate of 3,900 rounds per minute – in bursts of about 1.5 to 2.0 seconds from an ammunition drum holding 1,150 rounds), there is no high-amplitude vibration induced into the A-10 airframe. Each firing burst expends an average of approximately 110 rounds per burst.

The early A-10s had a problem with corrosive gun-gases entering into the access panel covering the forward avionics bay. The access panel was manufactured to closer sealing tolerances and the problem of avionics failures due to gun gas corrosion was resolved. Shown next is the A-10 in a gun-firing mode:

Fairchild Republic A-10 in GAU-8 Gun firing Mode (1972)
(Source: USAF Photographic Files)

[A street-language description of the sound of the GAU-8 gun-firing – with the Author's pre-apology if offending any Reader - is that it sounds like King Kong passing gas. Quite honestly, that is exactly how it sounds and the Author has heard it many times.] There is almost imperceptible vibration during gun firing, and there is a significant recoil force acting to decelerate the A-10 when gun firing. A recoil mechanism reduces the recoil forces into the airframe to be almost negligible, and the deceleration of the airplane does not create a flight safety problem. The A-10 speed deceleration is barely detectable by the pilot.

Originally, the reloading of ammunition in the ammunition drum was manually done; however, that time-consuming and hazardous practice was replaced with an automatic loading system. When rounds are fired from the ammunition drum, the expended casing is cycled back into the drum. In the downloading and reloading process, the empty casings are cycled out of the drum and the loaded rounds are cycled into the drum. In addition, the A-10 was designed to carry highly effective external loadings of air-to-ground missiles and bombs to 16,000 pounds as shown in the photograph next.

A-10 with large weapon load on wing hard-point stations
(Source: USAF Photographic files)

[Author's note: The conduct of the AX/A-10 program using competitive fly-off of two different prototypes - Northrop YA-9 and Fairchild Republic YA-10 - served to inspire the USAF to use essentially the same approach (to be described in detail later in this book) for the USAF Light Weight Fighter Prototype Program, which saw the invaluable fly-off between the General Dynamics YF-16 and Northrop YF-17 prototype fighters).

The A-10 was built with an extreme amount of survivability (and decreased vulnerability) protection to ground fire based on lessons learned from the heavy losses of F-4s (both USN/USMC and USAF variants) in the Vietnam War. Void-filler foam was installed in cavities in the fuselage to deter the accumulation of explosive fuel/air mixtures in the event of fuel leaking from any damage done to the fuel tanks. U-shaped channels were set in place beneath the flight control cable runs to protect the control system cables from ground fire. (Many of the "stuffing the sausage – installing components and equipment - for improved survivability/vulnerability" considerations learned from the A-10 were applied in the USAF Light Weight Fighter Program.)

Shown next is a photograph of an A-10 which was heavily damaged by ground fire from both small arms fire (and from apparently AAA-fire to the leading edge of the starboard horizontal tail), but survived to RTB (return To base) primarily because of its very carefully thought-out survivability/vulnerability design features.

A-10 Warthog with extensive ground gunfire damage
(Source: USAF Photographic files)

Added in the upper left corner of the photograph is a patch worn by those (including the Author) who were assigned to an A-10 Accident Investigation Team after five A-10s were lost when on training missions during ground support exercises – in a no-threat environment.

Several A-10 pilots complained in post-mission debriefings that on numerous occasions there were unusually high friction forces experienced in their control column such that there was loss of positive rapid control of control surfaces (and especially in ailerons) for roll angle control. During a flight, a pilot called in declaring an emergency such that he had to put one foot up on the instrument panel to achieve enough leverage to control the A-10 in pitch angle. The pilot's flight boot left a dent in the instrument panel and cracked the glass lens on one of the instruments in the flight panel. The pilot was able to land (at Barksdale AFB) and the aircraft was impounded for investigation.

A thorough investigation of the flight control system revealed that a large threaded bolt had somehow been introduced into the U-channel beneath the flight control cable (protecting the cable from ground fire) and had become lodged under a bell crank converting longitudinal (parallel to the fuselage axis) motion of the cable into lateral motion of the cable out to a wing aileron. The bolt was considered as being FOD – any object which causes foreign object damage (or loss of function) – but it was unknown where the bolt had come from. By inspecting the thread on the bolt - and matching thread markings on the bell crank and the length and head configuration of the bolt FOD – inspectors concluded there was no bolt of that configuration which had been installed anywhere in the A-10 during its manufacture (at Gaithersburg, Maryland). After much sleuthing and several interviews of ground maintenance personnel, one crew member identified the bolt as being of the same type that was used as a pin to support a rod that held the access panel over the avionics bay in an open position. Another crew member admitted that sometimes the rod support pin would be dropped down into the avionics bay and could not be found – with the result being that a replacement for the unusual configuration bolt had to be drawn from the Tool and Supply crib. The FOD bolt had jumped through pass-through holes in two bulkheads during the A-10 maneuvering and traveled about seventeen feet down through the U-shaped channel and became lodged under the bell-crank located near the wing root – fuselage junction. By changing maintenance instructions and a small redesign of the rod support attachment, there were no repeats of that type of FOD incident.

However, other pilots also reported friction being higher than normal in their control column and described that they had just descended into the ground attack mode when the increased friction was first sensed. As the control column was moved, eventually the friction decreased. In this case, the Accident Investigating Team found that considerable rain water (especially at bases where the A-10s were based and frequently exposed to rain storms such as Bent waters Air Base, England; Barksdale AFB, Louisiana; and Eglin AFB, Florida) was found internal to the A-10 fuselage. After much investigative analysis, it was found that rainwater, entering through openings between fuselage skin panels and around access panels, had soaked into the foam pieces that had been installed as void-fillers (to reduce explosions in cavities where an explosive ratio of fuel and air had accumulated).

When the aircraft were taken to altitude enroute to the target area, the water-soaked foam pieces froze and were breaking their adhesion to the inner skin surfaces, and chunks of the frozen foam were also dropping into those U-shaped channels protecting the flight control cables from ground fire. In effect, the ice blocks were FOD, and the flight control cable had to be moved much like a saw blade to cut through the ice block FOD eventually reducing the friction sensed by the pilot in his control column. The method of attaching the void-filler foam was changed and a recommendation made that the gaps allowing rainwater entering the fuselage of the A-10 be sealed (with a common fish tank fluorocarbon sealant). A recommendation that lightweight plastic cap plates be installed over the top of the U-shaped channels to enclose them against FOD and/or rain water intrusion was not approved.

It is recognized that the above lengthy details of FOD-induced damage and accidents (of any type and source) may not be of interest to all Readers, but they should be noted carefully by those Readers who are planning on a career in System Engineering. It is usually an unexpected small detail that creates big problems. It is the seemingly insignificant details, such as dropping a bolt inside an aircraft and give up searching for the bolt, that must always be considered as being significant until proven otherwise. While a technology feature may be of benefit to one application, it may become a problem to a different technology application.

Norman Augustine compiled an amusing set of Murphy's Laws, and while very tongue-in-cheek humorous, a Murphy's Law states: If it can happen, no matter how low a probability, it will. Another corollary to that Murphy's Law is: When the impossible problem happens, it will happen at the worst possible time. Such was the case with several A-10 accidents happening when the A-10s were in the attack mode only 50-60 feet above a target and rolling into a near 90-degrees bank to evade damage debris thrown up from the damaged target. High friction in the flight control system may cause the pilot to be unable to reduce from a 90-degrees roll maneuver close to the ground and fly into the ground when unable to reduce the roll angle command quickly enough.

For those Readers also who may at some point in their career be responsible for installing equipment items or components into the fuselage of a flight vehicle, always think about what might happen if an adjacent piece of equipment (someone else's responsibility) moves out of its position (due to a loose attachment bolt, or due to damage by a gunfire round) becoming a piece of FOD. Respect the Laws of Physics as well as those of Murphy. The Author fanatically observed them to the letter during the development of the Light Weight Fighter Prototype aircraft, and even more so during the transition of the advanced technology demonstrator YF-16 as it was being missionized into the operational F-16 Air Combat Fighter.

The effects the Cold War had on military aircraft design (airframe and propulsion systems) was accentuated starting in the late 1950's, continued throughout the decade of the 1960s, and lasted well into the decade of the 1970s. When the B-58A Hustler Supersonic Bomber program was in its development test phase in the late 1950s (being designed to replace the slower B-52 Bomber), and the Department of Defense had studied the need for what was to become the Mach 3 high-altitude XB-70 Valkyrie interceptor, on May 1, 1960, the Soviets shot down a high-altitude (70,000 feet) CIA U-2 reconnaissance aircraft (piloted by Francis Gary Powers) with an S-75 Dvina (SA-2 Guideline) surface-to-air missile (SAM). The US Government (President Eisenhower and Secretary of State Allen Dulles) tried to conceal the details of the U-2 aircraft mission and told the World it was a U-2 performing meteorological research. However, the pilot Gary Powers had ejected and was captured alive by the Soviets, and had confessed to being on a spy mission not meteorological research one. Remnants of the fuselage of the U-2 were photographically distributed World-wide as well as photographs of Gary Powers and his statement of confession.

While the circumstances of the U-2 shoot-down event were devastating to the credibility of the US Government, the realization by US military technology specialist (Foreign Technology at Wright-Patterson AFB and those at the Pentagon) that the Soviets had a SAM capability up to at least 70,000 feet altitude cast serious doubts on the military value of "faster and higher" being the path which future military aircraft should be taking. The emphasis swung toward a high-speed high-altitude aircraft being able to also have a subsonic low-altitude capability – sufficiently low that it could penetrate deeply into countries and go undetected. This is the origination of the variable sweep-wing technology for military aircraft – giving rise to both the F-111 fighter-bomber development (presented earlier in this book) and the B-1 bomber concept (which, being a bomber, is not included for detailed description in this book dealing primarily with technologies of fighter category aircraft).

While this change in the technology thrust for airframes and configurations capable of operating efficiently and safely in the diverse altitude and speed spectrums and in environments near the ground terrain (i.e., requiring terrain-following electronics technologies), an equal if not larger challenge was to the propulsion manufacturers to produce turbojet engines (albeit in multiple numbers different for bomber and fighter aircraft sizes and weights) of a thrust-producing capability matching the widespread operating environment. Engines operating at subsonic speeds at low altitude (where the air is dense and there are no inlet shock effects to be considered) may not be as challenging as operating at high altitude (where the air is very low density and at supersonic speeds in excess of Mach 2 where inlet shock effects are extremely challenging), but the necessity for having one engine cycle to accommodate the challenge of operating in both environments presented a formidable challenge to the engine manufacturers. At that time period, there were essentially only two major engine manufacturers of engines meeting the military demands for performance and safety – those being the General Electric Company in Lynn, Massachusetts, and the Pratt & Whitney (Military Engine Division) in West Palm Beach, Florida.

When any manufacturing company of any product enters into a competition with another manufacturing company developing a similar product, it is highly dependent on the upper management – having widely different experience base and thought-process – for the performance and cost of their respective products satisfying the Users need and affordability. That places emphasis on one of the underlying themes in this book. It is the personalities and the capabilities/experience of a few key people who determine the outcome of a technological advancement and for its future application. For the engine competition (a book titled *The Air Force and the Great Engine War* was written in 1987 by Robert W. Drewes (published by The National Defense University Press), and which is the most authoritative and factual account of the "engine war" between its two primary combatants – the General Electric development team led by Gerhard Neumann and the Pratt & Whitney development team led by Robert Carlson (under Harry Gray CEO of United Technologies Corp., Pratt & Whitney's parent corporation).

An outcome of the early phases of the engine war was that the General Electric Company began development of its F101-GE-102 engine, which would provide the power for the Advanced Manned Strategic Aircraft (AMSA – with tongue-in-cheek definition as America's Most Studied Aircraft), later to be developed as the B-1A Strategic Bomber. The B-1A had the variable sweep wings technology for the two challenging engine operating design regimes - high speed (Mach 2.5+) at high altitude, and subsonic speeds at low altitude, and with a terrain-following capability). There were four of the GE F101-GE-102 engines powering the B-1A. The GE F101-GE-102 was first run in early 1970 and was equipped with GE's first afterburning turbofan jet engine, generating 30,780 pounds of thrust in maximum afterburning. The F101 engine had an overall pressure ratio of 26.8:1 and an engine thrust- to-weight ratio of 7.04:1 (its weight was 4,400 pounds).

In 1967, the USAF and USN issued a joint engine request for proposal (RFP) for the Advanced Turbine Engine Gas Generator (ATEGG) Program. The ATEGG Program had as its goals to significantly improve thrust in combination with reduced weight to achieve an engine thrust-to-weight ratio of 9. In 1970, Pratt & Whitney was awarded a contract to produce the P&W F100-PW-100 engine for the USAF F-15 Eagle program and to produce the P&W F401-PW-400 engine for the USN F-14 Tomcat program. The specifications of the developed P&W F100-PW-100 engine are summarized in the following table:

F100-PW-100 Specifications:

Thrust: 23,770 (max no afterburner) –29,160 (max afterburner) lb.
Weight: 3,740 lb.
Engine Thrust-to-Weight Ratio: 7.8:1
Length: 191 in
Inlet Diameter: 34.8 in
Maximum Diameter: 46.5 in
Bypass Ratio: 0.36
Overall Pressure Ratio: 32 to 1

(Source: USAF Engineering Files)

The Pratt & Whitney F100-PW-100 engine is of modular construction. Shown next is a diagram identifying the assembled modules in a cutaway drawing.

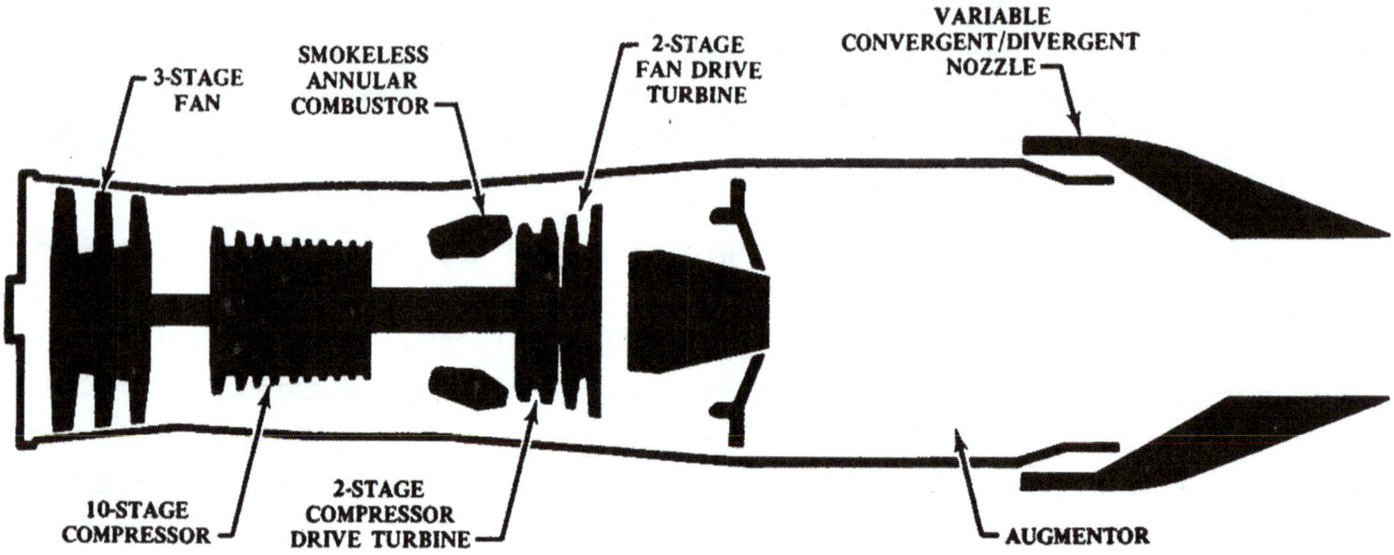

Major Components of the P&W F100-PW-100 Engine
(Source: USAF Engineering files)

When assembled, a cutaway diagram reveals the intricacy of the fully assembled P&W F100-PW-100 engine:

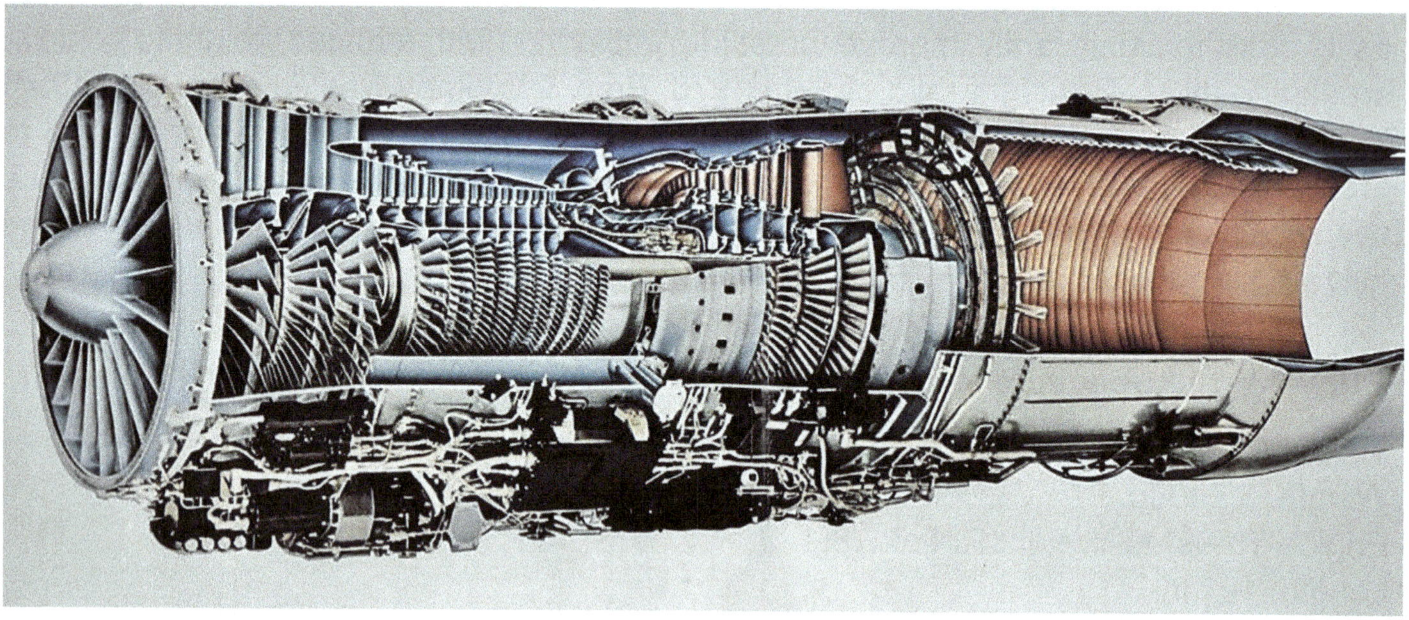

Cutaway of the Pratt & Whitney F100-PW-100 Engine
(Source: USAF Engineering Files)

At the bottom of the colored cutaway photograph are shown the external accessories which control the operation of the F100 engine. These include the Electronic Engine

Controller (EEC), the fuel, oil, and hydraulic pumps and feed lines, the electric generator for the engine combustion, the main electric generator for aircraft power, and the actuators for the setting of the afterburner leaves to control nozzle exit area. For the initial F-15 operations, the EEC was analog; however, the EEC was later upgraded to be a Digital Electronic Engine Controller (DEEC). There are also five ports in the outer engine case to allow borescope inspection of the turbine blades during scheduled (and unscheduled) inspections of the several hundred turbine blades (buckets). Although not obvious, the engine accessories are mounted on the underside of the engine such that they are the same on each engine being manufactured. There are no "left" or "right" engines for installation in the two-engine F-15. [This "no left or right" feature of the engine configuration was imposed on the installation of the F100-PW-100 engines in the YF-16 Light Weight Fighter (and subsequent F-16 Fighting Falcons), and gives rise to the description that the F-16 is literally wrapped around the engine. The location of each accessory on the engine dictated where many rapid-access panels had to be formed into the skin on the underside of the F-16 aft fuselage – for ease of access and for minimum time in the maintenance, inspection, servicing, and repair – of those engine accessories when needed.]

The modules comprising the F100-PW-100 engine are shown in the next series of photographs (contained in a Pratt & Whitney non-copyrighted brochure provided for the USAF Lightweight Fighter program):

F100 FAN SECTION

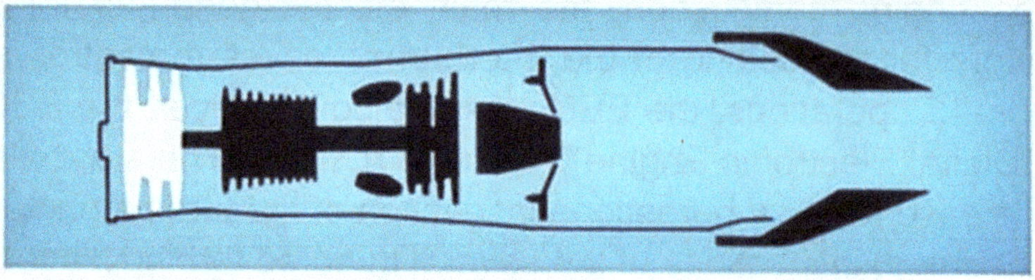

The F100 utilizes a three-stage fan with variable inlet guide vanes to provide optimum airflow scheduling and maximum operational stability throughout the flight envelope.

Advanced blade technology is coupled with a variable camber guide vane which has been amply proven in the J52 and J58 engine programs.

F-100 Fan Section Module
(Source: USAF Engineering files)

F100 COMPRESSOR SECTION

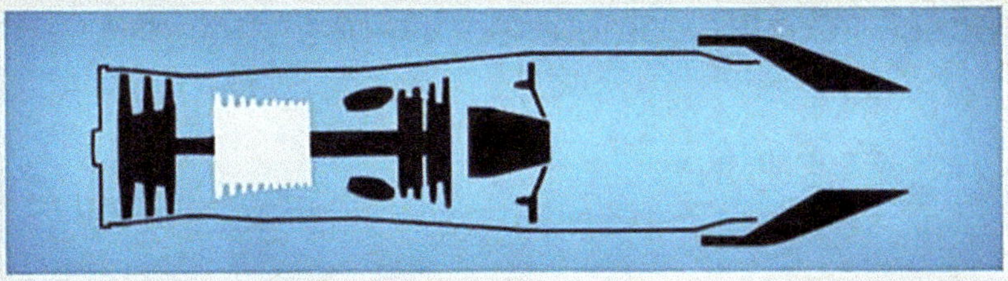

The F100 employs a 10-stage compressor with three stages of variable stators for optimum airflow scheduling.

F100 Compressor Section Module
(Source: USAF Engineering files)

The outer housing of the compressor module is shown in place covering the stators and turbine blades. The outer housing serves also for blade containment in case a blade fails during engine operation (especially at high RPM). At the 7 o'clock position is shown a bleed-airport that provides high-pressure air into the Environmental Control System (conditioned air into the pilot's crew station). Also shown are the ports enabling bore scoping inspection of the compressor stages of blades.

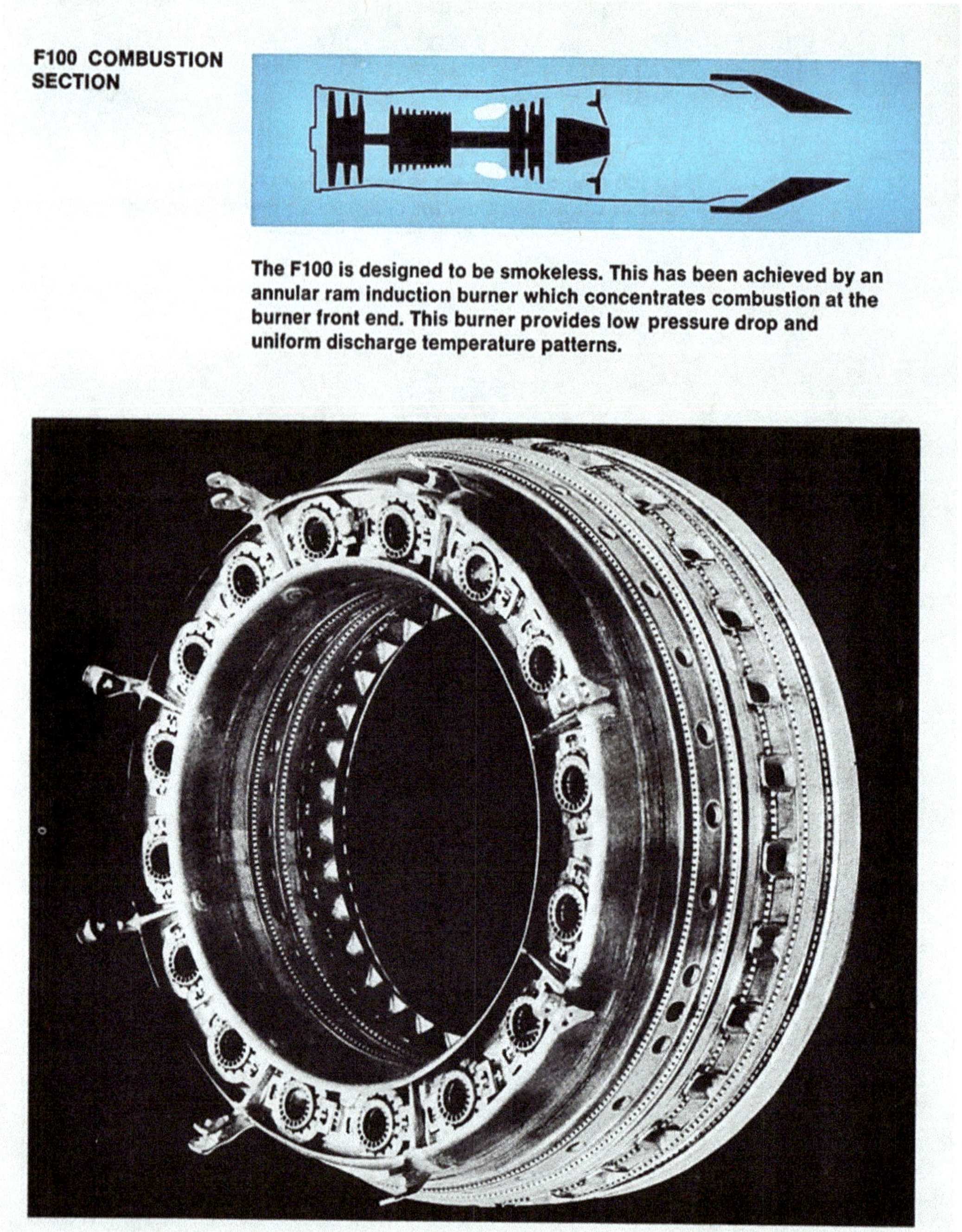

F100 Engine Combustor Section
(Source: USAF Engineering files)

Within this Combustor Section, the highest temperatures are experienced within the engine – about 3,600 F^0. It is through the annular series of holes that cooling air is passed around through the combustor case to cool the Combustor Section. The heated air from the cooling flow is introduced into the final modules and mixed with the heated air from the combustion process. The nomenclature that the F100 engine has a by-pass ratio (of 0.63:1) indicates that that fraction of the air exiting from the fan at the face of the engine will be used for cooling the combustor section (the remainder of air entering into the combustion process). As might be expected, the material used in the Combustor Section is of the highest grade on steel capable of the extremely high temperature and pressure generated within that section.

F100 TURBINE SECTION

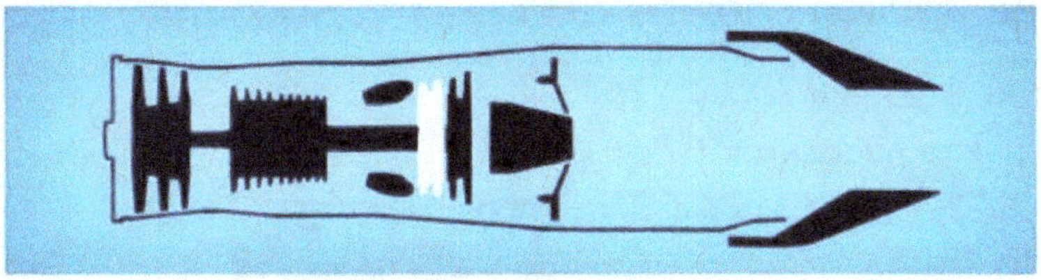

F100's high operating temperatures are achieved through the use of an air-cooled, two-stage high pressure turbine. Film and impingement cooling techniques are utilized to obtain maximum performance in the high pressure turbine vanes and blades as illustrated below.

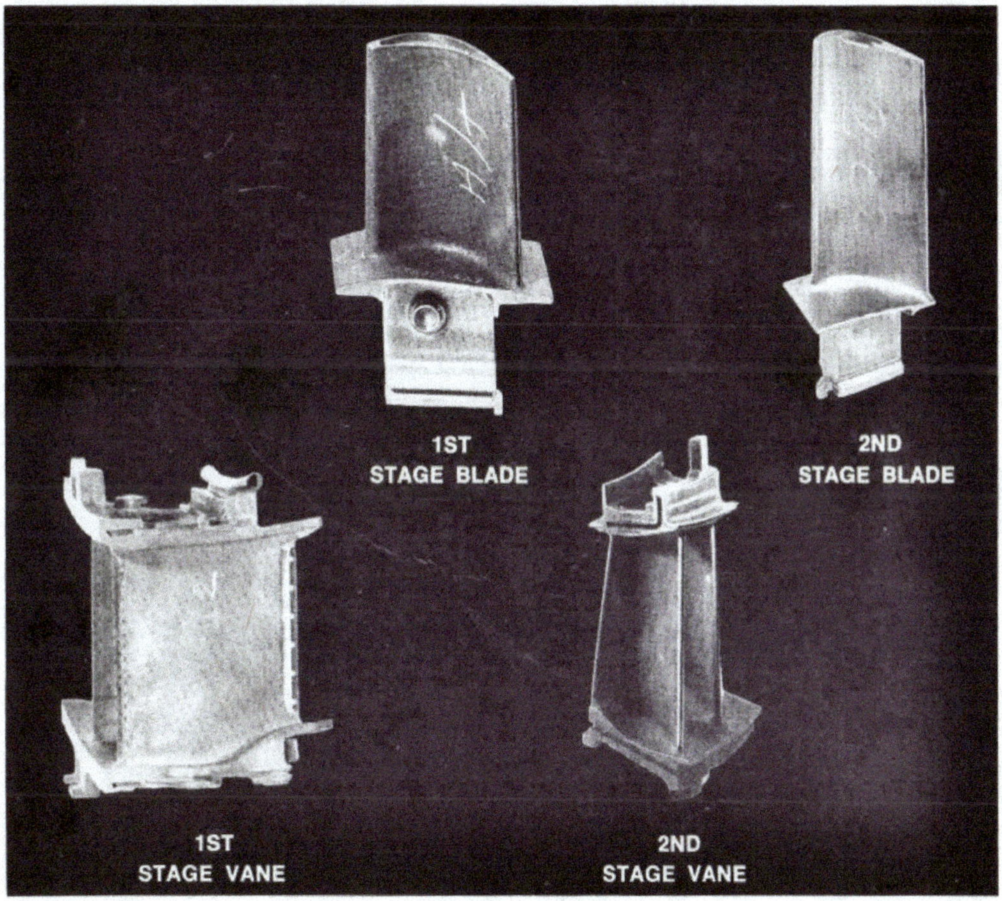

F100 Turbine Section
(Source: USAF Engineering files)

The heated, high-pressure air from the combustor section flows through the first stage vane, then onto the first stage blade, then past the second stage vane, and finally onto the second stage blade. The flow of air onto the first and second stage blades causes the turbine in the turbine section to rotate at high RPM. That rotation of the blades causes the shaft onto which the blades are attached to rotate, and it is that shaft rotation that drives the rotation of the front fan section and the compressor section. The flow of high temperature and high-pressure air that exits the second stage blades is introduced into the augmentor section shown next.

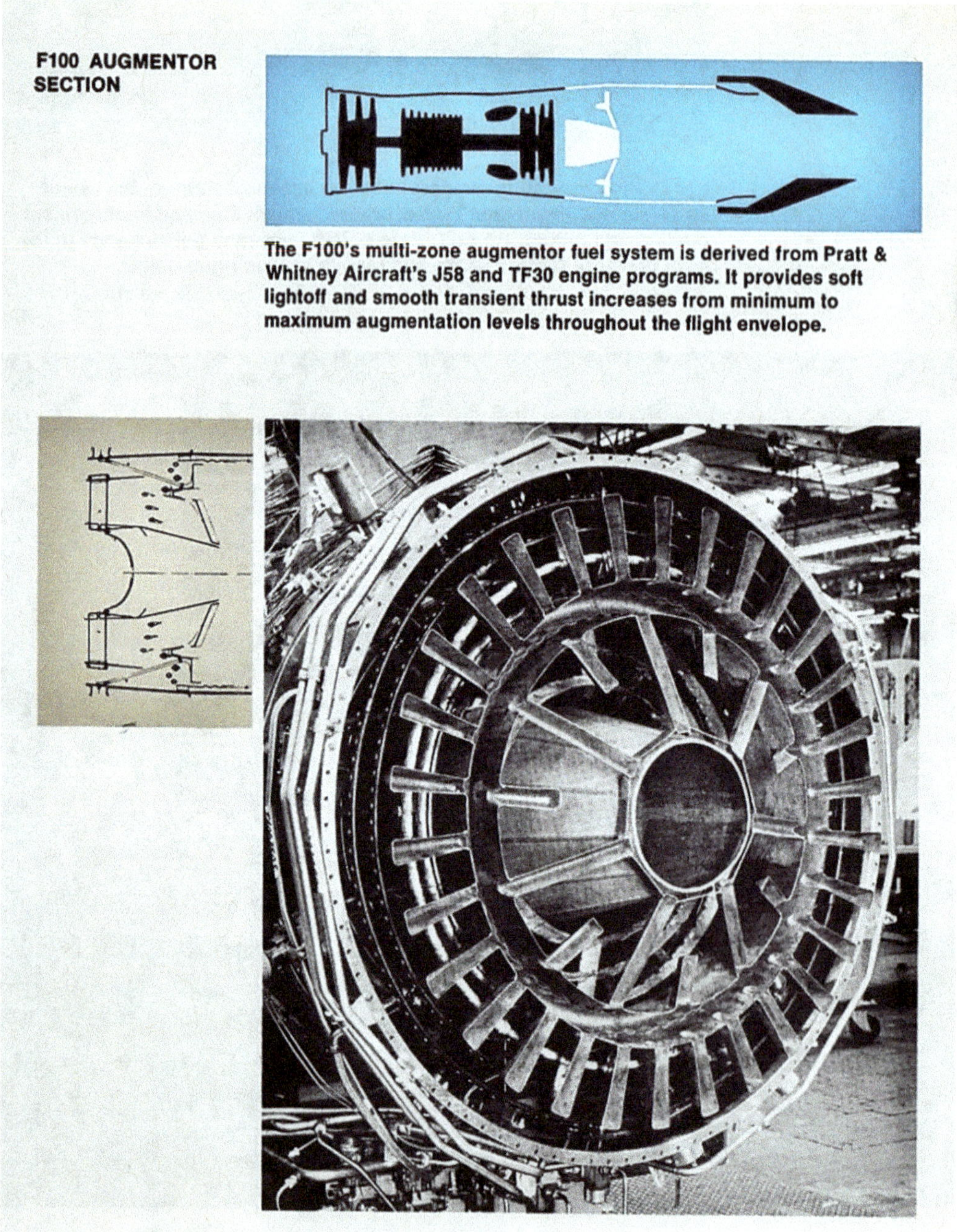

F100 Augmentor Section
(Source: USAF Engineering files)

The heated, high-pressure air exiting the turbine section and entering the augmentor section may be further heated and raised to a still higher temperature and pressure level by the introduction of fuel injected into that heated air and causing additional combustion of that airflow. The augmentor is designed with five rings and provides the pilot with commanding from one to five additive zones of augmentation of the thrust exiting downstream of the augmentor and exiting the nozzle through a series of external leaves that adjust the nozzle exit area to create the augmented thrust. The Augmentor Section is more commonly called the afterburner. The five zones of afterburning are selected in detents on the engine thrust column at the left of the pilot on the left console in the YF-16 cockpit.

F100 NOZZLE SECTION

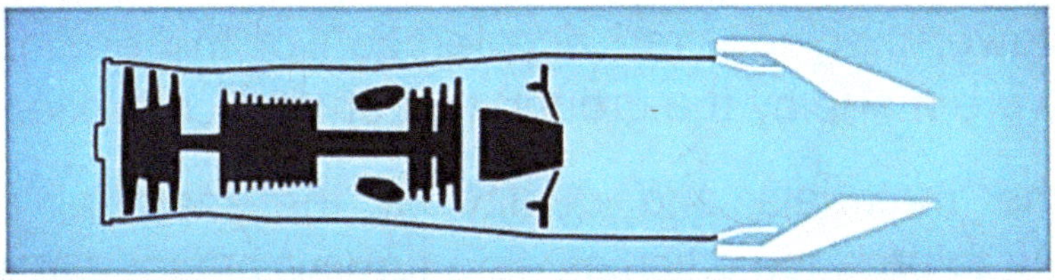

The *balanced beam* exhaust nozzle is a lightweight convergent/divergent nozzle which is fully variable for optimum performance. The nozzle walls are fabricated of a honeycomb sandwich material and require no airframe-supplied cooling air.

The *balanced beam* concept is simply the application of the principle of equal moments acting on a single pivot or hinge point. Nozzle pressure loads are transmitted through the pivot to a fixed ring.

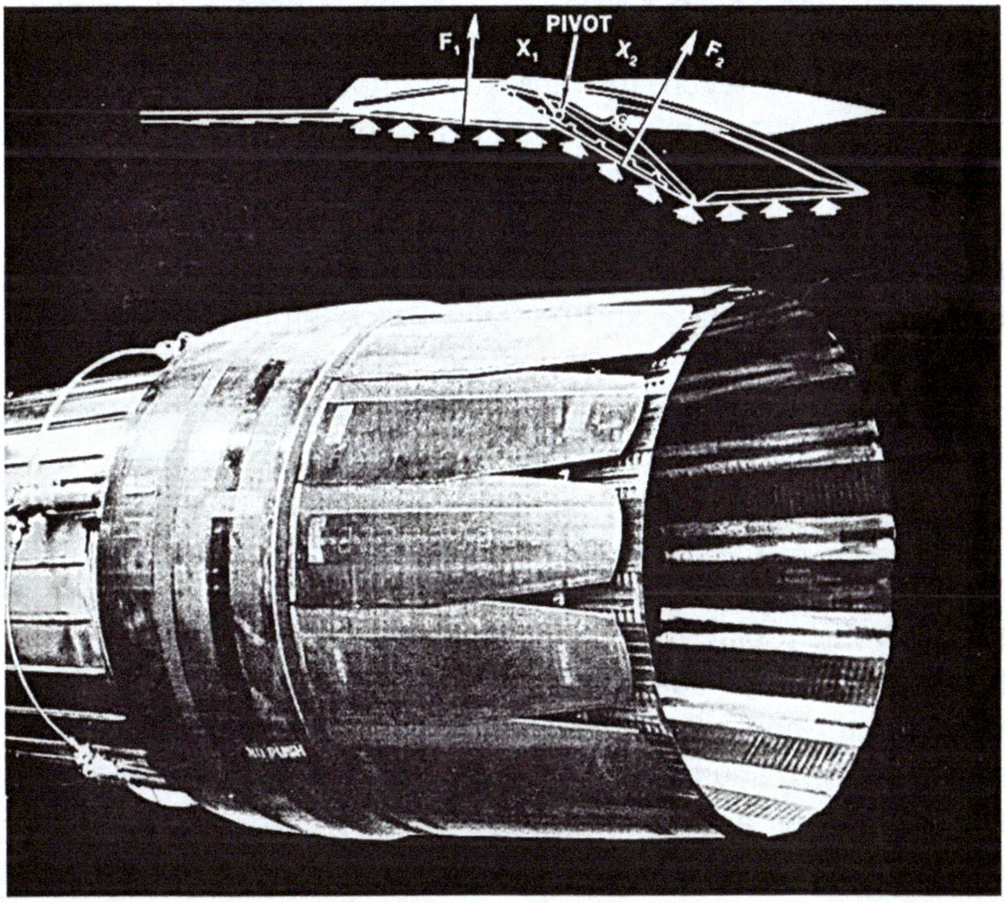

F100 Nozzle Section
(Source: USAF Engineering files)

As with almost every advanced technology development and most often those developments of advanced propulsion cycles there are likely to be experienced growing pains. These unexpected problems surface early in development testing. When the Pratt& Whitney F100-PW-100 was at 132 hours of a Congressionally-mandated operational test, the engine exploded. The cause of the engine disintegration was found to be the failure of turbine blades that had fatigue failures. The turbine blades in the F100 engine are subjected to the highest temperatures and pressures of all components within the engine. Further, the dynamic loading and decreased loading of the turbine blades as they rotate at high RPM past the exit from the combustor section are not a constant high stress loading but has a slight cyclical nature to it. A lesser significant problem was that there was slower-than- desired acceleration of the engine RPM as the throttle lever commands were made by the ground test engine controller.

Improving the materials and quality of the material used in the turbine blade manufacturing to produce efficient, safe turbine blades caused about a six-month slip in the F100 operational test schedule. A rescheduling of the electronic engine control to cause more responsive fuel flow into the combustor section resolved the slow engine acceleration problem to a tolerable level.

The nozzle leaves are subjected to very high acoustic noise level vibrations and also to the high temperatures and high pressures when in any Zone of afterburning – the higher the Zone number the more severe the vibration/temperature/pressure-induced stresses on those nozzle leaves. After a few hundred flight hours, early nozzle leaves developed cracks that propagated getting longer as the engine life increased. Maintenance personnel at forward bases attempted to stop the crack length from increasing by drilling a hole at the far end of the crack in an attempt to relieve the crack-inducing stress. That process was only very temporary and partially successful as the cracks started at the other end of the hole and continued propagating. The original nozzle leaves were rapidly replaced with a much higher crack-resistant steel material and the need for stop-drilling by maintenance ground crew personnel was obviated.

The FX / F-15 Development Activities (Pertinent to LWF)

In late1963, USAF planners had been planning the conduct of a study to develop a new fighter aircraft that would replace the Century Series fighters and the F-4 Phantom II that was not performing as well as desired in the Vietnam War especially against the Soviet MiG-21 Fish bed fighter being flown by pilots of the North Vietnamese Air Force. (That poor performance will be discussed in the last part of Chapter 5). Additionally, the USAF rejected the commonality approach desired by then Secretary of Defense Robert McNamara based on the problems (especially weight increases) experienced with the other two commonality attempts with the F-4 Phantom II and the F-111 programs. There were many lessons learned about the losses of F-4 Phantom IIs due to enemy ground gunfire bringing down the very visible smoke-stream producing F-4 Phantom II ("Smokey

Bear") at an unacceptable rate. Ground gunners did not need to visibly detect the F-4 target – as in duck-hunting, all they needed to do was fire at the approaching smoke column and with lead ahead of the front of the column with enough time for their ground-fired projectiles to intercept that smoke trail behind the F-4.

Named as the FX, the new fighter aircraft would be designed primarily as a USAF air superiority fighter and not burdened by being a common development with the USN/USMC carrier-based fighter aircraft. In a separate activity from that of the USAF, the USN/USMC planners were studying a fighter concept (VFX) that would feature the variable sweep angle of the wing technology developed for the F-111B (the USN version of the USAF F-111A). This VFX would lead to development of the Grumman F-14 Tomcat for the USN/USMC having carrier capability as a fleet-defense aircraft.

The new USAF FX air superiority fighter was to be designed with a high T/W (approximately T/W = 1.2 at start of air combat) surfacing from the energy – maneuverability work that USAF now-Major John Boyd and his cohort Tom Christie at Eglin AFB were championing. Major Boyd was instrumental in writing the maneuvering performance requirements for the FX air superiority fighter. To achieve the high T/W (1.2 at start of combat) requirement, there would necessarily be development of a propulsion system for the FX as there was no available afterburning turbofan jet engine yet developed by either General Electric or Pratt & Whitney matching the T/W = 1.2 at start of combat and being smokeless (a severe problem with the F-4 Phantom II in Vietnam), both of which requirements were specified for the FX by Boyd. Design of the new engine (later to become the Pratt & Whitney F100-PW-100) and with performance capability functionally matched with that of the design of the new FX aircraft configuration) began in 1965.

After Boyd's performance/maneuverability requirements were established for the FX request for proposals, by mutual agreement of Boyd and the FX Program Office, his further participation in the design development of the FX was essentially ended. That was primarily due to two reasons: first, Boyd had a personality clash with the Chief System Engineer (Harold Brown) and his Deputy Chief System Engineer (Fred Rall) – all three had tremendous egos and Boyd's was that of the alpha male. It was an environment of "No pilot is going to tell us engineers how an aircraft should be designed." vs "You dumb engineers don't know what kind of fighter aircraft we pilot truly need in air combat." Secondly, and more importantly, Boyd saw that those engineers did not grasp his concept of energy-maneuverability which relied on developing a new propulsion system for the FX. Boyd also saw there was no effective control over the contractor (later chosen to be the McDonnell Aircraft Company in St. Louis which had developed the F-4 Iron Horse) establishing the geometric size and especially the weight of the F-15 as an air superiority fighter.

Boyd was also aware of the emerging technology in the USAF Flight Dynamics Laboratory at Wright-Patterson AFB where the Author was serving (as the acting Chief of the Aeromechanics Branch in the Flight Mechanics Division), and developing the technology of computer-controlled leading-edge and trailing-edge wing flaps for enhancing the maneuverability of fighter aircraft. The Author had earlier approached Fred Rall in his office at the F-15 SPO and suggested that the technology of wing leading-edge and trailing-edge maneuvering flaps be incorporated in the RFP for Contractors to consider in their responses to the FX request for proposal. The response the Author received from Fred Rall was: *There's no God-damned way I will allow you technology toads to screw up my aircraft design. Don't let the door hit you in your ass on the way out.*

As a result, the F-15 wing design which resulted from the FX RFP design activities was a fixed, low-camber airfoil wing designed for highest efficiency at supersonic speeds – which was learned from the Korean and Vietnam Wars to being a low need for maneuvering against an aggressor aircraft at supersonic speeds (above Mach 1). The engagement may start at supersonic speeds but the ensuing maneuvering will drag (in all senses of that meaning) the engagement down to be at high subsonic speeds. The large increase in aircraft drag above Mach 0.9 (due to compressibility effects and shock wave formation) is adverse to the aerodynamic efficiency needed for high maneuverability.

The original thinking for the F-15A Eagle was that it was to be an air superiority fighter to counter the Soviet MiG-25 design (for which Foreign Technology activities had provided some measure of the MiG-25 speed and altitude capabilities). Being much of what is termed as being a point design, the F-15A had virtually no capability for support of ground troop activities. The buzz phrase in the F-15 System Program Office at the Aeronautical Systems Division, Wright-Patterson AFB, was "Not a pound for air-to-ground". That lack of an important capability was remedied several years later when the F-15E Strike Eagle was developed.

But it was not the F-15 configuration which contributed much technology to the development of a Lightweight Fighter (later to be developed into the F-16 Air Combat Fighter). By far, the propulsion system for the F-15A was the technology development which was the most important factor allowing the development of the F-16. There was much opposition to the development of the Lightweight Fighter Program – or any other program – which would require funding by the USAF, and detract funds from the F-15 development program. The F-15 program, like most new aircraft development programs, had encountered several unexpected circumstances to arise which had twofold effects. First, was that most of those problems were of an aerodynamic performance nature – i.e., the flight testing revealed that there was a significant difference in the drag of the F-15 from that obtained in the earlier wind tunnel testing done. The performance of the F-15 was classified as Secret, and therefore there was little (basically none) that the Flight

Dynamics Laboratory at Wright-Patterson could assist the F-15 System Program Office in resolving. The F-15 Program Office engineering management, as mentioned earlier, regarded the Flight Dynamics Laboratory engineers as being a bunch of technology toads, and that negative attitude persisted during the entire F-15 development activities. As a result, some of the emerging technologies for advanced technology fighter aircraft being pursued in depth at the Flight Dynamics Laboratory were ignored – intentionally – both by the F-15 Program Office engineers and by the design engineers at McDonnell Aircraft Company in St. Louis. (The McDonnell Aircraft Company was not allowed to reveal the aerodynamic problems emerging during F-15 development flight tests to anyone other than F-15 Program Office engineering).

In parallel with the airframe development of the F-15 to overcome the deficiencies identified during the Vietnam War, there was the development of the engine for the F-15A propulsion system. [Author's note: The engine part of the F-15 development program was initially denied by HQ USAF for use in the rapidly-following Lightweight Fighter Program, but that denial was reversed by Department of Defense officials and the engine became the largest asset allowing the Lightweight Fighter Program to progress past the paper design phase of its development history and into prototype hardware for flight test. This aspect will be described in greater detail later in this book. It is hoped that those Readers most interested in the development of propulsion systems for fighter aircraft (which has been traced in earlier chapters of this book) may gain an appreciation of how a highly advanced technology engine developed for one system – the two-engine F-15 – became the engine used in another program – the single-engine F-16.]

Next, the Author wishes to describe the circumstances of how the Pratt & Whitney F100-PW-100 engine continued its development for the USAF (in their development of the F-15 Eagle). Following that will be a description as to how that engine was transitioned from initially being denied (by the Commanding General of the Air Force Systems Command, Gen Alton D. Slay) for its use later (after Gen Slay's decision was reversed) in the Light Weight Fighter to becoming the key factor in successful development of the Lightweight Fighter Program. As with all technology advancements, there is usually one individual who by his/her leadership makes that technology advancement become a most momentous achievement in that technology domain. This book recognizes that singular contribution to the technology of high speed (Mach number greater than M 2.0) to have been by USAF officer Benjamin N. Bellis.

In the mid-1960s, Ben Bellis was the principal author of the Air Force Regulation AFR-375 series of regulations outlining how the development of military weapon systems should be managed. When the need for development of a new weapons system is defined and receives funding to proceed into development, a System Program Office (SPO) is established and manned at a government facility – for the USAF systems that was typically located at the then Aeronautical Systems Division (ASD) of the Air Force Systems

Command (AFSC). The Headquarters of ASD was at Wright-Patterson Air Force Base situated a few miles east of Dayton, Ohio. In the mid-1960s, Ben Bellis (then a USAF Lieutenant Colonel) lectured on that management process and which was followed by several AFSC flight vehicle programs then in the process of being developed. Colonel Bellis in 1965 was selected to be the SPO Director directing the development of the Lockheed SR-71 Blackbird, the high-speed, high-altitude strategic reconnaissance aircraft which set and still holds the worldwide speed and altitude records for a manned atmospheric flight vehicle.

The SR-71 was designed as a Mach 3+ interceptor of the Soviet Mach 3+ MiG-25 Foxbat. It was as the SR-71 SPO Director that Colonel Bellis' outstanding leadership was demonstrated and gained the highest respect for managing complex, challenging advanced technology flight vehicles and their integration into major weapon systems. When appointed to be the F-15 SPO Director, Col. Bellis was promoted to Brigadier General. B/G Bellis accepted the assignment with the condition that he head the F-15 SPO but also head the F-15 Joint Engine Program Office (JEPO) where the development of the Pratt & Whitney F100-PW-100 engine for the F-15 Eagle was being accomplished along with the development of the Pratt & Whitney F401-PW-400 being accomplished for the USN F-14 Tomcat. (The JEPO was a concession B/G Bellis was willing to give to Secretary of Defense Robert S. McNamara to implement "commonality" between USN and USAF weapons systems procurements.) The USN would later cut back and then completely cancel its order for P&W F401-PW-400 engines and continued to use the Pratt & Whitney TF-30 engine (from the F-111) for the F-14 Tomcat. The USN decision did not surprise or disappoint B/G Bellis as he could better concentrate then on the two development activities that were of highest interest to him – the F-15A Eagle aircraft and the P&W F100-PW-100 for the F-15A Eagle.

The Advanced Day Fighter (ADF) Study - Boyd and the LWF Mafia

In 1963, the USAF conducted a Project Forecast "to identify future weapons trends and certain high-priority areas for research and development (R&D) with recommendations based on the greatest potential payoff for the future." This document had a very polarizing effect throughout the Air Force setting up two camps – those who agreed with the recommendations (partially or in total) and those who disagreed with the recommendations (partially or in total). There was also mass confusion among the aircraft manufacturers as to really what were the weapons (aircraft – fighter, bomber, interceptor with high speed/high altitude or subsonic low altitude, terrain following or not, missiles – with air-to-air or air-to-ground capability and having short-range, medium- range, or long-range capability, and with or without nuclear capability, etc.) that the government contractors having some degree in confidence that they should invest company-sponsored and IR&D (independent research and development) funds to pursue a specific technology domain. The Project Forecast recommendations were referred to as being a wish list with something for everybody. In addition, the USAF budget could not afford to proceed with production of each and every one, but refused to give any indication of priority for which weapon system type to proceed with in earnest. Out of that chaos emerged the TFX and FX programs being given the nod to proceed with development while the other candidate capabilities were being sorted out and prioritized. [The foregoing observations were derived by the Author while assigned in the Advanced Concepts organization at the Aeronautical Systems Division, Wright-Patterson AFB.]

As the Vietnam War progressed and escalated, it became clear that the TFX (USAF F-111A and USN F-111B) aircraft would not be developed in time to salvage the F-4 Phantom II fleet (both that of the USAF and of the USN/USMC) from being totally destroyed by ground gunfire or by the equally capable MiG-21 fighter with its Atoll IR missile. The USN divorced itself from the F-111 Program and progressed independently on a fighter aircraft for its fleet defense, that being the genesis of the USN/USMC F-14 Tomcat. In retaliation, the USAF generated the need for an air superiority fighter that would *not* have any commonality with the F-14 Tomcat Program, and that was the genesis of the FX Program leading toward development of the USAF F-15 Eagle. Recognizing that the F-15 was to be a high-speed air superiority fighter and with "not a pound for air to ground", the USAF rapidly put together the AX program leading to the A-10 Thunderbolt II which was a large cannon with wings for anti-tank and anti-armored vehicles ground support missions.

Standing off to the side of all those circus activities, Lt. Col. John Boyd, who had distinguished himself with his *Aerial Attack Study* and had developed his energy-maneuverability theory, began to attract allies to his independent tactics proclaiming the need for a low-cost, advanced day fighter that could be built in large numbers at an affordable cost. Boyd had always held the belief, derived from his personal experience as an F-86 Sabre pilot in the Korean War that it is better to have a large number of fighters with moderately high tactical capabilities on the "flight-ready" line than to have only a third of that number of "wow" fighters having every bell and whistle known to aviation technology – but having no capability for supporting ground forces. John Boyd generated the Growth of the Fighter historical chart shown next.

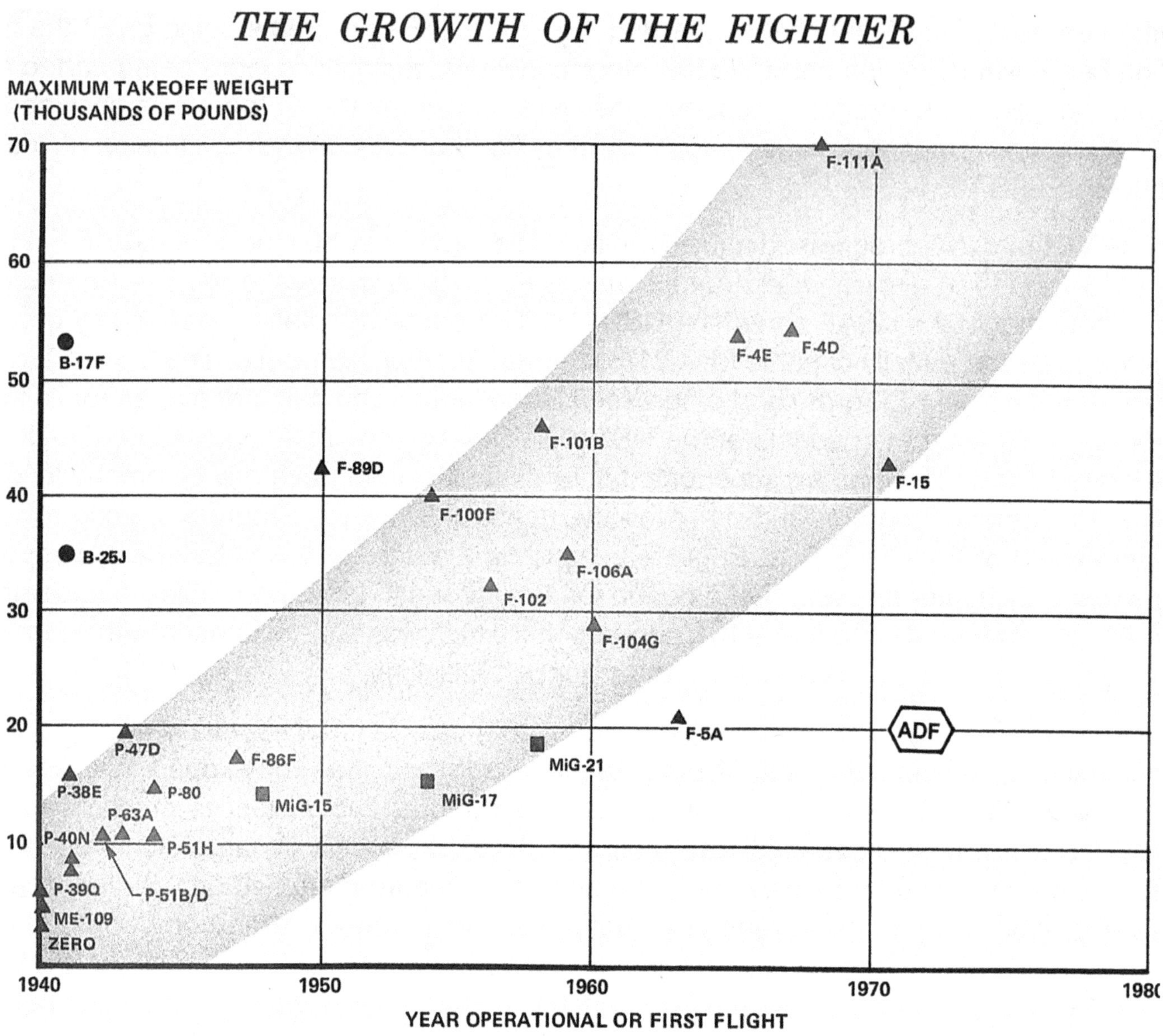

Boyd Growth of the Fighter Chart in ADF Study
(USAF Engineering files)

Boyd added a point on the growth trend chart labeled as ADF to emphasize the importance of minimizing the weight for an affordable Advanced Day Fighter. The goal weight for the ADF was to be in the 20,000 pounds class (about one-half that of the F-15A) but having a thrust- to-weight ratio greater than 1.2 at combat weight. Boyd reasoned that Northrop had developed the F-5A Tiger at a weight in those same 20,000 pounds class. Pierre Sprey (being a far more smooth-talking lobbyist than John Boyd) meanwhile was trolling the corridors of the Pentagon spreading the politically-correct message that the Advanced Day Fighter concept was founded on being an advanced technology fighter which could provide a complementary capability with that of the F-15, not as a substitute for it. Boyd projected the time period for introduction of the ADF into a high/low mix with the F-15 to be in the 1972 to1973 time period, shortly after the F-15 had passed its IOC (Initial Operating Capability) milestone.

Few in the USAF Development Planning offices in the Pentagon desired to pursue the concept of a high/low mix and did not support any funding to pursue the concept any further. Two however did. Colonel Everest "Rich" Riccioni in the DCS Plans organization and Pierre Sprey, a weapons system operational analyst for the Department of Defense began a dialogue with John Boyd and Tom Christie and they colluded to unofficially pursue the Advanced Day Fighter concept fostered by Boyd. In 1969, Col. Riccioni "found" (best not to inquire where and how) $149,000 to conduct a study titled "Study to Validate the Integration of Advanced Energy-Maneuverability Theory with Trade-Off Analysis". Boyd and Christie had been having unofficial under- the-table (actually on top of tables, and on bar napkins) discussions with two elements of the aircraft manufacturing companies who also shared the belief that Boyd's Advanced Day Fighter featuring high maneuverability and as a high/low mix with the USAF F-15 Eagle made great sense and should be pursued further. Those two companies and the individuals wishing a piece of the action where Lee Begin at the Northrop Aircraft Division in Hawthorne, California, and Harry Hillaker at the General Dynamics Convair Division in Fort Worth, Texas.

Based on where the two companies stood on developing a small, low-cost fighter (ADF) with in-house funding, the $149,000 from Riccioni was split, with Northrop receiving $100,000 and with General Dynamics receiving the balance of $49,000 for participating in the ADF study. Pierre Sprey provided his experience to the two companies in the form of "guidance to in-house efforts of applying emerging technologies to develop affordable new capabilities of potential interest to the government". The ADF study started rapidly – an advantage of not having a huge formal contract and not having to meet any official requirements and adherence to military specifications – and both Northrop and General Dynamics were off and running defining an advanced day fighter capability.

Recalling that John Boyd's energy-maneuverability theory is summarized in the following equation (derived and explained earlier in this book):

$$\text{Specific Excess Power, } P_S = V \times (T - D) / W$$

the three elements for maximizing P_S at combat conditions call for a maximization of thrust T and a minimization of drag D and weight W at the combat conditions (V and H).

Both General Dynamics and Northrop conceived and studied, including wind tunnel testing of models, the ADF concept with continual contact on progress with the Boyd/Christie/Riccioni/Sprey quartet. The technologies that were given prime attention were those of the propulsion system (to maximize T), the external aerodynamics of the configuration (to minimize D), structural materials (to minimize W), and low complexity of avionics consistent with a day clear-weather environment (to minimize cost and achieve affordability). Added to those four technology domains was the technology of the flight control system that would make the design not only highly maneuverable but rapidly responsive to pilot maneuvering commands. Not to be forgotten were the lessons learned from the F-4 Phantom II losses in Vietnam due to a flight control system that was highly vulnerable to ground gunfire.

Shown next is the evolution of the Northrop P-530-3 configuration resulting from the refinement of in-house efforts with meeting the ADF goals (maximization of P_S and small, low complexity – to minimize cost).

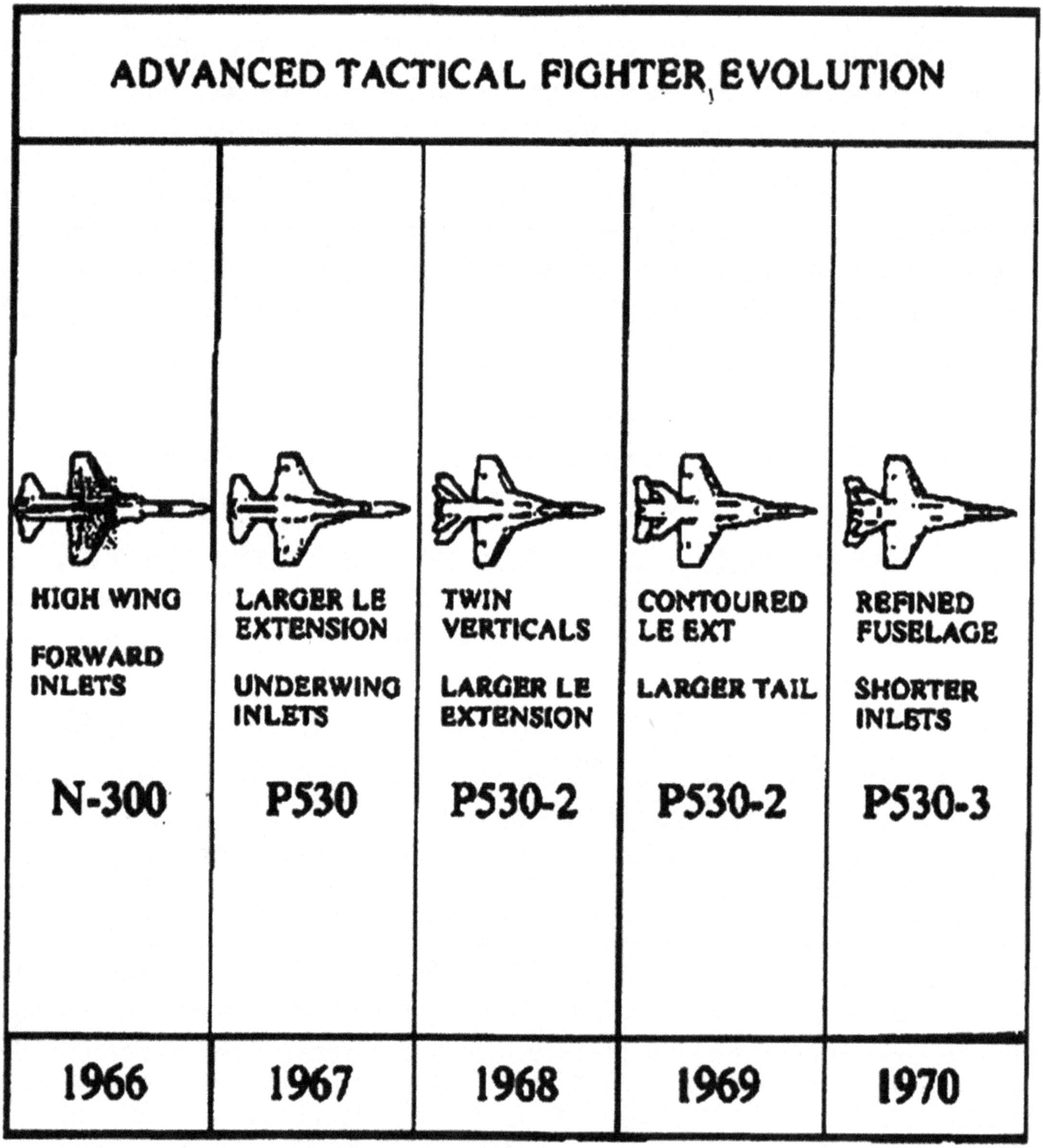

Evolution of Northrop P-530 Design in ADF Study
(Source: USAF Engineering files)

Northrop had started development of a small fighter in 1966 as a company-sponsored and funded venture and labeled it the N-300. In 1967, the design had been refined using USAF-provided IR&D funding and was renamed as the P530 Cobra. During 1968 and 1969, wind tunnel tests caused the P530 to be modified with the configuration changes shown on the chart. For the propulsion system, Northrop had chosen a two-engine configuration and had chosen the General Electric YJ101 turbojet engine, in the prototype stage at GE, as the engines for the N-300. Based on the efforts of the USAF Flight Dynamics Laboratory in the technology development of a survivable flight control system, Northrop had chosen a hybrid flight control system – hydromechanical for the primary control functions but with an electronic backup control augmentation system. In addition, Northrop had pursued the technology emerging from within the Flight Dynamics Laboratory of having a wing with leading-edge flaps and trailing-edge flaps which were computer-controlled. To better enable the pilot to endure the physiological effects of high-g maneuvering, the pilot's heel rests and rudder pedals were raised several inches. The pilot's control column was a conventional between-the- legs design. Emphasis was placed on pilot visibility in the entire upper hemisphere and a two-piece bubble canopy was selected by Northrop. Northrop constructed a full-scale mock-up of the P530 Cobra as shown in the next photograph.

Northrop P530 Cobra (full-scale mock-up), Hawthorne, CA, 1970.
(Source: USAF Engineering files)

With the $100,000 obtained from the USAF for the ADF study activity, additional configuration improvements for the Northrop ADF configurations were made during and as a result of wind tunnel testing at NASA Langley Research Center with the advice of Dr. Dick Whitcomb. Although the photograph gives the appearance of the wing planform seeming to be a straight (i.e., zero leading-edge sweep) wing, the small diagram in the previous evolution chart shows the P530-3 to have a clipped delta-wing planform suitable for Mach 2 operation. The Cobra mock-up shows only four missiles being carried on 2 wing stations on each wing; however, when the Cobra became the configuration later known as the USAF YF-17 Light Weight Fighter prototype, the number of external under wing external stores stations had been increased to four stations per wing.

Harry Hillaker at General Dynamics had also been developing configurations for a fighter of the type revealed to him in a chance meeting with John Boyd at Eglin AFB, Florida, some years before the funded start of the ADF Study. The $49,000 received from Col. Riccioni for the ADF Study went toward payment of wind tunnel tests also at the NASA Langley Research Center (at Dr. Dick Whitcomb's transonic wind tunnel.) Instead of refining one configuration as Northrop had done (with its Cobra configuration), the Hillaker configuration development team had explored several different wing, tail, and inlet placement designs, as well as having settled on a single- engine design (to minimize the weight of the concept). Shown next are four of the configurations developed by General Dynamics during Boyd/Riccioni's ADF Study:

ADF Configurations at Gen. Dynamics Ft. Worth, TX, 1970.
(Source: USAF Engineering files)

The configuration in the lower right corner was designated as the Model 401. It is this Model 401 which General Dynamics later offered to the USAF in response to the USAF Light Weight Fighter request for proposal.

General Dynamics had been highly responsive to what fighter pilots with combat experience had given as feedback for the features they would like to have in the design of an advanced cockpit. As shown in the photograph, all configurations of the GD ADF featured a one-piece bubble canopy – to eliminate the bow-frame which pilots had complained as obstructing their view of the enemy aircraft during maneuvering engagements. In a risky decision made by Harry Hillaker, the all-electric fly-by-wire (FBW) flight control system technology (an emerging technology from the USAF Flight Dynamics Laboratory Survivable Flight Control System activities) was selected.

At the completion of the Advanced Day Fighter Study (i.e., when the original $149,000 provided by Col. Riccioni had been expended), further attempts by John Boyd to gain interest in his Advanced Day Fighter concept were disappointingly disregarded as the USAF considered that there should be no funds expended on any activity other than continued development of the F-15 Eagle and continued development testing of the Pratt & Whitney F100-PW-100 engine. Not to be dissuaded, however, the Col. John Boyd mafia (Boyd, Christie, Riccioni, and Sprey) continued their lobbying for a lighter- weight, lower-cost fighter as the low end of the high/low mix with the high- end F-15 Eagle. [The identification of that group of four as "Boyd's mafia" was given by Col. Riccioni as a humorous reference to Riccioni's Italian heritage and culture and referred to Boyd's somewhat heavy-handed (and his dagger-tongued) approach conflicting with the USAF bureaucratic chain-of-command manner of accomplishing anything – or studying it to death.

Harry Hillaker at General Dynamics rightfully did not acknowledge his being a charter member of the Boyd mafia as that would constitute a conflict-of-interest in the event that General Dynamics was to be considered for a more legitimate procurement activity to conduct further official development of an advanced technology fighter. Harry Hillaker and John Boyd did remain in constant contact, with Harry frequently updating Boyd on what advanced technologies were being considered for the next phase of a light-weight, low-cost, high-T/W fighter. Principal among those technologies were that of a survivable flight control system as being a necessary technology for any fighter having a ground attack mode of capability. The losses of over 200 F-4 Phantom IIs in the Vietnam War due to flight control damage from ground gunfire (when hydraulic system components in the hydro-mechanical flight control system were damaged and caused loss of control of the F-4s). Harry Hillaker and his flight control team followed very closely the development of fly-by-wire technology with an all-electric (i.e., no hydro-mechanical links) flight control system with a flight control computer of the type being pursued by the Flight Control Division in the USAF Flight Dynamics Laboratory (AFFDL) at Wright-Patterson AFB. Based on results of the battle damage assessment and recovery (BDAR) activities on aircraft and aircrew losses in the Vietnam War, the AFFDL and McDonnell Aircraft Company had modified an F-4 Phantom II with a partial fly- by- wire flight control system to become the FBW F-4 demonstration flight test platform. Based on highly successful flight testing with the FBW F-4, the all-electric fly by wire flight control system technology was considered ready for application to an advanced technology fighter aircraft design. Because the General Dynamics ADF design was not as "cast in concrete" as was the Northrop P-530 Cobra ADF design, the all-electric fly by wire flight control design was the first of its kind to be applied to an aircraft that would eventually be transitioned into operational use. In USAF parlance, FBW was considered a "high- risk, high- payoff technology which promises increased effectiveness of future USAF fighter aircraft weapon systems".

Propulsion for the General Dynamics Model 401 version of the ADF was selected for weight, cost and availability reasons, as well as a somewhat programmatic reason. The latter reasoning being that if the ADF was being touted to be a high/low mix with the two-engine F-15 Eagle a most logical choice for the single-engine ADF would be the engine being developed for the F-15 Eagle – the Pratt & Whitney F100-PW-100 afterburning turbofan engine. That would simplify the procurement process as well as the maintenance process at bases where the high/low mix aircraft were to be co-located. The P&W F100 engine was well into its development testing and was a perfect thrust (25,000 pounds) class engine producing the T/W of 1.2 at start of combat weight set by Boyd as a goal for the ADF. [The use of the P&W F100-PW-100 engine from the F-15 development program was not an approved HQ USAF approach, but then the whole concept of an ADF had not been officially approved by HQ USAF. General Dynamics proceeded to wrap its Model 401 design around the P&W F100 engine on the speculation that it would later gain approval by the HQ USAF Commanding Officer. This proposed use of the P&W F100-PW-100 engine from F-15 engine resources became a major point of contention later in the formulation of the USAF Lightweight Fighter Program, as will be described in the next chapter.]

[Author's note: Harry Hillaker expressed surprise and disappointment to the Author that the McDonnell Aircraft Company had not incorporated an all-electric fly-by-wire flight control system in the F-15 Eagle which McDonnell Aircraft Company was developing under contract for the USAF. It was concluded the reason being that the F-15 Eagle was designed as an air superiority fighter and boasted of "not a pound for air to ground". Survivability and vulnerability of the F-15 to ground gunfire therefore was not a large issue. Harry was also interested in close contact with the AFFDL for the advanced technology thrust underway in the Author's Aeromechanics Branch with the use of the emerging technology of computer-controlled leading-edge and trailing-edge flaps. Earlier in this book, the Author described the incident where a colleague (Major Darryl Sandlin) and the Author had approached the Deputy Chief System Engineer (Mr. Fred Rall) in the F-15 System Program Office to gain his interest and support for McDonnell to consider integrating computer-controlled leading edge and trailing edge flaps for wing camber control during maneuvering mode of flight – and were invited to immediately leave his office as being technology toads pursuing an imaginary problem on the F-15. (The inside reference is commonly titled as a technology looking for a problem to be solved.) Readers who are planning to become members of a design team should always be aware that it is the personalities, mutual respect, and professional courtesy of upper team management that will have a direct impact on the success of a technology development program.]

The final portion of this chapter describes the activities within the USAF that were conducted as a result of the Project Forecast study earlier in the decade, or as a result of the Lessons Learned from the Vietnam War. In addition, there were a host of new technologies generated within the several laboratories of the USAF Systems Command at Wright-Patterson AFB. The most productive of these activities was the Advanced Prototypes Study, described next.

USAF Advanced Prototypes Study

On April 21, 1970, HQ AFSC (Headquarters, Air Force Systems Command at Andrews Air Force Base, Maryland) requested that all AFSC agencies submit candidates for FY-72 and FY-73 to support an AFSC program initiative of prototyping. [For Readers not familiar with the term FY – FY represents "fiscal year" which runs from October 1 of one calendar year to September 30 of the next calendar year in which that FY ends. FY-71 for example is the period from October 1, 1970, to September 30, 1971. All government agencies receive an approved budget from their Headquarters' budget office authorizing expenditure of the funds that have been allocated to them during that specified FY.] The definition of "prototype" included the spectrum from a complete system with its aerospace ground equipment to proving the technical feasibility of a technique or a piece of equipment. Some 220 proposed projects were received from the AFSC production divisions, centers, and laboratories. Of the 220 proposed projects received, 45 were candidate system projects and 175 were subsystem/equipment or technology items.

A letter dated May 7, 1971, from the Secretary of the Air Force to the Deputy Secretary of Defense David Packard recommended a program to advance technology, to reduce technical and strategy uncertainties, and to provide a variety of hardware options in anticipation of future military needs. In June 1971, the USAF Vice Chief of Staff recommended that the Secretary of the Air Force initiate a comprehensive prototype program and to form a study group of Air Force executive officers and senior civilians to define that program. A Prototype Study Group, headed by a Steering Group (comprised of Secretary of the Air Force Robert C. Seaman's Assistant for Research & Development Grant L. Hansen, Lt. Generals O'Neill and Glasser, and Major Generals Kent and Graham) was organized into a team directed by a Study Director Brigadier General Kenneth L. Chapman (AFSC). The specific objectives set for the study were to develop the rationale for prototyping, to determine what to prototype, and to determine how to manage the selected prototype programs.

Supporting the Study Director were four Task Groups having the functional titles of Integrating Group, Project Selection Task Group, Management Approach Task Group, and Procurement Approach Task Group. The four Task Groups were each headed by a Colonel and had 4-6 senior officers and senior civilians from USAF, AFSC, ASD, ESD, DOL, and SAMSO as members. [Of primary interest to Readers tracing the "genealogy" of the Lightweight Fighter, the Project Selection Task Group was headed by Col. Lyle W. Cameron and with Col. John R. Boyd as his assistant. Col. Cameron was named later to head the Prototype Programs Office at ASD Wright-Patterson AFB for which the Lightweight Fighter Prototype Program and three of the other five recommended Prototype Projects were later conducted. Col. John R. Boyd is the same John Boyd whose earlier classic activities (*Aerial Attack Study*, energy-maneuverability theory, and *Advanced Day Fighter Study*) have been described earlier in this book. Readers should feel free to be thinking "That's like putting a fox in the henhouse to guard the chickens."]

The Project Selection Task Group rationalized that the 175 proposed candidate projects in the subsystems/equipment and technology domains were normal activities that are funded annually by R&D budgeted funds and transferred these to other "home" organizations to be funded by other means than by the Prototype Study Group. The 45 candidate system projects were then compared to evaluate their relative degree of risk/payoff and potential for development under current ongoing development programs. In the final cut, the Project Selection Task Group recommended six programs be given further consideration, planning, and scheduling by the four Task Groups of the Prototypes Study Group including distribution of the total budget allocated by the Secretary of the Air Force.

The final Prototype Program Study Report produced by the Prototype Study Group was completed in late July 1971. It was one of the finest and most comprehensive reports ever compiled and written by a government study group. [Two commonly heard statements in the halls of government offices are (1) "When the paperwork equals the weight of the aircraft being designed, the construction of the design can begin"; and (2) "The camel is a horse designed by a committee." The Prototype Program Study Report" did not receive these comments.]

Each of the six projects selected for further study and planning were discussed in depth among the four Study sub-Groups (Integrating Group, Project Selection Task Group, Management Approach Task Group, and Procurement Approach Task Group). The length of time spent on each candidate project varied depending on its nature of technologies involved as well as the scope of development activities that were projected as being needed.

On August 5, 1971, Brigadier General Kenneth Chapman, the USAF Prototype Study Director, presented to Deputy Secretary of Defense David Packard a summary of the general concepts used by the Prototype Study Group: the objectives, the approach, the motivations used, the 6 projects recommended to be pursued further, the cost to pursue the six recommended projects, and a management summary. The presentation consisted of twenty-four charts for the 1-hour presentation. Rather than the Author presenting an interpretation of what was presented by B/G Chapman to Mr. Packard, a copy of the eight most pertinent briefing charts will be shown with a minimum of comments from the Author. The presentation charts were at the time classified as being Confidential; however, they have been declassified and marked as Approved for Public Release under authorization provisions in the Freedom of Information Act.

The first chart extracted from the B/G Chapman briefing to Mr. Packard (Chart #2 shown next) reveals the key strategy points of the presentation

STATEMENTS ON "PROTOTYPING" **Chart #2**

- Use "Fly-Before-Buy approach
  - Reduce risks through hardware demonstration
  - Build and test before production commitment
- Streamline management and procurement approaches
  - Less Paper
  - Less Reporting
  - Less Service Involvement
- Design to performance goals rather than specifications - keep price firm
- Stated military "requirement" not needed
- Interest in industry design team continuity
- Not all selection emphasis on "proposals"
- Competition not always needed

Chart #2 from B/G Chapman Presentation
(USAF Approved for Public Release - FOIA)

The next three charts (Charts #3, 4, and 5), also extracted from the Chapman/Packard presentation, summarize the study objectives, study approach, and study motivations.

STUDY OBJECTIVES — **Chart #3**

- Develop rationale for prototyping
- Determine what to prototype
- Determine how to manage and procure program
- Examine advisability of, and an approach to, preserving continuity of industry design teams

Chart #3 from B/G Chapman Presentation
(USAF Approved for Public Release - FOIA)

STUDY APPROACH — **Chart #4**

- Focus on advanced development program category
- Consider prototyping to be complementary to normal development process
- Stress aircraft and aeronautical technology prototyping at system level
- Review past and current USAF prototype programs for lessons learned
- Stress innovation and emphasize simplicity in management and procurement approaches

Chart #4 from B/G Chapman Presentation
(USAF Approved for Public Release - FOIA)

MOTIVATIONS FOR AIR FORCE PROTOTYPING PROGRAM — **Chart #5**

- Improve total acquisition process
  - Better product visibility at relatively low cost
  - Potentially lower costs in any subsequent development phase
  - Added flexibility in management and procurement
  - Evolutionary approach to satisfying force needs
- Provide a variety of demonstrated options for application to military hardware needs
- Add "confidence" factor in decision making
- Make possible project initiation without force structure justification
- Permit the system demonstration of new, high risk technology
- Foster a viable system design and modeling capability within industry

Chart #5 from B/G Chapman Presentation
(USAF Approved for Public Release - FOIA)

The six prototype projects recommended for further development are shown in Chart 7:

RECOMMENDED PROTOTYPE PROJECTS — **Chart #7**

- Advanced Medium STOL Transport
- Very Low Radar Cross Section (RCS) Vehicle
- Large Tanker Aircraft
- Lightweight Fighter (LWF) Aircraft
- Quiet Aircraft
- Remotely Piloted Vehicle (RPV)

Chart #7 from B/G Chapman Presentation
(USAF Approved for Public Release - FOIA)

For each of the six recommended Prototype Projects, a summary of the principal technical objectives of the project was presented. Shown next in Chart #9 were the principal technical objectives of the Lightweight Fighter Aircraft Project:

PRINCIPAL TECHNICAL OBJECTIVES (CONTD) Chart #9

- Large Tanker Aircraft
  - Controllability of receiver aircraft and boom
- Lightweight Fighter Aircraft
  - High maneuverability and performance
    - Neutral stability
    - High aspect ratio wing
    - Automatic variable camber
    - High thrust/weight
    - Thin wing
  - Controllability
    - High "G" cockpit/sidestick control
    - Fly-By-Wire

Chart #9 from B/G Chapman Presentation
(USAF Approved for Public Release - FOIA)

In government–sponsored projects, little if anything can be accomplished without the authorization for expending funds for accomplishing the project. The Procurement Approach Task Group of the Prototype Study Group developed a cost summary profile for each of the six recommended Prototype Projects for FY 1972 through FY 1975. For the Lightweight Fighter Prototype Project, the assumption used for estimating the costing was that there would be two contractors each designing, building, and flight testing two prototypes. Given an initial total budget from the Office of the Secretary of Defense, the Procurement Approach Task Group did its best to allocate that total approved budget to make sufficient funds available and responsive to the needs of each of the six recommended Prototype Projects. Chart #11 next summarizes that allocation. [In the next chapter of this book, it will be described how the prime contractors managed to augment the allocated funds from the Government with their own company funding and with the cooperation of component vendors providing their components at cost (i.e., no profit) to the prime contractors so as to have their component flight- tested.]

COST SUMMARY
RECOMMENDED PROTOTYPE PROJECTS

Chart #11

	FY-72	FY-73	FY-74	FY-75	TOTAL
• Adv Med STOL Transport (1 Cntr, 1 or 2 Vehicles)	(3.5)	10-15	10-56	0-15	20-86
• Very Low RCS Vehicle (1 Cntr, 3 Vehicles)	5	8	7	-	20
• Large Tanker Aircraft (1 Cntr, 1 Vehicle)	2	-	-	-	2
• Lightweight Fighter A/C (2 Cntr, 2 Vehicles ea)	8	46	16	-	70
• Quiet Aircraft (1 Cntr, 2 Vehicles)	4	8	3	-	15
• Remotely Piloted Vehicle (1 Cntr, 2 Vehicles)	-	6	6	3	15
TOTALS	19	78-83	42-88	3-15	142-208

Chart #11 from B/G Chapman Presentation
(USAF Approved for Public Release - FOIA)

At the start of this chapter, a description was given of the AX program from which the Fairchild Republic Company developed the A-10 Thunderbolt II (a.k.a., Warthog). The management approach used during the AX prototype phase was one of unconventional management by the USAF of the Prime Contractors Northrop and Fairchild Republic. The AX competitive fly-off and the cooperative management approach, used by Fairchild Republic and the USAF A_10 Program Office, served as the model for the Management Approach Task Group of the Prototype Study Group in formulating a similar management approach to be used with all six of the recommended prototype projects. There is necessarily a mutual trust developed between the contractor and the government when using the prototype management approach, but it was shown to work well on the AX and A-10 development programs.

Chart #16 next summarizes the management approach which the Management Approach Task Group of the Prototype Study Group formulated for use in each of the six recommended Prototype Projects. (Chapter 7 will describe how this was accomplished within the Lightweight Fighter Prototype Project, and how it proved to be a very successful method of program management.)

MANAGEMENT SUMMARY **Chart #16**

- Advanced Prototype Programs can be managed effectively with:
  - Small Program Offices
  - Maximum Use of Contractor Formatted Data
  - Tailored Application of Formal Procedures
  - Minimum Reporting

  - THROUGH

  - Increased Industry Responsibilities
  - Direct Air Force Participation
  - Integrated Flight Test Programs
  - Use of Contractor Control Systems

Chart #16 from B/G Chapman Presentation
(USAF Approved for Public Release - FOIA)

Shortly following the presentation by Brigadier General Kenneth Chapman, Deputy Secretary of Defense David Packard gave an indication that the funding level and FY profile for the 6 recommended projects seemed reasonable and that he would recommend to the Secretary of the Air Force that it be approved and provided to the Prototype Program Office (to be established expeditiously at the Aeronautical Systems Division at Wright-Patterson AFB in Dayton).

USAF Col. Lyle W. Cameron was named as Director of the Prototypes Program Office and chose Lt. Col. William E. Thurman as Deputy Director. Formation of the Prototypes Program Office included those six projects as shown in Charts #7 and #11.

Chapter 7 of this book describes the formation of the USAF Lightweight Fighter SPO, the Request for Proposals (RFP) sent to six US aircraft manufacturers and for evaluation of their responses to the RFP for their design of a Lightweight Fighter most capable of meeting USAF LWF goals. The Source Selection of the two most promising of those six RFP proposal responses resulted in selection of the General Dynamics Model 401 and the two-engine Northrop Model P-600 to be manufactured and competitively flown in a one-year flight-test program. The winner of the USAF Lightweight Fighter “fly-off” will be identified at the end of Chapter 7

CHAPTER 7
LWF PROGRAM OFFICE, LWF RFPS, LWF PROPOSALS, TECHNOLOGIES IN LWFS (1971 TO 1972)

The Office of the Secretary of Defense (OSD) acted quickly on the recommendation of Deputy Secretary of Defense David Packard, pursuant to the presentation by Brigadier General Kenneth Chapman to Deputy Secretary Packard on August 5, 1971, that a Prototypes Program Office be established within the Aeronautical Systems Division, Air Force Systems Command, Wright –Patterson AFB, Ohio. HQ USAF had mixed emotions about instituting a new program office that might siphon funding from the F-15 System Program Office (SPO) and/or the Pratt & Whitney F100-PW-100 Joint Engine Program Office (JEPO); however, the Office of Management and Budget (OMB) had been allocated funds from the OSD to pursue all six of the recommended prototype programs to be funded as advanced technology demonstrators with no official recognition that there would be any funding extended for those prototype advanced technology demonstrators after their initial programs had been completed per the presentation given by B/G Chapman to Deputy Secretary Packard. Readers are reminded that the six programs (and their funding profiles) recommended in the Chapman/Packard presentation were detailed in Chart #11 in Chapter 6. (Chart #11 is repeated next for Reader's convenience.)

COST SUMMARY
RECOMMENDED PROTOTYPE PROJECTS — **Chart #11**

	FY-72	FY-73	FY-74	FY-75	TOTAL
• Adv Med STOL Transport (1 Cntr, 1 or 2 Vehicles)	(3.5)	10-15	10-56	0-15	20-86
• Very Low RCS Vehicle (1 Cntr, 3 Vehicles)	5	8	7	-	20
• Large Tanker Aircraft (1 Cntr, 1 Vehicle)	2	-	-	-	2
• Lightweight Fighter A/C (2 Cntr, 2 Vehicles ea)	8	46	16	-	70
• Quiet Aircraft (1 Cntr, 2 Vehicles)	4	8	3	-	15
• Remotely Piloted Vehicle (1 Cntr, 2 Vehicles)	-	6	6	3	15
TOTALS	19	78-83	42-88	3-15	142-208

Chart #11 from B/G Chapman Presentation
(USAF Approved for Public Release - FOIA)

The then-Commander of the Aeronautical Systems Division Lieutenant General James T. Stewart (not the same Hollywood actor James M. Stewart, who incidentally was also a Brigadier General in the USAF Reserve), assigned Colonel Lyle W. Cameron as the Director of the new Prototype Programs Office. Col. Cameron was selected primarily due to his having served exceptionally well as the Group Leader of the Project Selection Task Group in the Prototypes Study Group, headed by the Steering Group Co-Leaders Grant Hansen from OSD and four Lt. Generals and Maj. Generals. Colonel Cameron proceeded rapidly with organizing the efforts and staffing of the approved Advanced Prototypes Projects at Area B of Wright-Patterson AFB with full support from the ASD Commander Lt. General Stewart and all staffing organizations within ASD.

To assist Col. Cameron, Lt. Col. William E. Thurman was assigned as Deputy Director of the Prototypes Program Office. Both Col. Cameron and Lt. Col. Thurman were also USAF fighter aircraft pilots with very commendable service records. Col. Cameron flew

F-84s in Vietnam and was shot down, spending the next two years as a prisoner of war in the Hanoi Hilton. His stories of what he endured at the hands of his Vietnamese captors were amazing, but a recount of them is beyond the scope of this book. Lt. Col. Thurman flew F-105 aircraft in Vietnam, and also was shot down, but was able to return his badly damaged F-105 to friendly territory and where he was rescued. With their experiences of having both been shot down by enemy ground gunfire, it is of little wonder why the technology areas of survivability and vulnerability were given a high priority during the conceptual design of the USAF Lightweight Fighter, so that future pilots of the new fighter aircraft would fare better when exposed to a hostile ground threat environment.

Not all of the six approved prototype projects, however, were considered as needing to be conducted within Col. Cameron's Prototypes Program Office. The two highest funded projects and given the highest priority were (1) the Advanced Medium STOL Transport (AMST, which was decided to be pursued with two contractors, not the single contractor shown on chart #11, and fabricating/testing two vehicles each; and (2) the Lightweight Fighter (LWF) to also be pursued by two contractors and fabricating/testing two vehicles each. The Quiet Aircraft was to be pursued under Col. Cameron's Prototypes Program Office but not with as high a priority as the AMST and LWF, and accordingly was not set up as a manned project office but handled more as a separate special project.

The Very Low RCS Vehicle project was transferred to the Defense Advanced Research Projects Agency (DARPA) for pursuit as a sole-source development with the Lockheed Skunk Works, because of its stealth technology nature (and having a very highly classified application). The Large Tanker Aircraft project was transferred to the USAF Logistics Command and pursued as a sole-source basis with Boeing in response to an unsolicited proposal received from Boeing. The project entailed mainly a major design modification of the refueling boom on a candidate Boeing 747 aircraft. The Remotely Piloted Vehicle (RPV) project was not funded for a FY-72 start, and therefore was also on a low priority level compared to AMST and LWF. Further, the RPV was more appropriate to be developed as an unmanned flight vehicle falling into the drone/target vehicle technology domain, for which the USAF Armament and Weapons Laboratory at Eglin AFB, Florida, was responsible.

The management approach approved by OSD for the Prototype Projects was that as presented by B/G Chapman to Deputy Secretary Packard in the August 5, 1971, presentation. The charts outlining that management approach approved were identified in the presentation charts (chart #2 and chart #16) shown in chapter 6 related to management approach, which Readers are invited to recall for reference. A specific chart not shown in chapter 6 but which is more appropriate to show in this chapter 7 displays a somewhat more visually useful outline for how the AMST and LWF Prototype Projects were to be managed on a day-to-day working level management process. Chart #13 from the Chapman/Packard presentation is shown next.

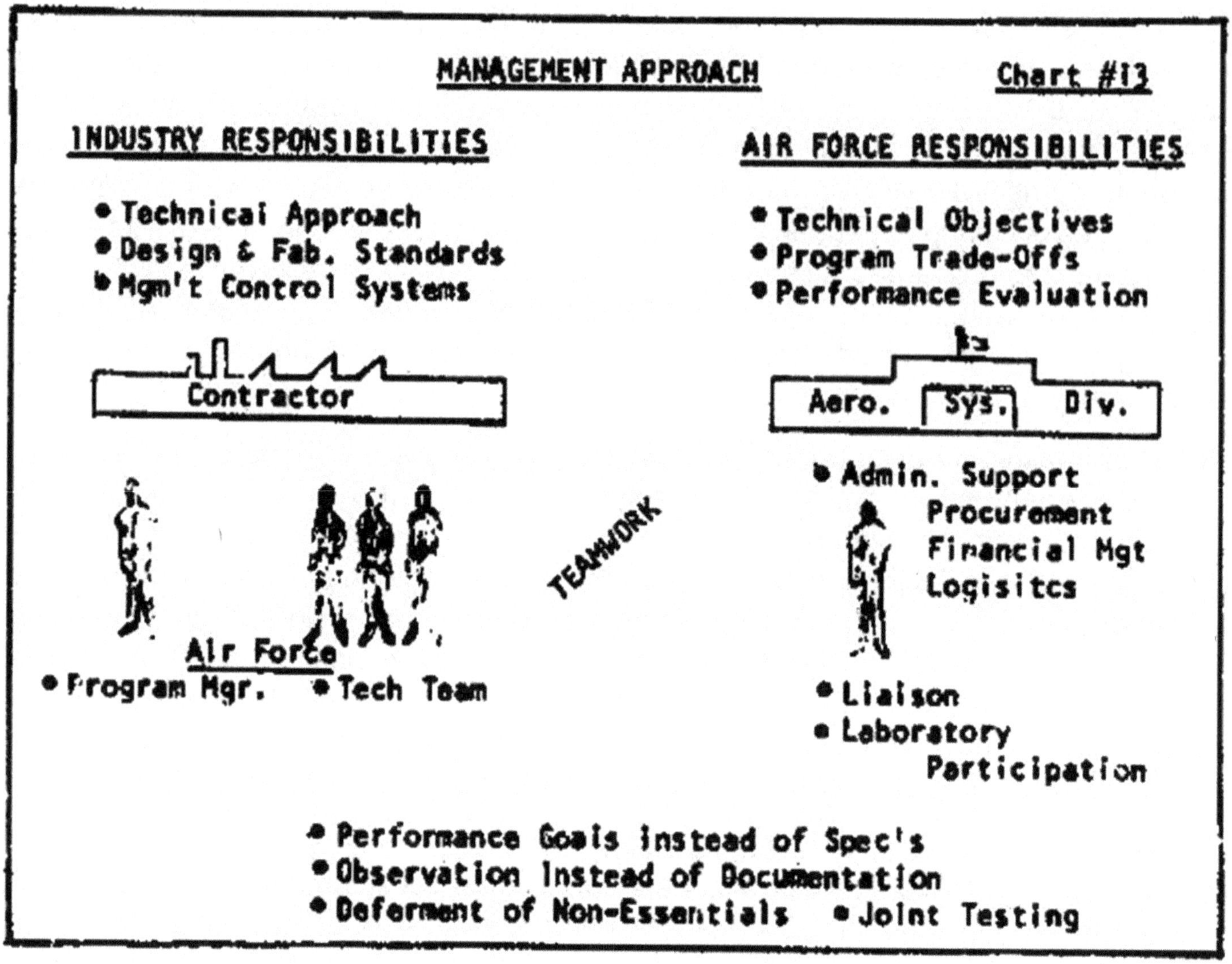

Chart #13 from B/G Chapman Presentation
(USAF Approved for Public Release - FOIA)

When B/G Chapman presented Chart #13 to Dep Sec Def Packard in August 1971, the text he used for his narrative was included with a copy of Chart #13. That text clarifying the management approach to be taken by the Prototype Project Office teams is presented next verbatim:

The management concept is based on the Air Force retaining responsibility for establishing technical objectives, for maintaining a proper balance between these objectives and the program progress, and for evaluating the final results of the project. Responsibilities for establishing the technical approach, for selecting design and fabrication standards, and for exercising adequate management control of the project are assigned to the Contractor. Because we advocate replacing paperwork with people, it is desirable to have the Air Force Program Manager and a small technical team working closely with the Contractor. This team could on a selective basis be located at the plant with the liaison function at home where the additional support that a Program Manager requires can be provided on an "as-needed" basis. Laboratory participation will be obtained, perhaps to the extent of "managing the project" in some cases. Laboratory involvement will assure that the full spectrum of appropriate technologies is infused into the program. Each program management team would consist of a small number of highly qualified Air Force people separated organizationally from the Division Commander by not more than one level of authority. In summary as shown on the chart, Air Force specified performance goals would be used for establishing the program rather than firm requirements and rigid specifications, direct Air Force observation and participation would be substituted for contractor documentation. All matters not essential to the advanced prototype demonstration would be deferred, and the method of testing would be planned to satisfy both the Contractor and the Air Force's interest at minimum cost and with minimum resources.

As woven into the previous chapters of this book, it is the talents, personalities, and experience of the persons directing the advancement of a technology/technologies or its/their integration with a set of other advanced technologies comprising a weapon system. Since the Prototype Program Office was located within the Aeronautical Systems Division at Wright-Patterson AFB, Col. Cameron and L/C Thurman had a very large pool of professional talent, both civilian and military, from which to choose who were assigned in the several organizations comprising the Aeronautical Systems Division and the Air Force Research Laboratories resident at Wright-Patterson AFB.

The remainder of this chapter 7 will address only the formation of the Lightweight Fighter Project, the preparation of the project goals/requirements and statement of work for the LWF request for proposals (RFP), and the evaluation of the Proposal Responses to the LWF RFP, including the award of contracts to two aircraft manufacturing companies to each fabricate two flight test articles to be flight-tested.

Col. Cameron and L/C Thurman selected 5 persons to form the nucleus of the team managing the LWF Project. Chart #13 shown earlier was marked (in red) to show who and what function the Project team members would perform in management of and in teamwork with the future two contractors designing, fabricating, and flight testing the two LWF advanced technology demonstrator prototypes. A similar selection of personnel from within the Aeronautical System Division and AFSC Laboratory organizations at Wright-Patterson AFB was done for the AMST Prototype Project, but is not shown here.

A copy of that edited Chart #13 for the staffing of the Lightweight Fighter Project is shown next.

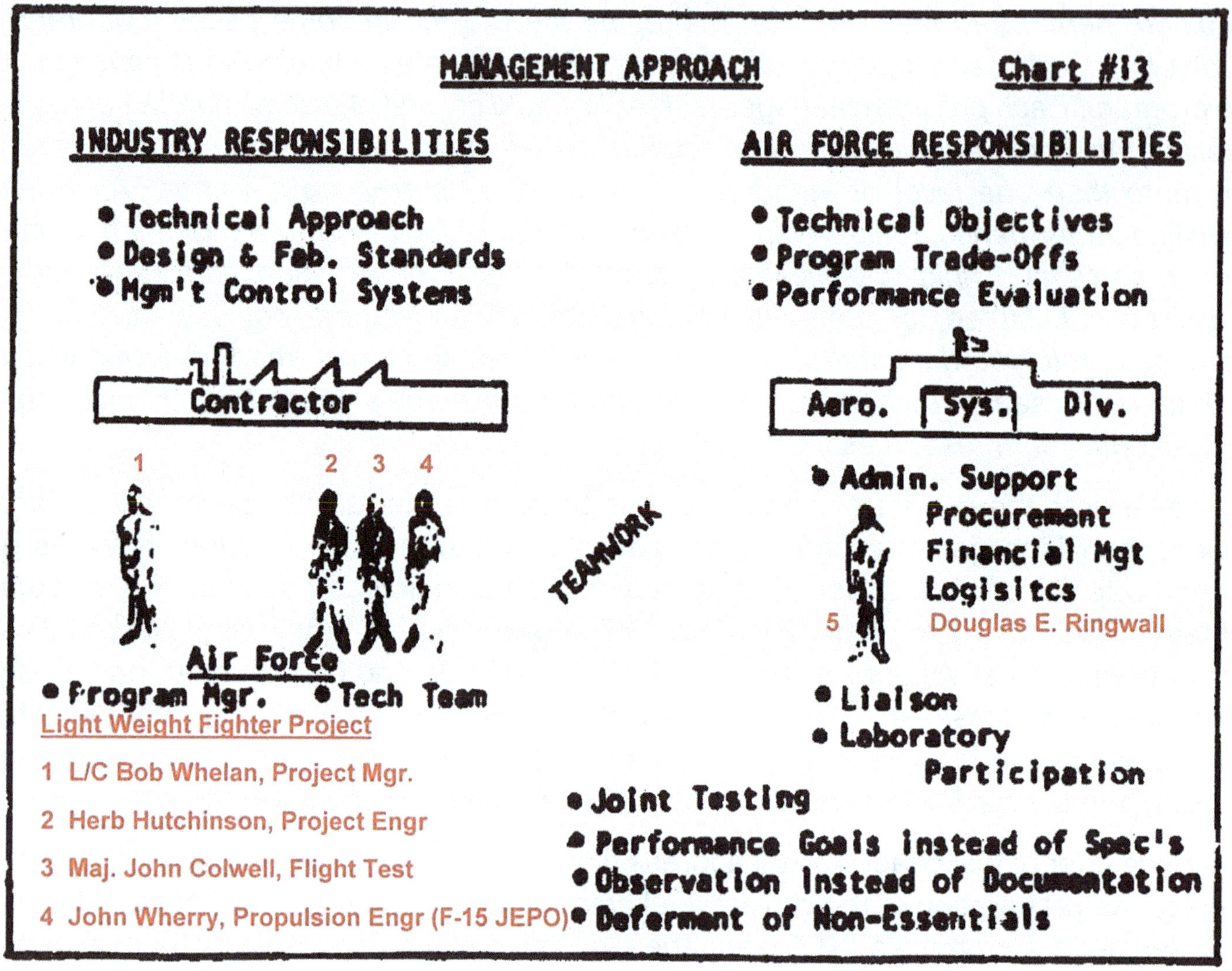

Chart #13 Mod. from B/G Chapman Presentation
(USAF Approved for Public Release - FOIA)

[Author's note: Four of the five persons named to form the charter LWF Prototype Project Office were reassigned from ASD organizations. The Project Engineer Herb Hutchinson (the Author) was assigned from the Air Force Flight Dynamics Laboratory due to the advanced technologies to be tested on the LWF prototypes – many of those advanced technologies having emerged from AFFDL technology development activities.

A traditional Dinner, hosted by the Commander of ASD, Lt. Gen Stewart, was held at the Wright-Patterson AFB Officers Club in September 1971 to introduce the new Prototypes Program Office staff to the other existing System Program Offices staffs (F-15 SPO, F100 Engine JEPO, A-10 SPO, B-1 SPO, etc.). Col. Cameron introduced his entire LWF project staff as being all of those asked to stand - his "five Young Turks". Proudly among them was the Author.

The Program Plan for the Lightweight Fighter Prototype, developed by the Prototype Study Group, (summarized on pages 4-13, 4-14, and 4-15 in Appendix 4 extracted from the Final Report of the Prototype Study Group) is shown next.

Air Force Advanced Prototype Program Plan (U)

1. **Program Element Number:** 63XXXF

2. **Project Title:** Lightweight Fighter Aircraft (U)

3. **Program Summary:**

a. Objective: To demonstrate the high maneuverability and controllability of a lightweight fighter aircraft.

b. Description:

(1) Background: Because of the technical risks involved the Air Force has been unable to include an optimum combination of new aerodynamic concepts and design ideas in approved development programs of fighter aircraft. By restricting any potential mission applications to day air-to-air capability a lightweight fighter design with high performance should be achieved by combining these new technologies.

(2) Program Brief:

(a) Design Goals: A dual source (two aircraft each) prototype effort is planned to evaluate two different technical approaches. Significantly different proposals are expected from industry due to the large number of contractors expressing interest in pursuing this program. The design and performance objectives will be goals in order to allow design tradeoffs which will result in the best overall system. Guidelines for design and performance goals will include: low weight - less than 20,000 lbs., unequalled performance and maneuverability in the transonic (8 - 1.5 Mach), high-g, combat arena; combat radius on internal fuel of about 225 NM, with increases up to 700 NM with external tanks; limited supersonic capability at sea level (Mach 1 - 1.2), increasing to about Mach 2 at altitude as constrained by fixed geometry inlets optimized for the transonic battle arena; in being or late developmental propulsion; mission essential avionics; armament to include a representative state of the art/high muzzle velocity gun, and an effective low cost air-to-air missile; hardpoints and systems to give a credible air/ground capability; outstanding visibility; and excellent handling qualities optimized for the combat arena.

(b) Test Program: The test program will be limited to demonstrating the performance and operational capabilities of the aircraft as a prototype vehicle rather than as a production model. As

Atch 4

such, full scale Category I, II, and III tests will not be performed. Instead, a joint airworthiness and flight performance evaluation will be conducted with the Air Force entering the program at the earliest possible point in time. The contractor will retain his traditional role in that he will demonstrate the basic airworthiness and initial expansion of the flight envelope, with emphasis on flight loads, flutter, and structural demonstrations. This will allow the contractor to incorporate design refinements in the vehicle during the flight test portion of the program. The Air Force will then conduct limited performance and operational suitability tests to include: stability and control, systems, performance/propulsion, and operational effectiveness capabilities. The tests are envisioned to be accomplished in approximately 300 flight hours, with about 100 to be flown initially by the contractor, 100 to demonstrate performance, systems, and stability goals, and 100 to investigate operational capabilities.

(3) Technological Assessment:

(a) Uncertainties involved in the prototype effort vary with the individual proposals. While no formal request for proposal has been issued, contractor data presently available indicate generally reasonable chances of technological/operational success. Representative areas of possible risk include: technological advances in high acceleration cockpits; sidestick/fly-by-wire control; automatic variable camber; neutral stability; and flutter, lift, and drag associated with high aspect ratio thin wings. Operational risks encompass the effectiveness of a small, lightweight, simply designed, relatively inexpensive air-to-air fighter aircraft with exceptional maneuverability and austere mission essential avionics. Growth options for follow-on technological investigation may include: composite structures; side force, direct lift, and task oriented control; integrated stores; and attack subsystems investigations.

(b) Potential combat effectiveness gains of the lightweight fighter concept appear to be exceptional. The most advanced of the available proposals indicate combat performance capabilities far superior to the present and postulated fighter threat. Attainment of these goals in the prototype aircraft appears to be reasonably probable.

4. RDT&E Funds: ($ Millions)

	FY 72	FY 73	FY 74	Total at Completion
Proposed	8.0	46.0	16.0	70.0
FYDP (F&FP)	0	0	0	0
Net Change	8.0	46.0	16.0	70.0

5. Funding Levels:

a. FY 72

(1) Complete conceptual design.

(2) Initiate preliminary design.

b. FY 73

(1) Complete wind tunnel testing.

(2) Complete detail design.

(3) Initiate fabrication.

c. FY 74

(1) Complete fabrication.

(2) Conduct flight testing.

6. Management: Management of the Lightweight Fighter program will follow the Adaptive Management concept as defined in Appendix 2.

7. Procurement Approach: The approach for procurement of the LWF will be in accordance with the Advanced Procurement Plan in Appendix 3, Sec 5.

8. Program Schedule and Milestones:

Jan 72 - Program start

Feb 72 - Initiate conceptual design

Apr 72 - Complete conceptual design

Jul 72 - Initiate preliminary design

Jan 73 - Initiate fabrication

Aug 73 - First flight

Aug 74 - Complete flight testing

9. Support: Support requirements will be dependent upon the contractor selected.

In late September 1971, Col. Lyle Cameron assembled the staffs (all nine persons – John Wherry serving as the Propulsion engineer for both programs) of the AMST and LWF prototype projects in the Program Office conference room and started with the question: *Do you know what Adam said to Eve that first time in the Garden of Eden?" "You'd better stand back. I don't have any idea how big this thing is going to get.* He then authorized and tasked the LWF Prototype Program Staff (five "young Turks") to begin putting together the design and performance goals and the statement of work for the request for proposals to be issued to the aircraft manufacturers eagerly awaiting those RFPs. Col. Cameron referenced the applicable portions of the Prototype Study Group Final Report (the preceding extracted three-page program plan shown for the LWF prototype) as the place to start.

Based on the results of the Advanced Day Fighter Study conducted by Col. John Boyd, it was anticipated that the General Dynamics proposal response would feature the Model 401 design and having the Pratt & Whitney F100 engine as the source for meeting the desired T / W of about 1.2 at the start of combat. It was known also that the Northrop proposal response would feature the Model P-530 (3) design featuring the use of two General Electric YJ101-GE-100 turbojet engines. It was not known if any of the other proposal responses from the responding aircraft manufacturers would also be using the P&W F100 turbofan engine. It would therefore be possible that the two awards for the LWF Project would be to at least one and possibly two contractors proposing use of the P&W F100 engine. The ASD Commander, Lt. Gen. Stewart considered it absolutely necessary to obtain Gen. Slay's approval and commitment that up to potentially six of the Pratt & Whitney F100 engines would be made available as GFE (government-furnished equipment) – one for each of the four flying prototypes and two spare engines for the two competing designs in the event of damage to one engine each contractor. Less than that number of GFE engines could jeopardize the completion of the prototype demonstration flight-testing.

Approximately two weeks later (early October 1971), L/C Bill Thurman, LWF Project Manager L/C Bob Whelan, LWF Project Engineer Herb Hutchinson (the Author), and F-15 Engine JEPO Engineer John Wherry, traveled to Andrews AFB for a presentation to the Commander of the Air Force Systems Command Alton D. Slay, with three primary objectives for the presentation being as follows:

1) To assure Gen. Slay and the AFSC Staff that the LWF Project Office well understands the directions given by HQ USAF to proceed with the LWF Project in accordance with the stated program plan;

2) To assure Gen Slay and the AFSC Staff that the two Contractors to be selected for the LWF prototype activities will be made fully aware that when their proposal responses have been submitted for the LWF advanced technology demonstrators, the proposals have been evaluated, and two contractors have been awarded the

two-years contracts for designing/fabricating/testing of the four-prototype aircraft, there was no plan established for either of those prototypes to enter into a production;

3) To solicit the AFSC preapproval for the use of the Pratt & Whitney F100-PW-100 turbofan engine (nearing completion of development for the F-15 Eagle) in the proposal responses as being the propulsion source for meeting the high T / W goal for the propulsion system in the Lightweight Fighter.

It was known before the presentation to Gen Slay that he was much opposed to the use of any F-15 resources for support of the LWF Prototype Project. Col. Cameron and L/C Thurman had developed the strategy to convince Gen. Slay that it would be to the advantage of the F-15 Program (which was faced with higher than initially planned cost per unit F100 engine due to some unexpected problems in fuel control experienced in engine ground test runs at P&W in West Palm Beach) in the event the LWF Prototype Project were to go into any level of production as a high/low mix with the F-15. If it did, there would be a greater number of F100 engines to be produced which would reduce the per-unit cost due to the larger number of engines to be produced. If that should occur, then both the F-15 Program and the LWF F-XX Program would benefit from the lower cost per engine unit.

That strategy worked as hoped, and after the presentation to Gen. Slay, he gave approval (with three stipulations) for up to six Pratt & Whitney F100 engines to be made available as GFE for use in the LWF prototype flight-testing phase of the LWF project. The three stipulations imposed by Gen. Slay's approval were: (1) the up-to-six GFE F100 engines would not be modified in any way such that when all six were returned to the F-15 JEPO they could be used exactly as they were originally configured for use in the F-15 Program; (2) that if the LWF F-XX Program would result in a production run increase of the F100 engine, there would be no requirement to have a configuration difference on the F100 engine production line as to the engine being configured for use in the F-15 or configured differently for an LWF F-XX; and (3) that every precaution be taken to neither FOD (foreign object damage) the loaned F100 engines nor to subject them to any exceedance of their known operating limit during the maneuvering portions of the LWF flight-test phase.

With that major potential obstacle avoided (i.e., having Gen Slay's approval to use up-to- six GFE Pratt & Whitney engines), the activities to prepare the requests for proposals for the LWF prototype were able to proceed expeditiously.

A key document for the RFP (as in any Government contracted work effort) is the Statement of Work which details the work to be performed by the Contractor. The Statement of Work generated for the LWF contract effort was a tailored version of the Statement of Work shown in example format in the RFP/Model Contract for the Advanced Medium STOL Transport (AMST) Prototype Project – as contained in

Appendix 7 to the Prototype Study Group Final Report. The Model Contract shown in Appendix 7 had also been tailored to account for the stream-lined management approach as shown in earlier extracts from the Prototype Study Group final report (issued in September 1971). Appendix 7 to that Prototype Study Group final report included models of a Letter of Invitation, a model RFP, and a Model Contract – shown for the AMST Prototype Contract. It was that RFP and Model Contract that was further tailored for the LWF Prototype Project (i.e., substituting "LWF" for "AMST" and substituting alternate descriptions for the Work Required, Proposal Data Requirements, Deliverables, and for the GFE and Test Facility requirements unique to the LWF Prototype Program Plan). Additional boilerplate forms and instructions were included in the model contract for the AMST but again were minimized so as to retain only those regulations needed to protect the safety of humans and high-cost equipment items engaged in the development and testing process planned for the LWF Project.

Copies of the actual Request for Proposal sent to potential responding aircraft manufacturing companies were not found in the Author's pack rat files; therefore, the Statement of Work portion of the RFP package has been reconstructed in the best- effort nature of the work requested to be performed.

The Statement of Work was developed using the example Statement of Work contained in the Prototype Study Group Final Report released in September 1971. A reconstruction of the Statement of Work (segmented in Tasks and Deliverables) of the Contract Effort to be accomplished by the LWF Project Contractors is shown below:

Statement of Work Tasks & Deliverables

Task 1 – Contractor shall design, develop, and fabricate two (2) prototype Lightweight Fighter aircraft based upon the Performance/Design Goals of Attachment A. Contractor shall also provide the equipment necessary for support of the aircraft with maximum use of standard hardware.

Task 2 – Contractor shall conduct sufficient ground testing and systems check-out to assure that the aircraft is ready and safe for flight. Provide a Safety of Flight Certification prior to first flight to a USAF Executive Independent Review Team (EIRT) for review and approval of first flight. Provide an updated version of that Safety of Flight Certification to the Prototype Program Office prior to the aircraft delivery to the Air Force for testing by Government pilots. The Safety of Flight Certification shall include a matrix of major components, safety critical functions and verification methods, a definition of the flight profiles within which the aircraft can be safely operated and any limitations or restrictions required to assure that no unsafe conditions occur during flight testing.

Task 3 – Contractor shall conduct a flight test program to verify the capability of the aircraft to satisfy the design requirements as established by the Contractor considering the Air Force goals defined in Attachment A. Based upon the flight test program, the

Contractor shall provide the Air Force with the test data generated by the Contractor (in Contractor format) defining the aircraft flight characteristics.

Task 4 – Contractor shall provide the necessary instruction and data for ground and flight training of six (6) Air Force test pilots in Contractor developed ground simulators and cockpit mock-ups.

Task 5 – Contractor shall provide logistics and maintenance support of the two (2) prototype aircraft during the Air Force flight testing. The Air Force flight test evaluation program, evaluating the flight performance of the prototype aircraft relative to achievement of the Performance Goals of Attachment A, will consist of approximately 200 flight hours accumulated in the two LWF Prototype aircraft over a 12-month period at the Edwards AFB Flight Test Center.

Task 6 – Contractor shall support and conduct any repair to or unscheduled maintenance of aircraft structure or equipment (Contractor-furnished or GFE) accidentally caused during ground and/or flight-testing during conduct of Task 2 through Task 5. (Costs for this contingency will be addressed if/when damage is incurred).

Task 7 - Contractor shall frequently interface with the Air Force to assure that sufficient information, in Contractor format and content, concerning progress of the program is available when necessary.

Task 8 – Contractor shall generate design of a two-seat variant of the LWF Prototype final configuration during, and in parallel with, the flight testing of the single-seat aircraft undergoing testing.

Additional Deliverables with RFP

1) Offeror shall deliver at same time of submittal of Proposal Responses to the RFP a wind tunnel model of the final flight configuration of the LWF Prototype aircraft for testing by Government personnel at Government wind tunnel facilities (to be arranged for and funds transferred to cover testing by the Prototype Program Office);

2) Offeror shall deliver an IBM Card Deck(s) containing the aerodynamics of the aircraft and of the installed propulsion system net thrust throughout the flight envelope for the LWF aircraft configuration proposed;

Offeror shall submit cost estimates for accomplishment of the seven (7) planned tasks identified above and for the Additional Deliverables on a DD Form 633.

3) Offerors will comply with the instructions on the DD Form 633 and furnish the necessary supporting details.

Deliverables at Completion of Contract

1) Contractor shall submit Final Test Reports (in Contractor format) for Contractor-Conducted Tests (Laboratory, Ground, and Flight) at end of Contract.

2) Contractor shall submit Final Report (in Contractor format) for Final Designs (single-seat and two-seat variant) at end of Contract.

3) Contractor shall deliver the two LWF Prototype aircraft to the Air Force for continued testing as deemed desirable by the Air Force.

The Statement of Work for the LWF Program references Attachment A which contains the Design and Performance Goals prepared for inclusion in the LWF RFP. The specific Design and Performance Goals generated for the LWF were extracted from the Prototype Study Program Plan for the Lightweight Fighter Project shown on the three pages (numbered 4-13, 4-14, and 4-14 extracted from the Prototype Study Group Final Report) earlier in this chapter 7. The Design and Performance Goals are recognized as having been largely generated by Col. John Boyd and are based on his *Aerial Attack Study*, his development of the *Energy-Maneuverability Theory* and the *Specific Excess Energy* comparisons of aircraft. It also capitalized on his leadership in the Advanced Day Fighter Study done with General Dynamics and Northrop – using funds found by his lightweight fighter mafia associate Col. Everett "Rich" Riccioni, and from his own talents/experiences as an F-86 Sabre and F-100 Super Sabre pilot (and as an instructor at the USAF Fighter Weapons School).

For Readers convenience, the LWF Design and Performance Goals have been condensed into the format of a briefing chart and are as presented on the next page. The listing of the design and performance goals are not in any order of priority although those that appear early in the listing tend to have more significant effect on the low-weight, low-cost, highly- maneuvering features of the LWF design, whereas those that are listed later in the listing add to the capability of the LWF's fighter capabilities and not being drivers on the LWF weight, cost or maneuverability.

DESIGN AND PERFORMANCE GOALS FOR LWF

(No Priority Order)

- Low weight – less than 20,000 pounds at start of combat;
- Unequalled performance and maneuverability in the transonic (Mach 0.8 – 1.5), high- g, combat arena (especially at altitudes between 20,000 and 40,000 feet);
- Use of in-being or late developmental stage propulsion system to provide T/W NLT 1.2 at Start of Combat (P&W F100-PW-100 turbofan engine in single engine application preferred);
- Combat radius on internal fuel of about 225 NM, with increase up to 700 NM with external tanks (jettisonable);
- Limited supersonic capability at Sea Level (Mach 1.0 – 1.2), increasing to about Mach 2 at altitude as constrained by fixed geometry inlets optimized for the transonic battle arena;
- Mission essential avionics (clear weather, day conditions) with capability to be upgraded later to all-weather & all-light conditions;
- Outstanding pilot visibility (bubble canopy preferred);
- Excellent handling qualities optimized for the air combat arena;
- Armament to include a representative state of the art/ high muzzle velocity gun (20 mm M61-A1 Vulcan 6-barrel Gatling preferred with minimum of 500 rounds of ammunition in a reloadable drum), and effective low-cost air-to-air missiles;
- Structural hard-points and systems to give a credible air-to-ground support capability.

Design and Performance Goals for LWF RFP Proposal
(Chart Created by the Author)

The USAF Lightweight Fighter request for proposal (RFP) and model contract package with a Letter of Invitation to submit a proposal in response to the invitation was sent on 6 January 1972 to nine aircraft design and manufacturing companies in the United States and with a time span of six weeks in which to respond with proposal responses fully responsive to the RFP issued, the Statement of Work and Deliverables, the Design and Performance Goals (Attachment A), and a Model Contract. Those nine Companies to whom the RFP package was sent included (in alphabetical order):

Boeing Aircraft Company (Military Products Division), Wichita, Kansas
Fairchild Industries, Hagerstown, Maryland
General Dynamics Corporation, Fort Worth, Texas
Grumman Aircraft Company, Bethpage, New York
Ling-Temco-Vought (LTV), Dallas, Texas
Lockheed Aircraft Company, Burbank, California
McDonnell Aircraft Company, St. Louis, Missouri

Northrop Corporation (Aircraft Division), Hawthorne, California
North American Rockwell Aviation, El Segundo, California

The Letter of Invitation to potential proposal responders from Lt. General James T. Stewart, ASD Commander, was worded as shown in the example next

(EXAMPLE)

DEPARTMENT OF THE AIR FORCE
HEADQUARTERS AERONAUTICAL SYSTEMS DIVISION (AFSC)
WRIGHT-PATTERSON AIR FORCE BASE, OHIO 45433

FROM: ASD/Office of the Commander 6 January 1972

SUBJECT: Lightweight Fighter Request for Proposal

TO: XYZ Corporation

1. You are invited to submit a proposal to the Air Force for furnishing two (2) prototype Lightweight Fighter aircraft as described in the Request for Proposal attached hereto. A test program and certain data will also be required. Two Contractor sources are planned to be selected to perform the Lightweight Fighter Prototype program. Funds per Contractor will be provided to not exceed $4.0M in FY 72, $23.0M in FY 73, and $8.0M in FY 74 ($35M per Contractor for the Total Contract effort planned to be completed by 30 September 1974). This is to be a "best effort", "design to goals", and "adaptive management" type of Contract. No follow-on effort is planned at present. Once contract awards are made, further funding will not be accomplished.

2. The primary objective of the Lightweight Fighter program is to investigate and demonstrate advanced technologies which when integrated in a light-weight, low-cost fighter aircraft will result in a fighter having unequaled performance and maneuverability in the transonic, high – g combat arena when being powered by an existing or in-development propulsion system. The aircraft must have a credible air-to-ground capability, have outstanding visibility for the pilot, and have excellent handling qualities optimized for multi-modes of operation in the combat arena. All of these beneficial performance characteristics are to be accompanied in an aircraft having outstanding maintainability, supportability, survivability, and availability characteristics when being operated in a clear- weather, day environment with mission-essential avionics and armament.

3. Please submit, for evaluation only, that data requested by this RFP. Any data not directly pertinent to the RFP will be either returned to you immediately, or discarded. The attachments to this letter describe the proposal requirements. Your proposal must be responsive to the RFP attached hereto, and be limited to 50 pages of technical information, 10 pages of management data, and 10 pages in your cost proposal. Required certifications and cost data format are specified in the DD Form 633 attached.

4. Your proposal (seven copies) must be furnished to Hq ASD/YP, Bldg 14, Area B, Wright-Patterson AFB, by 1300 on 12 February 1972.

/S/ James T. Stewart
Lt. General, USAF
Commander

1 Atch
RFP

Example USAF Letter of Invitation to Propose
(From USAF LWF PPO File – FOIA)

Four of the above nine companies advised the Air Force that they would not be submitting proposal responses to the LWF RFP Invitation to propose. Those four companies planning not to respond (for reasons not required to be explained) were Fairchild Industries, Grumman Aircraft Company, McDonnell Aircraft Company, and North American Rockwell Aviation.

The five companies which submitted responses to the LWF RFP to ASD on/before the due date of 12 February 1972 were the Boeing (Military Aircraft Division), General Dynamics/Fort Worth Division, LTV, Lockheed (Military Aircraft Division), and Northrop (Aircraft Division).

In the case of the Northrop proposal response, Northrop had earlier notified the Prototype Program Office that it would comply with the requirements of the RFP in which the USAF had stated a preference that the design of the offered LWF prototype be a single-engine configuration (to achieve the low-weight goal) and that the preferred single-engine being of the status "existing or in development" with again a preference stated that the engine be the Pratt & Whitney F100 engine being developed for the F-15 Eagle program. Northrop pointed to the fact that the Northrop Aircraft Division had been pursuing development of a two-engine design (its Model P-530 Cobra) since the conduct of the Advanced Day Fighter Study sponsored by Cols. Boyd and Riccioni and had a strong desire to see that their two-engine (using General Electric YJ101 engines being developed by the GE Company in Lynn, Massachusetts) design be considered for inclusion in the evaluation of the other LWF Program proposal responses.

Shown next is a photograph of a full-scale mock-up of the P-530 Cobra at the Northrop Aircraft Division in Hawthorne, California.

Northrop P-530 Cobra Mock-Up Hawthorne, CA, 1970.
(Source: USAF Engineering files)

The Prototype Program Office gave careful consideration to the Northrop request and concluded that the best way, being fair to the other Offerors would be for Northrop to submit a single engine design to be in responsiveness to the issued RFP, and in parallel to submit an alternate Proposal Response which also was fully responsive to the original issued RFP except for the design being a two-engine design and using the General Electric YJ101 prototype engines. The rationale used by the Prototype Program Office was that since the objectives of the LWF Prototype Program were to explore and demonstrate advanced technologies for a new fighter having low weight, low cost, and unequaled maneuvering capability in the combat arena – and not limiting the designs proposed to be only single engine designs to meet those goals. Accordingly, Northrop was allowed to submit two responses to the LWF RFP – the single engine design and the two-engine design as an alternate proposal. Both the Northrop single engine proposal (Model P-610) and the two-engine design proposal (Model P-600) would therefore be fully evaluated along with those submitted by the other Offerors.

For Readers clarification of the decision for the Prototype Program Office to evaluate both the single engine and two engine designs Northrop proposed for the LWF RFP responses, there had been strong interest expressed by the US Navy that it wished to closely follow the USAF LWF Program developments, as it also had been exploring designs in a Navy study for a low-cost aircraft in the fighter fleet-defense technology domain to complement the Navy F-14 Tomcat in operation. The US Navy has long-favored their carrier-based aircraft inventory to have two engines for safety when operating far out at sea. That strong desire is well understood – losing one engine of a two-engines aircraft permits return to the carrier, whereas losing one engine on a one engine aircraft is virtually a certain loss of that aircraft and its crew (unless the single engine aircraft with no power crashes into the ocean and within a distance that the crew can be rescued by carrier-based search-and-rescue helicopter aircraft).

A second point of interest to Readers may be that there has been a long-standing difference in perspective (often very heated debates at Happy Hour at the Officers Club bar) within the Air Force as well as between the Air Force and Navy pilots of which is better in air combat – a single engine fighter or a two-engine fighter? Being given the opportunity to address the long-standing "one versus two" question under controlled tests such as that in a LWF Prototype Program being proposed was one which was attractive to both the USAF and the USN. Therefore, when the Prototype Program Office allowed Northrop to submit a primary and an alternate proposal response, the Prototype Program Office also advised the other Offerors (Boeing, General Dynamics, LTV, and Lockheed) that if they wished to submit an alternate two-engine proposal for USAF evaluation of Proposal Responses, they were welcome to do so.

In the end result, only Northrop chose to submit both a single engine design and a two-engine design response to the LWF RFP. As a means of focusing on their best candidate for the USAF LWF Proposal Evaluation, Northrop did not fully respond to the LWF RFP in delivering a wind tunnel model (to be tested at NASA facilities by USAF engineers) for both their single engine design and their two-engine design. Northrop did however, deliver (as per the Deliverables Requirement of the RFP) an IBM card deck containing their estimate of the aerodynamics and installed YJ101 engine thrust for evaluation of the maneuvering performance and energy maneuverability properties of a single-engine design. A wind tunnel model of the Northrop proposed two-engine P-600 (derivative of the P-530 Cobra) design was submitted per the RFP Deliverables requirement.

LTV submitted a response to proposal in accordance with the invitation to offerors placing extreme emphasis on the low cost for developing a prototype of a new fighter, and based their design ((Model V-1100) on the use of many components from their USN F8 Crusader and A-7 Avenger aircraft designs. The LTV proposal response was later withdrawn by LTV from further consideration by the USAF as LTV considered it technically non-competitive. The proposed configuration (Model V-1100) did not sufficiently advance the new technologies needed in order to be fully responsive to the design and performance goals sought by the USAF.

The Lockheed CL-1200 Lancer capabilities formed the basis for the proposal response submitted by Lockheed in response to the USAF LWF RFP, and was in an already- flying status, making it the most rapidly available candidate for the USAF LWF Program. Shown next is a photograph of the Lockheed CL-1200 Lancer.

Lockheed Model CL-1200 Lancer (Mod. for LWF RFP)
(Source: USAF Engineering files)

For much the same reasoning as that used by LTV, Lockheed elected to withdraw their proposal from further consideration in the USAF LWF prototype program as it did not feature any substantially advanced flight technologies that were then emerging from the Air Force research and development laboratories. The general perception held by the LWF Program Office of the CL-1200 was that it was “only a luke-warm version of a warmed-over F-104”.

That narrowed the LWF RFP evaluation to four proposed designs: the Boeing Model-908 design, the General Dynamics Model 401 design, and the two proposed designs from Northrop – the single-engine Model P-610 and the alternate proposed two-engines Model P-600. Northrop chose those designations as the two-engines Model P-600 had two-holes in its end, and the single-engine Model P-610 had only one hole in its end.

Shown in the photograph next is a full-scale mock-up of the Boeing Model 908 submitted as the Boeing proposal response design to the USAF LWF RFP.

Boeing Model 908 (USAF LWF RFP)
(Source: USAF Engineering files)

Initially, the Boeing Model 908 was considered to be a very competitive design as it utilized the Pratt & Whitney F100 as desired by the USAF LWF Program Office, featured a forward fuselage having a pair of chines to provide the resistance to departure (an adverse stability property of fighter aircraft) of the nose at high angle of attack, and featured extensive use of composite materials to achieve a very low empty weight condition. There was concern however that there was insufficient volume in the internal fuselage for fuel tanks capable of achieving long enough legs (flight distance) to meet the combat radius goal for the aircraft.

From an external aerodynamics consideration, the juncture of the wing to the fuselage did not reflect any wing-body blending by which NASA had shown in their configuration research not only improved the external dynamics (since air molecules do not like to flow into 90-degrees corners) but also added substantial volume for internal fuel. There were no leading-edge extensions which NASA had developed to produce vortex lift. As can also be seen, there was insufficient room in the nose section for a potential future search-and-track fire control radar system, although the RFP did not require such an advanced radar system. The visibility in the cockpit was awesome, despite there being a bow-frame in the two-piece bubble canopy.

Shown next is the General Dynamics Model 401-16B submitted in response to the USAF LWF RFP:

General Dynamics Model 401-16B Ft. Worth, TX (1971)
(Source: USAF Engineering Files)

Comparing the external contours and features of the two configurations, it can be seen that the Boeing Model 908 was a rather close-configuration to the General Dynamics Model 401, differing mostly in the shape of the chin inlet. The Boeing Model 908 had a more rectangular shape inlet compared to that of the General Dynamics Model 401 which featured a warped-elliptical "smiley" chin inlet... The Boeing Model 908 had no wing-body blending – reflecting that Boeing had not utilized much, if any, of the aerodynamics technology advancements being made by Dr. Dick Whitcomb at NASA in the domain of transonic and supersonic aerodynamics. The GD Model 401 canopy was close to the same bubble-shape which gave pilots outstanding visibility in the upper hemisphere of their field of vision, General Dynamics apparently having decided to eliminate the troublesome bow-frame which pilots generally disliked in air-to-air combat engagements.

The primary Response to Proposal (responsive to the LWF RFP Invitation and RFP Design and Performance Goals) from Northrop featured a derivative of the two-engine P-530 Cobra configuration but altered to be a single-engine configuration powered by the P&W F100 turbofan engine, and designated as the Northrop P-610. As allowed by the Prototypes Program Office, Northrop submitted an alternate proposal response closely similar to that of the P-530 Cobra in that it featured a two-engine configuration powered by two General Electric YJ101 turbojet engines (in development by GE as a speculative development in the advanced turbojet engine technology domain). That latter configuration was designated as the Northrop P-600.

Shown next is a planform comparison of the Northrop Model P-600 (two engines GE YJ101 turbojet) and the Northrop Model P-610 (single engine P&W F100 turbofan).

Planforms of Northrop Model P-600 (L) & Model P-610 (R)
(Source: USAF Engineering files)

Not shown in the photo is that the inlet configuration of the Northrop Model 600 was virtually identical to that of the P-530 Cobra (a photograph of the P-530 Cobra was shown earlier). Northrop designers had chosen the inlet for the single-engine Model P-610 to also be a single, chin- inlet, which was an aerodynamically efficient type of inlet for single-engine aircraft designs.

Wind tunnel models of the Boeing Model 908, the General Dynamics Model 401, and the Northrop Model P-600 were submitted by the three design teams. The Boeing Model 908 and General Dynamics Model 401 wind tunnel models were trans-delivered for testing at the NASA Langley Research Center (Dr. Whitcomb's transonic wind tunnel) virtually in parallel with their paper proposals. The Northrop Model P-600 wind tunnel model was trans-delivered to the NASA Ames Research Center (transonic wind tunnel where Stuart Treon was the wind tunnel facility Manager). Prime responsibility for the wind tunnel testing aspect of the LWF proposal evaluation was assigned to the Author where all the pre-test briefings and test planning was literally a three-ring circus. The testing schedule at USAF-sponsored wind tunnels were all saturated with testing of other on-going acquisition programs (F-15, A-10, several classified cruise missiles, etc.), and the burden fell on getting NASA to test the three models at the Langley Research Center (Hampton, Virginia) and at the Ames Research Center (Mountain View, California). NASA was agreeable to scheduling of the wind tunnel testing of the Northrop Model P600 at the Ames Research Center, and the wind tunnel testing of the Boeing Model 908 and General Dynamics Model 401 at the Langley Research Center.

There was a string attached to that agreement, however. A NASA general policy was to not test USAF models for development testing objectives; however, by allowing NASA to stencil the NASA emblem on the models, they could be tested for aerodynamics and configuration aerodynamic research purposes. In addition, NASA required that the wind tunnel test data obtained in NASA facilities could be written into NASA technical research reports and made available freely to the public at any educational or research facility without USAF prior approval of the NASA documentation. As the USAF representative, the Author agreed to those NASA conditions, but since the models and their data was competition sensitive, the Author negotiated an agreement with NASA to not release any of the wind tunnel data to the public for the three models until the completion of the USAF LWF flight testing, scheduled for the end of 1974. Release of the NASA wind tunnel data generated for the three LWF proposal models to the USAF source selection activity was accomplished rapidly and met the USAF LWF source selection evaluation schedule needs. The Author was, and remains, grateful to NASA and the staff of the Ames and Langley Research centers for their outstanding cooperation in accomplishing a challenging technical and schedule –driven activity as professionally and as rapidly as they did.

IBM card decks were correspondingly submitted by the three proposing companies for their respective three designs (Boeing Model 908, General Dynamics Model 401, and

Northrop Model P-600) external aerodynamic configurations and engines. These three IBM card decks were trans-delivered to Tom Christie's computation facilities at Eglin AFB, Florida, for analysis of the respective models' specific excess power characteristics and associated energy-maneuverability comparisons against threat aircraft. The energy-maneuverability comparisons followed the process developed earlier by Col. John Boyd, and were the only security classified material in the LWF Program. The comparisons were therefore provided only to those with a need-to-know (the Source Selection Committee of USAF senior officers and high grade federal civil service civilians appointed by the Commander of the Aeronautical Systems Division, Lt. General James T. Stewart). Membership on the Source Selection Evaluation Factor panels was drawn from the functional organizations (ASD, AFLC, Technology R&D Laboratories, etc.) resident at Wright-Patterson AFB.

The Proposal Evaluation Team diligently evaluated the four proposal responses for their responsiveness and adherence to the LWF RFP Statement of Work tasks and the design and performance goals in addition to adherence to the officially- required USAF regulations and safety specifications tailored for the prototype technical approach and program adaptive management approach. The four evaluation factors teams were in the domains of management, technical, test planning, and cost. The evaluation process used also followed all USAF rules and regulations for professional ethics and policies of conduct. All material generated during the source selection evaluation was considered competition sensitive and was strictly protected as such. Classified material (a small amount) was handled in strict adherence to security material handling regulations.

It is beyond the scope of this book, as well as it would be ethically inappropriate, to reveal the myriad of details in the three finalist proposal responses, the results of the wind tunnel tests on the three models, or the results of the digital analysis of the three finalist IBM card decks (specific excess power and energy-maneuverability comparisons). By a process of ranking of the technology advancements offered and according to the degree of their meeting the design and performance goals, the four proposal responses eventually were ranked as #1 General Dynamics Model 401 followed by #2 Boeing Model 908 and #3 the Northrop single-engine Model P-610 as being the best all-around proposals for the single-engine LWF designs. The Northrop P-600 was ranked #1 as the best of the two-engine LWF designs, not based on it being the only two-engine design proposal offered, but more importantly because it was evaluated (after adjusting its aerodynamics data based on the government wind tunnel tests) to also have the second highest capability toward meeting the RFP design and performance goals.

After evaluating the single-engine designs of both the Boeing Model 908 and the General Dynamics Model 401, the USAF LWF Proposal Selection Committee rated the single engine design of the Northrop P610 to be much less capable of performing in an operational theater at bases in forward area locations where FOD ingestion probability is always higher than it is at USAF bases in the CONUS. Simultaneously, however, the

Proposal Selection Committee had seen the value of the LWF Fly-Off Competition being enhanced if the two competitors were a single-engine design and a two-engine design to answer the often-argued question of "Which is better for a fighter design – having one large engine or two smaller engines?" Using that rationale and since Northrop had been the only company to alternately propose the two-engine P600 for its LWF Proposal Response, the USAF LWF Selection Committee chose their best evaluated single-engine design (the General Dynamics Model 401) and the only two-engine design offered (the Northrop Model P600). The Northrop Model 600 had a drag problem which was unexpected but correctable, and that will be discussed later in this book.

In late March 1972 when the LWF source selection evaluation was completed, the Source Selection Committee confirmed their final selections to be the General Dynamics (Model 401) and Northrop (Model P-600) to perform the Tasks and deliver the Deliverables identified in the LWF RFP. Funded contracts were awarded to General Dynamics and Northrop on 14 April 1972 to commence "best-effort" work on the USAF Lightweight Fighter program.

Approximately a week after the LWF contract awards were made to General Dynamics and Northrop, Col. Cameron arranged for travel (using Mil-Air, to reflect the low-cost program approach being adhered to) of the LWF Prototype Project staff – the five "young Turks" identified earlier – to have face-to-face meetings with their counterparts at both General Dynamics (in Fort Worth, Texas) and Northrop (in Hawthorne, California). In effect, this was the kickoff of the USAF LWF Prototype Program. [The Mil-Air four-prop engine Luxury Liner with rearward-facing seats took almost eight hours to fly from Dayton to Los Angeles on a Sunday. The best part of the trip was stopping to refuel at the small airfield in Pueblo, Colorado, and having a Buffalo Burger for lunch at the small cafeteria which also doubled as the Base Operations.]

The kickoff meetings at Northrop (on Monday and Tuesday) and then at General Dynamics (from noon Wednesday to noon Friday) went very well and established the teamwork for conducting the program on its "adaptive management", "design to cost", and "design to goals" basic nature. At Northrop, the kick-off meeting was started with a tour of the full-scale mock-up of the Northrop P-530 Cobra which Northrop had developed following the John Boyd and LWF mafia-sponsored Advanced Day Fighter study activities in 1970 and 1971.

The full-scale mock-up seen at Northrop created two somewhat diverse impressions among the visitors from the Lightweight Fighter Prototype Project Office. The first impression was that the mock-up revealed that Northrop had created a very credible fighter aircraft design and was much in synchrony with the John Boyd concept of what a lightweight day fighter should be like in three-dimensional full-scale form. The second impression was that the design had been created without consideration of the "design to performance goals" and "design to cost" goal and "best effort" towards development of a lightweight, low-cost, advanced technology demonstrator as described in the LWF RFP. Also very noticeable was that it was a two-engine design and not fully responsive to the RFP which preferred a single-engine design and preferably designed for the Pratt & Whitney F100 engine already well-along in development for the F-15 Eagle weapon system. In fact, however, it was because it did feature a two-engine approach that was the basis for it having been selected as providing very competitive capabilities to contrast with the best of the single- engine fighter designs that had been proposed in responses to the USAF LWF RFP. More will be described later in this book how the two different General Dynamics and Northrop LWF proposed designs were evaluated during the ensuing flight test program as to how well each had designed having the capabilities for meeting the design and performance goals outlined in the USAF LWF RFP.

Working-team relationships were established, the design approach in all major functions (aerodynamics, propulsion integration, structures, avionics, and especially in-flight control and flight safety, ground and flight test planning, and manufacturing approach) was initiated. An invitation for USAF personnel to attend any and all meetings at both contractors' facilities related to the LWF Program was extended and gratefully accepted by Col. Cameron on behalf of the entire LWF Prototype Project team. During the two-days visit at each contractor, Northrop in Hawthorne, California, and General Dynamics in Fort Worth, Texas, the first day was spent generally on learning how each company was organized and planned to conduct the LWF program based on what was submitted in their response proposal. By contract agreement, each contractor's response to the LWF RFP was considered as being the contractor plan (design, fabrication, ground test, flight test, and with schedule and cost for accomplishing the Tasks enumerated in the Statement of Work) to which the contractor proposed to follow as a best effort.

The first morning of the two-days visit was a joint meeting with all visitors from the Project Office spent mainly with the contractor describing the planning approach that would be used, and by whom, to conduct the work described in each task. That was done in order to assure no one could be unclear about what the objectives and work tasks were for the contract program. As an old adage the whole symphony must play from the same musical score, otherwise the result will be noise – well organized noise, perhaps, but noise rather than harmonic music. The first afternoon of the two-days visit was to organize into the main factors of the LWF prototype project visiting staff at the meeting – Management (L/C Bob Whelan), Cost and Contract (Mr. Doug Ring wall), Technical (Mr. Herb Hutchinson – the Author), Propulsion (Mr. John Wherry), and Test and Evaluation (Major John Colwell).

In a similar manner to that at Northrop, the team visiting at the General Dynamics facility were given a short tour of the forward fuselage full scale mock-up which General Dynamics had also created at about the time of the Advanced Day Fighter Study in 1970 and1971. A photograph of the General Dynamics forwards fuselage mock-up featuring the efforts General Dynamics had expended in making the Lightweight Fighter design concept approach as being "a fighter-pilot's fighter" is shown next.

YF-16 Forward Fuselage/Cockpit Module Ft. Worth, TX, Feb 1972.
(Source: USAF Engineering files)

In an adjacent area of the General Dynamics manufacturing facility, there were activities that were most interesting. General Dynamics had been developing a soft (i.e., non-metallic) mock-up of their center fuselage and aft modules of their YF-16 design – and these were not nearly as far along as the forward fuselage mock-up. In a contrasting manner, where Northrop had a complete three-dimensional full-scale mock-up of the external P-530/YF-17 configuration, the General Dynamics approach was that positioning of the internal fuselage components were to be located (stuffed) in the center fuselage module for survivability/vulnerability considerations as well as for the other associated "ilities" (maintainability, accessibility, replaceability, functionality, etc.) considerations. The approach planned was to use the cut-and- try technique. For the soft mock-up, the skeleton of the mock-up was made from plywood bulkheads (for ease of locating pass-through cutouts), thick cardboard skin and access panels, Styrofoam blocks shaped to the exact dimensions as their metallic component counterparts, and ropes of various diameters to represent wire bundles/ hydraulic/pneumatic cables/pipes, mailing tubes to simulate ducts (ECS, etc.). Shown next is the status of the center module at the time of the kickoff meeting of the General Dynamics Lightweight Fighter contract activities (March 1972):

YF-16 Center Module Soft Mockup
(Source: USAF Engineering files)

It is worth mentioning that there was a very close working relationship between the technicians on the drawing boards (very little computer aided design capabilities existed at that time [1972]) and those technicians in the mock-up area cutting the wood and Styrofoam pieces representing the internal functional components. With very little expenditure of effort, material, and cost, the resulting mock-up clearly avoided a great number of drawing changes and expensive modification to aircraft structure and configuration – a pitfall for aircraft programs developed with concurrency and no prototype (either mock-up or aircraft).

Shown next is the soft mockup further along in the stuffing process which was started soon after the larger equipment manufacturers had been selected:

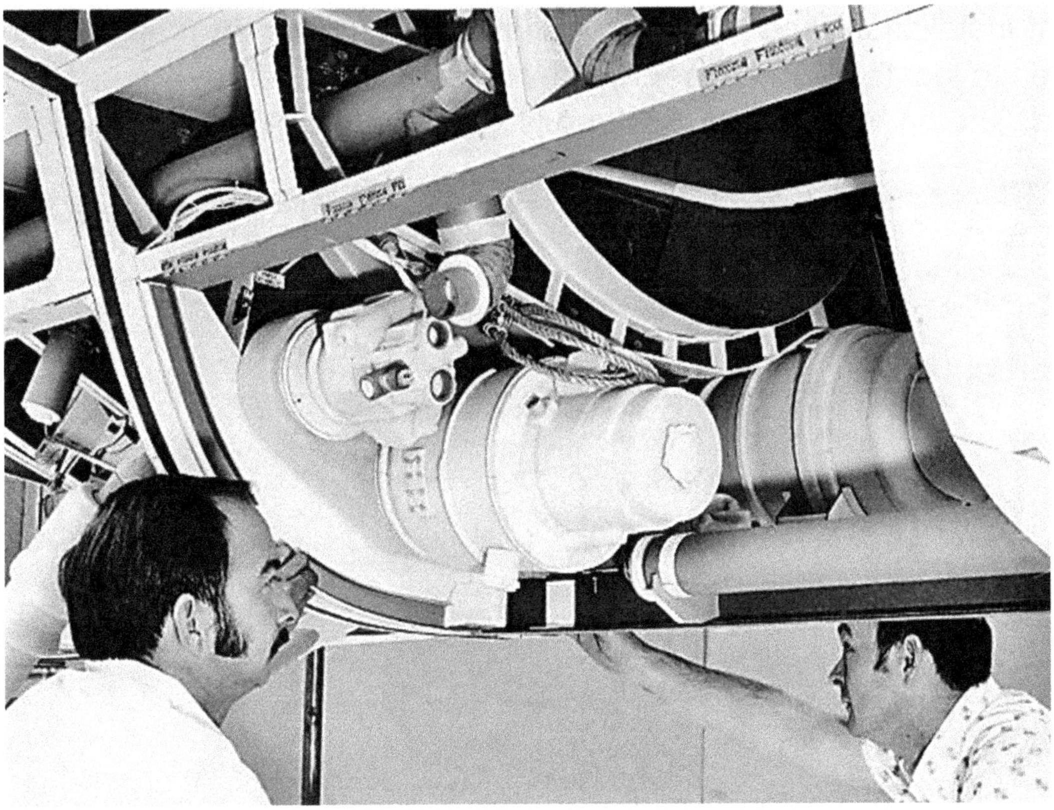

Styrofoam Components being fitted YF-16 mock-up
(Source: USAF Engineering files)

Shown next, a third view of the soft mockup illustrates the components being arranged within the soft mockup for survivability reasons as well as maintainability/accessibility reasons:

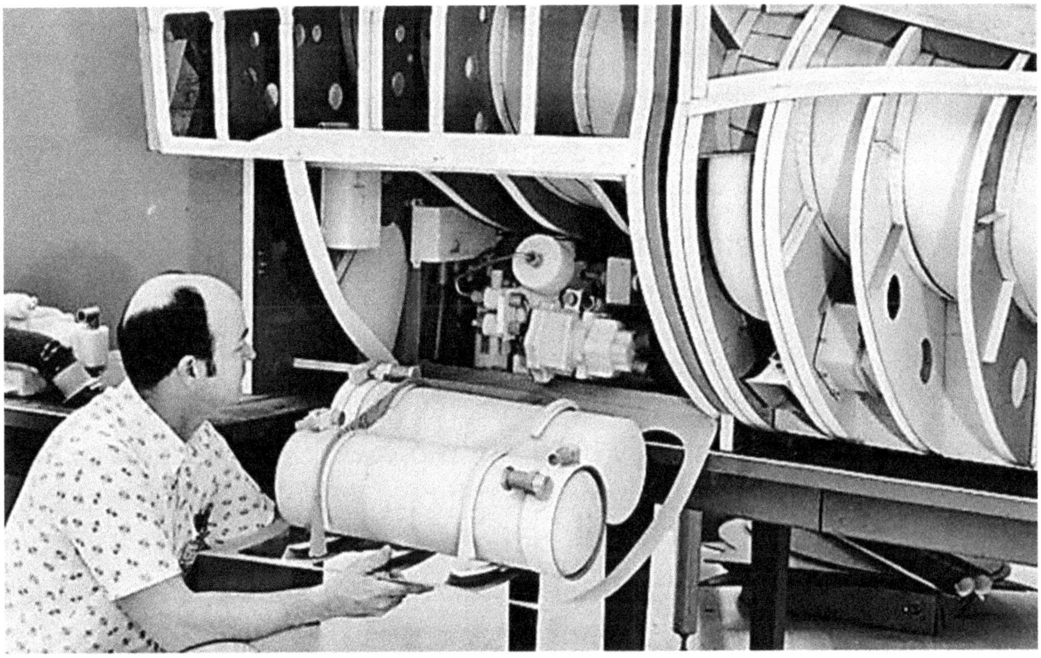

Two-tier layering of components beneath inlet duct
(Source: USAF Engineering files)

For ease of installation (and removal when necessary), the aft fuselage module mock-up was used to locate the rails upon which the engine would travel when being installed or removed. The aft module soft mock-up is shown in the photograph next.

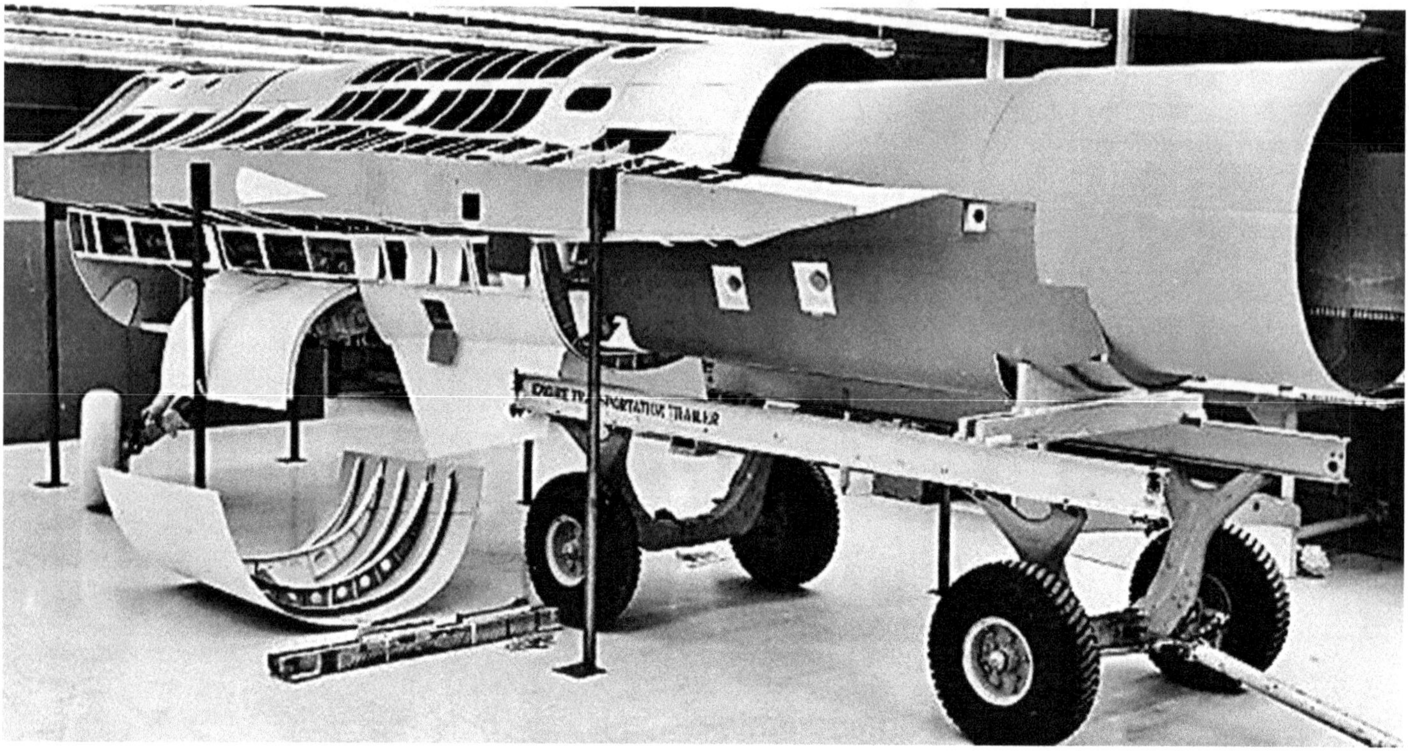

YF-16 Aft fuselage mockup used for locating engine rails
(Source: USAF Engineering files)

It should be noted that the mock-up engine is being transported on a four-wheel replica of a transporter cart that was developed as government-furnished equipment (GFE) for the F100 engine for the F-15 Eagle program. As a very wise decision on the part of the Air Force Logistics Command, three senior ground crew "blue-suiters" had been assigned to the USAF LWF program to assure that the "ilities" associated with both LWF designs were considered for design of especially the YF-16 early in its development phase. This was another illustration of the philosophy expressed in the metaphor - As the twig is bent, so shall the tree grow.

Ease of maintainability and servicing/rearming was designed into the YF-16 from the time the soft mock-up was being created. An essential part of that maintainability was that it was to be performed eventually by ground crew blue-suiters at operational bases using GFE extensively. One of the more difficult maintenance tasks performed at operational bases is the removal of the large engine for scheduled inspection and/or maintenance and repair; therefore, the fighter aircraft designs must feature a large number of access panels with a variety of fasteners depending on the frequency and urgency of the maintenance to be performed on the equipment items and components stuffed primarily in the fuselage. The Northrop P-600 design was also developed with all of those desired "ilities" in mind, but did not have the benefit of the three AFLC senior blue-suiters to provide suggestions based on their invaluable experience at operational bases under far-less than ideal environmental conditions.

For orientation of the technical team to the LWF designs, the technologies selected for the respective designs were presented by both contractors. On the chart next is a list of those more significant technologies that were identified by General Dynamics for integration into the YF-16. The Author has reconstructed the chart from General Dynamics presentations and design reports submitted during the LWF Program activities and stored in the Author's memorabilia file.

Advanced Technologies Considered/Selected
(For YF-16 LWF Prototype Design and Construction)

- **Near Fully Developed Propulsion System** (P&W F100-PW-100 from F-15) - 25,000 lbs Thrust SLS Max A/B [T/W~1.2 @ Start Combat Wt.]; Smokeless
- Clipped Delta Wing; Blended Wing Body Design
- Wing Root Leading Edge Extension For Vortex Lift
- Bubble Canopy (No Bow Frame for Pilot 360^0 Visibility)
- Ventral Fins For Improved Directional Stability
- **All Electric Fly-By-Wire Flight Control System** (Computer Controlled)
- Side-Stick Pilot's Flight Controller – Small Displacement 2 Axis
- All-Movable One-Piece Horizontal Tail Surfaces (Stabilators)
- Slight Negative Static Stability Margin
- Automatic Leading Edge & Trailing Edge Flaps (Setting Per Flight Mode)
- 9 g's Maneuvering Load Factor Limiter
- Seat-Back Angle Reclined 30 Degrees and Raised Heel Rest

<u>Internal Equipment</u>

- Emergency Power Unit (Safety – Loss of Main Generator)
- Jet Fuel Starter for Rapid Engine Start
- 9 External Stores Stations (**Multi-Mode Capability**)
- 20 mm 6-Barrel Gatling Gun (580 Rounds)
- **Survivability and Maintainability Features Designed-In At Start**

Advanced Technologies Proposed for the YF-16
(Chart by the Author)

For this book, concentration is on those technologies listed which were selected for integration into the YF-16 design.

A similar list, but with far fewer advanced technologies described, was presented by Northrop for the YF-17 design. These included:

- **Two engines for safety-of-flight redundancy**
- **Large leading-edge extension ("Cobra hood") (for vortex lift)**
- **Fly-by-wire ailerons**
- Automatic leading edge and trailing edge flap settings
- Seat-back angle reclined 18-degrees (vs 11-degrees conventional)
- Raised heel rest
- Boundary layer air diversion slots (for improved inlet airflow)
- Two-piece bubble canopy for excellent visibility in upper hemisphere

Advanced Technologies Proposed for YF-17 (Compiled by the Author)

The second morning of the kickoff meeting at the General Dynamics and Northrop facilities was spent conducting technical sub-meetings and in which each visitor from the Project Office discussed with his contractor counterpart what, if anything, may have been unclear in the contractor proposal either in function, schedule, plan, and/or cost to provide guidance back to the contractor with suggestions how that observation might be improved, or with compliments how that observation was found to be totally responsive to the USAF "desirements". During the last afternoon, sessions were conducted in arriving at agreements as to what were the near-term plans for future progress meetings, design reviews, test planning reviews, and/or what action items needed to be addressed and when.

Based on flight safety concerns identified during the Government review of the General Dynamics and Northrop Proposal Responses to the USAF LWF RFP, ground tests were identified to be needed for the canopy of the YF-16 (e.g., its ability to withstand bird-strikes), to more fully develop and test the emergency power unit for the single engine YF-16 (in the event of power loss to the flight control computer and flight control system), to conduct wind tunnel tests of both YF-16 and YF-17 in the NASA Spin Test Facility (vertical wind tunnel for spin susceptibility entry modes and recovery) and in the NASA Free-Flight Test Facility (for dynamic stability of both the YF-16 and YF-17 at high angle of attack), both specialized test facilities located at the NASA Langley Research Center. These tests were to be planned by the two contractors but conducted in government facilities provided by the USAF and NASA at government expense for use of the test facilities. Additionally, the USAF would independently conduct wind tunnel tests of the basic aerodynamics models provided by both contractors as part of their proposal responses to the USAF LWF RFP (to verify the aerodynamics of the LWF designs contained in the paper proposal responses).

Because it was not feasible to have someone from Wright-Patterson AFB at the two contractor facilities 100% of the time, arrangements were made with the AFPRO (Air Force Plant Representatives' Office) assigned at both facilities in Fort Worth and Hawthorne to represent the USAF interest and to serve as the traffic cop in the flow of information between and control of visitors at the facilities of the contractors from the LWF Prototype Project Office. Spurious visits by inquisitive personnel outside the Program Office were severely restricted so as to not impede the contractors' work activities.

Officially, the AFPRO functions at all USAF Contractor facilities are intended to closely monitor the progress and adherence of USAF Contractors to the safety and schedule/cost terms required by the Contract – but with the adaptive management and design to goals, not specifications atmosphere, the AFPRO also had to adjust its daily modus operandi to conform with their modified responsibilities. The deep trust placed in the AFPRO members to be the eyes and ears when Project Office members could not be at the Contractors facilities was more than amply rewarded. Much of the success for the LWF Project accomplishing all it did in the short schedule and within the low funds allotted to the program is attributed to the assistance received from the AFPROs at both the General Dynamics and Northrop facilities.

Because both contractors had essentially frozen (but not cast in concrete) their external configurations to be that of the General Dynamics Model 401 (YF-16) and Northrop Model P-600 (YF-17) resulting from the Advanced Day Fighter studies, the initial activities at both General Dynamics and Northrop focused on selecting the myriad of internal equipment items and functional components that needed to be stuffed into those small designs and so as not to create any undesirable trespasses of the skin lines of the external configuration. By virtue of Northrop having built the P-530 Cobra full-scale hard mock-up circa 1970, Northrop was ahead of General Dynamics in selecting all the primary pumps, valves, generators, actuators, connectors, fasteners, etc. that were off-the-shelf and qualified for safety of flight. To be accurate, the Northrop integration of their internal arrangement of the P-600 was driven more by cost, weight, availability, and size/shape constraints of all the equipment and components for reliability, maintainability, and supportability but with vulnerability to ground gunfire being a low priority on its selection and location of those items in their internal arrangement of the P-600. By way of comparison and as described earlier, General Dynamics elected to build a soft mock-up constructed with plywood bulkheads and stringers with wrapped cardboard skin to cut, fit, and try the design of their internal components and equipment items as well as to support the decisions for the "make or buy" option for the internal components.

In an unexpected but most welcome manner, the suppliers of the aviation items of equipment and components were all vying for their products to be included in the LWF competition as the smell of a potentially large market was in the air. Northrop made its selection of their full suite of internal equipment items and functional components substantially before General Dynamics had come to that point in their selection process. There is much intrigue among aircraft manufacturers in knowing what other aircraft manufacturers (domestic and foreign) are producing.

There is even a much more intriguing atmosphere existing between equipment items and component suppliers when it is known what their competitors are developing and producing. Few things are a more powerful motivation than is the spirit of competition. When the Northrop selection of a major equipment item or component was made known to the suppliers' community, the competitors for that same type equipment item or functional component approached General Dynamics and offered to provide, at low or no cost, those same functional items so as to be given the opportunity for a fly-off competition with their competitors. With the availability of small funds available to both Northrop and General Dynamics, suffice it to say that the availability of low-cost equipment items and components of high quality and with emerging technology improvements was an unexpected cost-avoidance surprise to the LWF Prototype Program. (Programs such as the sole source USAF F-15, SR-71, F-117, and F-35 which do not have competition between suppliers to motivate their selection of equipment items and functional internal components usually end up paying more for the same items and components due to the absence of the motivation of a fly-off competition potentially leading to a large market for suppliers' equipment items.)

The purpose of the government conducting independent wind tunnel tests of the prototype designs proposed was to evaluate if the aerodynamic characteristics of the design configuration matches with the drawings, photographs, and analyses included in their response to RFP, and does the total aircraft design function as a system to achieve the design and performance goals (outlined in the customer's RFP) in a three-dimensional realistic dynamic environment.

The General Dynamics Model 401 (YF-16) external design had undergone extensive wind tunnel testing during and after the Advanced Day Fighter study described in the previous chapter. So too had the Northrop Model P530 Cobra external design, but for a different motivation. The Northrop Cobra followed a company strategy of designing a fighter aircraft to replace its F-5 family of light-weight, low-cost fighter designs attractive to other countries – i.e., the Northrop Cobra was begun largely as a foreign military sales venture. Because the Model 401 (YF-16) had been tested in the NASA Langley Wind Tunnel (by Dr. Dick Whitcomb), it had been configured to feature the wing-body-lift and blended-wing-fuselage which NASA had recommended to Harry Hillaker and the General Dynamics design team. Northrop had elected to develop their P-530 Cobra design configuration more from testing at commercial and engineering college (e.g., Cornell University) wind tunnels not having the diverse expertise possessed at NASA.

Many outsiders (those not privy to what was actually going on within the corridors and rings of the Pentagon and within the Prototype Program Office in Building 14 at Wright-Patterson AFB) misunderstood the LWF Prototype Project objective, and many news media articles declared the Lightweight Fighter Program to be a head-to-head competition between the General Dynamics LWF design and the Northrop LWF design. From the very start, the USAF Prototype Project Office planned the ensuing flight-test program to be one in which both designs would be tested to evaluate how well each independent design came in demonstrating achievement (or missed achieving) the design and performance goals as stated in the LWF RFP. There was never a USAF plan to have a head-to-head fly-off of the General Dynamics Model 401 (YF-16) design against the Northrop P-600 (YF-17) design – it was initially approved to be and was an advanced technology demonstration program. (There were a few jokers in the USAF who claimed that the best way to determine which was the better fighter would be to load the 20 mm guns with live ammo and with live AIM-9 IR missiles instead of inert missiles on both YF #2s and have them go “fight’s on” at Nellis AFB. Whoever showed up for happy hour at the Nellis AFB Officers Club Bar that night would demonstrate which was the better air combat fighter. None of those jokers were assigned in the USAF LWF Prototype Project Office.]

Shortly after the LWF Prototype advanced technology demonstration contract awards were awarded to General Dynamics and Northrop, and the fact that the Northrop two-engine P-600 was to be flight evaluated for its capabilities in meeting the USAF LWF Design and Performance Goals (as stated in the USAF LWF RFP), the USN saw that the Northrop P-600 had a potential for winning the USAF contract (if any) for its next fighter as a high/low mix with the F-15. The USN had in parallel started an NACF study of its own to explore a new USN/USMC fighter aircraft with closely similar performance capabilities as the USAF LWF but which could be upgraded to have a carrier-based capability and serve as the high/low mix with the USN F-14 Tomcat. Accordingly, the USN (with strong urging from OSD) sent written correspondence to the USAF ASD Commander expressing a strong interest in the outcome of the USAF LWF competition to the extent that if the decision were to be made by OSD for the winner of the LWF competition to proceed into full-scale development, the USN would give “strong consideration” to adapting the winning design (with recognized required modifications for having a carrier-based operational capability).

The USAF ASD Commander Lt. Gen Stewart thanked USN for their expression of interest and invited the USN to assign a senior USN pilot to be present as often as deemed useful to serve as a liaison between the USAF LWF program and the USN NACF Program. A USN Lt. Commander (Bob Thalman) came to the USAF Prototype Program Office at ASD to observe what progress was being made on the USAF LWF

Project. It was not exactly a two-way street for the USAF to collaterally this be made aware of what progress the USN was making on its USN NACF Program. (Readers are asked to hold in mind this presence of the USN Lt. Commander as the USAF/USN liaison function in the USAF LWF Program as this liaison function was later to play a very key role in the outcome of the USAF LWF Program.)

To constrain the size of this book to being reasonable for its stated objectives, only the most major technology features selected for the General Dynamics YF-16 and Northrop YF-17 designs are able to be discussed, and with most coverage of those advanced technologies selected for integration into the YF-16. Following the propulsion system technology selection description (shown next), the ground testing aspects to validate the flight safety of both the YF-16 and YF-17 designs will be photographically summarized (canopy bird-strike tests, spin susceptibility and recovery characteristics, and free-flight dynamic stability at high angles of attack).

Propulsion System Technology Selection

The selection of the propulsion systems for the two different LWF configurations was a most important influence on the design of each respective fighter design as an advanced technology demonstrator. The fact that they were being manufactured by different engine manufacturers had less to do than the fact that they were in two different thrust classes – the P&W F100 being in the 25,000 pounds thrust size for incorporation in the General Dynamics Model 401 (YF-16) application whereas the General Electric YJ-101 engines were in the 15,000-pound class for incorporation in the Northrop Model P-530 (YF-17) application.

If Northrop had chosen to use two P&W F100 engines in their two-engines basic design, the 50,000 pounds of thrust would either be far too much power than was needed or would enable the size of their P-600 configuration to grow to be a rival to the F-15 Eagle program. The far-too-much power would also reflect in a more visible, heavier weight and higher cost design so as to not even be close to that of the General Dynamics Model 401 (YF-16) design. Also, the spirit of competition was less a factor in that Northrop had selected the use of the General Electric YJ-101 engines, which were referred to as leaky turbojets, for their two-engine design based on a long-standing working relationship that Northrop and General Electric had shared on preceding aircraft programs – the most notable and most recent being that of the Northrop F-5 series of light weight, low-cost fighters. The “leaky” nomenclature was given as the YJ-101 was a turbojet but diverted a small portion of the air taken in through the inlet to be flowed over the combustor section of the turbojet engine.

The technical specifications for the YJ-101 functional characteristics were as summarized listed next. For comparison, the specifications for the P&W F100-PW-200 which powered the F-15 Eagle and early F-16 Air Combat Fighter are also shown.

	General Electric YJ101 Engine	P&W F100-PW-200 Engine
Type:	Turbojet	Turbofan
By Pass Ratio:	0.20 : 1	0.36 : 1
Pressure Ratio:	21	25
Thrust (full afterburning):	14,960 lbs	23,770
Thrust (max dry):	9,460 lbs.	14,590
Mass flow (max dry):	123.2 lbs / min	43.8
Length:	11.58 ft.	15.92
Diameter:	32.63 in	34.8
Weight:	1,804 lbs.	3,234
Thrust/Weight Ratio:	8.29	7.35

Comparison of GE and P&W Engines Specifications (Source: USAF Engineering files)

In its proposal response to the USAF LWF prototype RFP, the Northrop Corporation offered a single engine (with the Pratt & Whitney F-100 turbofan) configuration as the Northrop Model P-610. More than one proposal evaluator (especially Col. John Boyd) noted a deficiency against the P-610 design for having the nose gear placed directly ahead of the single chin type inlet. The concern was not so much as for the aerodynamic effect of the nose gear strut on inlet flow distortion, but for the nose wheel potentially slinging foreign objects (FOD) up into the inlet. Depending on the nature of the FOD, there was a potential for major engine damage. That potential for engine FOD damage from the nose wheel is much decreased with "armpit" inlets not in line of FOD trajectory.

Northrop also offered as an alternate proposal response to the USAF LWF RFP, a two – engine Model P-600 which had been derived from the longtime Northrop venture into the design of their P-530 Cobra design (developed under company funding as an IRAD (independent research and development) effort and with a potential FMS (foreign military sales) market.)

Northrop did a commendable job of installing the equipment items in the YF-17 and especially in installing the two GE YJ101 engines so that their accessories would be easily accessible for engine inspection, maintenance, and repair. Another cost-savings feature is that the GE YJ101 engines were all produced in one configuration on the engine manufacturing and assembly line. There was no need for a left engine and a right engine – and as seen in the next photograph, the accessories are of the same singular arrangement at the easily accessible bottom of the engines.

Underside of YF-17 showing two GE YJ-101 engines/accessories
(USAF Engineering files)

While there was much discussion during the proposal evaluation about the high thrust-to-weight ratio achieved in both the General Electric YJ101 turbojet (a leaky turbojet and technically a turbofan cycle with a low 0.20:1 BPR) and its 8.29 T/W ratio, and the Pratt & Whitney F100 turbofan (with a nearly twice higher 0.39:1 BPR) and its 7.35 T/W ratio, the GE YJ101 engine with the higher (8.29) T/W had the least amount of validation run time. Comparatively, the Pratt & Whitney F100-PW-100 engine with the slightly lower (7.35) T/W was much further along in its demonstrated performance run time. That was due to the P&W F100 engine having been developed as the most advanced technology item the USAF developed for the F-15 Eagle fighter aircraft.

The existing P&W F100 engine was integrated into the General Dynamics YF-16 design and was installed from the rear end into the large cavity in the aft fuselage. By direction from HQ AFSC, the engines borrowed from the F-15 JEPO were to have *zero* modifications done to them for the General Dynamics YF-16 Lightweight Fighter. That was the reason that program insiders described the YF-16 as "a design wrapped around the engine". And to large extent, that is a factual description. When the team first went to the General Dynamics facility in Fort Worth after the announcement that the two Contractors in the USAF Lightweight Fighter Program, the team was shown an activity that General Dynamics engineers had started perhaps four or five months prior to the official announcement that General Dynamics was to be one of the two Lightweight Fighter competitors in a later flight- test fly-off competition.

The chart next indicates that by the planned first flight date of the YF-16 with the P&W F100 engine in early 1974, the engine test run hours would be about 16,000 and the F100 engine would have passed its important Qualification Test milestone for the F-15 Eagle program.

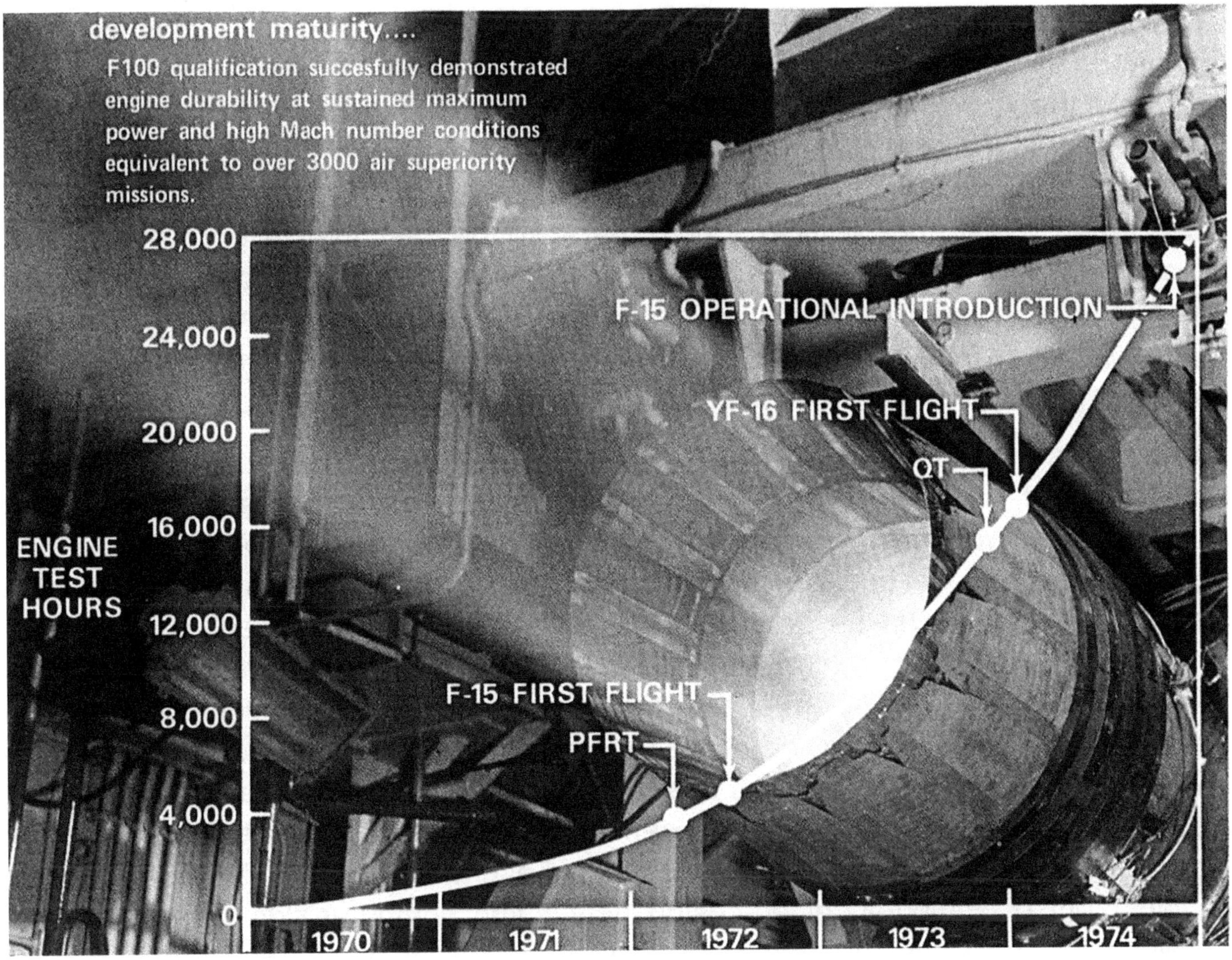

Development Status of P & W F100-PW-100
(Source: USAF Engineering files)

By July 1972, General Dynamics was well along in construction of a full-scale soft mock-up of the forward, center, and aft modules of the Model 401 external contour lines. The aft module mock-up was used extensively in designing the to-be-manufactured version of the aft fuselage for the installation and removal of the Pratt & Whitney F100 engine. The engine, on a manually-operated GFE ground-cart when being installed, engages the mounting rail at the top inside of the engine cavity, and is pushed along that rail until the engine front face matches up with the main bulkhead ring and which is the exit of intake airflow from the inlet.

Readers should note that three of the five features that dominate the internal arrangement within a small aircraft fuselage with a large engine installed (such as the General Dynamics YF-16) are propulsion system related. The engine itself is by far the largest volume that must be contained within the fuselage exterior contours established for aerodynamic efficiency. Second is generally the inlet ducting configured to efficiently deliver air from the exterior of the fuselage to the front face of the engine, with usually a significant distance from the inlet lip to the front face of the engine. Third is the sizing and arrangement of the fuel tanks to contain sufficient fuel for accomplishing the design mission of the aircraft and installing the associated fuel system components and controls for providing that fuel to the engine. The fourth area, occupying a large percentage of volume of that available within the fuselage, is that of the pilot and the cockpit arrangement. That cockpit integration aspect will be described later in this chapter separate from the first three high-volume aspects which are propulsion system related. The fifth area requiring large volume occupation in the fuselage is that of the M61 A1 Vulcan 20mm cannon system and the addition of wing hard-points for a credible air-to-ground capability, both of which will be described later in this chapter.)

This book cannot include the myriad of design challenges that design engineers face in the design of an aircraft, and Readers are left to research the level of details commensurate with their individual level of interest in the design process. There have been a large number of books printed recently and over the past decades instructing the steps that are taken in an iterative process to develop an efficient aerodynamics configuration of an aircraft (depending on the user requirements for speed/altitude/payload/mission) and then stuffing all the necessary components within the confines of the space-limited fuselage to achieve the mission requirements safely and efficiently. The design process is also a very iterative one that affects the scaling (usually to larger dimensions) of the exterior configuration to integrate all the required internal components inside the skin lines of the aircraft, especially the fuselage). The development of computer- aided-design (CAD) and computer-aided –manufacturing (CAM) digital tools and programs certainly assist design engineers throughout that iterative design process, but as with any tool, the operator of the tool applies experience and acquired skills in the final arrangement of the interior and exterior configuration of the aircraft.

Addressing the three propulsion items dominating the integration and arrangement of the interior fuselage of the YF-16 designs and configurations, for the purposes of this book only the final configuration will be briefly presented next, recognizing that the final configuration has resulted from a large number of iterations using mock-ups and CAD tools.

The engine cavity in the aft fuselage of the YF-16 has been described sufficiently verbally earlier in this chapter. Shown next are two diagrams describing the interior arrangement of the YF-16. The first diagram shows the front ¾ of the fuselage and the second diagram shows the aft ¾ of the fuselage interior (a reduced scale diagram showing the total fuselage length resulted in titles and identifiers being too small for legibility). At the top of each chart are shown cross-sectional orthogonal cuts through the fuselage at several locations of the fuselage longitudinal axis. Cross section A-A indicates the arrangement of the electronic components housed in the nose of the F-16A aircraft. (The YF-16s did not have a radar capability installed and the nose contour was longer and more pointed than that shown below for the F-16A.) Cross sections B-B, C-C, and D-D reveal the features of the inlet duct as it transforms from the warped elliptical shape into a more symmetrical elliptical shape and then into a circular shape matching the face of the engine. With regard to the inlet duct transformation from the warped elliptical (smiley) shape and to the circular shape having the same diameter as the face of the round cylindrical-shaped engine. The duct also makes a gradual upward curvature as it progresses aft toward the engine so that the air being passed into the circular engine face is in alignment with the engine central axis of rotation.

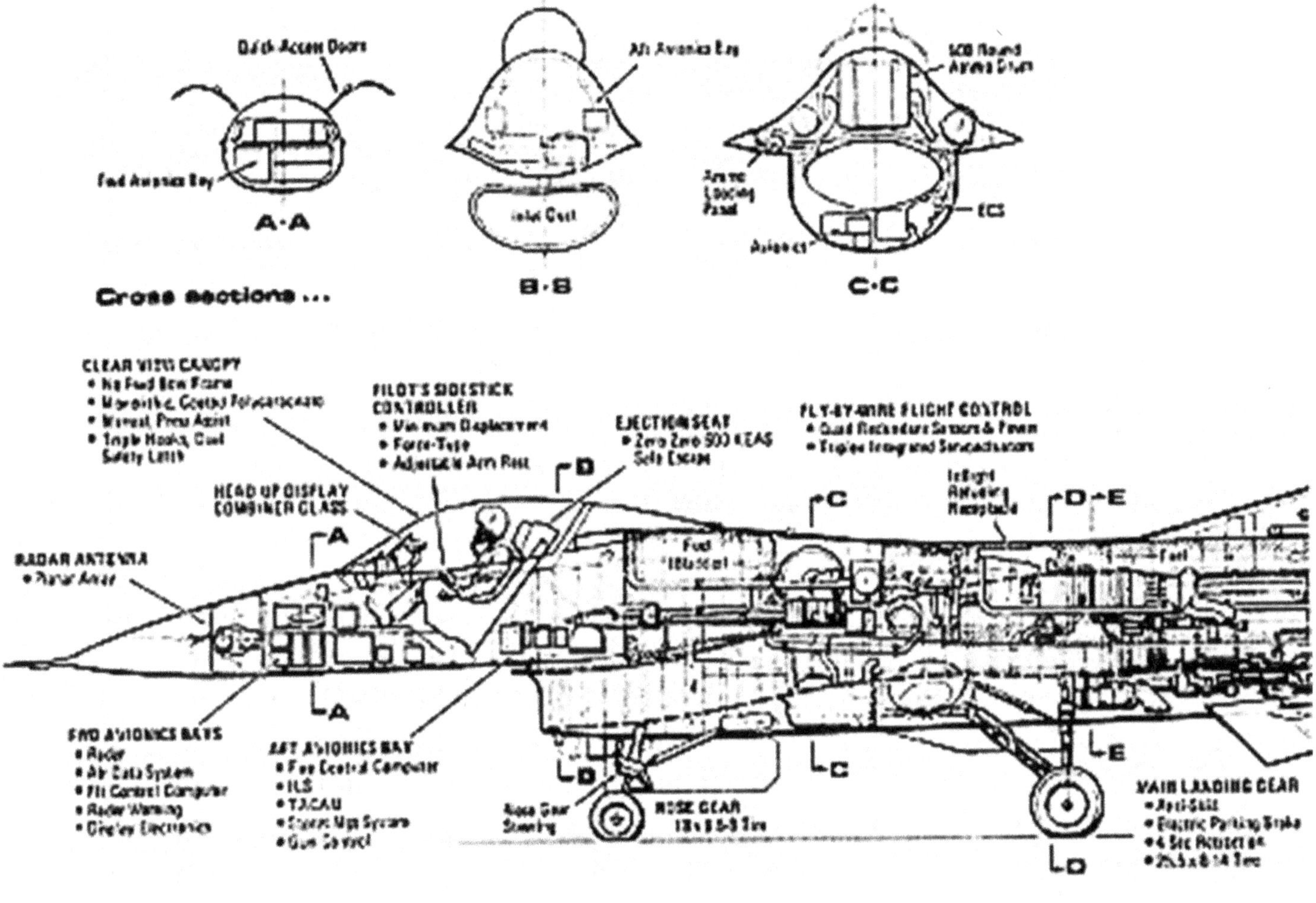

YF-16 Forward Fuselage Interior; Cross-Sections A-A, B-B, C-C
(Source: USAF Engineering files)

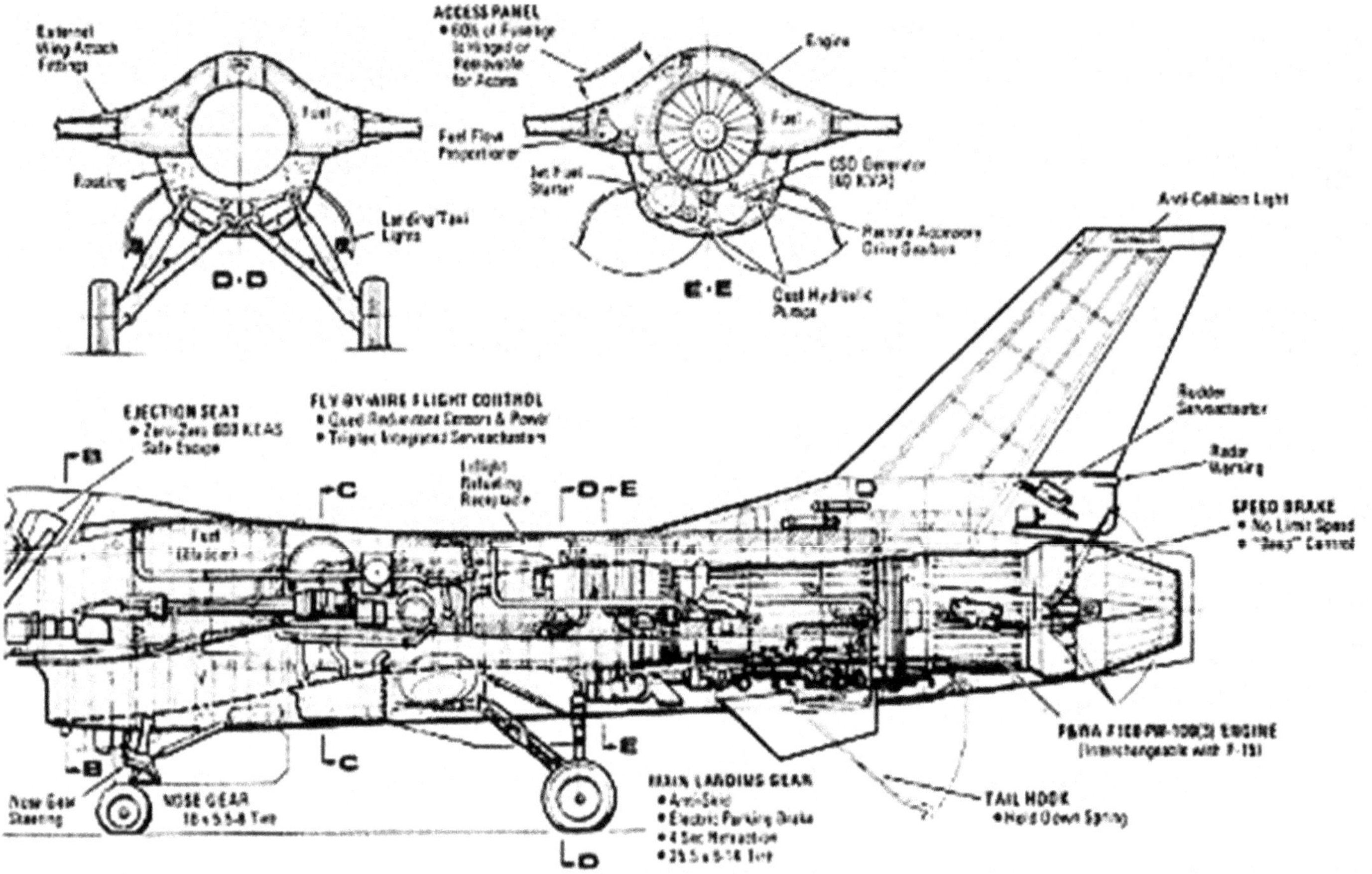

YF-16 Aft Fuselage Interior; Cross Sections D-D, E-E
(Source: USAF Engineering files)

The interior arrangements in the two preceding diagrams indicate the location and relative shapes/sizes of the internal fuel tanks within the YF-16 and F-16A FSD fuselages. The most forward fuel tank is shown to contain a bladder within the tank. The bladder and outer metallic tank provide a double barrier for pilot safety. When a tank is partially full and is subjected to a high deceleration (as in a crash), the remaining fuel in the tank will tend to slosh forward and if the deceleration is high enough, as they are in crashes, the fuel creates a hydraulic ram effect that can rupture the metallic tank structure. Without the bladder, the fuel would splash into the pilot's compartment and add a fire hazard to the pilot's other woes due to the crash. The other fuselage tanks are in a saddle-shaped configuration wrapping over the top of the fuselage and extending down into the blended wing-body area where the planar wing mates to the fuselage in five attachment fittings.

The blended wing-body configuration was designed primarily for aerodynamic reasons, but the added volume and added structural nature of the blending synergistically provides additional volume for fuel and adds to the structural strength of the wing-body juncture especially in wing bending-moment capability. Another frequent question posed in recent years was whether the LWF prototypes should have been configured to have a low radar-cross section in all aspects related to stealth. For several reasons, most importantly the design of the YF-16 being a multi-mode, low end of the high/low mix with the F-15 Eagle and having an outstanding external stores carrying capability associated with its multi-mode capability, stealth capability per se was considered but dismissed from being integrated into the YF-16 design. A second reason for not having stealth (that which is obtained by special radar-absorbing materials, or having to apply special tapes over seams where panels of the exterior skin are mated together, was counter to the low cost and "ilities" goals for the Lightweight Fighter concept. Low detectability was strongly considered in the design, and was achieved by keeping the size of the YF-16 very small, and with their not being any sharp corners for example where the wing abuts the fuselage. (The Boeing Model 908 submitted as the Boeing design response to the USAF LWF RFP was adversely evaluated for having that sharp corner juncture with the wing and fuselage primarily for aerodynamics and fuel volume reasons, however, it was also considered to not be advantageous as it produces a large radar cross-section (RCS) signature from a lateral perspective.)

The blended wing-body of the YF-16 does not have a large radar cross-section from the wing region but the largest contributor to RCS in the side perspective from a co-altitude radar seeker was the single vertical tail of the YF-16. A two-vertical canted tails configuration for the YF-16 was studied in the Advanced Day Fighter and showed that in the absence of a stealth requirement, the single vertical tail was a much lighter weight and low-cost design feature for the YF-16. The cost consideration was in the additional cost of manufacturing (adverse to the affordability goals set by the USAF Prototype Program Office, but more importantly in the aspect of operating and support (O&S) costs associated with the special materials and stealth features maintenance. A goal for the YF-16 was to have a high number (above five) of turn-arounds per day if in a combat theater, and in fact it will be later shown in this book that during the LWF flight test, the YF-16 demonstrated having the capability of eight turn-arounds in one day.

Cross-sections C-C, D-D, and E-E in the two diagrams shown on preceding pages indicate the saddle shape of the middle and aft fuselage fuel tanks. All fuel tanks are purposely located on the top of the fuselage to minimize the tanks being hit by enemy ground gunfire. In addition, non-safety-of-flight critical components which are less susceptible to loss of function if hit by ground gunfire are located nearer the bottom of the fuselage and act as a shield for the fuel tanks. The main landing gear is a good example of that interior arrangement (stuffing) approach for using lower criticality components to shield higher criticality components.

Cross-sections B-B, C-C, and D-D on the preceding two diagrams also reveal the inlet duct cross-sectional shape at that cut location and reveals both the transformation from the warped elliptical ("smiley") face of the inlet to a more symmetrical elliptical shape and at a higher elevation, to the circular cross-sectional shape at the elevation of the engine axial centerline.

The YF-16 fuselage had over-240 access panels to access internal equipment and components for ease of inspection, maintenance, repair, replacement, and servicing. Of those over-240 access panels, 8 access panels are located (ports 1, 2, 3, 4, 6, and 8 on the bottom and ports 5 and 7 on the top) are to allow bore scoping of the engine while the engine is installed. (It is unknown whether this bore scoping technique was developed before or after the medical profession had developed the same type fiber-optic instrument for performing laparoscopies during human surgical procedures and/or before the ever-popular colonoscopy or endoscopy procedures.)

The borescope ports on the GE YJ101 engines in the two-engine YF-17 were also located top and bottom of the engine and accessed through access panels for borescope inspection of its fan, compressor and turbine blades, and combustion chamber.

To inspect an installed engine requires two crew members: one to crawl into the inlet and slowly rotate the engine with a torque-handle attached to the exposed axle hub of the front fan while another crew member inspects the blades passing by with a borescope instrument inserted into the engine through one of the eight ports. Shown below is a photograph of a ground crew inspecting the front face of the F100 engine while installed in the aircraft.

Inspecting the installed engine front fan blades
(USAF Photographic files)

The inlet for the General Dynamics Lightweight Fighter was developed to provide the required flow of air to the Pratt & Whitney F100 engine with the minimum amount of distortion in the flow of air through the inlet. A common statement made among the inlet design community is that "Round engines like round air" which would suggest that the inlet should be of circular cross-section shape. However, circular-shaped inlets are not easy to be integrated into the external shape of fighter aircraft configurations, especially those with a single engine. A single circular nose inlet (as popular in the early Soviet MiGs) can be achieved if the nose of the aircraft is of a circular cross-section and the nose cone is not an acceptable location for the forward-looking front hemisphere radar detection and tracking radar system. Also known to be avoided in inlet design is the existence of sharp corners inside the inlet near the face of the inlet, as those sharp corners cause local pressure distortion of the inlet air and result in a distorted flow field being presented to the engine front blades.

Very soon after the Advanced Day Fighter study was completed, the basic warped-elliptical (smiley) inlet shape was developed by General Dynamics in a series of wind tunnel tests – conducted at the NASA Lewis Research Center near Cleveland, Ohio. For inlet wind tunnel testing, usually only the front portion of the forward fuselage of the aircraft and containing the inlet(s) are commonly tested so that as large an inlet can be used for the testing. A photograph shown next illustrates one of the near-final inlet configurations for the single-engine General Dynamics Model 401 undergoing wind tunnel testing to map out the flow field pressure distribution of the inlet at different angles of pitch and low angles of yaw.

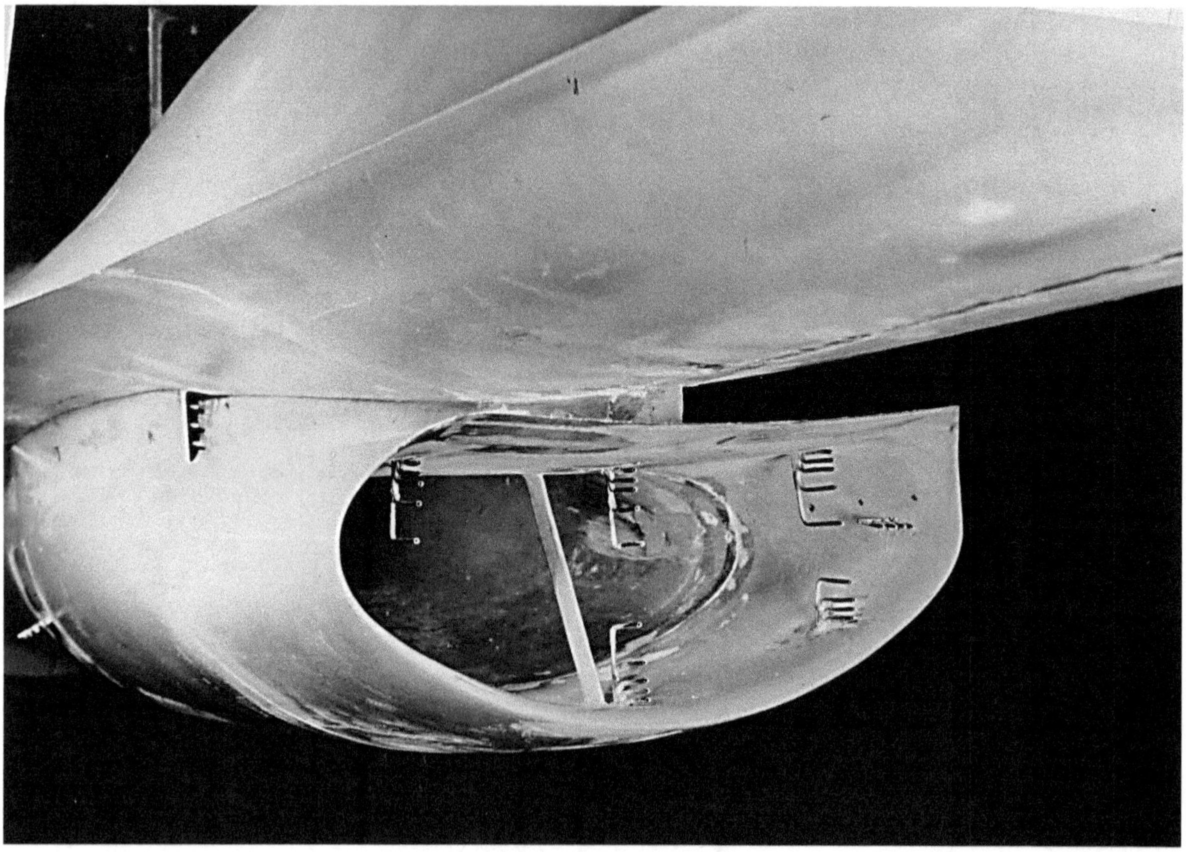

YF-16 inlet model undergoing flow distortion testing at NASA, 1973.
(Source: USAF Engineering files)

Seen installed in the inlet are five of a set of six pressure rakes to obtain test data for the distortion mapping of the inlet flow field. It should be noticed that the upper lip of the inlet extends a significant distance ahead of the location where the upper inlet joins the fuselage. Turbulent air beneath the forward section of the nose of the aircraft is split away from the laminar air entering the inlet and diverted laterally downstream from the inlet. The swept-back rectangular vertical structural piece on the centerline of the inlet was found to be necessary to keep the warped elliptical shape of the inlet when the inlet is at very high air pressure tending to circularize the inlet. (Ground crews sometimes humorously inform visitors at ground displays of the F-16 asking "What is that strut in the inlet for?" with the response that "it gives them something to grab onto if they get sucked into the inlet". A second tongue-in-cheek response is that "it cuts the bird being ingested in the inlet into two smaller pieces so the engine can survive the bird ingestion". A third non-factual response also given sometimes is that "the strut gives off a nice humming sound at high speeds and makes the inlet smile". None of those responses are factual although the visitors are typically amused by the candid answers.]

The vertical strut is a means of avoiding circularization of the inlet (a one-piece aluminum casting) when the inlet is subjected to the very high-pressure loadings inside the inlet at supersonic speeds. The vertical strap is the lowest weight approach to maintain the warped elliptical shape of the smiley inlet. The vertical strut could be eliminated by designing thicker walls for the inlet casting but the weight savings by installing the vertical strut were found by structural design analysis to be significant. In a later chapter describing unexpected problems encountered during flight testing of the YF-16 prototype aircraft, it will be explained why it was necessary to not mount the vertical strut in a fixed manner at both ends within the inlet.

Another distinctive external feature of the YF-16/F-16 configuration is the clipped delta planform of the wings – and to a smaller scale – the clipped delta planform of the one-piece all-movable horizontal tails (stabilators). In its operational configuration, the F-16 will always be seen to have a launcher rail for the tip missiles (typically AIM-9 Sidewinder-type missiles) installed on the wing tips (stations 1 and 9). The launcher rails on the wing tip stations are there as a result of flutter testing that was done on the YF-16 wing models at the NASA High Speed Flutter Test Facility at Langley Research Center. The wing section chosen for the YF-16 wings was an NACA 64A204 airfoil section where the last digit 4 indicates the percent thickness of the airfoil being 4- percent thick relative to the chord of the wing. As the chord length shortens in the spanwise direction, so does the physical depth of the airfoil, such that near the tip of the 4-percent thick wing the physical depth of the wing is extremely small. For that reason, the YF-16 clipped-delta wing planform was in fact clipped to have the wing tip with a structurally sufficient physical depth to avoid a flutter problem. However, even though clipped, flutter-testing of the YF-16 wing experienced a catastrophic flutter as shown in a photograph of the wing flutter test model next:

Failed wing flutter test model (in Flutter Wind Tunnel Tests)
(Source: USAF Engineering files)

To stiffen the outboard trailing edge portion of the wing, General Dynamics considered various options (aerodynamic, materials, and structural) and decided to make the close-out tip rib of the wing to be the launcher rail for the tip-mounted missiles. That brought the flutter margin to acceptable levels for safety of flight throughout most of the flight envelope for the YF-16/F-16. When the tip missile is mounted on the launcher rail, it creates an even higher flutter margin throughout the entire operational flight envelope. Flutter wind tunnel tests confirmed the safety of flight margin with the launcher rail and missile (in blue color) mounted as shown in the photograph next.

YF-16 wing with launcher rail undergoing flutter tests
(Source: USAF Engineering files)

Fly-By-Wire Flight Control System Technology

If there was one single factor that drew most attention during the evaluation of the Responses to the LWF RFP, it would be the difference in the technology domain of the flight control system proposed by General Dynamics for the Model 401 (YF-16) design. As described earlier, the huge number of losses of F-4 aircraft in the Vietnam War to ground-gunfire (small arms, automatic weapons, AAA artillery, and SAMs) were found to be caused by damage to critical elements of the F-4 flight control system. The most prevalent form of damage causing loss of the F-4 and their crews while still over enemy territory was identified in BDA (Battle Damage Assessment) statistics to be due to the severing of hydraulic lines transferring F-4 pilot commands to control surfaces aft of the engine nozzles.

This BDA assessment, done in view of the high operational losses of the F-4s in the Vietnam War, caused the Air Force Flight Dynamics Laboratory (AFFDL) at Wright-Patterson AFB near Dayton, Ohio, to accelerate their technology development activity known as the Survivable Flight Control System (SFCS) which had been started conceptually in 1959. To a moderate degree, the AFFDL capitalized on earlier work done by NASA in the modification of an F-8 aircraft having a flight control system prototype for use by astronauts in exploring how to most effectively control the motion of their spacecraft when conventional aerodynamic forces on control surfaces were not present (such as during extra-atmospheric modes of flight). NASA had been developing the use of reaction force rockets at significant locations on their space vehicles, and pointed to the need for an all-electric flight control system to be the most effective means of controlling the space vehicle with rocket reaction forces.

The potential and advantages of a fly-by-wire system had been demonstrated in several exploratory development programs, beginning in 1959 at the Air Force Flight Dynamics Laboratory. At that time, the comparison of the FBW concept with conventional mechanical flight control systems had the main objective of improving the reliability of flight control systems designed for military fighter type aircraft. That objective was centered on the premise that electrical and electronic components technology had been improved to the point that their reliability was better than that of physical components which were failing either structurally or mechanically and required much maintenance to remain in calibration (to provide the pilots with consistent stick feel.

Shown next is a comparison of the significant differences in the type components comprising a mechanical flight control system with that of an all-electrical fly-by-wire flight control system.

MECHANICAL FLIGHT CONTROLS

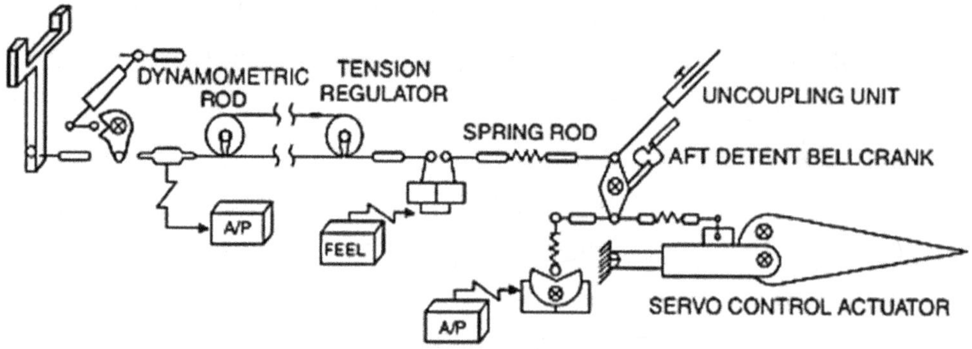

ELECTRICAL FLIGHT CONTROLS (FLY BY WIRE)

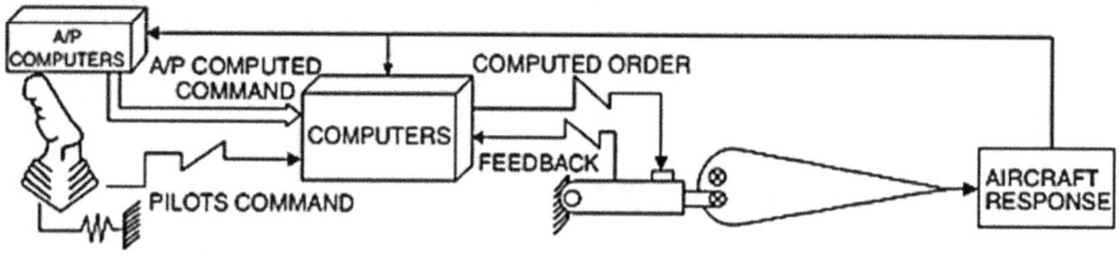

Comparison of Mechanical and Electrical Flt Control System
(Courtesy: LWF Engineering files)

The exploratory efforts initiated within the Flight Control Division of the AFFDL in 1959 led to a larger funded Advanced Development Program in the AFFDL in the 1967 to 1971 time period. A report "Survivable Flight Control System Fly-By-Wire Development and Flight Test", AFFDL-TR-71-20, authored by James E. Hunter of the Flight Control Division of AFFDL contains the Summary and initial paragraphs of an Introduction and are extracted from AFFDL-TR-71-20 as follows (verbatim):

SUMMARY

The United States Air Force initiated the Survivable Flight Control System (SFCS) Advanced Development Program in July 1969 in which one of the major objectives was to establish the practicality of the Fly-by-Wire (FBW) concept for use in military fighter aircraft. This advanced development program provided for the design, fabrication, qualification, and highly successful flight test evaluation of a quad-redundant FBW primary flight control system in an F-4 aircraft without conventional mechanical controls.

Results from this intensive FBW advanced development effort indicate significant improvements in overall flight control system performance, reliability, safety and maintainability. Additionally, the strong and credible FBW technology development base developed as a result of this program has paved the way for further exploitation through the application of advanced concepts such as Control Configured Vehicles and Multi-Mode Controls.

INTRODUCTION

The Air Force has accomplished a significant major milestone in advanced fighter flight control design and reliability with successful completion of its fly-by-wire flight test under the Survivable Flight Control System (SFCS) Program. This portion of the SFCS Program has consisted of a four-years, $16.5 million development program with primary objective for developing and flight-testing control elements to improve combat survivability of aircraft weapon systems. The development program was performed by McDonnell Douglas under contract to the Air Force Flight Dynamics Laboratory, with Sperry Rand, General Electric, and Lear Siegler as the principal equipment suppliers. Although the primary purpose of this program was to develop a highly reliable flight control system designed to improve survivability, major improvements in handling qualities, stability and performance, and weapon delivery accuracy were also achieved goals.

The SFCS mechanized in a YF-4E test aircraft uses a quadruple redundant (two-fail operate), dispersed. Three-axis, analog, fly-by-wire (FBW) primary flight control system allowing the pilot to command aircraft motion rather than the conventional control surface position. This is the first high performance fighter aircraft ever to fly using a futuristic "all electric" system."

Summary & Introduction of SFCS in AFFDL TR-71-20 (USAF Approved for Public Release – FOIA)

Readers who may wish to see the myriad of technical details and control laws contained in AFFDL-TR-71-20 will find the full AFFDL-TR-71-20 and its several supplements on the internet.

When John Boyd and the Lightweight Fighter mafia created the Advanced Day Fighter study program described in the previous chapter, only Harry Hillaker and his team of flight control system specialists (essentially the whole team who had developed the flight control system for the USAF/General Dynamics F-111 Aardvark Program) at General Dynamics included the advanced technology feature of an all-electric Fly-By-Wire flight control system in the Model 401 (YF-16) configuration, which was developed during wind tunnel tests for the Advanced Day Fighter study program. By contrast, the Northrop Aircraft Division had used their P-530 Cobra design (developed as a potential replacement for its F-5 small, lightweight fighter) with a conventional primary hydro-mechanical flight control system with an electronic Control Augmentation System. Hillaker had also thoroughly followed and studied the results of the F-4E/F Battle Damage Assessment studies done to identify and categorize the causes for loss of about 200 F-4s in the Vietnam War to ground gunfire.

In an off-the-record conversation the Author had with Harry Hillaker of General Dynamics during a meeting at the AFFDL to discuss emerging technologies, Harry expressed his opinion that he was both surprised and disappointed that the F-15 Eagle had not been designed with a fly-by-wire flight control system, especially since the flight test work on the YF-4E, modified to have a fly-by-wire flight control system, had been accomplished by the McDonnell Douglas Company, who also had designed and were manufacturing the F-15 Eagle. The Author explained to Harry Hillaker that the Chief Engineer of the F-15 Eagle Program and his Assistant Chief Engineer did not have a reputation at Wright-Patterson AFB for applying emerging technologies (even one as advanced in development as the fly-by-wire system on the YF-4E, that had undergone a comprehensive flight test program) and regarded the technology development AFFDL community as being technology toads.

It is beyond the scope of this book to fully describe all of the salient features and functions of the fly-by-wire flight control system developed for both the YF-16 and subsequent F-16 designs. For a complete description of that system, Readers should consult the document "The General Dynamics Case Study on the F-16 Fly-By-Wire Flight Control System" by Carl S. Droste and James E. Walker, written for the AIAA Professional Study Series (ISBN 978-1-56347-307-4). This case study was prepared in the 1981 time period and reflects the design features of the FBW FCS as originally designed for the Model 401 (YF-16) proposal response to the USAF LWF RFP, with modifications to that design for the YF-16 LWF prototype aircraft (which underwent extensive flight testing during the General Dynamics YF-16 and Northrop YF-17 flight demonstration program in 1974).

Before attempting to describe the function each one of the internal components of the F-16 all-electric fly-by-wire flight control system performs in control of the motion of the aircraft, it is considered necessary to first summarize the aerodynamic and configuration technologies of the external configuration in the YF-16 and subsequent F-16 aircraft. The technology of Relaxed Static Stability is one that is easily understood by a pilot of any aircraft, but for other Readers may sound somewhat too high tech to be understood. In any aircraft in flight, the weight (W) of the aircraft may be considered to be acting as a downward force vector acting through the center of gravity (c.g.) of the total aircraft. In order to lift an aircraft off the ground and become airborne, there must be a lift (L) force vector opposite to the weight vector and that lift vector must be greater in magnitude than the weight vector. The lift force is generated by virtue of applying a thrust (T) force from the propulsion system to increase the forward speed of the aircraft and for the wing surfaces to aerodynamically generate the lift vector.

The fourth vector opposing the forward motion of the aircraft is that of aerodynamic drag (D) resulting from both the friction of air molecules over the exposed skin of the aircraft and from a result of rotating the lift vector backward as the angle-of-attack is increased producing a "drag-due-to-lift" component. The four force vectors (L, W, T, D) combine with the force vectors produced by control surfaces deflection in determining the flight path (heading, speed, and rotation rates and radii of flight path curvature in three-dimensional space) of the aircraft as well as the orientation of the aircraft with respect to its three axes.

The diagram next illustrates for a conventional aircraft how the relative location of the lift and weight force vectors affect the static stability of the aircraft about the pitch axis. This is applicable to the subsonic and supersonic flight regimes.

CONVENTIONAL STATIC STABILITY

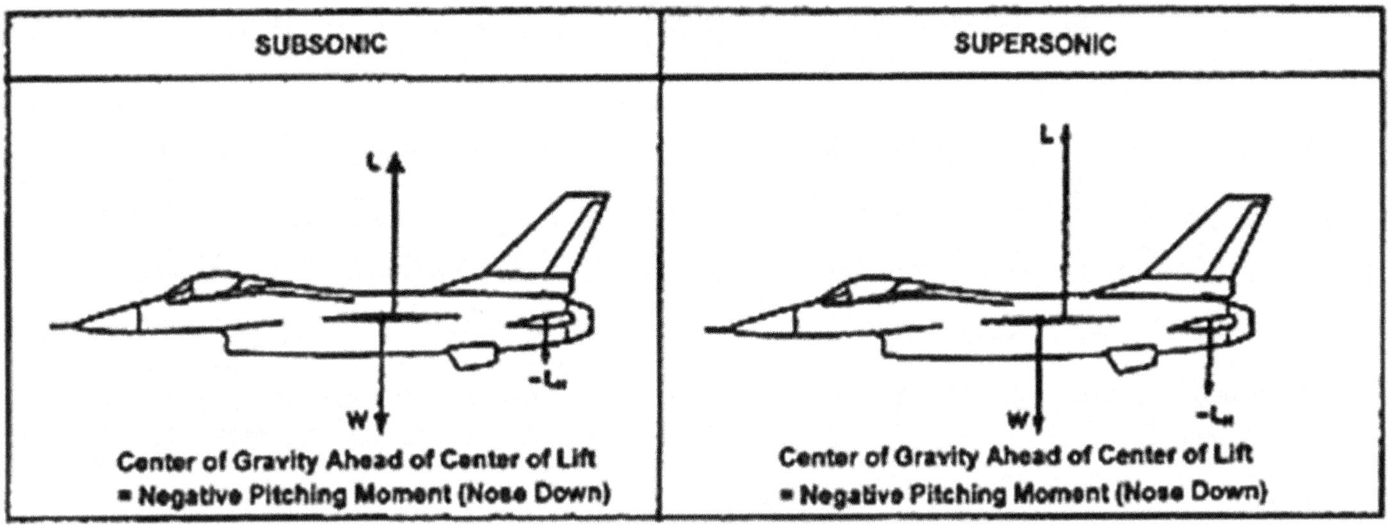

Conventional Static Stability (subsonic and supersonic)
(Source: USAF Engineering files)

As shown in the diagram, when at subsonic speeds, conventional designs will require a force in the downward direction from the horizontal tail to trim out the negative (nose down) pitching moment caused by the lift L vector being aft of the c.g. of the aircraft. Because the resulting forces acting upward in the lift direction are reduced by the amount of downward force ($-L_H$) of the horizontal tail, the angle-of-attack of the aircraft must be increased slightly to increase the basic wing-body lift of the aircraft in order to maintain level flight conditions with the conventional static margin.

When the aircraft is accelerated from subsonic speeds to supersonic speeds, the center of lift is shifted further backward (approximately a distance of 15-20 percent of the mean aerodynamic chord (m.a.c.). At supersonic speeds, there is required an even larger downward force required at the horizontal tail to trim out the increased negative (nose down) pitching moment. Once again, the angle-of-attack of the lifting surface (wing body) must then be increased to cause additional lift to overcome the larger downward force at the tail to maintain flight altitude and in a stable condition. It should also be noted then that with conventional static stability, the nose-down pitching moment providing static stability is opposite to the nose-up pitching moment commanded by the pilot when initiating a high-G maneuver. In effect then, conventional static stability is not favorable to the agility of fighter aircraft. To improve the adverse effect of conventional static stability, the approach of relaxed static stability has emerged in the stability and control technology domain.

To assist Readers in understanding the relaxed static stability concept, a diagram is shown next.

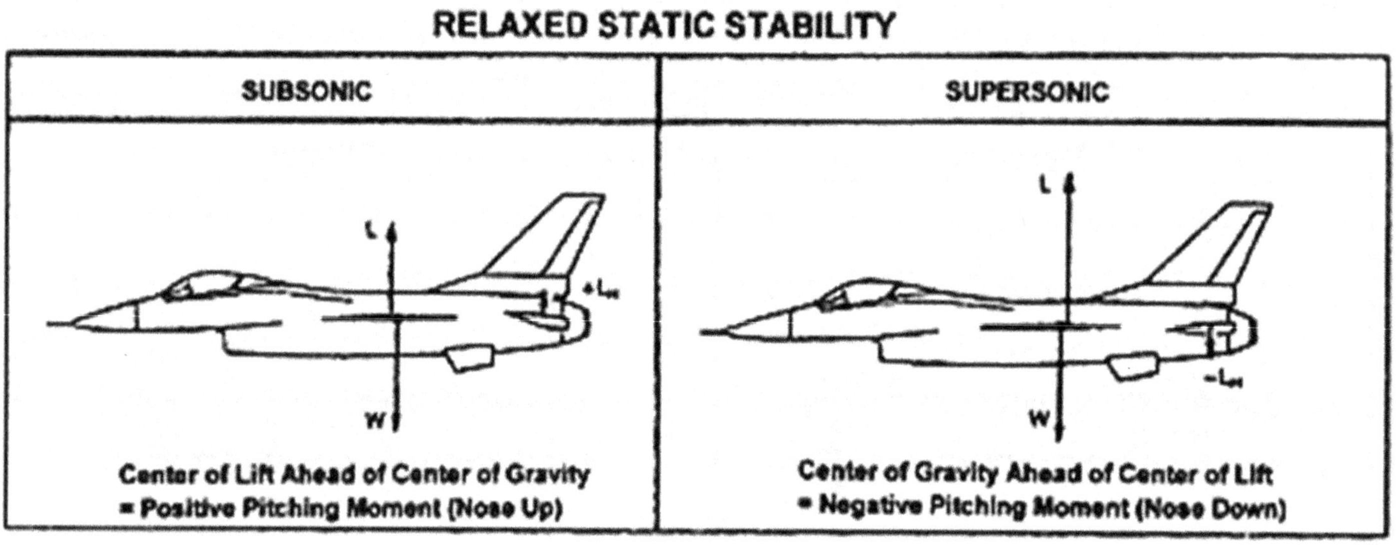

Relaxed Static Stability (subsonic and supersonic)
(Source: USAF Engineering files)

With relaxed static stability, the lift vector at subsonic speeds is slightly forward of the weight vector such that a nose-up (unstable) pitching moment exists on the aircraft in flight. To trim out the slight instability, the force at the horizontal tail is reversed from that in conventional static stability and requires an upward force (i.e., in the lift direction) to trim the aircraft. The positive aspect of relaxed static stability is that the pilot when commanding a nose-up maneuver (such as when commanding a high-G maneuver) will more rapidly achieve the increase in angle of attack and will have the additional upward force vector ($+L_H$) at the horizontal tail acting with the direction of the maneuver rather than opposing it. This increases the agility of the fighter aircraft to perform pitch maneuvers more rapidly both in onset rate of the maneuver as well as taking advantage of the reduced induced drag to be used positively for sustaining the maneuver.

Also with relaxed static stability, when the aircraft is accelerated to supersonic speed and the center of lift translates backward an additional 15-20 percent of the mean aerodynamic chord (m.a.c.), the amount the static margin changes (from a nose-up pitching moment to a slightly nose-down pitching moment) is smaller than that when that shift in static margin is on a conventional stability aircraft. As a result, then, at supersonic speeds the static stability is a much lower percent of the m.a.c. and requires a lower downward force at the horizontal tail to trim out the smaller (less nose-down pitching moment) static stability.

A chart extracted from the General Dynamics F-16 Fly-By-Wire Flight Control System database provided to the USAF Lightweight Fighter Program Office is shown next to indicate the percent changes in the lift/drag (CL / CD) graphical representations (referred to as polars) for the F-16 by incorporating the relaxed static stability (RSS) technology.

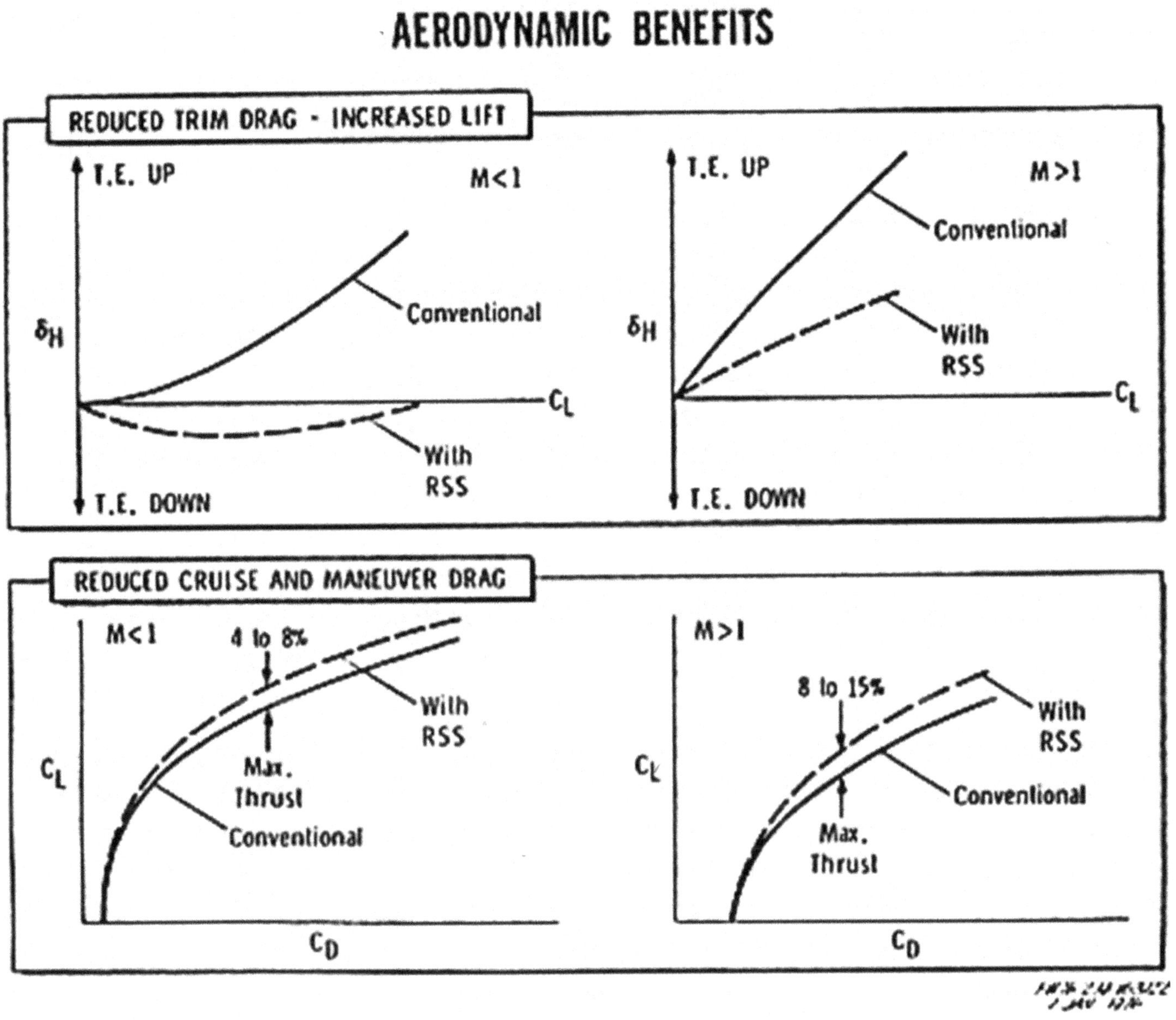

Aerodynamic benefits of relaxed static stability
(Source: USAF Engineering files)

The upper two portions of the diagram show the effect on trim drag with horizontal tail deflection degrees between the conventional and the relaxed static stability (subsonic speeds on the left quadrants of the chart, and supersonic speeds on the right quadrants of the chart). The effect on cruise drags and on maneuvering drag between the comparisons of the variations attributed to the conventional and the relaxed static stability concepts are shown on each quadrant of the chart.

The capability for designing an aircraft with a relaxed static stability characteristic is possible only with some form of electronic stability augmentation system, as requiring the pilot to achieve the correct deflections of the control surfaces on the aircraft to perform aggressive maneuvers and without loss of control of the aircraft would be exceedingly demanding and hazardous to flight safety. With the use of the FBW flight control computer managing the proper deflections of the control surfaces of the aircraft, the achievement of unexcelled agility of the F-16 is made possible.

[Author's note: Col. John Boyd was ecstatic when provided with the information contained in the General Dynamics proposal response with the Model 401 (YF-16) design showing the agility of the Model 401 (YF-16). During his evaluation of the General Dynamics response to the LWF RFP, more than three times he is quoted as booming out to the Source Selection Evaluation Team, "This baby will really roll your socks down." A more typical roar by John Boyd was emanated by the Roar Master while watching a film *The Battle of Britain* at a theater in Gaithersburg, MD (outside Andrews AFB). The Author was told by Boyd's theater companion (preferring anonymity) that in one scene an RAF Spitfire pilot was in the process of gunning down a German Messerschmitt, Boyd became totally engaged in the action (as he felt obligated to do) when he was on flight status in his F-86 Sabre Jet in the MiG Alley of Korea) and roared "*Hose* that son of a bitch!". Theater patrons changed nearby seats to be as far from the Roar Master as possible. In the great biography *Boyd* by Robert Coram, description of an excursion made by Boyd in chasing a MiG-15 north of the Yalu River is given in greater unauthorized detail (i.e. prohibited by the then Commander-in-Chief President Truman)].

Another flight technology that was made possible on the YF-16 by virtue of it having a flight control computer was that of the variable wing camber technology. This technology was a major technology development activity which started at the Air Force Flight Dynamics Laboratory circa 1966. The Author had received a merit promotion within the Federal Civil Service System to transfer from the Advanced Concepts Division of the Aeronautical Systems Division to the Flight Mechanics Division of the Air Force Flight Dynamics Laboratory, all located at Wright-Patterson AFB, Ohio. Known at first as the Wing High Lift Devices Program and then as the Variable Wing Camber Technology program, the objective of the technology activity was to adapt the use of a flight control computer to the varying of wing leading-edge and trailing-edge control surfaces to vary the wing camber to be optimized for the Mach number and mode of flight for fighter aircraft. Harry Hillaker of General Dynamics and Lee Begin of Northrop were highly interested visitors at the Flight Mechanics Division of the AFFDL and both had expressed high interest in the Wing High Lift Devices technology as being applicable in their company IRAD programs for a new fighter aircraft capability.

Northrop had designed a somewhat conventional hydro mechanical primary flight control system for their Cobra design but with a computer-controlled control augmentation system. The Stability and Control Department at Northrop Aircraft Division in Hawthorne,

California, opted to apply the use of the Cobra CAS computer to vary the leading edge and trailing edge flaps of the Cobra to vary both LEF and TEF to optimum wing camber configurations depending on the flight mode and Mach number conditions.

The resulting control scheduling of the LEF and TEF for the P-530 family derived from the basic Cobra development activity (a few years before General Dynamics considered the variable camber wing technology) is shown in the diagram next.

LEF and TEF Scheduling for Northrop YF-17

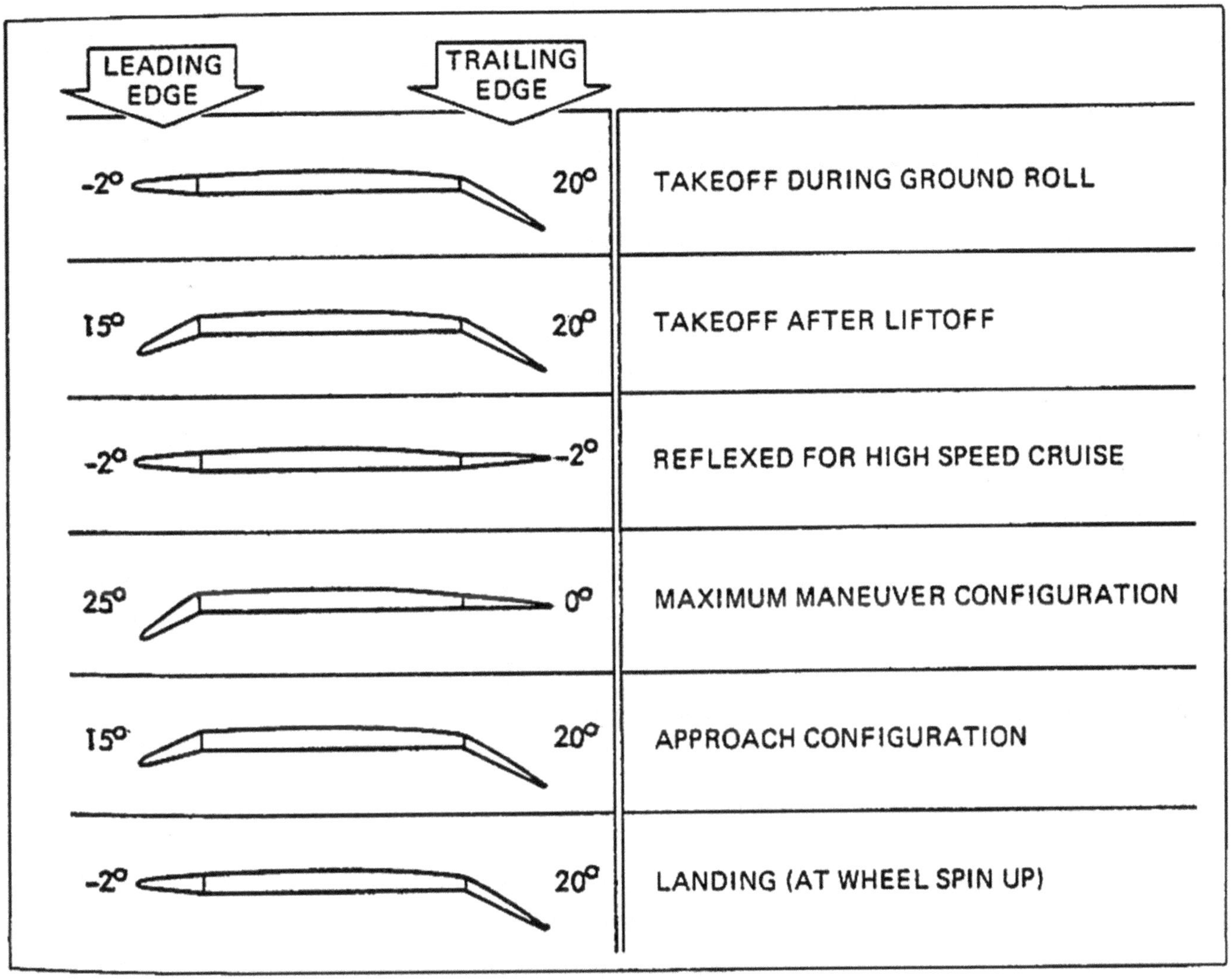

LEF and TEF deflection schedule for YF-17
(Source: USAF Engineering files)

General Dynamics soon after (during the Boyd LWF mafia Advanced Day Fighter study in the 1968 time period) configured their Model 401 (YF-16) to also have variable wing camber with the LEFs being continuously varying under control of their FBW FCS flight control computer. The TEF was not a continuously varying optimization but relied on the position of the landing gear position handle and a weight-on-wheels (WOW) switch for deflecting to a fixed position for takeoff and Landing. The TEF deflection schedule of the

Model 401 (YF-16) when at supersonic speeds (and obviously with landing gear up) was set to a TEF deflection of -2 degrees.

The chart next (extracted from "The General Dynamics Case Study of the Fly-By-Wire Flight Control System on the F-16") reflects the General Dynamics approach to variable wing camber depending on flight mode for their Model 401 (YF-16) proposal response and YF-16 configuration.

Variable Camber Flap Scheduling on General Dynamics YF-16

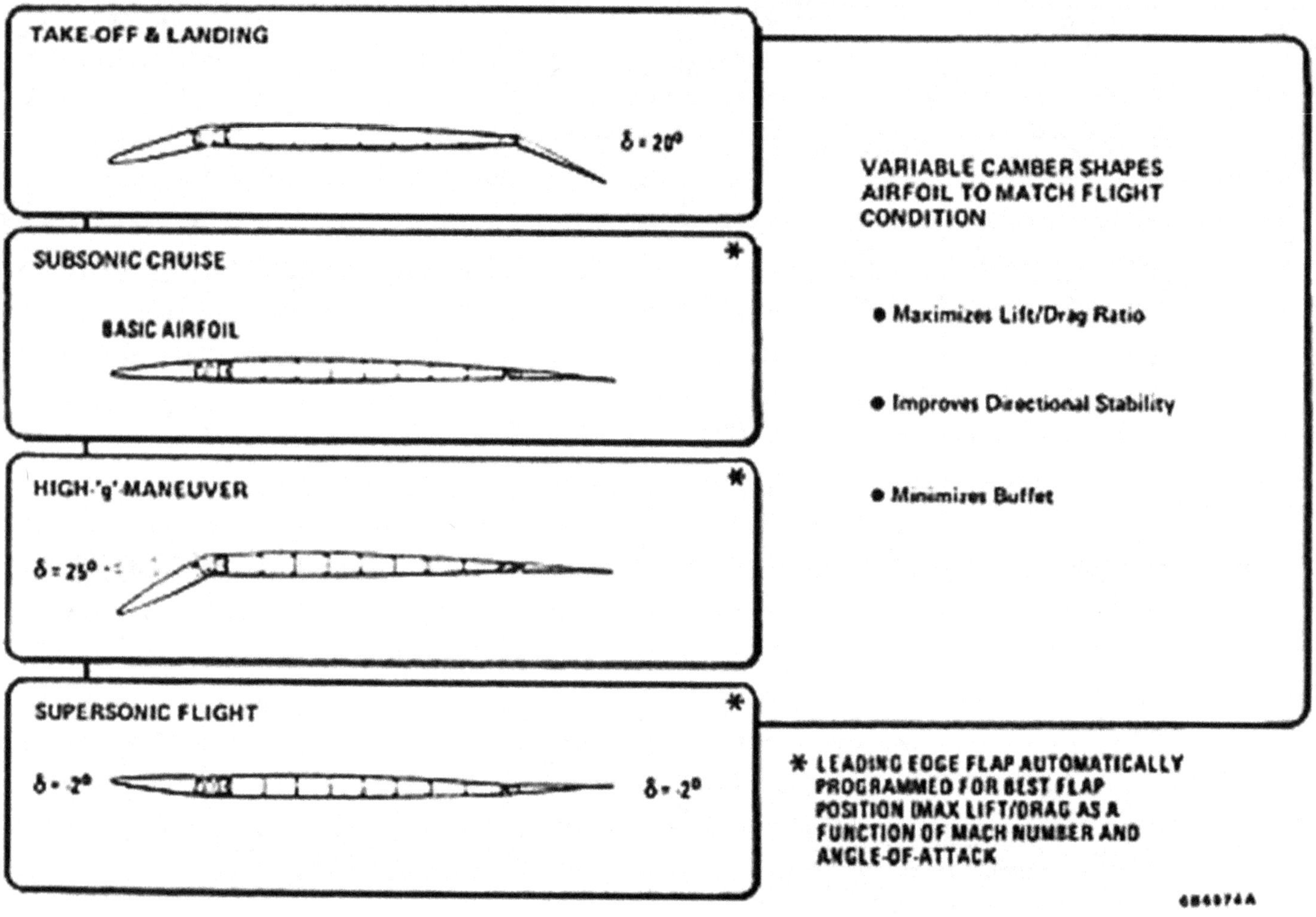

Variable camber flap schedule on YF-16 (per mode of flight)
(Source: USAF Engineering files)

Another external configuration feature present on both the Northrop P-530 Cobra family of fighter aircraft and on the General Dynamics Model 401 (YF-16) and YF-16 are the forward wing leading edge extensions (LEX). It is the size and shape of the LEXs on the Northrop P-530 Cobra which appear much like the hood of the cobra snake when preparing for a strike of a victim (human or animal) that gave the well-related name Cobra to the P-530 Cobra aircraft.

The LEXs on the General Dynamics Model 401 (YF-16) and YF-16 advanced technology features were of a smaller area and of a much more subtle geometry than those on the P-530 Cobra. A sketch of the LEXs (more commonly referred to as the strakes by General Dynamics aerodynamicists) is shown next and peeling off vortex pairs that flow aft over the inboard wing juncture with the blended fuselage and grow in diameter as the vortices progress rearward. To take advantage of that additional volume within the wing-body blended juncture, General Dynamics used the additional volume for internal fuel storage.

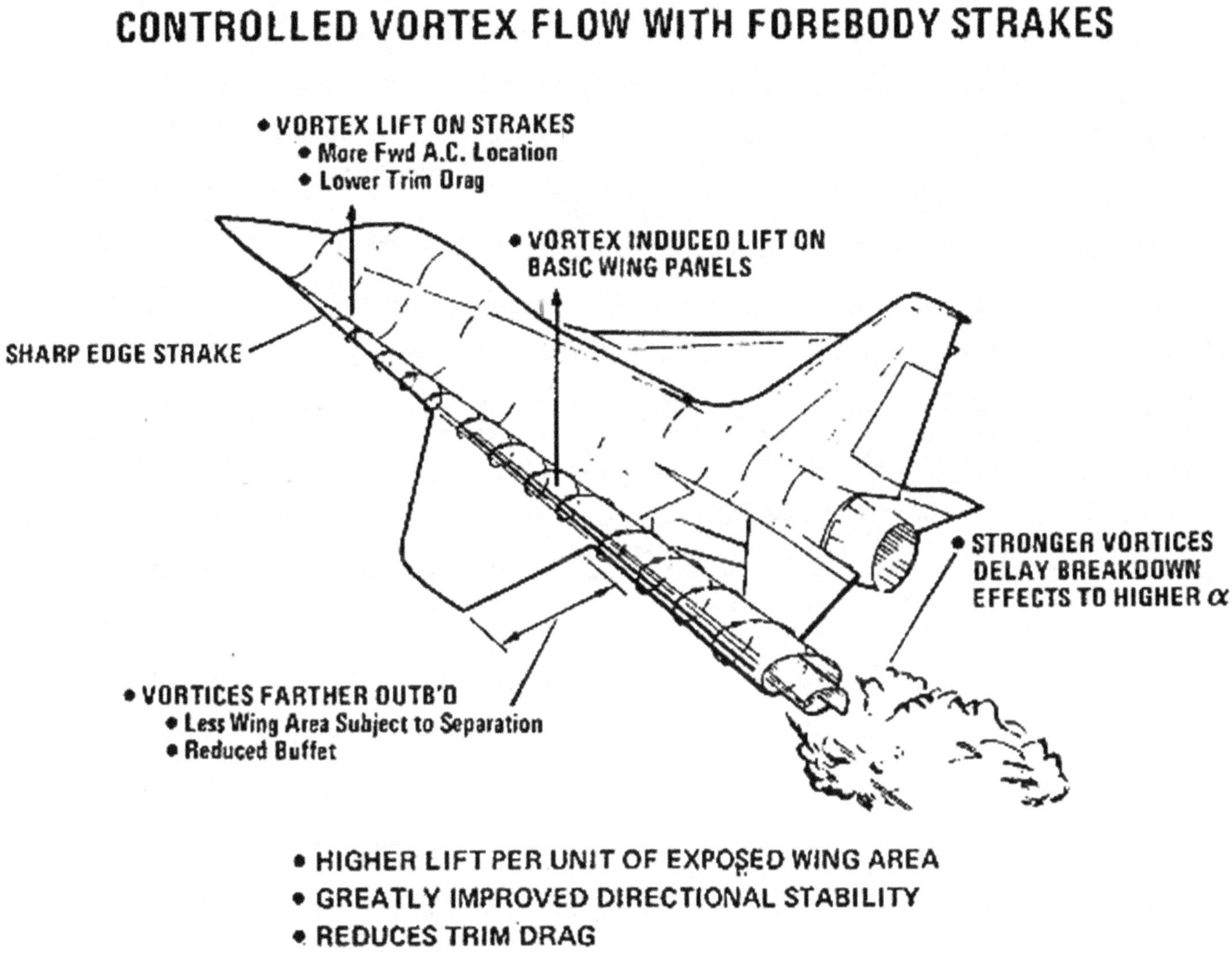

Controlled Vortex Flow with Forebody Strakes YF-16
(Source: USAF Engineering files)

Actual photographs of the vortices emanating from the YF-16 during flight maneuvers by the Joint Test Team in 1974 at Edwards AFB are shown later in this book.

Never missing an opportunity to make an off-the-wall comment, in this case about the YF-16 wing-body blending feature, Boyd felt certain that Harry Hillaker had designed-in that extra volume for cool fuel to be used in keeping the wing attachment fittings (five fittings in each wing root) from getting overheated from all that bending moment heat energy created at nine Gs which the Lightweight Fighter pulled so often.

As mentioned earlier, the wing body blending and the size/shape of the strakes on the General Dynamics Model 401 (YF-16) and USAF YF-16 had been largely developed by the Harry Hillaker team during extensive wind tunnel testing (at the NASA Langley Research Center transonic wind tunnel under the talented guidance of Dr. Richard T. Whitcomb) for the USAF Advanced Day Fighter study sponsored by then Lt. Col. John Boyd and Col. Everett “Rich” Riccioni.

A pair of ventral fins (trapezoidal in shape) were installed below the aft fuselage of the YF-16 starting approximately at the trailing edge of the clipped delta wing. Those ventral fins in combination with the variable camber wing (computer controlled LEFs) and controlled vortex flow from the leading-edge extensions (strakes) combine to have a strong beneficial effect on the high lift coefficient, to improve directional stability, and to reduce buffet intensity of the YF-16.

For reference purposes, the Leading-Edge Extensions (LEXs) on the Northrop YF-17 are much larger in area than those on the YF-16. To be factual, Northrop has employed leading-edge extensions on prior fighter aircraft especially in their T-38 and F-5 series of aircraft and had developed the larger Cobra leading edge extensions after many hundreds of wind tunnel test hours at both NASA and commercial wind tunnels. However, for the P-530 series of configurations culminating in the P-600 design that was submitted in response to the USAF LWF RFP, Northrop added the boundary layer diversion slots also seen in the photo of a YF-17 shown next.

YF-17 Leading-edge extensions and boundary layer diversion slots
(Source: USAF Engineering files)

Although there were no incidents of injury reported when any USAF, USN, or USMC pilots were flying the YF-17, the Boundary Layer Diversion Slots were described by some of those pilots (not in the USAF or USN) as being ankle busters.

For unknown reason, Northrop did not detect that the flow through the boundary layer diversion slots had an adverse interaction with both the vortex flow (i.e., positive pressure was being diverted into the lower pressure flow field atop the LEXs) and having an adverse effect on the aerodynamics of the aft fuselage above the wing, which had a reflex in curvature downstream from the slots. The flow through the slots acted to create a region of aspiration of the flow over the aft fuselage above the wing and created additional, though not huge, drag downstream from that region. In the next chapter, a photograph taken during phosphorescent oil wind tunnel testing of the flow field above the wing and over the upper aft fuselage of the YF-17 configuration will be shown to enable Readers to see the complex flow field due to the boundary layer diversion slots.

Returning to the General Dynamics YF-16 technology features and their effects, a chart is shown next providing flight-derived quantification of the beneficial effects of the advanced aerodynamic technologies on the YF-16 performance at high coefficients of lift and compared with a conventional configured (external geometry and static stability) fighter aircraft.

COMBINED EFFECTS OF CONTROLLED VORTEX FLOW AND VARIABLE CAMBER SHOW SIGNIFICANT PERFORMANCE GAINS

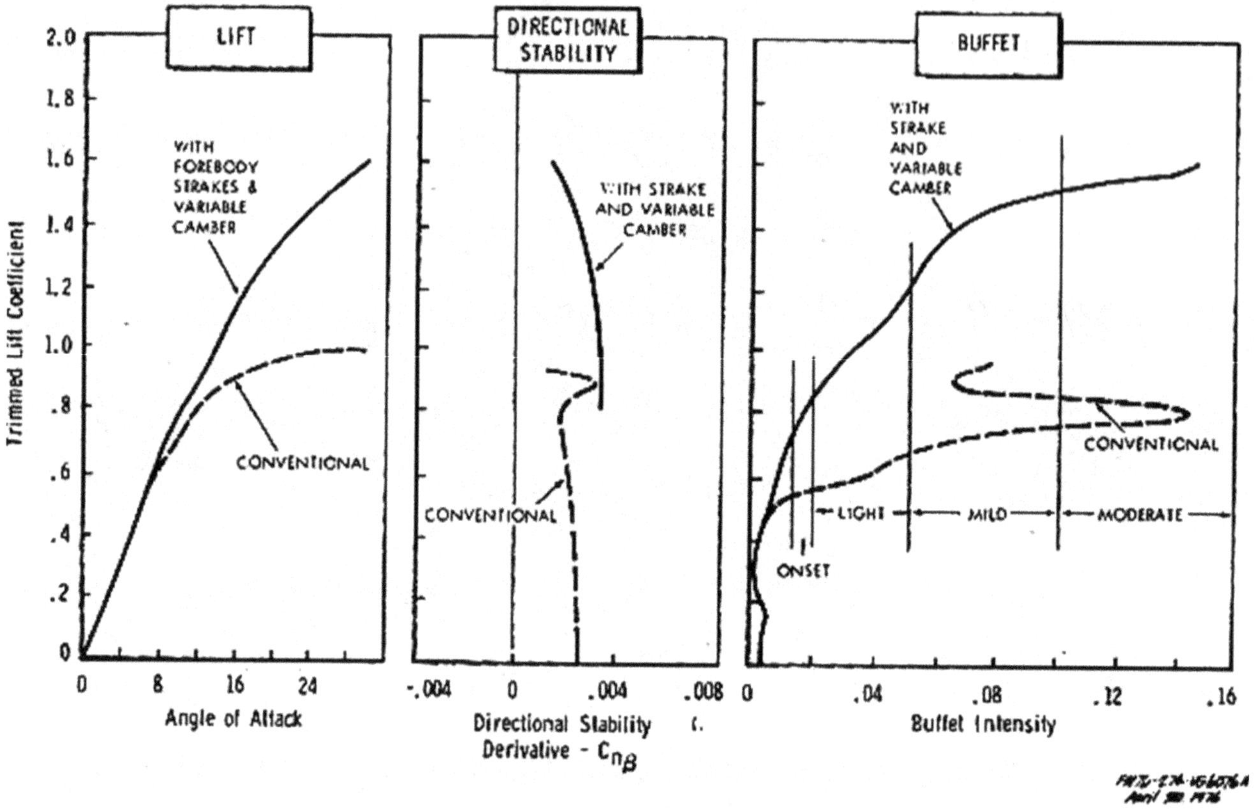

Combined Aerodynamic Effects on YF-16
(Source: USAF Engineering files)

In viewing the lift curve above, it is seen that for a conventional mode of aerodynamics (i.e., no variable camber wing mechanisms and no beneficial effect of forebody strakes (a.k.a., leading edge extensions), the maximum lift coefficient is obtained in the region of angle of attack between 24-degrees and 30-degrees. In that same region of angle attack, there is also seen to be rapidly changing directional stability magnitude and buffet intensity. However, wind tunnel tests as confirmed by flight tests indicate that the combined interactive effect of having variable camber wing control (with the flight control computer optimizing the LEF and TEF for the flight mode – without pilot command) and with the existence of the forebody strakes/LEXs, the trimmed lift coefficient continues to exhibit the ability for producing high lift at angles of attack in the 24-degrees to 30-degrees region. Both the directional stability and buffet intensity are also seen to have more benign and beneficial variation with angle-of-attack in that region.

It may be of assistance to Readers to point out that in a program such as the Lightweight Fighter program in which a best-effort type of contract was awarded by the Government to two contractors, there will generally be no formal government reviews of the design progress of the prototype aircraft. In a conventional aircraft development program that has a planned full scale development phase leading to a rate production of the aircraft, the design of the aircraft would normally be reviewed by the Program Office in a two-review process entailing a Preliminary Design Review (PDR), and a more comprehensive Critical Design Review (CDR) approximately five to six months later. The Lightweight Fighter prototype aircraft were originally designed as being advanced technology demonstrators but having some inherent capability for performing multi-mode military missions. Accordingly, the need for a formal PDR and CDR of the Lightweight Fighter prototypes was waived by the government.

The technical information presented about the USAF Lightweight Fighters to this point in this chapter 7 was generated in a joint cooperative manner in which the government gave two contractors a set of design and performance goals to be met and without the government applying numerous military specifications (MIL-Specs) that detailed how the aircraft was to be built to fully comply with military-quality standards of construction. Military-quality standards are much more stringent than are commercial-quality of construction due to the military aircraft being exposed to, and needing to survive, the harsher operational environment associated with combat operations.

In lieu of a formal official PDR and CDR, there were instead a series of informal design reviews held at the contractors' facilities on an approximately monthly basis. Design problems were surfaced at the monthly progress reviews, and resolution was generally achieved by the contractor deciding on resolution consistent with the resources provided in the best-effort contract and consistent with government safety standards. As a design feature was being defined, if there was some ground testing needed to confirm the validity of the design approach, the government made arrangements with government-operated test facilities (NASA or USAF) to provide those specialized test facilities to collect the data needed by the contractor.

For the Lightweight Fighter program, the two contractors – General Dynamics and Northrop – were contractually required to, and did, submit wind tunnel models of their respective YF-16 and YF-17 LWF designs for wind tunnel testing by the Government. (The Y at the start of the YF-XX designation is used by the USAF to distinguish a prototype aircraft from an aircraft being developed in quantity for operational use.) In that manner, the Government independently verified how well the contractors' estimates, contained in their proposal responses to the LWF RFP for the design and performance of their prototype aircraft, met the design and performance goals expressed by the Government in the LWF RFP. Additional analyses and ground tests were then conducted by the contractors to verify the safety for flight of their proposed designs. The contractor ground tests were conducted in technical design areas such as structural strength of the design, flutter margin for lifting surfaces, the stability and control aspects for safe control of the prototype aircraft in a planned flight test program, certain other safety tests such as canopy bird-strike tests to verify the canopy strength for pilot safety, and electromagnetic interference/lightning-strike tests on critical flight control components.

At the discretion of the commander of the Aeronautical System Division, an Executive Independent Review Team (EIRT) may be convened to review the adequacy of the ground tests performed and reviewed by the Program Office staff in an independent, second-opinion manner as to whether a prototype aircraft was safe to be flown by government pilots and at government flight test facilities. Because the LWF prototype designs featured some high-risk design features (e.g., the first all-electric fly-by-wire flight control system, and a canopy with no structural bow frame), an EIRT team and a well-defined safety review process was established for the USAF LWF Prototype Project by the ASD Commander, Lt. Gen. James Stewart.

The next chapter describes the additional tests that were conducted in response to EIRT items of concern for safety of flight, as well as items of residual concern for safety of flight generated by the LWF Prototype Project Office. After satisfactory accomplishment of the EIRT and LWF Project Office ground tests and safety analyses, the manufacture and assembly of the approved LWF prototype aircraft designs were allowed to begin. The next chapter provides photographs illustrating progress as it was made in the manufacture of the two-prototype aircraft that both contractors – General Dynamics and its YF-16, and Northrop and its YF-17 – delivered to the USAF at Edwards AFB for the competitive and combat-relevant flight testing by a ten- member Joint Test Team.

CHAPTER 8
SAFETY GROUND TESTS, EIRT SOF, MANUFACTURE OF LWFS (1972 TO 1973)

When General Dynamics, Northrop and the others submitted their responses to the USAF LWF RFP, they submitted designs for both a single seat variant and a two-seat variant, but with the contract requirement to each construct and flight-test only two of the single- seat variants of their designs. Only General Dynamics offered the canopy with no bow-frame design, but had the capability of adding the bow frame if bird strike testing and sled testing showed that a fixed portion of the canopy was dictated to be necessary for flight safety reasons.

Using the S-3 canopy bird-strike test facility at the USAF Arnold Engineering Development Center, Tullahoma, Tennessee, where testing of the F-15 Eagle canopy bird-strike vulnerability and structural integrity had recently been completed, the YF-16 canopy (with no structural bow frame) was tested for its flight safety. At the AEDC S-3 facility, a somewhat unique chicken gun was developed specifically for the testing of bird strikes on the canopies of aircraft. The gun features a nitrogen filled gas chamber which can be pumped to higher levels of pressure depending on the force to be applied to the projectile – a carcass of a bird weighing approximately four pounds. The cannon barrel is a tube approximately four inches in diameter and made of high strength steel to contain the pressure of the propellant (high-pressure compressed nitrogen) firing the projectile. The projectile carcass is aimed at the canopy being tested so as to impact the canopy in a low impact, mid-impact or, a high impact and at an impact speed typically of 400 to 450 knots.

The original design for the YF-16 bubble canopy was of a uniform thickness (5/8 inch is in the Author's memory) polycarbonate material formed by a thermoforming process developed by Texstars Inc of Grand Prairie, Texas. The thermoforming process essentially was/is to uniformly heat the polycarbonate sheet to an elevated temperature, allowing it to be deformed without changing thickness and/or optical quality, and the heated sheet forming the bubble shape by being drawn by suction pressure through a porous metal heated mandrill shaped to the desired bubble contour. The deformation of the polycarbonate is slowly achieved so as to not cause any thin or thick spots causing optical distortion in the polycarbonate canopy.

The F-15 canopy tests at AEDC S-3 utilized dead chickens obtained from a nearby farm as their projectile. General Dynamics was not allowed (by the Humane Society) to slaughter chickens except for use as a food supply. Therefore, the first tests utilized a four-pounds slug of soft clay to simulate the chicken; however, the slug of clay did not disintegrate the same as does a feather-covered chicken carcass and when the clay slug impacted the polycarbonate carcass, the clay slug drove a wave back through the deforming polycarbonate bubble and the inside surface of the wave impacted the dummy pilot, breaking the neck of the dummy. The canopy itself did not sustain any damage during its deformation, but the test was declared a failure inasmuch as the dummy pilot had endured fatal injury. General Dynamics purchased dressed chickens (not frozen) at the local market as an alternate source of the chicken carcasses, and subsequent bird strike tests were deemed to be realistic.

The next series of bird-strike tests at S-3 used the commercially bought dressed chicken carcasses and the bird strike tests of the YF-16 canopy were completed as shown in the photograph next.

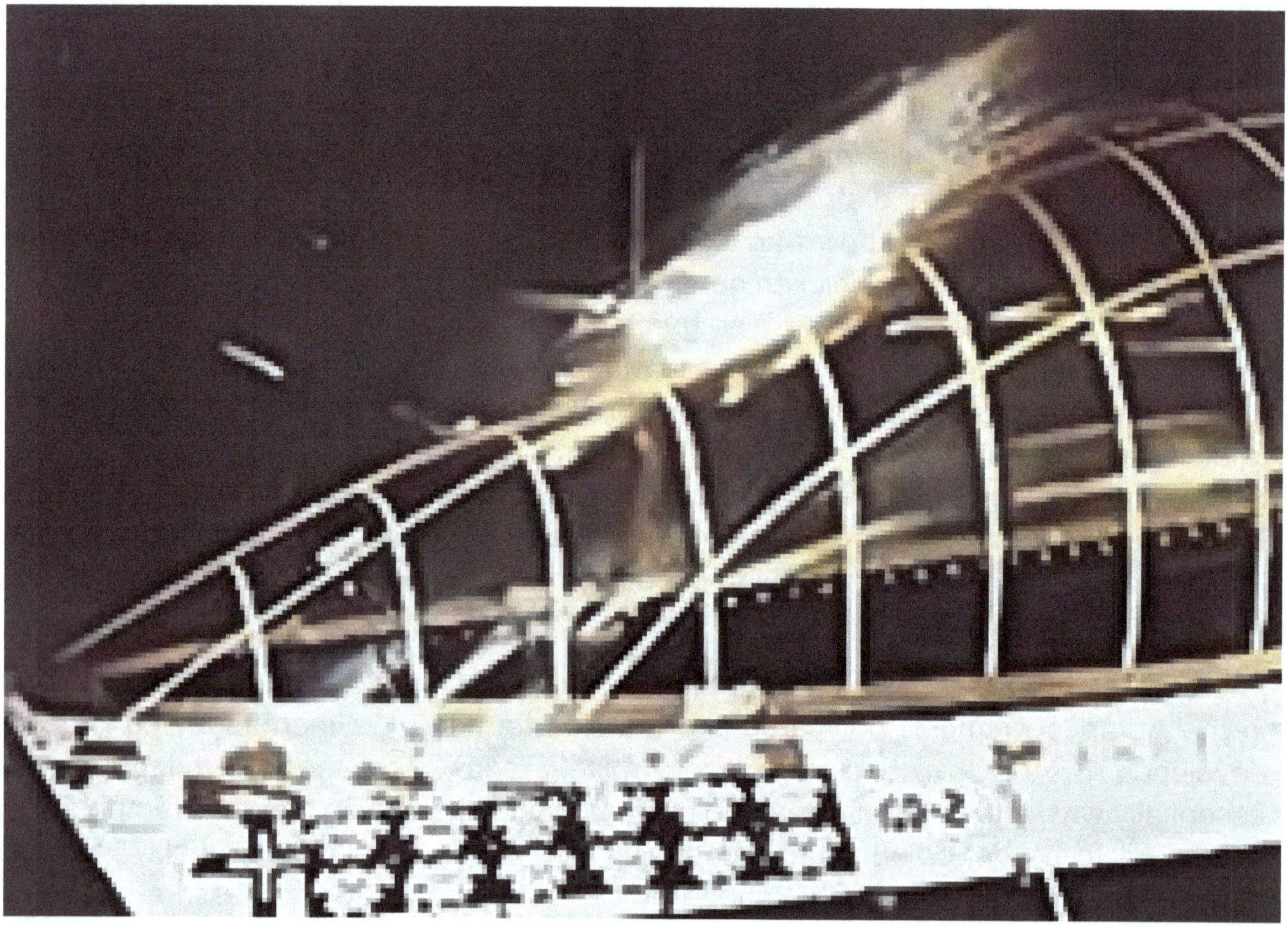

Bird-strike Tests at AEDC S-3 Test Facility of YF-16 Canopy
(Source: USAF Engineering files)

As seen painted on the side of the fuselage nose module and below the canopy was the accounting for the canopy bird strike encounters (not counting the test in the photograph) as GD canopy suffering two defeats and the chickens suffering 13 defeats. While this test series considered the original YF-16 canopy to be adequate from a flight safety aspect at AFFTC, it was recorded that if there was a production run made of the bubble polycarbonate material, the thickness of the front portion of the canopy would necessarily be constructed stronger so as to have less distortion by a bird strike event. It was also recommended that the production canopy be of a laminated construction, with the outer layer being of a harder acrylic material that is much higher scratch-resistant in an operational environment.

To validate the stability of flight and spin susceptibility being adequate for flight safety of both the YF-16 and YF-17, special spin susceptibility and spin mode testing was done in the vertical wind tunnel and large free-flight wind tunnel at the NASA Langley Research Center. The spin mode and recovery testing will be described first.

The NASA Spin Tunnel was originally used as a test facility for parachute design over 75 years ago. The test section is in a vertical orientation, is twenty feet in diameter, open at the top, driven by a large fan at the bottom and does not operate at very high speed since a parachute descends at very low speed. However, in recent years it has been used mainly for testing small light (typically balsa wood) models of aircraft to develop their proclivity for entering into a mode of spin and for developing control surface deflections or deploying spin chutes to effect recovery from the spin it has entered into. Control of the vertical speed of the wind tunnel is between zero and 85 feet per second to maintain the spin model at a constant height above the recovery screen located below, in order for the spin characteristics of the model to be photographed with high-speed photography.

The test model is launched by hand by a skilled launch member of the facility staff (often referred to as the Golden Wrist). Photographs are shown next describing the overall testing process and include a photograph of the YF-16 model being spin tested at the NASA Spin Test Facility (vertical wind tunnel).

View From Below of NASA Spin Test Facility Arrangement
(Source: USAF Engineering files)

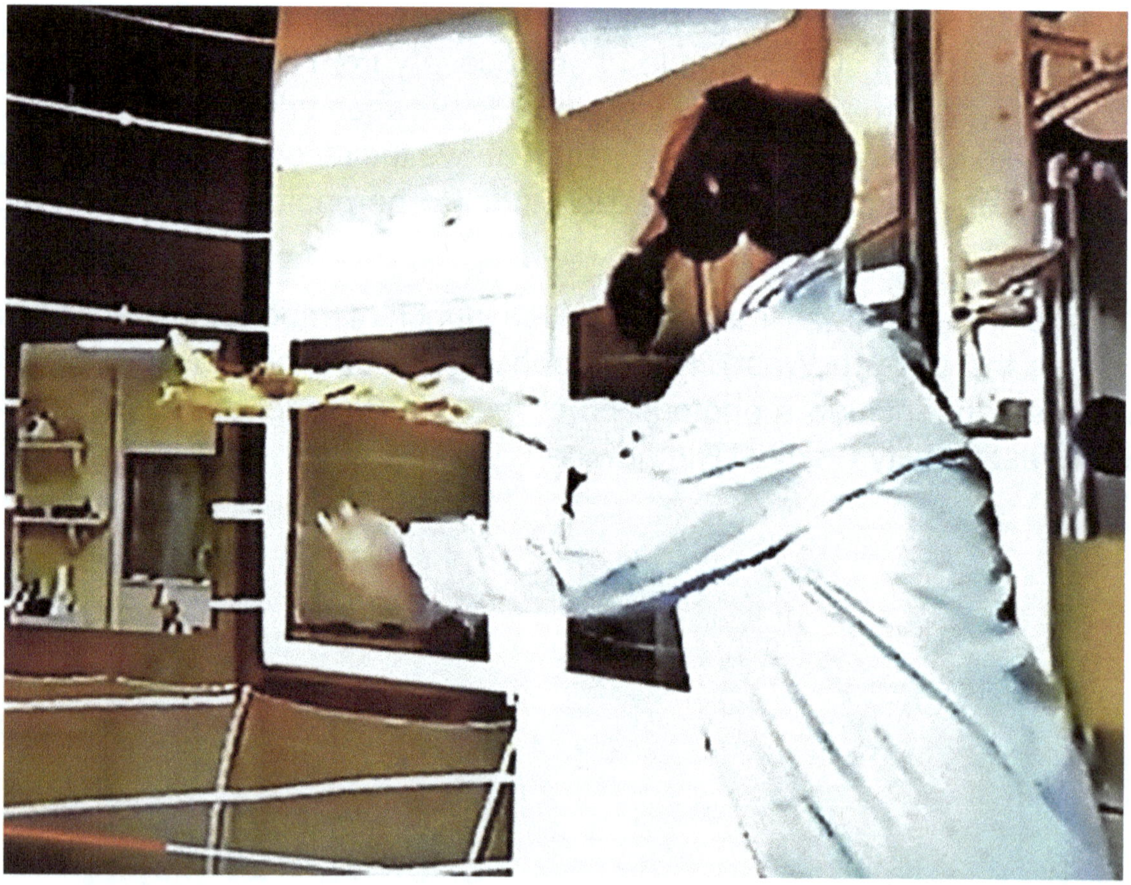

NASA Spin Test Model Launched by Golden Wrist in Spin Mode
(Source: Engineering files)

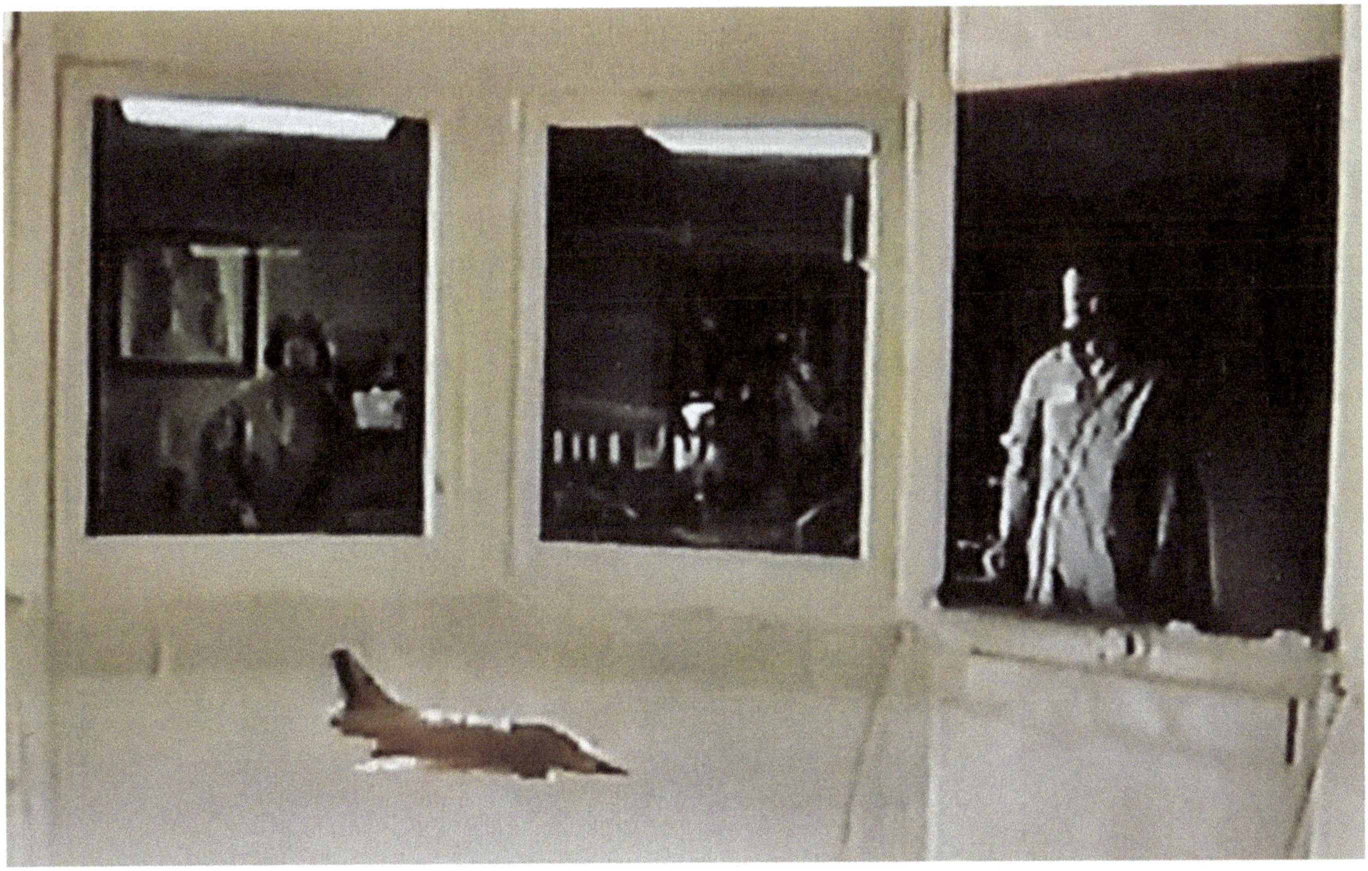

YF-16 Spin Recovery Testing in the NASA Spin Facility
(Source: USAF Engineering files)

As seen in the above photograph, the model has been launched by the launcher shown in the open window at the right. A second test staff member, shown behind the window at the left, controls the vertical speed of the airflow of the wind tunnel to maintain the spinning free-flight model at a relatively constant height (at about window level height) such that high speed photography (through the window in the middle) can capture the dynamics of the test model very accurately for post-test analysis.

Two spin-recovery modes are possible to be tested in the NASA vertical wind tunnel. These two modes are by (1) deflecting the appropriate control surfaces on the test model to effect recovery from the spin mode; and/or (2) deploying a spin recovery parachute (spin chute). If recovery is by control surface deflection (1), the test model has electrically-operated control surface deflection – such that the model is deployed in the spin mode with no control surface deflections. Then by sending a control signal to the test model, the control surfaces are deflected to the test deflection angles in various combinations that anti-spin recovery analysis has indicated effective. The high-speed photography captures the entire sequence from established spin mode until spin recovery has been obtained. If recovery is to be obtained with the use of a spin recovery parachute, a small-scale spin chute will have been attached to the test model prior to launch in the spin mode. Again, by radio-controlled signal, the spin chute can be deployed to recover the test model. The vertical speed operator gradually will gradually decrease the vertical velocity of the model and spin chute until it settles on the recovery screen below (just above the fan blades). A photograph next shows a recovery of a test model from a spin using the spin-chute technique. In a later chapter describing the flight test program for the YF-16, a segment of that testing included a spin chute to be installed on YF-16 #1 before the aircraft was tested for high angle-of-attack flight characteristics and spin susceptibility.

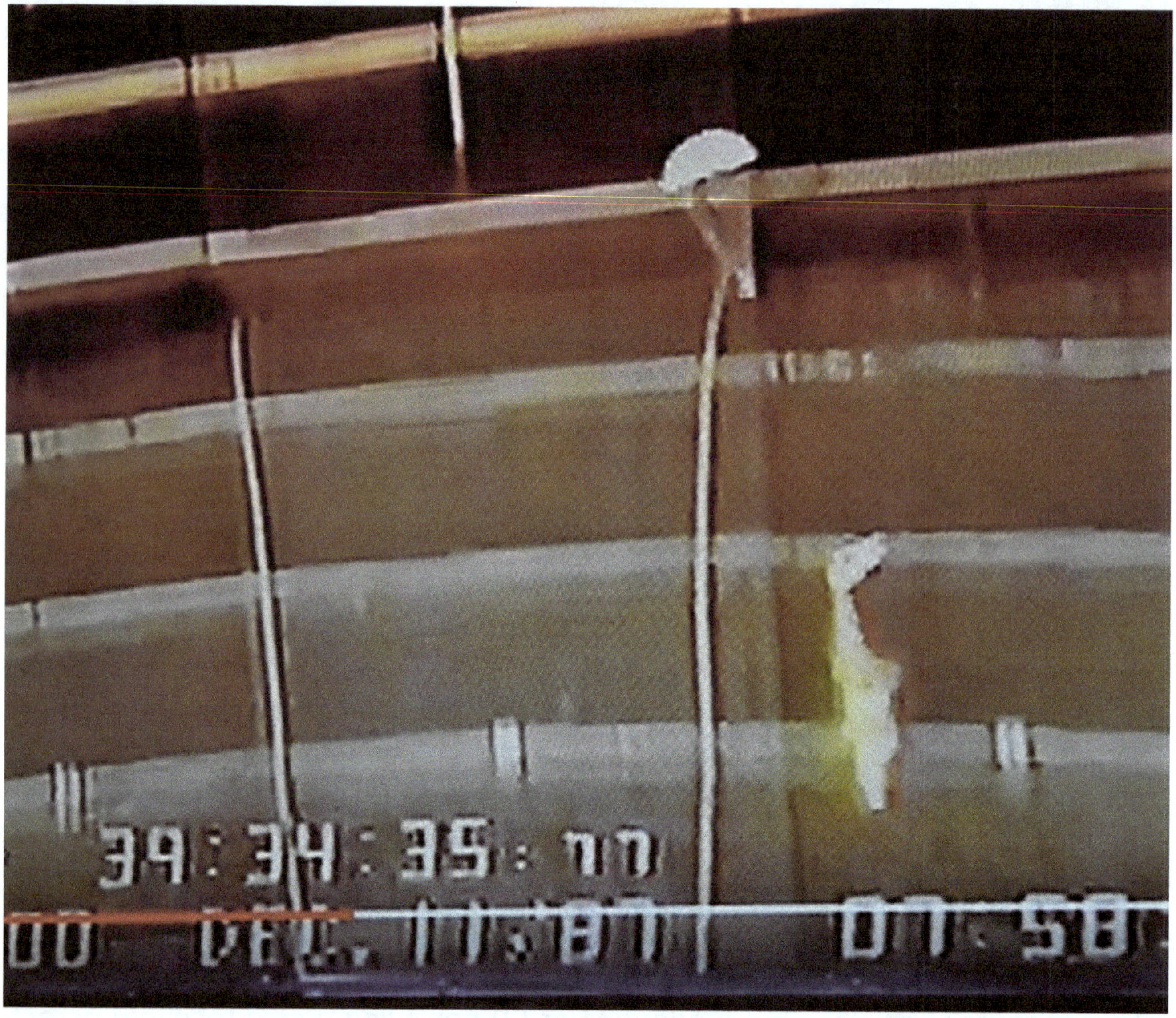

Recovery of a Spin Test Model by Recovery Parachute
(Source: USAF Engineering files)

The horizontal white stripes on the circumference of the test section are vertically spaced one foot apart and assist the photography analysis in correlating the rate of descent of the test model during the recovery mode (and accounting for the vertical speed of the airflow through the wind tunnel). A long pole with a grasper (similar to that used to retrieve grocery items from high out-of-reach shelves) then is inserted through the open launch window to retrieve the test model on the bottom screen.

The NASA wind tunnel test facilities at the Langley Research Center (near Hampton, Virginia) also includes a free-flight wind tunnel for testing of tethered wind tunnel test models. This type of testing is also for safety of flight to determine the dynamic flight characteristics of aircraft configurations operating in high angle-of-attack portions of the maneuvering flight envelope. A schematic of the NASA Free-Flight Wind Tunnel facility is shown next.

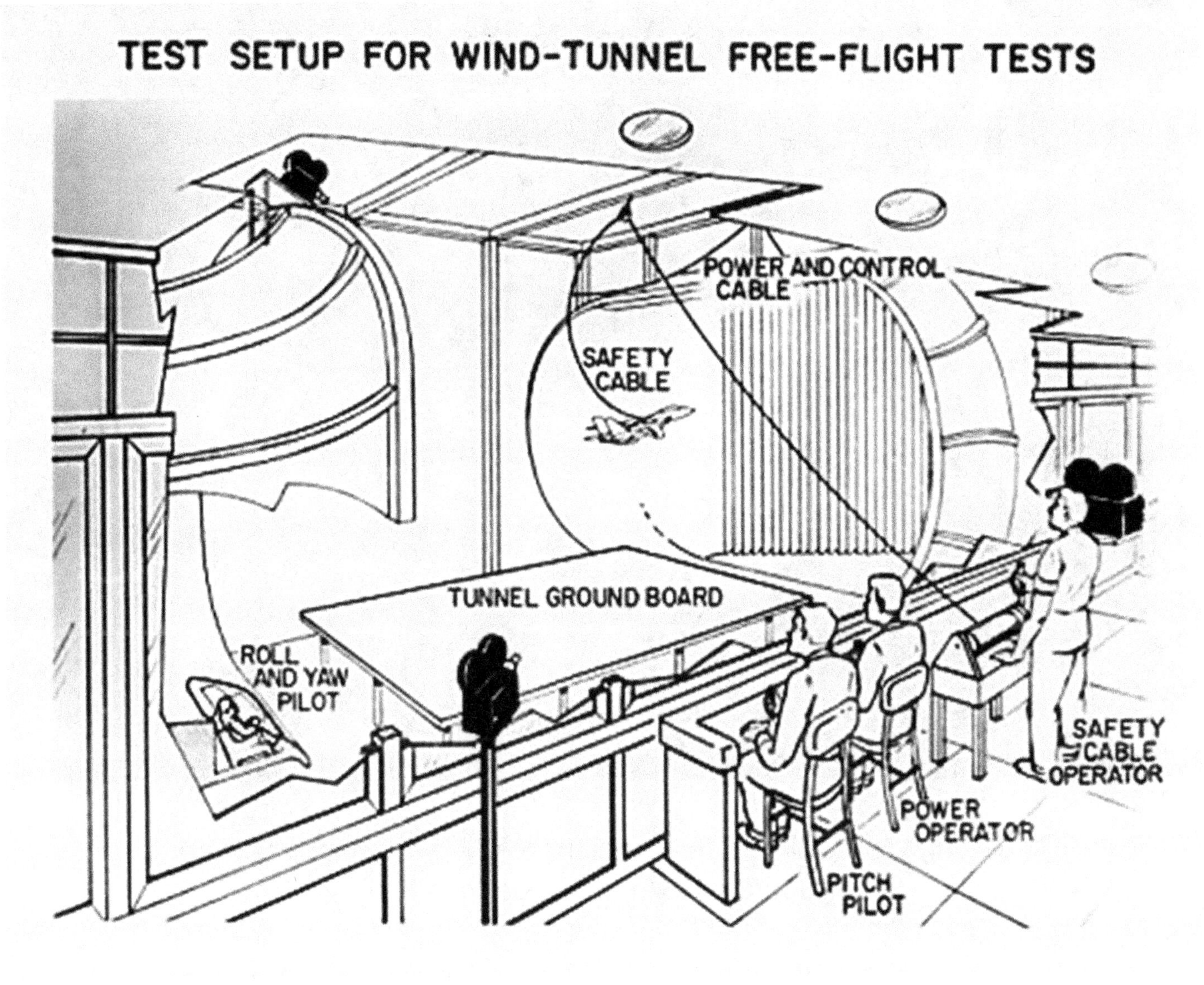

NASA Facility Set Up for Free-Flight Wind Tunnel
(Source: USAF Engineering files)

Both the YF-16 and YF-17 configurations were tested in the NASA Free-Flight Wind Tunnel to obtain their dynamic stability derivatives when operating at high angle-of-attack in their maneuvering flight regime. Shown in the photograph next is the large-scale model of the YF-16 being prepared for testing in the free-flight wind tunnel.

Setting Up USAF YF-16 Model for NASA Free-Flight Testing
(Source: USAF Engineering files)

In the photograph on the next page, the YF-17 model is shown in a high angle-of-attack attitude undergoing free-flight dynamic testing in the free-flight wind tunnel with the tether cable and control surface signal control cables attached.

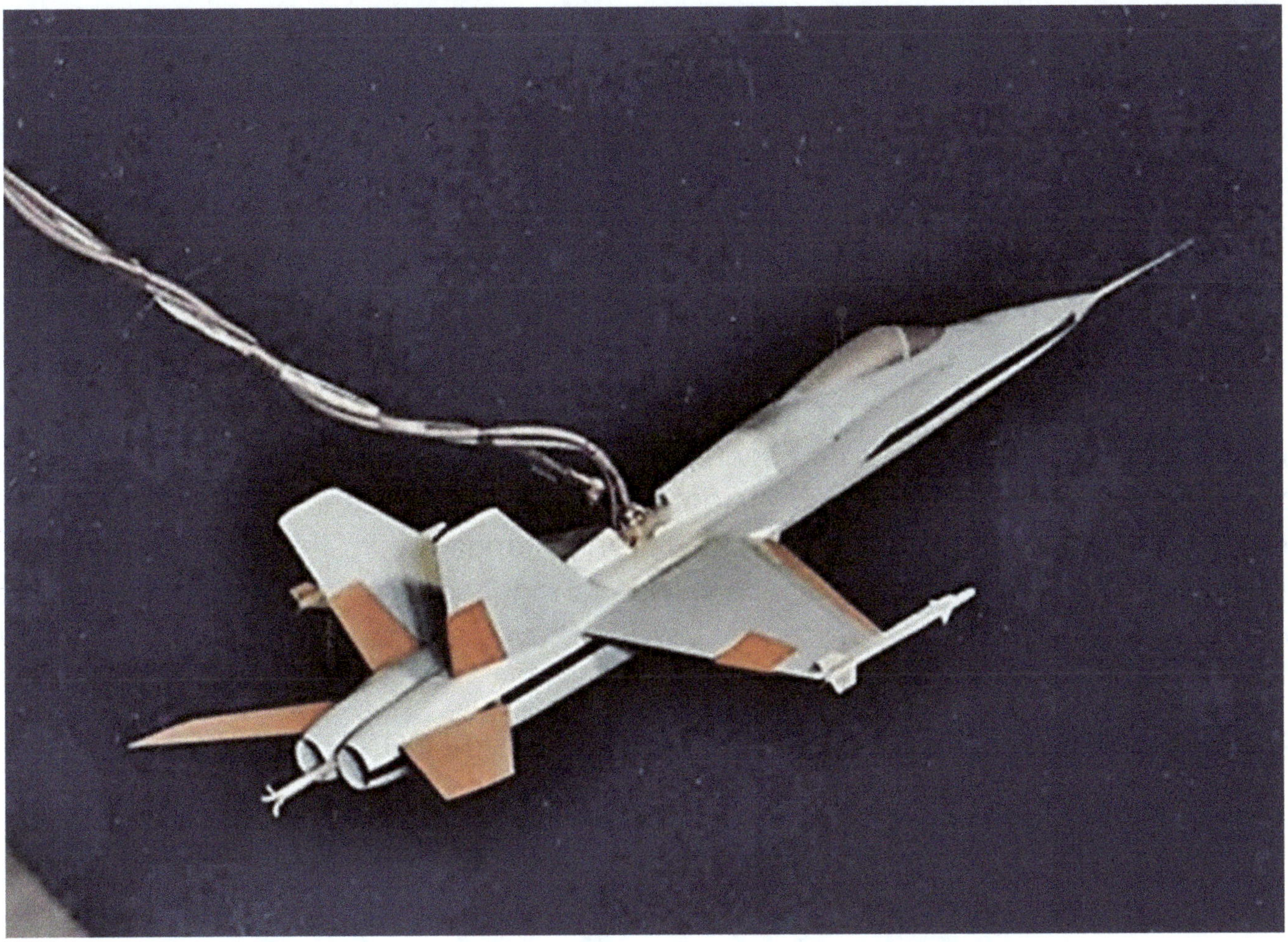

YF-17 Model Being Free-Flight Tested at High AOA
(Source: USAF Engineering files)

Propulsion of free-flight models is provided by high-pressure air being supplied through one of the hoses attached to the tether cable shown in the photograph above. The high-pressure air is directed aft from the nozzle device shown mounted between the nozzles of the YF-17 model and is controlled by a facility staff member to maintain flight of the model within the test section. Other staff members control the flight of the model by commanding deflection of the model control surfaces (shown in the photographs in contrasting orange red color). The flight control surfaces of the model are painted in the high visibility red/orange color for dynamic photography purposes as well as to assist the facility pilots controlling the dynamic motion of the free-flight model being tested.

It should be apparent to Readers that in relying almost entirely on the fly-by-wire flight control system flight control computer to relieve the pilot from all the seemingly mundane but highly critical duties of maintaining control of the USAF F-16 aircraft (and for deploying spin recovery control surface deflections), there is extremely high dependence that the flight control computer will perform without failure. That dependence on the flight control computer controlling the aircraft while being shot at by enemy ground fire or being attacked by surface-to-air missiles (SAMs) causes the flight control computer to have much redundancy in order to survive in hostile environments (hence its origin in the Air Force Flight Dynamics Laboratory as being a Survivable Flight Control System).

Readers are reminded that earlier in this book, the difference between aircraft with fly-by-wire systems is that the pilot commands the motion of the aircraft by sending commands directly to the flight control computer which in turn converts those commands to electrical signals and relaying those signals (interpreted from the pilot commands) to the control surfaces of the aircraft that provide in an optimized manner which control surface to deflect to what degree of deflection and at what rate of deflection. It is considered appropriate to first include here a diagram showing what components comprise the YF-16 Fly-By-Wire Flight Control System to give Readers an introduction to the hardware aspects of that system before discussing the functions the various components perform to relieve as many of the manual tasks the pilot would have to perform to control the aircraft with a conventional flight control system. The diagram next was extracted from a set of charts provided by General Dynamics to the USAF Lightweight Fighter Project office:

COMPONENTS OF THE YF-16 FLY-BY-WIRE FLIGHT CONTROL SYSTEM

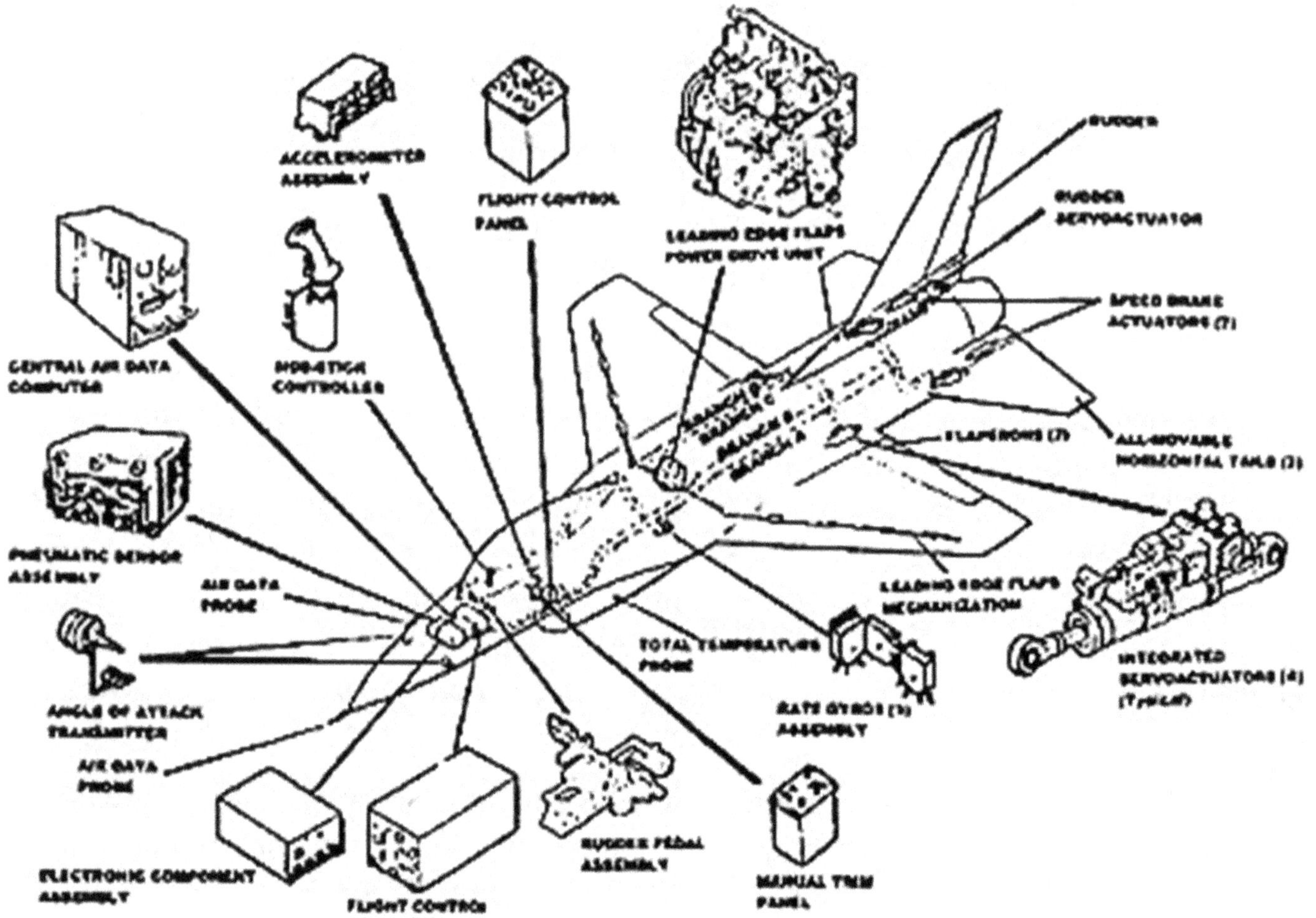

Components in the YF-16 Fly-By-Wire Flight Control System
(Source USAF Engineering files)

The pilot of the General Dynamics YF-16 applies a pressure to the side stick controller on the right console inside the cockpit, signaling the rotational motion of the aircraft and direction he/she wishes the YF-16 to move in the pitch plane and the roll plane. Located at the base of the side stick controller is a set of strain gages which senses and resolves the pressure applied to the side stick controller into the x-axis and y-axis components of the force (F_x and F_y) applied into the side stick controller. If the pilot also wishes to achieve a rotational motion of the aircraft having a component about the z-axis (contained in the yawing plane), he/she will apply feet pressure into the left and right rudder pedals. Similarly, there are strain gages at the base of the rudder pedal assembly to determine the magnitude and direction of the moment the feet pressure is commanding the aircraft to rotate about the z-axis. These three rotational commands are relayed such as to be received at the flight control computer in signals (direct current voltage levels). Simultaneously, there are external sensors that are sensing the ambient and total pressures and temperatures located at various locations on the external configuration, which are converted into the aircraft speed and altitude. The central air data processes the sensed pressures and temperatures into speed and altitude and relays that information (voltage signals) into the flight control computer.

It is beyond the scope of this book to describe all of the laws of motion in six degrees of freedom (6 DOF) and containing all of the significant dynamic variables and aircraft inertial terms in those multiple 6 DOF equations that are embodied in the YF-16 FBW FCS. Similarly, it is beyond the scope of this book to describe the effort and hundreds of thousands of man-hours expended in the development of those equations of motion and tailored control laws that were originated when development of the fly-by-wire concept began for application in a lightweight fighter as created by Harry Hillaker and John Boyd. There were many thousands of wind tunnel testing done to define the aerodynamics coefficients and stability derivatives providing inputs into the development of the 6 DOF and control laws, as well as many hundreds of hours of simulator time refining them. The highest of credit was earned by the General Dynamics team in development of the Fly-By-Wire technology, and the Author can attest that every person associated with the development of the all-electric FBW FCS was fully aware of the importance this technology potentially had on future fighter aircraft.

A high degree of credit is justifiably shared with the HQ USAF Advanced Prototype Study team headed by Brig. Gen. Kent for fostering the management concept of providing a limited amount of funds to two contractors (General Dynamics and Northrop) and allowing them to conduct their best effort toward the creation of the YF-16 and YF-17 advanced technology demonstrator prototypes.

While a Flight Control Computer cannot think as does a human pilot, the computer does have a memory capacity and a processing capability to convert, at the speed of electrons, what the pilot manually commands the pitch and roll motion of the aircraft to be and with yaw motion (if any is commanded) through foot pressure on rudder pedals.

Internally located are rate gyros sensing the heading changes being commanded by the pilot. Also internally located is a three-axis accelerometers assembly component sensing the acceleration of the aircraft in its three planes of flight (accelerations in the x, y, and z direction). The heading rate changes and the three axes of accelerations are used as a feedback loop into the flight control computer and are compared with what the pilot commanded to what motion the aircraft is actually taking. If there is significant difference between the pilot commanded motion and the aircraft actual motion, those sensed differences will be processed by the flight control computer into the appropriate additional deflections of the various control surfaces needed to bring the commanded and achieved aircraft motions into equivocality.

With his/her left hand on a throttle quadrant, the pilot controls the forward speed of the aircraft by adjusting the amount of thrust being produced by the propulsion system. The forward speed (i.e. proportional to the level of thrust being commanded) is controlled by electrical signals from the throttle quadrant to the electronic engine control at the engine and is achieved by adjusting the rate of fuel flow to the engine combustion process. In the design of the YF-16 cockpit, the principle of "hands on throttle and stick" (HOTAS) was created such that the pilot could manage the motion and speed of the aircraft directly by coordinated movements of the throttle and pressures on the stick and rudder pedals to fly the aircraft with near complete abandon – and not having to mentally process what deflections of primary flight control surfaces would be needed to achieve that motion.

To further illustrate the HOTAS objective, both the sidestick controller and the throttle quadrant have several other functions which are controlled by finger selection (both hands). [Author's note: It was often heard among human factors engineers that pilot of the YF-16 should have extraordinary eye-hand-finger coordination similar to clarinet players in a symphony orchestra in order to be able to read the musical score (on the HUD), to interpret what note is being identified to be played, and almost without thinking to move the appropriate finger to activate the correct switch on the sidestick controller or throttle quadrant – and do so while perhaps at elevated Gs. To do so harmoniously with all the others in the orchestra (or flight formation) requires constant practice and training.]

The sidestick controller configuration is as shown next.

Side Stick Controller

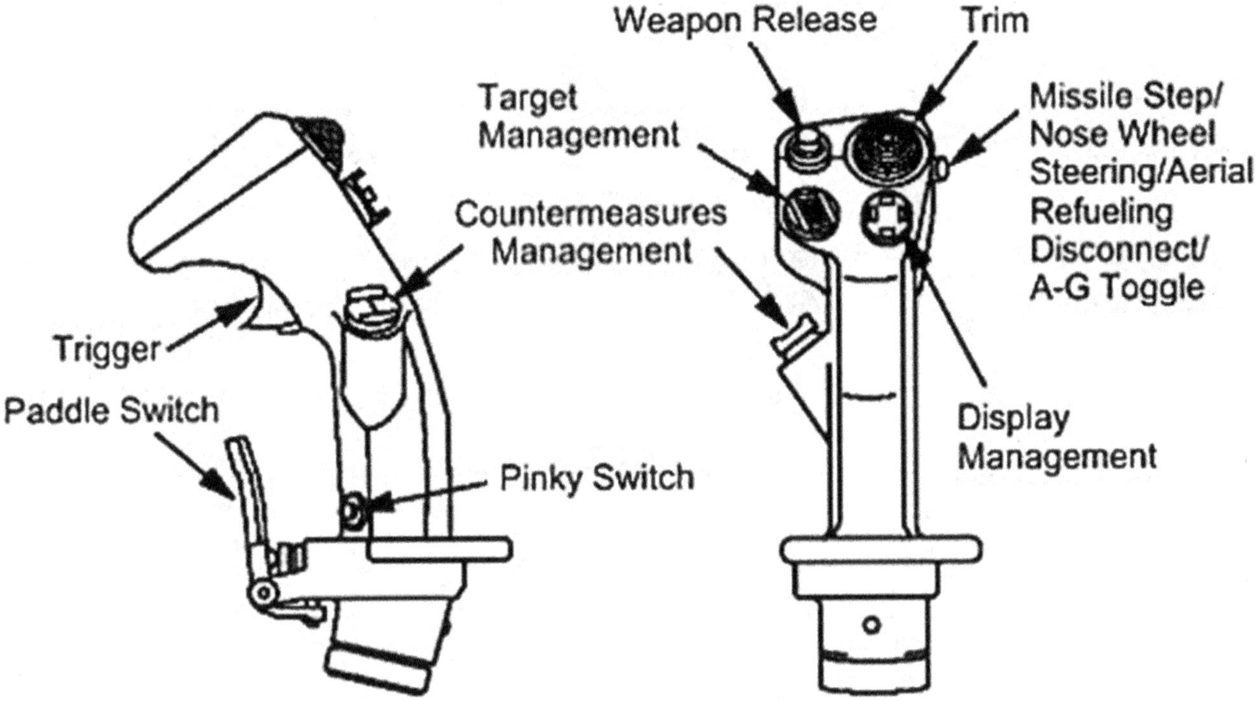

Type Side Stick Controller in YF-16 (Model in F-16C Shown)
(Source: USAF Engineering files)

The first side stick controller on the YF-16 LWF Prototypes was a fixed stick – there was virtually zero displacement. Although the test pilots did not evaluate the stick not being movable and not providing visual displacement clues for commanding the aircraft motion, there were unofficial comments made that the stick should have some displacement. Later in this book will be a description of a near-accident caused by the side stick controller having no displacement and which resulted in there being approximately +/- 1/4" displacement of the stick from its centered position in the plane defining the x and y axis.

[Readers may perhaps enjoy reading a humorous account regarding the original no-displacement side stick controller for the YF-16. At a briefing being given in early 1973 by Col. William E. Thurman to then-AFSC Commanding General Alton D. Slay and that had the objective of familiarizing General Slay with the technologies and capabilities of the YF-16 advanced technology demonstrator, Col. Thurman gave a summary description of the function of the side stick controller. It is necessary that Readers be aware that both Gen Slay and Col. Thurman wore Command Pilot wings and both with several thousand flight hours such that it was a pilot briefing another pilot on how the aircraft flew. When Col. Thurman stated "To make the aircraft increase its angle-of-attack and rotate its nose upward into a climb, all the pilot need do is to exert a force into the side stick controller in a rearward motion and the nose will pitch up." At that point in the briefing, Gen. Slay remarked: "Colonel, I really do not need to be reminded how to fly a fighter aircraft. I know that in order to increase my aircraft angle-of-attack into a nose-up attitude, I have to pull back on the stick." Col. Cameron recovered well by stating that unlike the other fighter aircraft in the inventory, it is the pressure force on the non-moving stick that is the mode of commanding aircraft pitch motion not the displacement of the conventional floor-mounted stick between-the-legs control column.

The Author was in attendance at the briefing and suppressed an incredibly strong urge to grin like a Cheshire cat while witnessing the exchange of comments between the AFSC Commander and the Author's LWF Prototype Project Director. When the memorable moment was re-told to a colleague, the advice given was that "When elephants make love, the monkeys should head for the trees."]

The throttle quadrant is a somewhat more conventional throttle used on aircraft with afterburning turbojet engines to control the thrust level being produced. The throttle quadrant has motion of travel permitting modulation of thrust from flight idle to full non-after-burning thrust and further movement through five detents to select thrust level from minimum afterburner to maximum afterburner in five zones of afterburner operation.

A diagram of the throttle quadrant is shown next (in the configuration of the throttle quadrant integrated later into the F-16C cockpit) and is seen to have additional finger switches for performing other functions while adhering to the HOTAS objective:

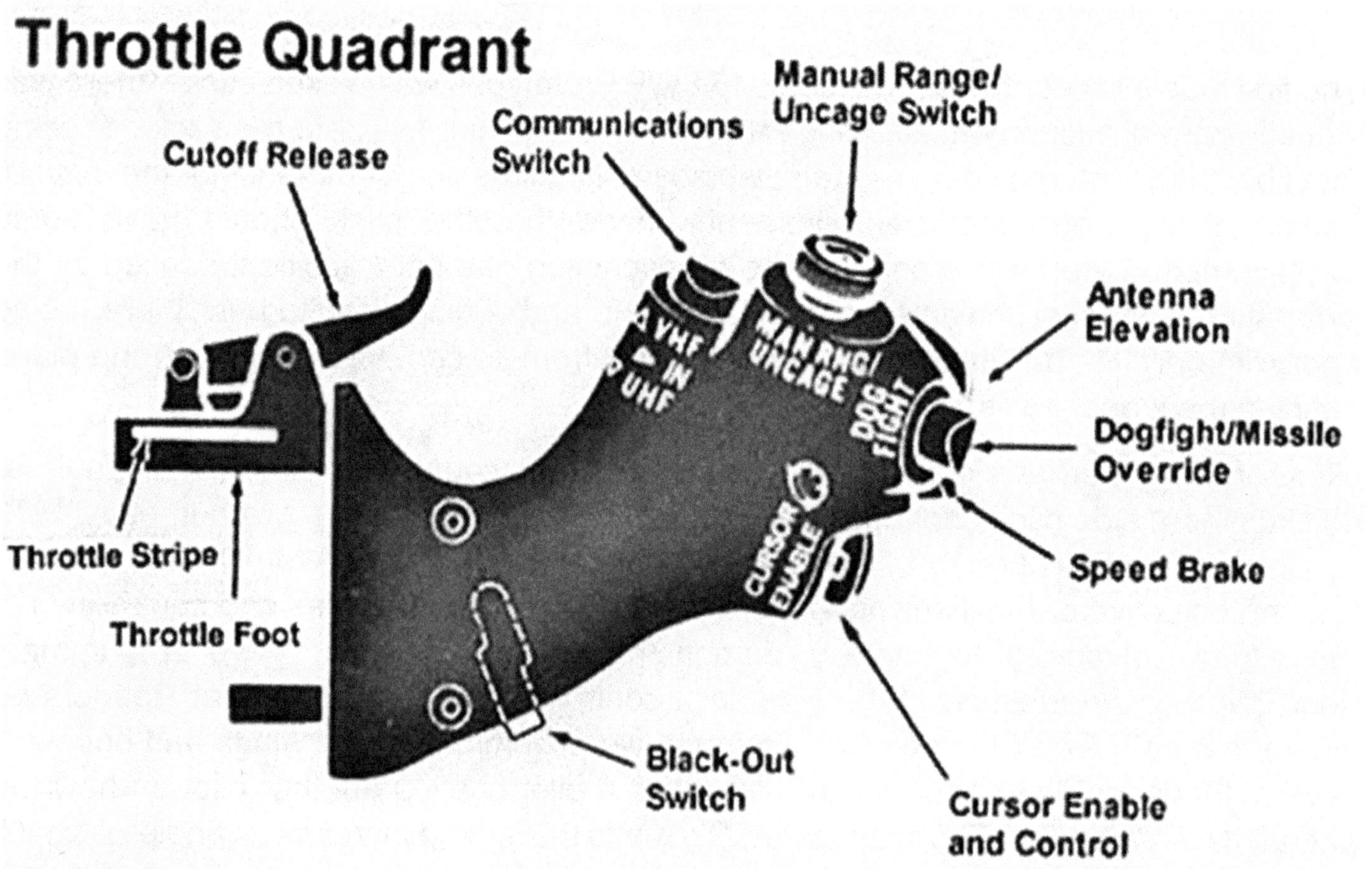

Type Throttle Quadrant in YF-16 (Model in F-16C Shown)
(Source USAF Engineering files)

The FBW FCC is of a quadraplex design and has a voting function in the event of failures (from any cause) in up to three of the four branches that carry the selected "best" signal from the computer to the control surface actuators to achieve the commanded motion by the pilot. The four branches (A, B, C, and D), installed for redundancy purposes are shown in the previous diagram – branches A and B installed internally on the upper port side of the fuselage and branches C and D on the upper starboard side of the fuselage. There is a significant space between each of the pairs of branches varying from approximately 4 inches to approximately thirteen inches depending on the local routing to pass by other installed components. These separated routings were done for survivability reasons and provide redundancy in the case of failure of up to three of the four branches from any cause.

The Flight Control Computer being quadraplex uses a voting process that routes the same best signal through each of the four branches aft to (1) the Leading Edge Flaps Power Control Unit; (2) the first two integrated servo actuators controlling deflection of the left and right trailing edge flaps (flaperons); (3) the third and fourth integrated servo actuators controlling deflection of the left and right all-movable horizontal tails (stabilators) and (4) the servo actuator controlling deflection of the rudder on the vertical tail. The signal to deflect the speed brakes from their normally closed position to full deflection is transmitted also through the four branches (A, B, C, and D) but that signal is seen to be commanded by the pilot operating the speed brake switch on the throttle quadrant as it is a speed control function and not a flight control function.

That best signal selection process is graphically summarized on the diagram next (extracted from sets of charts provided by General Dynamics to the USAF Lightweight Fighter Project office):

Quadruplex Flight Control Computer Signal Selector Logic

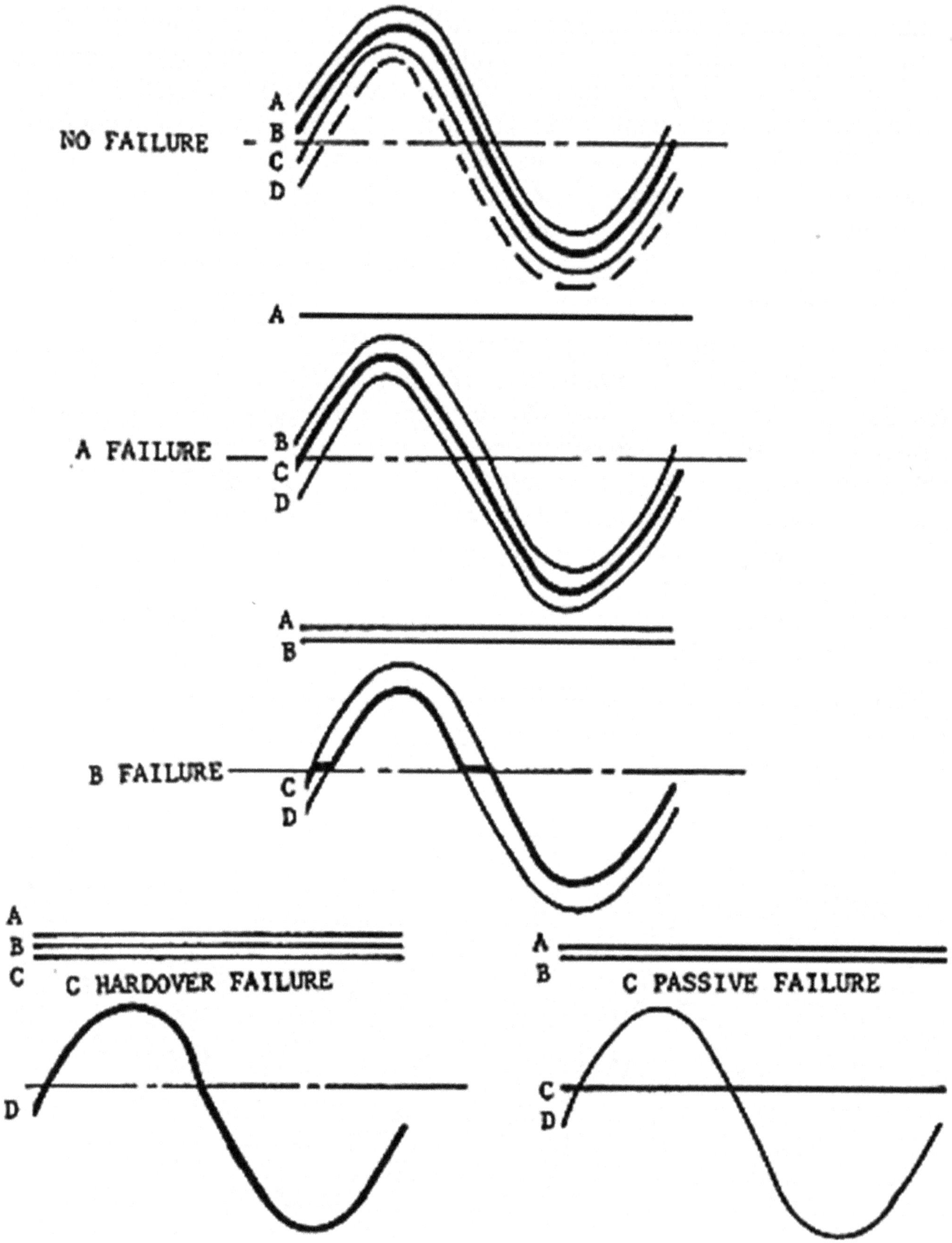

Quadraplex Flight Control Computer Signal Selection Modes
(Source: USAF Engineering files)

The above several signal selection modes indicate (in **bold** line) the Flight Control Computer logic used in selection of the best signal depending on the number of failures (if any) that are detected by the feedback of the control surfaces deflections in the feedback loop to the flight control computer. The uppermost selection diagram reflects the normal (no-failure) mode of the selection process. When three failures have been detected in branches, A, B, and C (highly improbable), the lowermost diagram indicates that if there is a hard-over type failure in branch C, the stand-by gains embedded in the flight control computer will be transmitted to the control surfaces' servo actuators permitting the aircraft to be flown safely, but with no capability for aggressive maneuvering, on a return to nearest base (RTB) heading.

As developed for the USAF LWF Prototype Program, the first configuration of the YF-16 flight control computers was configured to be suitable for safe flight control in the daylight and most often clear weather environment of the flight testing to be done at the AFFTC at Edwards Air Force Base, California. It was recognized that the all-electric FBW FCS had the Achilles heel safety symptom of relying on the supply of electric power to operate the all-electric FBW FCS. The two most major safety concerns were (1) loss of the electric power from the main generator driven by the single engine; and (2) the disruption of the quality of the electric power into the all-electric FBW FCS (especially from a lightning strike on the aircraft).

The first major safety concern was the loss of the engine power (for any cause) and resultant loss of the engine gear-driven single main electric generator as an accessory driven by the engine. A hydrazine-fueled emergency power unit (EPU) was added into the fuselage of the two YF-16 configurations directly aft of the bulkhead forming the rear wall of the pilot's cockpit. An engine RPM sensor was added into the emergency power subsystem that initiated operation of the EPU when the engine RPM was detected to be spooling-down at an abnormal rate and was approaching the lower limit of RPM needed to operate the main electric generator. When that dangerous engine spool-down situation was detected, the RPM sensor caused the ignition of the hydrazine fueled EPU to start generating sufficient power for safely operating the flight control system (but with all other unnecessary electric users for flight safety) being disabled. The EPU is essentially a rocket motor driving an electric generator, and uses a hydrogen and hydrazine (an ammonium-based chemical) reaction to produce the rotational power for the electric generator powered by the EPU. The hydrazine tank was sized to permit approximately ten minutes of operation, which was estimated to be sufficient time for the pilot to depart operation at Mach 0.9 at 30,000 feet and be able to descend to a safe, adequately controlled, gear-down, landing.

The second major safety concern was that of the possibility that the aircraft could sustain a lightning strike that would damage the flight control computer and other critical electronic components in the FBW FCS rendering them partially or totally inoperative. This concern caused the design and construction of the aircraft skin (that which is metallic) to serve as a conduction path for the large voltage electrical path for the lightning strike without that path damaging any internal safety-critical electrical components. For the two YF-16 prototype aircraft, their construction did not embody this lightning strike protection feature as the planned flight test program was to be conducted only in clear weather daylight environmental conditions. That did not preclude the FBW FCS from being shielded from electromagnetic fields created by internal fuselage electrical and electronic components and affecting the aircraft electrical control signals being routed through the four branches (A, B, C, and D) almost the entire length of the fuselage – from the flight control computer in the forward nose module and aft to the rudder servo actuator. [Later in this book, description will be given of a few instances of human error of oversight in the design of electrical systems interfacing with the flight control computer occurring that caused a few test flights to be aborted (fortunately with no damage to aircraft or human life). There was no incidence of lightning strike during the LWF flight test program in 1974 to either the two General Dynamics YF-16s or the two Northrop YF-17s.]

From the outset of the Lightweight Fighter concept, as envisioned by John Boyd, the objective was to develop an advanced technology fighter that was a fighter pilot's fighter. The first portion of this chapter described the major technologies of the Lightweight Fighter, and the next portion of this chapter describes the front office – its cockpit - of the General Dynamics YF-16 and, to a lesser extent, the Northrop YF-17.

Because of the limited funds available for designing, manufacturing, and testing (ground and flight) of the YF-16 but with the large number of advanced technologies incorporated in the YF-16 design, it was necessary to limit the controls and displays on the instrument panel and side consoles in the YF-16 cockpit to existing circular shape display dials with needle pointers and conventional toggle switches and press buttons. A photograph next shows the more or less conventional display and instrument panel (and shows the absence of the between-the-legs flight control column) in the YF-16s:

YF-16 HUD and Instrument Panel (1974)
(Source: USAF Engineering files)

By comparison, the external configuration features that existed on the YF-16 have not changed much at all relative to the equipment and component upgrading changes made internally in the F-16 design. [Vast improvements in HUD design and in instruments and displays technologies rapidly followed after the decision was made to produce the YF-16 design in quantity as the F-16A Air Combat Fighter, and continued in improvements during the decades of the continued large production of the F-16 Fighting Falcon Multi-National Weapon System Program.]

It is hastened to be clarified that the cockpit in the Northrop YF-17 advanced technology demonstrator also did not have many advanced technologies displays and controls incorporated in the cockpit at the time of their flight-testing but which were adequate for flight demonstration of the YF-17 advanced technology design features. Shown next is a photograph of the instrument panel as it was flown in the Northrop YF-17 #1 and #2 aircraft (tail numbers 01569 and 01570):

Instrument Panel in the Northrop YF-17 LWF Prototypes
(Source: USAF Engineering files)

When the Boyd LWF mafia-sponsored Advanced Day Fighter Study was being conducted in the late 1960s, Harry Hillaker and his team at General Dynamics did a thorough job of identifying advanced flight technologies that were in the emerging state of status at the several laboratories of the US Air Force Systems Command. A technology that Harry Hillaker saw being conducted in the Human Resources Laboratory (AFHRL) at Brooks AFB in Texas, and with participation by the Aerospace Medical Research Laboratory (AMRL) at Wright-Patterson AFB in Ohio was that of a recline angle of the pilot seat greater than the conventional eleven to twelve degrees seatback angle in fighter aircraft for the past several decades. The Aeromedicine personnel at both AFHRL and AFAMRL had both been exploring ways to enable the pilot community to sustain the physiological tolerance to highly maneuvering aircraft and had found in centrifuge testing that increasing the angle of the G-induced vector to the torso of the human test subjects gave an increased tolerance to the G effects with that angle between the G-vector and the spine increasing from the conventional 15-degrees upward to 40-degrees. Human factors engineers and the test pilots, especially Phil Oestricher at General Dynamics, constructed a rudimentary pilot seat to gain further physiological information on the ability of the pilot to perform manual tasks in the advanced configuration cockpit being designed and featuring the HOTAS objective. Shown next is that early variable seatback angle mock-up developed at General Dynamics:

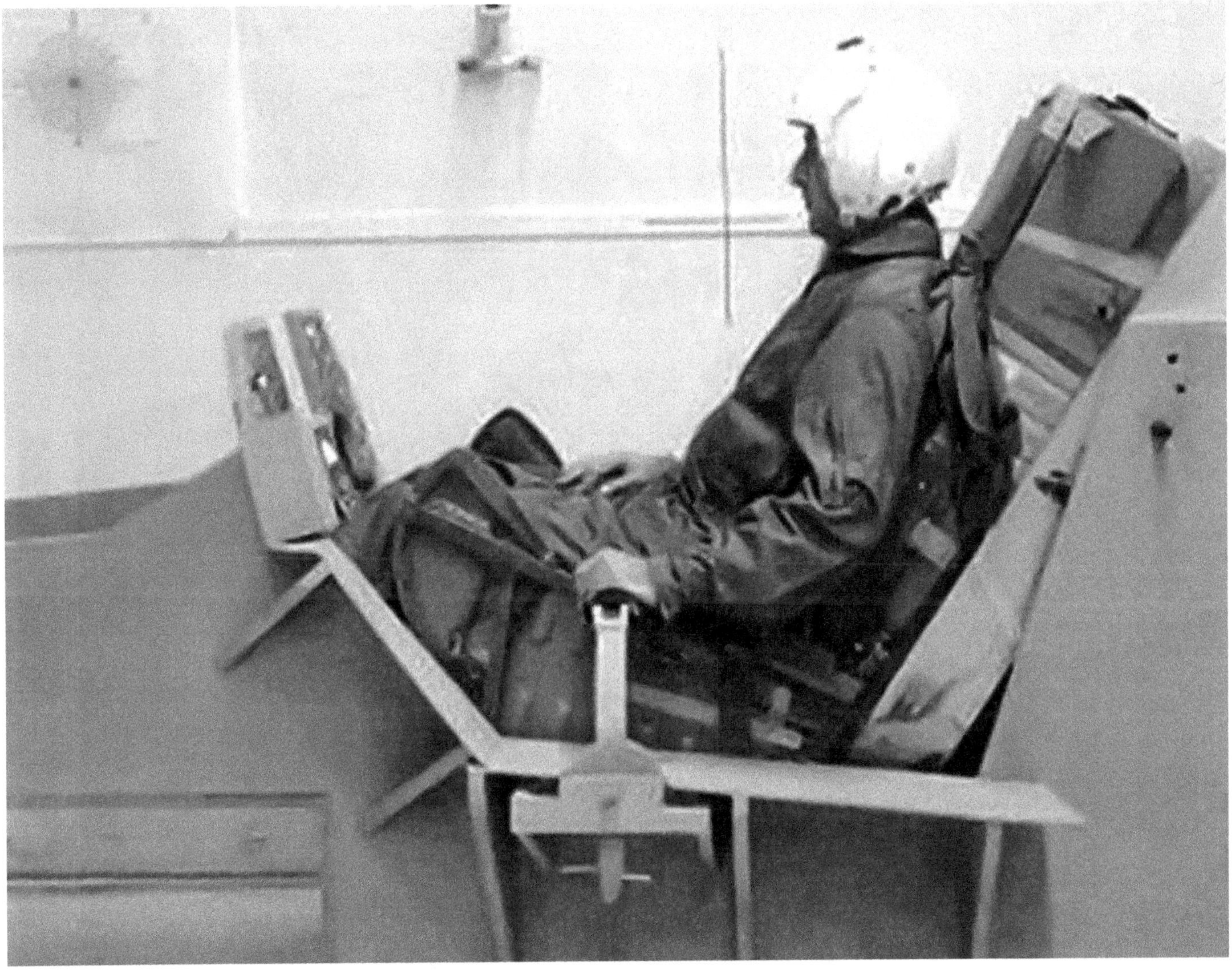

YF-16 variable seatback angle mock-up (at 30-degrees angle)
(Source: USAF Engineering files)

In the YF-16 aircraft, the Escapac escape seat was mounted in the 30-degrees reclined angle as shown in the photograph next.

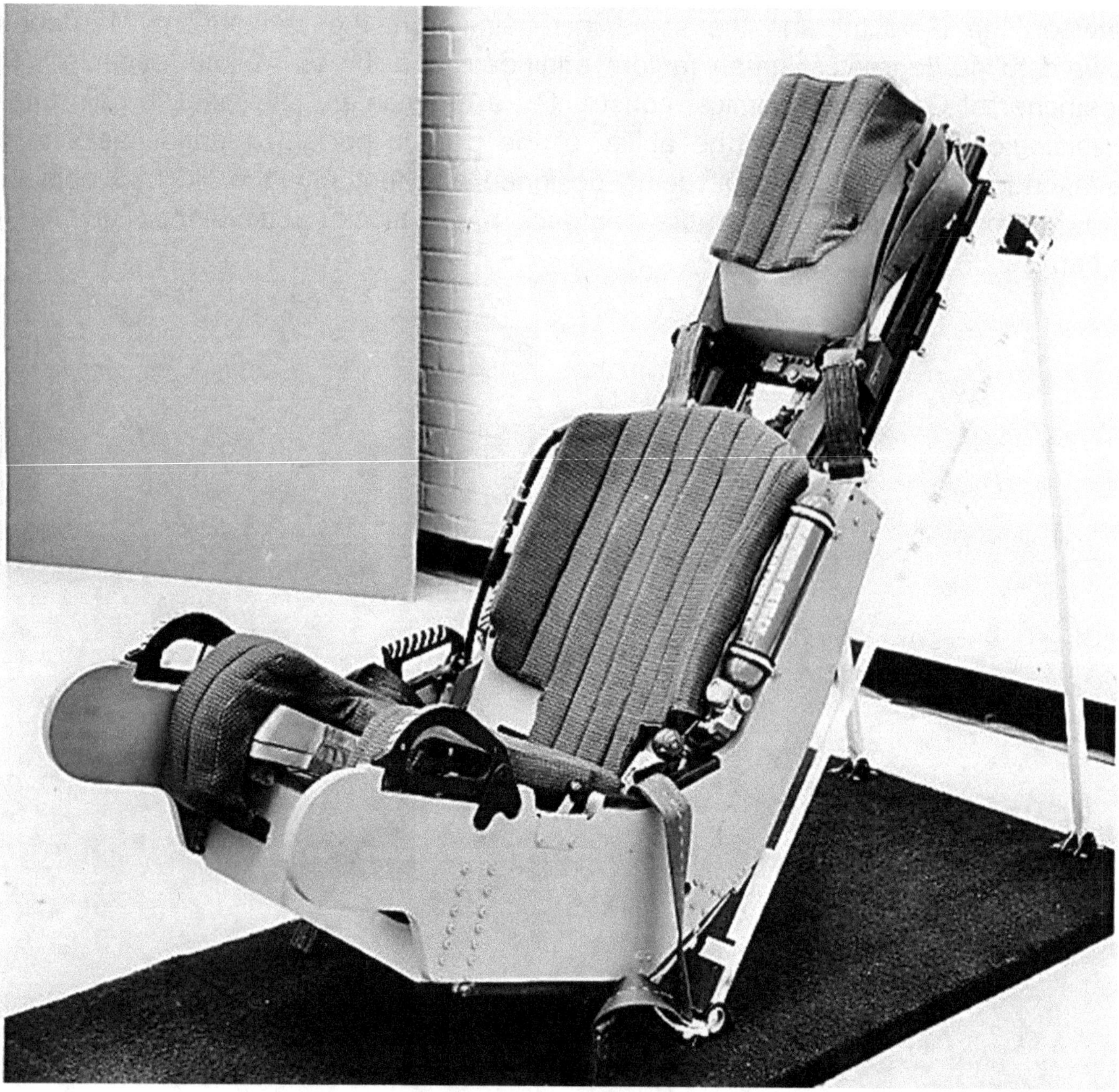

Escapac Model IH-8 seat installed at 30^0 angles in YF-16
(Source: USAF Engineering files)

Using the logic that the purpose of the USAF LWF Prototype Program was to gain flight assessment of the utility of the several candidate flight technologies on the pilot's ability to fly the LWF at near-complete abandon, the Hillaker cockpit design team arrived at three decision factors: (1) the operability of manual flight tasks and comfortability of movement appeared to degrade when seated in a seat back angle above 30^0; (2) the most reliable way to assess the value of a pilot's efficiency performing high G maneuvers and conducting the manual tasks for safe, controlled flight would be to conduct actual flight-tests with the seat back angle reclined to 30^0; and (3) it would be a much easier and faster modification to the test aircraft if the 30^0 seatback angle needed to be rotated to a lower angle (i.e. closer to the conventional 12^0) rather than to start flight test with a conventional seatback angle and modify the test aircraft later during the test program to explore test flying with a seat back angle of 30^0. Hillaker decided to propose the 30^0-seatback angle for the YF-16 test aircraft configuration in the General Dynamics response to the USAF LWF RFP, but had made provisions in the seat installation structure that it could be rotated to a lower inclination angle (in the 15^0-18^0 range of tilt angle) if during the flight-test evaluation of the cockpit arrangement pilots felt that the 30^0 reclined angle seat was not comfortable to fly in when not at elevated Gs. [No such comments were made in the LWF flight-test final report.]

The IH-8 version of the fully-developed Douglas Escapac family of ejection seats were proposed to be used for the YF-16 prototypes; this was for expediency, as the IC version of the Douglas Escapac ejection seats were used on the General Dynamics F-111A and B and were the ejection seats most familiar (and readily available) for the YF-16 design and program schedule. The initial production contract for the McDonnell Douglas ACES II ejection seat system was awarded in November 1976 and was the escape seat system preferred by Harry Hillaker, but the ACES II system, being developed for the McDonnell Douglas F-15 Eagle, was not fully qualified yet in the 1973 time period when equipment selections were being made for inclusion in the General Dynamics Model 401 configuration being prepared for the proposal response to the USAF LWF Project RFP

Using a similar set of logic and considered risks, the Hillaker cockpit design team also decided to gain flight evaluation of the technology of a canopy with no structural bow frame that partially blocks the pilot's forward visibility. Designing the canopy without a structural bow frame was based primarily on complaints of F-4 pilots in the Vietnam War (as well as the personal flight experiences of John Boyd while at the USAF Fighter Weapons School at Nellis AFB) that the Soviet MiGs are small size aircraft and too often were lost in visual contact during high maneuvering engagements by obstruction of the large bow frame in the canopy of their aircraft. Negative comments were also being surfaced about the poor out-of-cockpit visibility of the General Dynamics F-111A Aardvark and especially when performing low-altitude terrain-following flights.

The two greatest hazards with having a canopy with no structural bow frame are: (1) an inadvertent loss during flight of the one-piece, no bow frame canopy and (2) a bird-strike when at high speed/high altitude flight that could potentially cause a major breach of the canopy. Either event could seriously endanger the pilot since the only component in the cockpit that would be in the direct line of the high-speed airflow impinging on the visor of the pilot's helmet would be the HUD glass assembly.

Armament and Wing Hard Points for External Stores

Included in the LWF Design and Performance Goals section of the USAF LWF RFP (and shown graphically on the chart in Chapter 6 of Volume 1) was the goal statement: *"armament to include a representative state of the art/high muzzle velocity gun, and an effective low cost air-to-air missile; and hard-points and systems to give a credible air/ground capability".* Included elsewhere in the design and performance goals was a goal for *"combat radius on internal fuel of about 225 NM, with increases up to 700 NM with external tanks". Those goals were considered and accommodated in the design of the General Dynamics Model 401 external design to a significant extent. However, to actually integrate those capabilities for (1) fuel storage of the volume needed for the design missions goals; (2) carriage of the Model61 Al Vulcan 20mm cannon and ammunition drum/loading & unloading system; (3) hard-points in the wing and on the lower fuselage to carry external fuel; and (4) additional hard-points on the wing to carry an effective low cost air-to-air missile (e.g. the AIM-9 Sparrow infrared seeker type) and to carry external weapons giving a "credible air/ground capability" each presented unique challenges for including those capabilities for keeping the size and weight of the aircraft to a small size and with low cost. Those four areas of design challenge are discussed next – in the order shown above.*

1) Fuel Tanks   The interior arrangements in the two diagrams located earlier in this chapter containing cross-sections (A-A, B-B, C-C, D-D, and E-E) through the YF-16 fuselage (to show the inlet duct shape), also indicate the location and relative shapes/sizes of the internal fuel tanks within the YF-16 fuselage. The most forward fuel tank is shown to contain a bladder within the tank. The bladder and outer metallic tank provide a double barrier for pilot safety. When a tank is partially full and is subjected to a high deceleration (as in most crashes), the remaining fuel in the tank will tend to slosh forward and if the deceleration is high enough, as they are in most crashes, the fuel creates a "hydraulic ram" effect which can rupture the metallic tank structure. Without the bladder, the fuel would splash into the pilot's compartment and add a fire hazard to the pilot's other woes due to the crash.

The other fuselage tanks are in a saddle shaped configuration wrapping over the top of the fuselage and extending down into the blended wing-body area where the planar wing mates to the fuselage in 5 attachment fittings. The blended wing body configuration was designed primarily for aerodynamic reasons, but the added volume and added structural nature of the blending synergistically provides additional volume for fuel and adds to the structural strength of the wing body juncture especially in wing bending moment capability. Cross-sections C-C, D-D, and E-E indicate the saddle shaped of the middle and aft fuselage fuel tanks. All fuel tanks are purposely located on the top of the fuselage to minimize the probability of the tanks being hit by enemy ground gunfire. In addition, components which are less susceptible to loss of function from ground gunfire are located nearer the bottom of the fuselage and act as a shield for the fuel tanks.

2) Gun

Only the gun portion of the YF-16 weapons complement was carried internal to the fuselage. The cross-sectional cut C-C of the earlier shows the gun ammunition loading chute service panel (on the starboard exterior side of the fuselage) then the ammunition drum (capable of maximum of 511 rounds of 20 mm shells), and then the chute transporting rounds between the ammunition drum to the gun located at the left (port) side of the fuselage. Shown next is a photograph of the M61 A1 20-mm cannon system in an uninstalled configuration

M61 A1 20mm (Gatling) Vulcan Cannon Uninstalled (Source: USAF Engineering files)

The six barrels are seen located near the outer edge of the circular plate which supports the barrels and prevents them from oscillating out of boresight calibration during gun firing. Shown at the center of the circular plate is an axial support rod that performs two functions: (1) it attaches the circular plate to the barrel rotation drive shaft; and (2) it drives the six barrels in a steady rotational pattern to minimize the dispersal of the rounds emanating from the six barrels.

The firing rate (rounds per minute) of the M61 A1 20mm cannon is selectable to be 4,000 rpm (best for air-ground) or 6,000 rpm (100 rps, best for air-air). The ammunition drum can hold a maximum of 511 rounds of link less ammunition; however, as there are six barrels, there are six rounds not fired to clear the gun between firing bursts. Gun firing is done in up to 7 bursts of firing of approximately 1 second each. The post-firing rounds (expended and live) are recycled back into the ammunition drum for post-mission downloading and re-loading servicing.

The relative size of the 20-mm rounds in the M61-A1 Vulcan gun is shown next, compared to the 30-mm rounds in the A-10 (largest round at the right) and with other common hunting gun size rounds (smaller rounds at the left in the photograph) ...

Comparison of gun round sizes (20-mm Vulcan second from right)
(Source: USAF Engineering files)

There are three basic types of rounds of ammunition comprising the projectile portion of the round, (1) high-explosive; (2) armor-piercing; and (3) tracer rounds. The high explosive rounds are the most common rounds used in air-to-air gunnery. Armor-piercing rounds are used most often in gun firing against mobile armored vehicles on the ground or when targets are naval shipping. Tracer rounds are no longer of high use as they have an adverse property of visually showing where the aircraft is that is firing those tracer rounds. Technology advances in gun-firing electronic components and gun rounds trajectory-computing equipment obviate the need for the pilot to actually see the stream of rounds being gun-fired.

A diagram describing those essential elements of the mechanism and ammunition of the gun system illustrating the complete gun system when installed aft of the YF-16 cockpit is shown next.

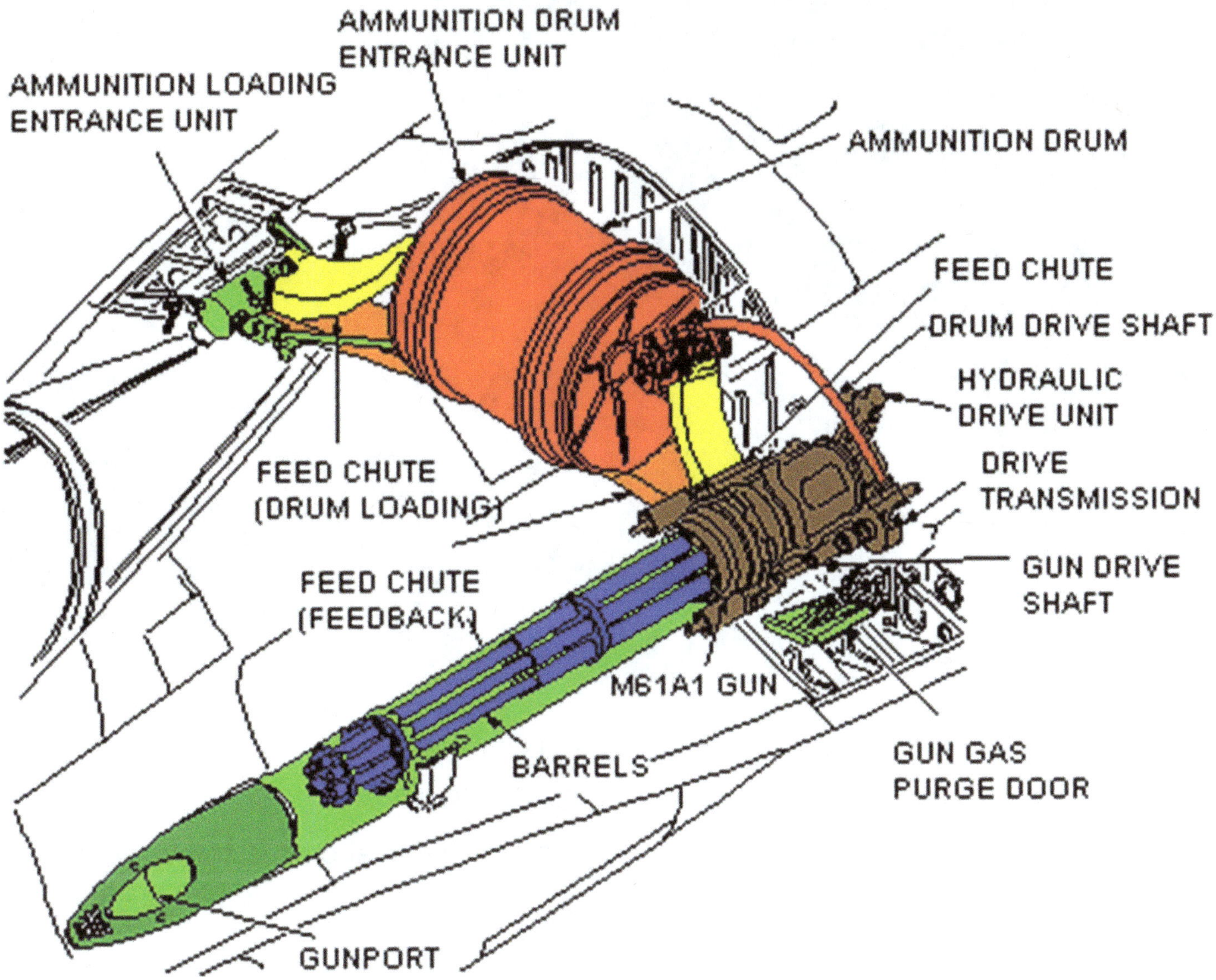

Installation Features of the M61 A1 20mm Cannon in F-16
(Source: USAF Engineering files)

The gun gas blast deflector was an aluminum casting and featured four rectangular slots that deflected the corrosive gun gasses behind the projectile away from the fuselage. A photograph of the gun gas blast deflector on YF-16 #2 is shown next.

YF-16 #1 Gun gas blast deflector
(Source: USAF Engineering files)

A frontal view of the gun port located on the port side of the aircraft is shown next.

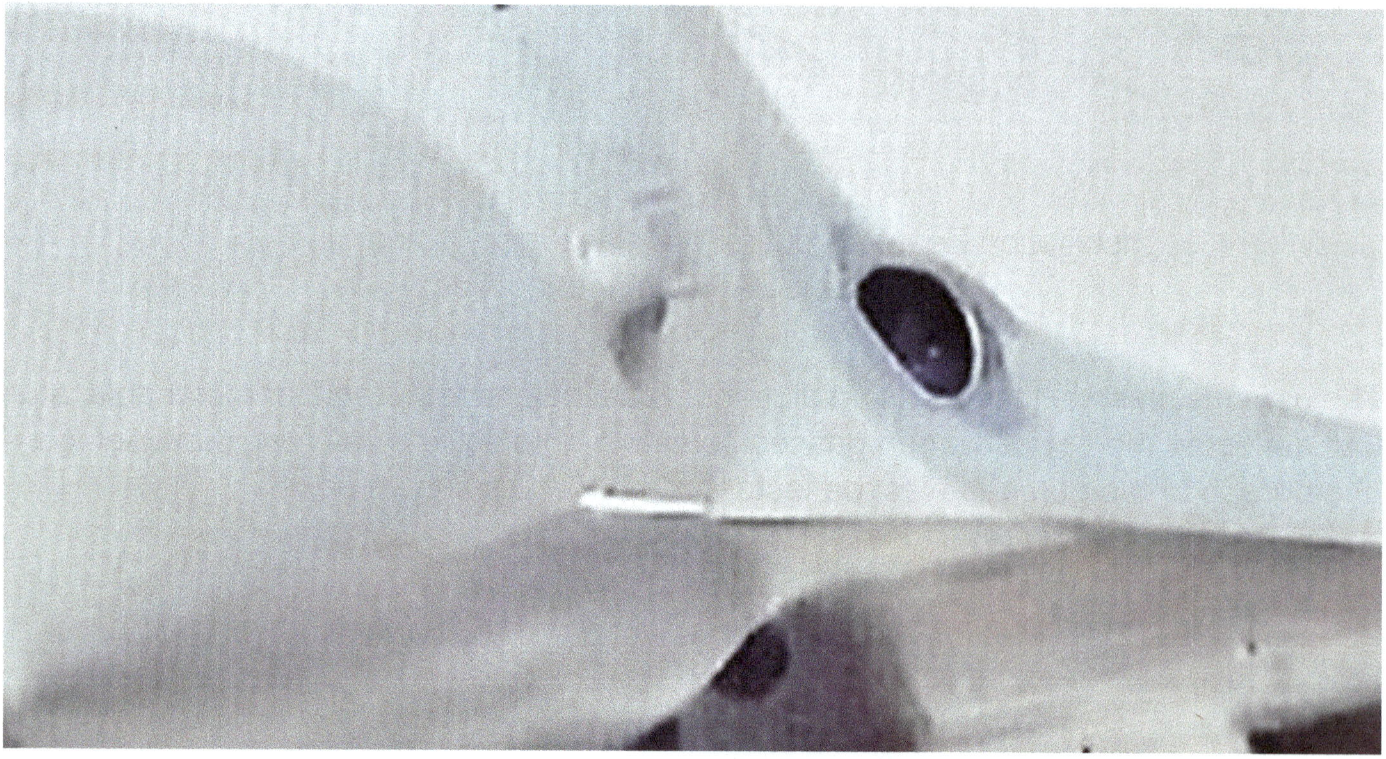

Frontal view of gun port (left wing root) on the YF-16
(Source: USAF Engineering files)

To address (3) hard-points and (4) credible air-to-ground capability, the diagram shown next should be of probable interest to Readers as it discloses the locations of the wing and fuselage hard-points, their load capacity at their Load Factor (G) limiting condition, and with an illustration of the locations where external fuel tanks can be carried (and jettisoned if/when necessary), or where alternatively external launched weapons (guided and gravity controlled) are capable of being carried.

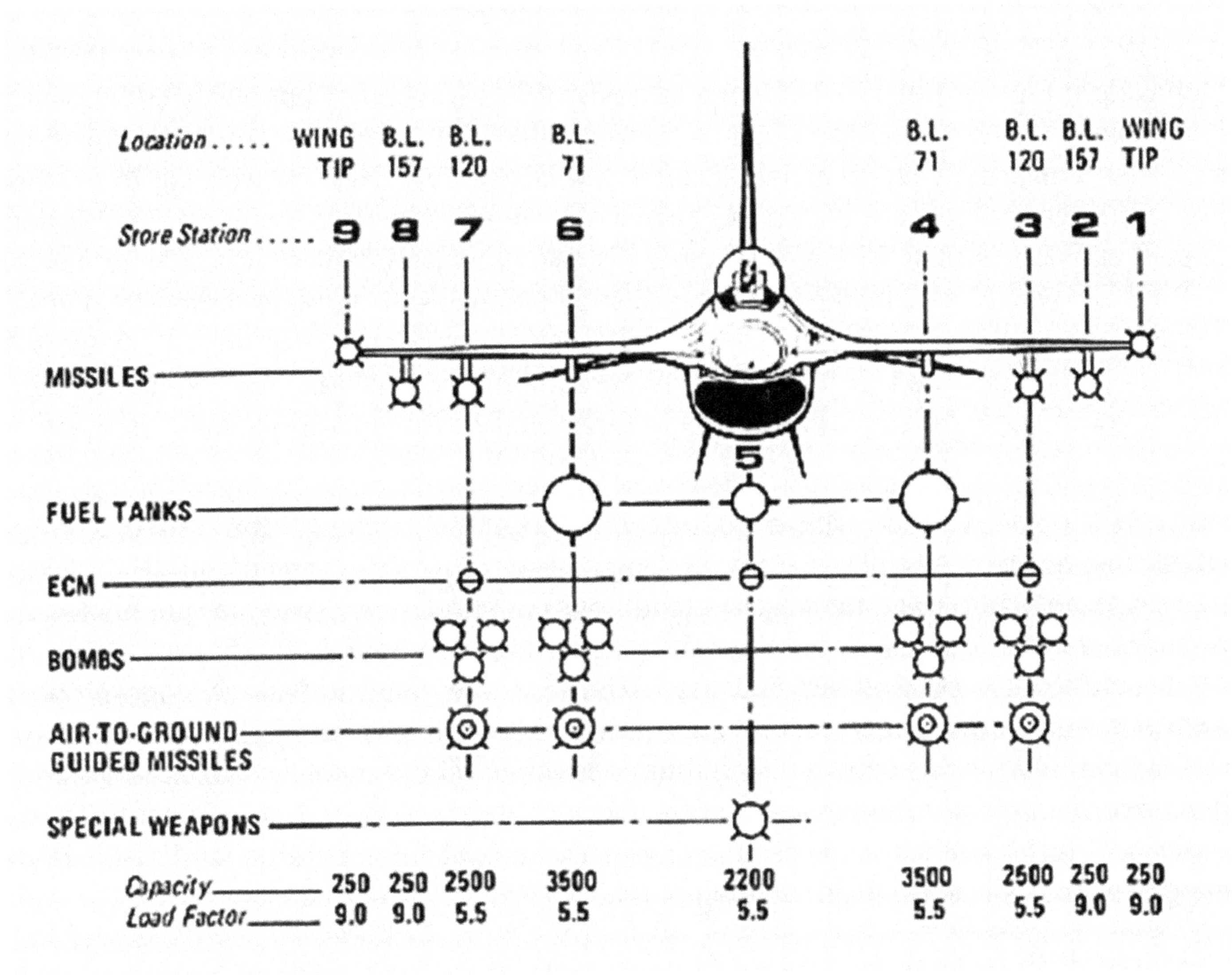

Wing hard-points in the F-16 design, 1974.
(Source: USAF Engineering files)

The YF-16 LWF prototype had only seven of these wing external hard-points (no stations 2 or 8) for expediency and being sufficient for meeting only the "combat relevant" flight-test goals.

Readers may have noticed that there has been little written about the avionics on the USAF LWF (both the YF-16 and YF-17). That is not to be construed as an oversight.

One of the ten commandments of John Boyd was "There shall be no bells and whistles on the Lightweight Fighter.", and Boyd was not an advocate for anything more than the basic avionics needed to fly and fight. While that was an important facet in the approach used to design-to-cost, and with that cost needing to be low in order to make the fight weight fighter an affordable one such that if the Lightweight Fighter were to be produced for the USAF inventory, many of them could be purchased. That conflicted only slightly with the General Dynamics desire to design with flexibility for growth – even if that was growth in avionics.

Another facet of that low-cost approach championed by John Boyd was that the Lightweight Fighter need to have only a daylight, clear weather capability (VFR, visual flight rules) and be used for close-in dog fighting. The need for a radar under those simplified constraints led to its exclusion in the USAF LWF RFP by wording the design and performance goals to cause the avionics on the Lightweight Fighter to be mission-essential avionics only. There were certainly two camps of thinking about whether the USAF LWF was to be configured as an advanced technology demonstrator with inherent capabilities that come with having a thrust/weight ratio greater than 1.0 at start of combat, or should it be a prototype of an operational fighter (albeit the low end in a high-low mix with the F-15 Eagle) having high technologies that provide it with unparalleled agility.

Harry Hillaker was faced with a quandary – should he configure the YF-16 with the USAF-desired M61 A1 Vulcan 20mm Gatling gun occupying the nose of the YF-16 (since there was no stated need for a radar in the nose) or should he locate the gun elsewhere and reserve the nose cavity for a potential radar at some point in the future? When his Avionics Engineer showed him how the F-15 radar was installed (see photograph next) and told Harry that USAF maintenance crews spoke highly of the swing-nose radar installation and hinged panels giving them access to all the avionics boxes in the F-15 nose, Harry decided to reserve the nose and compartment ahead of the cockpit for the radar and all the other avionics equipment that might later emerge and need to be integrated into a "missionized" version of the YF-16.

Radar System Access in the F-15 Eagle
(Source: USAF Photography files - FOIA)

As a result, Harry decided to locate the gun system installation for the YF-16 configuration in the space behind the cockpit and with the gun muzzles in the left-wing root (as shown earlier in this chapter). Recoil from the gun muzzles being 30 inches to the left of the fuselage centerline was expected to cause a nose-left yawing moment of motion during gun firing, but it was described earlier that a gun/rudder interconnect (GRI) function in the flight control system design to cancel that adverse yawing moment was to be added. Harry had called the shot right once again for designing the YF-16 with modular construction and to have the flexibility for growth.

Harry Hillaker further projected the implications of reserving the nose for a yet-to-be-determined radar system and directed that the nose material of both YF-16s be formed as elliptical fiberglass shells. It might be noticed that the radomes on the YF-16 #1 and #2 are painted in the overall paint schemes of the rest of the aircraft. Had there been radar equipment installed in their noses, the radomes would necessarily have remained in the black fiberglass color since black fiberglass was the state-of-the-art for radome construction material.

Executive Independent Review Team and Safety for First Flight

Flight safety engineers at both General Dynamics and at Northrop were advised by their counterparts within the LWF Prototype Project Office that one of the few mandated tasks imposed under the LWF Contract was that an Executive Independent Review Team (EIRT), composed of independent senior USAF safety analysts, must be conducted and any action items successfully resolved to assure that the four prototype test aircraft (two General Dynamics YF-16s and two Northrop YF-17s) would in all aspects be safe to fly a first flight and ensuing series of rigorous flight- tests at the Air Force Flight Test Center at Edwards AFB near Rosamond, California.

With the proposed first flight of the General Dynamics YF-16 #1 being January 1974, that gave approximately 12 months for General Dynamics to conduct safety analyses and safety ground tests evidencing the flight safety of the YF-16 prototype aircraft to the EIRT. Typically, a major concern for flight safety lies in the safety of the propulsion system for any new prototype aircraft, but inasmuch as the single engine YF-16 aircraft was to use the Pratt & Whitney F100-PW-100 engine developed and flight qualified for the McDonnell Douglas F-15 Eagle, the airworthiness safety aspects of that engine had already been proven to the USAF

First flight of the Northrop YF-17 #1 was projected to be April 1974 (that date driven by the planned ground testing and subsequent airworthiness certification of six flight-worthy prototype General Electric YJ-101 turbojet engines - two engines for YF-17 #1, two for YF-17 #2, and two spares for one ship-set in the event of FOD damage or inherent engine failure on either of the two YF-17 aircraft).

A high emphasis was placed by USAF EIRT flight safety review teams on the crew escape/ejection seat provisions and design in the event of a mandated pilot ejection from the prototype aircraft during the rigorous aggressive flight-test program to be conducted. General Dynamics selected the use of the Douglas IH-8 Escapac ejection seat for use in the YF-16 and Northrop selected the Martin Baker (MB SR-1) ejection seat for use in the YF-17.

The Escapac seat systems were developed primarily for use on USN aircraft (A-4 A, B, C, and E and for the A-7) and featured a pull-down-the-screen-over- the-face system. The IH-8 version of the Escapac seat, modified for the YF-16 LWF prototype application, did not have the pull-down-screen feature but instead had a between-the-knees ejection handle.

Another significant difference between the seat selected for the YF-16 and YF-17 test program is that the YF-16 escape ejection sequence requires that the bubble canopy be separated from the forward fuselage before the pilot and seat start up the ejection rails. To the contrary, the YF-17 escape ejection sequence utilizes a canopy penetrating and shattering mechanism atop the Martin-Baker ejection seat head rest assembly which shatters the canopy before the pilot and MB SR-1 seat moving up the ejection rails can pass through the shattered canopy material.

The General Dynamics YF-16s had an Escapac Model 1H8 ejection seat (photo shown earlier) and the Northrop YF-17s had a Stencel ejection seat – both seats being chosen because of their quick availability. A technology that was emerging starting in the 1960s which had great potential for manufacturing lightweight aircraft in both the commercial and military world of aviation was that of composite materials for use in the aircraft structural components and in aircraft panels and skin (typically made of aluminum, titanium, or high strength steel for aircraft with exposure to very high aerothermodynamic heating associated with Mach number flight above about Mach 2.2). A very beneficial aspect of composite materials is that they are used for applications where weight of the aircraft is critical and/or where its stiffness and strength/weight ratio is high. The use of composite materials for the USAF Lightweight Fighter design would at first consideration seem to be an ideal choice for application in a high percentage of the LWF structure.

A major drawback with extensive use of composite materials, which has high strength and light weight, in a fighter aircraft is that it also has high cost of manufacture, maintainability, reparability, and survival from penetrations of shrapnel or rounds from ground gunfire when exposed to hostile environments. The management approach of design for light weight and design for low cost have aspects that require careful trade-offs and compromise in realistically achieving light weight at low cost in military aircraft applications.

With regard to the structural strength of the YF-16 Lightweight Fighter, the Author cannot reword the description of the YF-16 strength to be any better than as stated by Harry Hillaker, the "father of the F-16" himself.

Harry Hillaker stated: *It is not generally appreciated that the aircraft's "light weight" was not achieved by resorting to high strength-to-weight materials, like titanium or composites, or other exotic materials. Nor was structural integrity sacrificed for the sake of lighter weight. The structural criteria were fully compatible with the aerodynamic capacity of the airplane and the pilot's physical tolerance. The structural design load factor was set at 9.0 g's at start of combat with full internal fuel, as contrasted to the normal military specification of 7.33 g's with only 60 percent internal fuel. Our* (General Dynamics) *approach to the structural design ensured that we could demonstrate the maneuvering capability of the airplane. Air Force ground rules dictate (appropriately) that an airplane be limited to 80 percent of its design takeoff weight until a structural static test has been completed. Because a structural test was not possible within our $37 million contract, we decided to design the two prototypes to 125 percent (a 25-percent margin) of the design loads (80 percent of 125 percent is 100 percent). The weight penalty was acceptably small (418 pounds or 3.5 percent). This was an* *unappreciated decision that contributed indirectly to the success of the YF-16 prototype program.*

With regard to the use of composite materials on the YF-16 aircraft, two additional considerations caused the minimal use of them on the YF-16. High quality of the carbon fibers embedded in a plastic-like matrix material is vital to the strength of the composite material component. When two US companies (MMM and Hercules) were approached in the 1973 time period and questioned about the availability of high-quality carbon fibers for aircraft application - but only in a small quantity for the two YF-16s and a spare set- in case of damage. Both of the companies considered that it was not a good investment for the companies to produce the small quantities of high-quality fibers being sought. It was more profitable to produce carbon fibers of a lesser but sufficient quality for the growing lucrative public market for composite-shaft golf clubs and fishing rods.

In a continued search for a small quantity of high-quality carbon fibers, it was discovered that a company in Japan did in fact have a production process that was producing high quality carbon fibers. After convincing those in the USAF who were on a Buy American crusade that Japan was the only source of quality fibers and that otherwise the parts intended to be made with high-strength, reliable composite components would continue to be made from conventional aluminum materials, the order was placed for an expedited delivery of the small quantity of high-quality fibers. The fibers thus obtained fro Japan were provided to the Hercules Manufacturing Company and the desired components were made for the two YF-16s.

Commercial aircraft manufacturing companies also indicated strong interest in using composite materials extensively in their commercial airliners where there is no concern for damage to the composite components due to the hazardous operational environments to which military aircraft with composite components would be expected to be frequently exposed. This was another strong incentive for materials manufacturers in the USA to later produce the necessary high quality carbon fibers (i.e., fibers which have a consistent circular cross-section and with no flaws).

Shown in the chart next indicating what material was used in what component and module of the YF-16 aircraft being produced at the General Dynamics Fort Worth facilities.

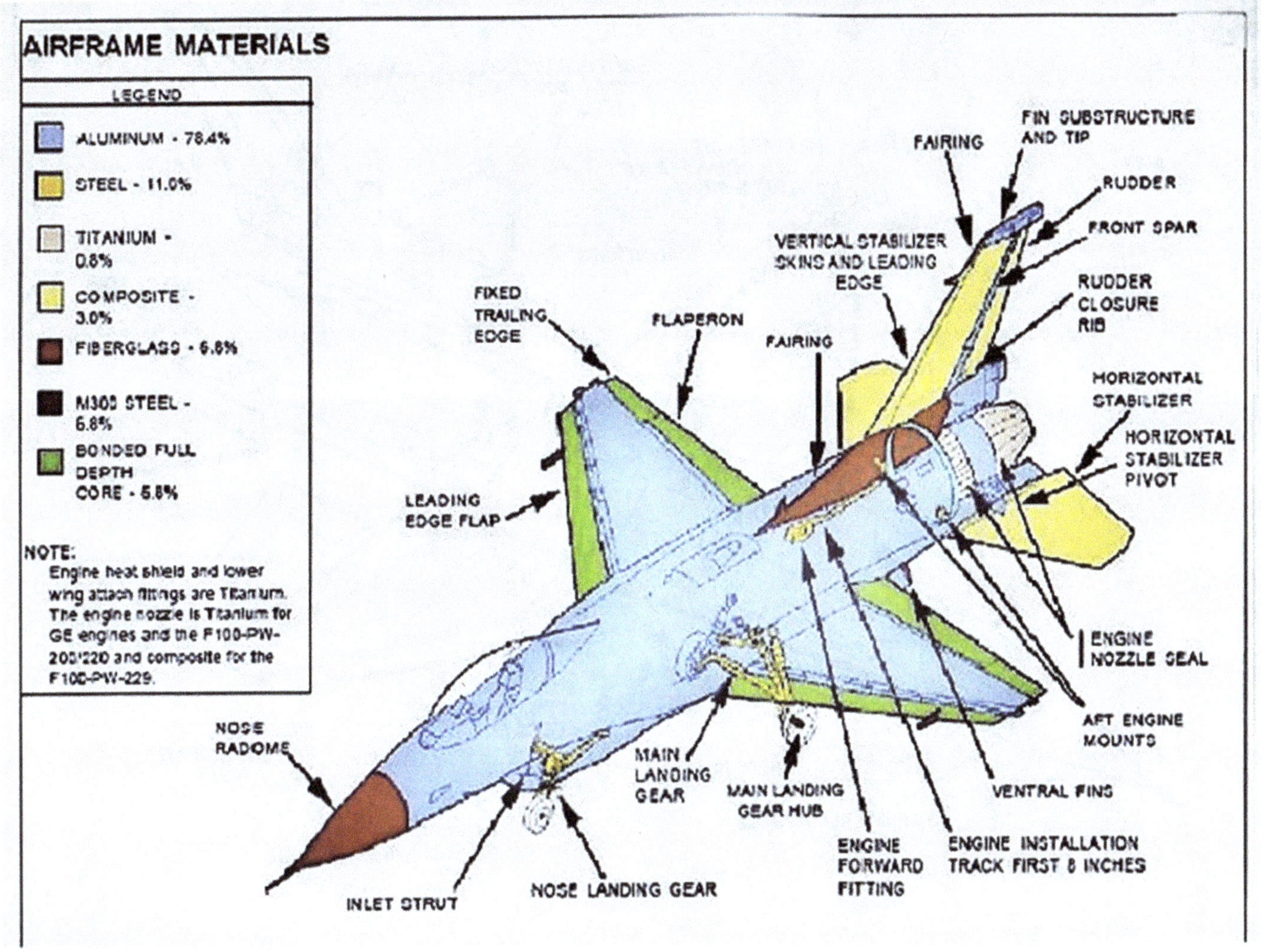

Materials used in the YF-16s by type, percent, and location.
(Source: USAF Engineering files)

The Manufacturing team at General Dynamics applied the design-to-cost goals set for the USAF for the Lightweight Fighter Project and developed a modular approach to manufacturing the aircraft. Their effort resulted as summarized in the chart shown next.

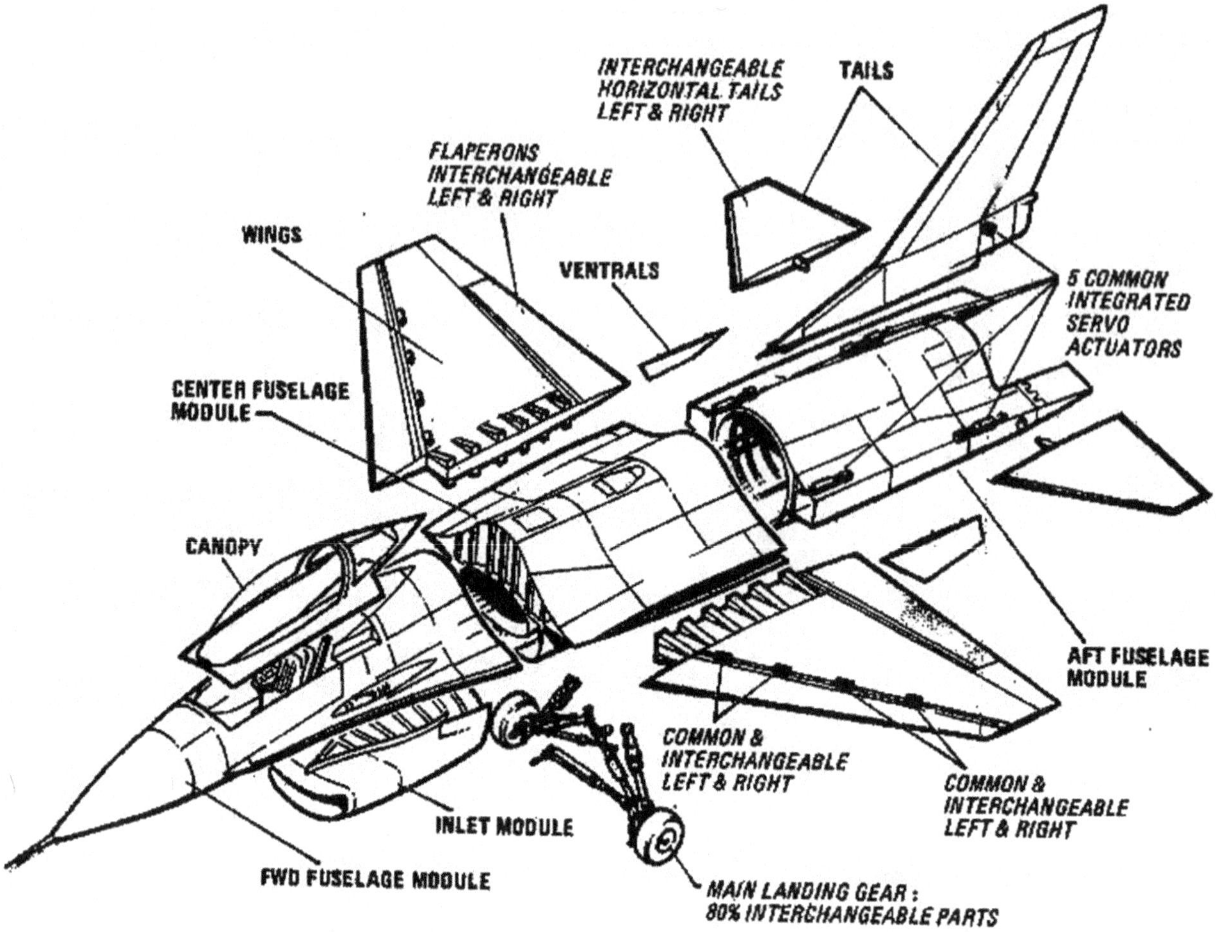

Modularity and interchangeability of YF-16 components
(Courtesy: USAF Engineering files)

As can be seen, the manufacturing concept of the YF-16 centered around the use of modularity, and the major modules are as shown in the diagram. There are several components made with an interchangeability principle (installation either left or right) to keep the manufacturing costs as low as possible for the number of different type components necessary to be manufactured. The larger of the interchangeable parts are: horizontal tails (stabilators); wing trailing edge flaperons; ventral fins; integrated servo actuators for primary control surfaces; actuators (powered rotary hinge) for wing leading edge flaps); main landing gear; and pilot's rudder pedals.

Another cost reduction approach in the manufacturing of internal components known as "make or buy" was used for both cost and schedule purposes. In general, if an equipment item was available off the shelf it was bought ("buy") at low cost because of its large production runs and low per unit cost. The most major item which General Dynamics decided to make in-house was the flight control computer, probably the most complex component and unquestionably of the highest importance to the successful demonstration of the all-electric fly-by-wire system to be flown in the YF-16s as the first USAF fighter aircraft designed with that type of flight control system.

Some assemblies (e.g., the three-axis accelerometer assembly and three-axis rate gyro assembly in the flight control system) were constructed by a combination of make and buy – the three accelerometers and the three rate gyros were bought individually and assembled in a mounting box made at General Dynamics. The canopy (one-piece bubble – with no bow frame) was subcontracted to be made by Texstar Industries of Texas who met the high-risk technology advancement in a fully professional manner. Similarly, the front portion of the smiley inlet was a one-piece aluminum casting also to be made under subcontract to Alcoa (the Aluminum Company of America). The highly important integrated servo-actuators (ISA) had been under research and development by National Water lift (NWL) and NWL also outstandingly performed their subcontract for providing approximately 18 of those ISAs to General Dynamics at low cost and well before actually needed for the tight manufacturing schedule of the YF-16s.

As a general comment, when the equipment manufacturers see that a prototype is being developed under design- to-cost and design- to- goals management and as a best effort type of contract in a "fly before buy" program with the USG, they provide their products at lowest possible cost (some at no-cost) in order to be regarded as being a part of the program team – and earning a lucrative potential by being selected to be the supplier of that equipment in a potentially very large production program.

Augmenting those parts and components that were either made or bought by General Dynamics for manufacture of the YF-16s was the GFAE (Government Furnished Aeronautical Equipment) as agreed to in the original USAF Lightweight Fighter Contract. This GFAE included

1. The loan of two Pratt & Whitney F100-PW-100 engines from the F-15 Program resources and being provided as GFAE to General Dynamics for their use in the flight test program of YF-16s #1 and #2 at Edwards AFB. (An additional two P&W F100 engines were to be made rapidly available if needed as spares for the first two engines.)

2. An M61 A1 Vulcan 20mm gun and ammunition was to be made available when needed for ground gun-firing tests and for installation in YF-16 #2 prior to its demonstration and flight evaluation of YF-16 gun-firing accuracy tests.

3. An unusual situation arose regarding the Main Landing Gear tires for the YF-16s in that the size chosen by General Dynamics was no longer available from the original supplier Goodyear (for the nose landing gear on the earlier General Dynamics B-58 Hustler Program). The tires of that size needed for the YF-16s were provided by cannibalization of the nose wheels on B-58s in mothball at Davis-Monthan AFB in Tucson, Arizona. Although not new tires, the best six tires that were cannibalized from the many B-58s in mothballs were provided to General Dynamics in excellent wear condition on a "return if requested by USAF" basis.

4. Most of the pilot's flight instruments were obtained from the F-111 program stock left over as surplus at the General Dynamics facility.

5. Two Douglas Escapac IH-8 seats (ejection qualified and modified for installation in the YF-16 LWF prototypes).

In new aircraft developments, there are special moments that are not recognized formally as Program Progress milestones but which are nevertheless highly meaningful to those honored to be "in the trenches" of the program. One such event was when General Dynamics started the first machining on the main bulkhead (at FS 253) of the YF-16 #1 airframe which joins the Forward Fuselage Module to the Center Fuselage Module of YF-16 #1 (tail number 72-01567).

Shown next is a small memento created by General Dynamics for those team members who endured all the many briefings, meetings, discussions, arguments, negotiations, travel inconveniences, inner concerns and fears for safety, etc. that somehow coalesce into the final design of the aircraft - on paper.

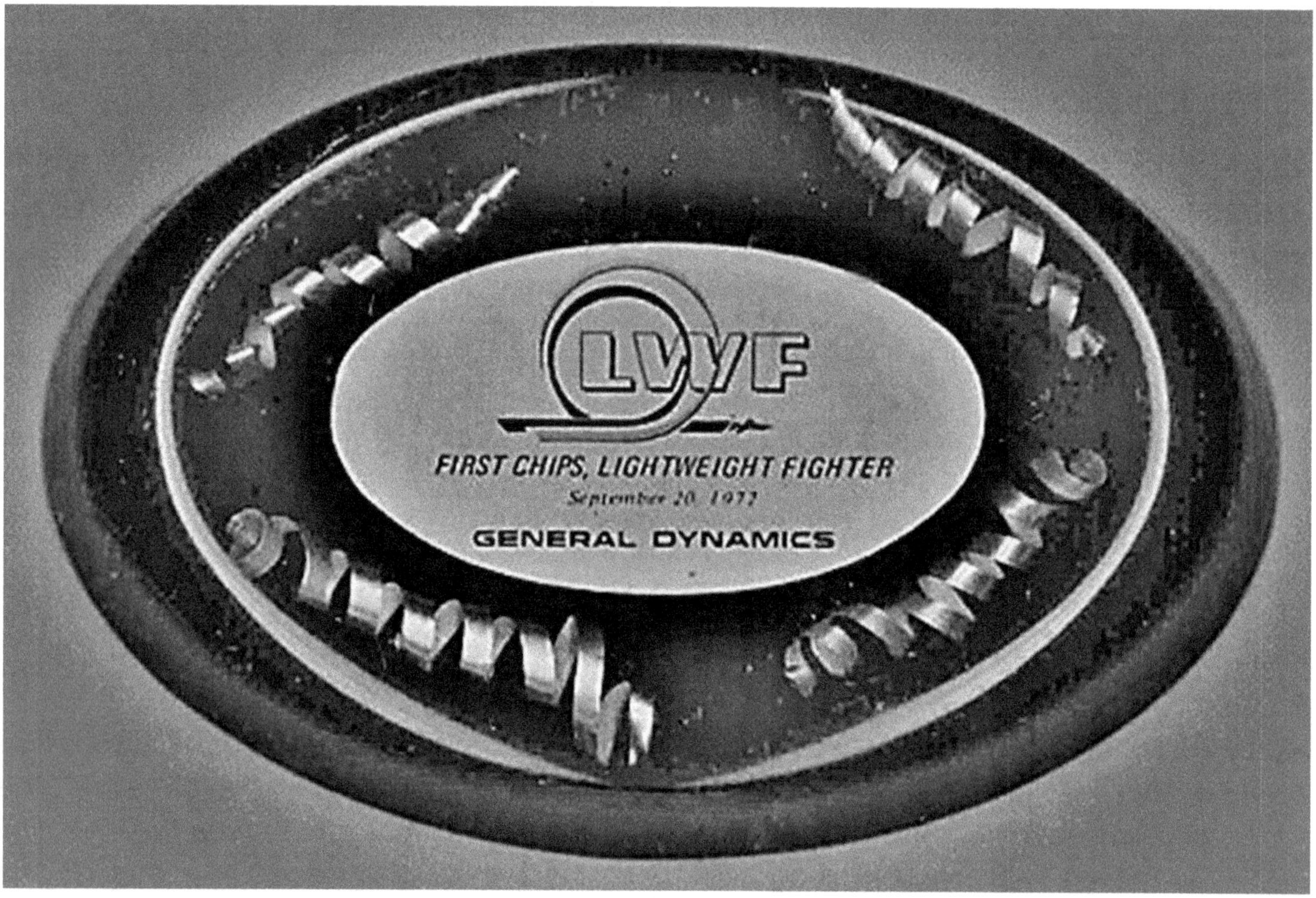

First Machining Chips of FS 253 Bulkhead (for YF-16 #1 Tail Number 72-01567) Sept. 20, 1972. (Source: Author's Memorabilia file – photo by Author)

Bill Dietz at General Dynamics witnessed the first machining event and it was he who remarked that the spiral chips were analogous to the vortices peeling off the leading-edge extensions of the YF-16 forward fuselage when pulling high Gs. Readers may detect on the first chips memento that the first-machining date was on September 20, 1972, but that the LWF contracts were not awarded to General Dynamics and Northrop until February 1973. General Dynamics had the highest of confidence that their Model 401 design concept would be chosen by the USAF as one of the two winners of the USAF LWF RFP contract awards, and had half-frozen their proposed Model 401 design. Using company resources and funding, General Dynamics had started the machining based on that high speculation. Readers may also note on the memento that the logo is shown as being for the LWF – as the designation of YF-16 for their Model 401 configuration was not assigned until after their Model 401 design proposed had been selected by the LWF RFP Proposal Response evaluation team.

By contrast of differences in company design approach and strategies, Northrop had not frozen the configuration of their proposed P-601 single- engine (P&W F-100) design at the time of their submittal of their proposal response to the USAF LWF RFP. Northrop had speculated that their alternate proposal of the P-600 two-engine (GE YJ-101 prototype) design was a better design for achieving the goals stated by the USAF in the LWF RFP. Northrop was well aware that the USAF had preferred a single-engine design and with that single engine being the Pratt & Whitney F100-PW-100 nearing completion of being fully developed and qualified for the F-15 program. As a result of the uncertainty as to which Northrop design (P-601 or P-600) might be selected by the USAF LWF Source Selection Evaluation Committee (headed by Lt. Gen. James T. Stewart at ASD), Northrop delayed the start of any machining of the airframes for their two YF-17s. (With a several month difference in delay between General Dynamics and that of Northrop in the start of manufacturing/machining, those three to four months of delay coupled with a parallel delay in getting six flight-worthy prototype YJ-101 engines from General Electric manifested itself in a four months interval from delivery of the General Dynamics YF-16 #1 and the Northrop YF-17 #1 to AFFTC for the flight-test phase. This interval of four months would become of critical importance as later events in the LWF Program flexible schedule would emerge.)

Three experienced senior USAF Maintenance Crew Chiefs (six plus stripes) were temporarily assigned to the LWF Program Office at Wright-Patterson AFB but who were in fact collocated at General Dynamics/Ft. Worth to provide advice and suggestions (but not direction) regarding maintainability, serviceability, reparability, safety, vulnerability, survivability (and seemingly every other "ility" known to man) of the Lightweight Fighter design while still in the formative stage of its development. (Readers are once again reminded of the metaphor woven into this book: "As the twig is bent, so shall the tree grow". It was most interesting to watch that growth of the Lightweight Fighter prototypes in real time – just as it is for parents watching a newborn baby grow to take his/her first step.)

Over the next several months into mid-1973, it was awesome to witness the manufacturing activities on the Assembly Station for the YF-16 #1 and see the modules of the airframe getting skin on their bones and starting to resemble in full scale the lines drawn on paper or assembled on a much smaller scale computer screen. Because of the need to build only two flight-worthy prototypes, those two prototypes would necessarily be hand-built - i.e., not using high-rate quantity production capabilities for manufacturing the components/parts/pieces. Only experienced senior workers (male and female) were chosen to apply their experienced talent on hand-making the prototypes. That fact was observed both at the General Dynamics facility in Fort Worth, Texas, and at the Northrop Aircraft Division in Hawthorne, California.

The photograph shown next captures construction of the wings for the YF-16 #1 underway at the General Dynamics facility in Fort Worth, Texas (the same facility used in manufacture of the B-58 Hustler Strategic Supersonic Bomber, and the F-111 Aardvark Tactical Fighter Bomber with variable sweep wings).

Manufacturing the wings for the YF-16 #1 LWF, July 1973.
(Source: USAF Engineering files)

Despite the manufacture and assembly of the structural elements of the YF-16 components being hand-carved (with the assistance of machine tools, when the modular components of the YF-16s were mated to their adjacent component, the fit was amazingly close. The small inaccuracy in fit between two components in YF-16 #1 were corrected in the assembly of components in YF-16 #2 and proved once again that the prototype approach of "fly before buy" expounded by Dep. Sec. Def. Packard might also be applied to the early manufacture of a prototype in a "fit before build" design approach.

The photograph next defined the moment when the forward fuselage module was mated with the center fuselage module for the first fit – which was a totally smooth mating.

First Mating of Fuselage Modules of YF-16 #1, July 1973.)
(Source: USAF Engineering files)

Within two weeks after the first mating shown above, all of the fuselage modules for YF-16 #1had been assembled and stuffing the sausage began.

In the late July 1973 time period, YF-16 final assembly had progressed to the installation of the many instruments and controls in the cockpit region of the forward fuselage module. The Human Factors engineers served much like a team of surgeons performing a surgery never before conducted. If the four Human Factors engineers could not form a consensus plan where to install a specific instrument, they called in the specialist who cast the swing vote. That specialist was most often chief flight test pilot Phil Oestricher or, if Phil was not available, flight test pilot Neil Anderson. As the real estate within the YF-16 cockpit was incredible tight, many compromises had to be made as to where something should be placed and since this was a fighter pilot's fighter, it was left to the experienced pilots to decide the best location for a number of reasons – flight safety, need for the pilot to be able to perform combat functions efficiently and effectively, and consistency with the HOTAS (Hands On Throttle And Stick) principle as the common goal of the YF-16 flight test pilots and Human Factors engineers team.

A photograph, shown next, was taken during a Human Factors engineers hands-on conference in the YF-16 #1 front office.

YF-16 #1 Fuselage 80% complete, late July 1973.
(Source: USAF Engineering files)

After the major airframe modules comprising the fully assembled airframe were mated, the all-important process of installing and/or connecting the wiring and plumbing lines (fuel, hydraulic fluid, compressed air, oil, ECS bleed, etc.) on the assembled modules of the YF-16 #1 was then accomplished. The cockpit is where all of the functions converge that are presented to the pilot for observation, orientation, decision, and commanding the motion of the aircraft and for functional status of all subsystems needed during the multi-mode flight. Placement of each instrument and display on the two side consoles –

to the left of the pilot, to the right of the pilot - and on the front instrument panel were the greatest challenge to be accomplished per all of the decisions made during numerous design meetings and reviews by Human Factors engineers and company test pilots. The role played by General Dynamics Chief Test Pilot Phil Oestricher in the selection and arrangement of all the instruments and displays/controls cannot be overrated for his invaluable contributions to the design of the YF-16 cockpit.

The next critical step in the manufacturing and assembly process was that of installing the many functional mechanical, electrical, hydraulic, and actuation components, as well as installing the pilot seat escape system, the HUD, and the canopy. As described in earlier portions of this book, high emphasis was placed on survivability during the technology and design development of the all-electric fly-by-wire flight control system - when/if exposed to hostile environments during operational engagements with enemy aircraft or during exposure to enemy ground gun-fire (small arms, automatic weapons, anti-aircraft artillery fire, surface-to-air missiles, etc.). That same high emphasis was placed on installing all other safety of flight critical components throughout the F-16 interior arrangement. Equipment installers referred to that as "stuffing the sausage".

Readers may recall the earlier section of this book that described General Dynamics constructing a full-scale mockup of the YF-16 fuselage made of plywood (simulating bulkheads with holes for pass-through services), cardboard (simulating skin with access holes cut out), rope (simulating wire bundle runs and fuel/hydraulic/oil lines). The components were made from Styrofoam blocks cut (by heated wire) to the exact shape, and size as the metallic component to be installed into the YF-16 test aircraft. That mock-up was of immeasurable benefit to assure that survivability and vulnerability factors had been properly considered as well as those of maintainability, serviceability, and reparability if/when the aircraft were to be used in an operational base environment.

Shown next is the interior arrangement of the F-16 indicating the placement location for the critical safety of flight components – derived from the YF-16 original design.

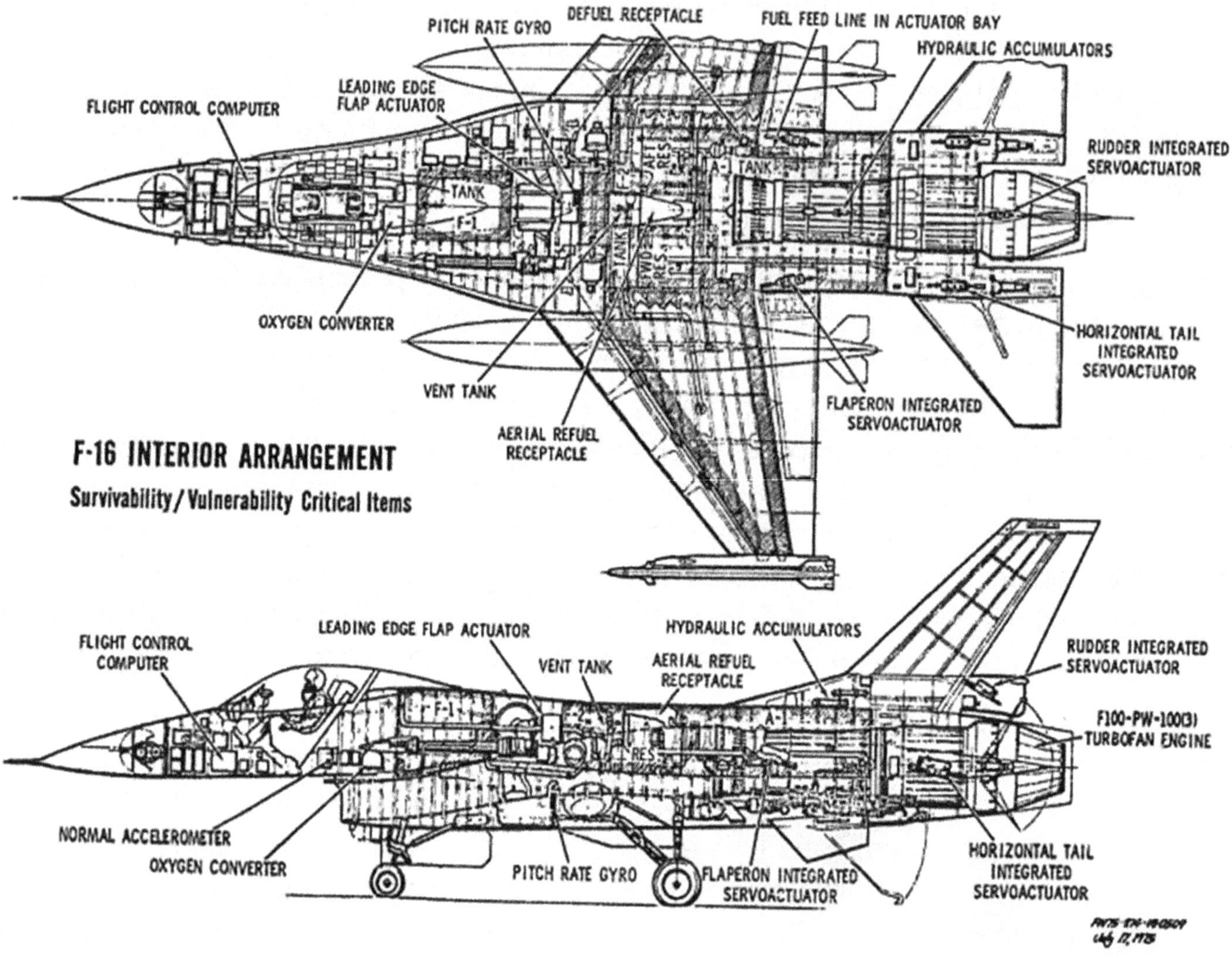

Positioning of Critical S/V and Flight Safety Components
(Source: USAF Engineering files)

In general, those safety-critical and/or vulnerable components are installed, when possible, to be in the upper locations of the fuselage so as to be shielded by less vulnerable and safety-critical components from ground gunfire hazards. For example, when the landing gear is in the up-and-stowed position, the relatively strong structural members (being constructed of high strength steel) of the landing gear system provides some shielding for the fuel tanks located above the stowed landing gear. Too many military weapon systems have been developed without adequate consideration of the high likelihood that the weapon system will be one day fielded at a remote operational base and operated in a far more severe environment than that for which the weapon system may have been originally designed to operate. Coupled with that operational safety concern for internally arranging the internal components to minimize their vulnerability to battle damage is that the aircraft is required to be maintained, serviced, and repaired in the event of battle-incurred damage by ground support personnel with standard tools and GSE (ground support equipment) at the forward bases.

When the fastener choices were first being made for the YF-16 prototype panels locations, there was historical data in the fastener database maintained and provided by the Air Force Logistics Command which reflected unclear criteria for use of the different approved type fasteners. To clarify the selection criteria for fasteners, General Dynamics elected to use fasteners for many panels on the YF-16 #1 and YF-16 #2 to be of different type for the upcoming flight testing in order to gain some flight test experience directly as to which fastener type would be the best choice for each panel. (A humorous comment made by one of the General Dynamics assembly workers was that "If this program is a fly-off competition, then should we call the fly-off between using different fastener types on #1 and #2 a screw-off?).

In the photograph next is shown the completely assembled and functionally tested airframe of the General Dynamics YF-16 #1readied for the first fit-and-function installation of a Pratt & Whitney F100-PW-100 engine.

First Mating of P&W F100 Engine in YF-16 #1, September 1973.
(Source: USAF Engineering files)

Readers who have an interest in the avionics equipment and components that were built into the YF-16 aircraft should be aware that YF-16 #1 was designed as a technology demonstrator aircraft, internally configured and instrumented to demonstrate the airworthiness of the YF-16 design and to establish its safe flight envelope. Accordingly, YF-16 #1 had neither a radar system typically mounted in the nose of the aircraft nor a 20 mm gun system installed. As seen in the photograph above, the radome constituting the nose of the aircraft is constructed of unpainted black fiberglass for radar transmissibility purposes although there was no plan to install a radar fire control system radar in the nose of YF-16 #1.

YF-16 #2 was configured with less internal instrumentation but with internal components that would later be needed when YF-16 #2 was being flight tested for its operational suitability (air combat maneuvering, gun firing, weapons launching, maintainability demonstration, etc.). However, YF-16 #2 also initially had no radar installed as that was not needed since the primary objective of the USAF LWF Project was, at least initially, to demonstrate the efficiency and effectiveness of advanced flight technologies of a light-weight, low-cost, fighter prototype to be flown under VFR conditions and only at the Air Force Flight Test Center at Edwards AFB, that features an enormous size dry lake bed for safer emergency landings if/when necessary. Nevertheless, for manufacturing commonality reasons, YF-16 #2 also was constructed with an unpainted black fiberglass radome as its nose structure – with only a rudimentary gunfire radar installed for later gun firing tests.

A “color #1” opportunity was given to the employees at General Dynamics for the paint scheme that was used on YF-16 #1, and the final paint scheme was chosen from one of those suggested by the employees.

The paint scheme for YF-16 #2 was chosen to be one of three candidate paint schemes to reduce the detectability of the small size fighter. The three candidates planned to be chosen later during the flight test program were (1) a blue and white camouflage to determine how well the aircraft would blend in to a blue sky dotted with white stratus clouds;(2) a medium grey for most of the external skin area but with a lighter shade of grey to fill in junctures where shadows might occur; and (3) a light to medium grey for the entire exterior skin. Photos of the 3 different paint schemes of YF-16 #2 (two paint schemes for low visual detectability purposes and one paint scheme – patriotic red, white, and blue - for planned demonstration flights at the 1975 Paris Air Show) will be shown in the next chapter.

After all final assembly tests and pre-first flight tests were completed on the YF-16 #1 (with the engine installed), the aircraft was moved to the paint station and painted in a most patriotic red, white, and blue paint color scheme. Shown on the next page is a photograph taken in early December 1973 of USAF/GD YF-16 #1 in its patriotic red, white, and blue colors, just a few days prior to its official roll-out debut to the open public and press media. It is also common to use a very visible paint scheme, with red being the dominant color, on aircraft that are to undergo a flight-test evaluation at AFFTC, Edwards AFB, as that makes its orientation much more visible for ease of being chased by an accompanying chase aircraft (or if inadvertently entering into a spin).

Although not readily visible in the photograph, the upper and lower surfaces of the wings are painted solid red, more for flight test purposes rather than for patriotic purposes.

YF-16 #1 with RWB Paint Ft. Worth, TX, December 1973.
(Source: USAF Engineering files)

There were a range of comments about the paint scheme chosen for YF-16 #1 ranging from being "as inspirational as the American flag" down to "looks like a flying whore-house". As always, beauty is in the eye of the beholder.

Those who have a knowledge of radar installations in an aircraft will note that the black fiberglass radome constituting the nose structure of the YF-16 #1 has been painted over with white paint which would detract significantly from acceptable radar transmissibility through the painted radome. Later when YF-16 #2 had been assembled and readied for painting, the decision was again to paint over the black fiberglass radome inasmuch as there initially was no radar system planned to be installed in YF-16 #2 either. Readers are asked to hold that thought until later in the next chapter when some very unexpected events happened during the flight test program that would indirectly affect even the paint scheme of YF-16 #2.

Meanwhile, back at the ranch, in parallel with the manufacturing and assembly activities out on the floor, the engineers and others at the General Dynamics facility, and to a lesser degree of activity at Northrop facilities which projected a first flight of their YF-17 four months after that of the General Dynamics YF-16, were engaged in planning for a USAF LWF Project Office-conducted flight readiness review (FRR) of their prototype aircraft designs. The FRR was contained in the USAF LWF RFP as being a contractually required task to assure that the completed prototype advanced technology demonstrator aircraft (two YF-16s and two YF-17s) could be flown safely when delivered to the Air Force Flight Test Center at Edwards AFB, California.

From about March 1973 to about June 1973, there was a growing interest expressed in the forthcoming "fighter pilot's fighter" competition between General Dynamics and Northrop and aviation magazines (e.g., *Aviation Week*, *Flight International*, *AIAA Journal*, etc.) as well as the major news media bureaus portrayed the program as being a fly-off competition with major implications for being the low end of the high-low mix with the F-15 Eagle. The F-15 Program had been very closed door about the progress being made in the F-15 development and test phase and the fact that costs and number of flight problems had escalated in the F-15 Program – such that the open nature of the Lightweight Fighter Project was being advertised (not entirely correctly) as filling in the short-comings suspected to be occurring in the F-15 program. It was not publicly well known that the F-15 had "not a pound for air-to-ground", but it was well-known within the LWF project engineering community that the F-15 Deputy Chief Engineer (Fred Rall) had crowed that mantra in closed-office meetings at Wright-Patterson AFB many times.

A major management change in the USAF Prototype Programs Office' Director assignment was made in June 1973, not long after the LWF contracts were awarded to General Dynamics and Northrop. The change entailed Col. Lyle Cameron being promoted to Brigadier General and being given an assignment to Hanscom Field, Massachusetts, to be the Director of the Airborne Command Post Program (ABCP). The ABCP program was already underway with Boeing as the prime contractor, but was experiencing cost and schedule problems as well as program management problems. Col. Cameron's Deputy Prototype Programs Manager, then-Lt. Col. William E. Thurman, was promoted to full Colonel and given responsibility for direction of the Lightweight Fighter Project and the parallel Advanced Medium STOL Transport Prototype Project.

Accompanying that management change was also a change directed from the AFSC Commander that the planned Flight Readiness Review of the YF-16 (and YF-17) planned to be headed by the USAF Prototype Program Manager Col. William E. Thurman was to be escalated into an Executive Independent Review Team (EIRT) assessment and headed by a Brigadier General (not assigned to the LWF Program to preserve the independent nature of the pre-first flight safety review. It was interpreted that this change was a recognition and acceptance at HQ AFSC of direction made by Dep. Sec. Def. Packard to the Secretary of the Air Force that the advanced technology demonstrator prototypes had increased fo a high level of importance in USAF acquisition planning. This change added a significant amount of preparation above that necessary for a FRR activity because it necessitated providing in-depth safety information to a general officer and his team who were not familiar with the designs of the YF-16 and YF-17. (Hence, the purposeful independent terminology in the EIRT charter). It was soon identified that the General officer heading the EIRTs was to be Brigadier General George H. Sylvester, and initially of the General Dynamics YF-16s progressing towards its first planned flight in December 1973. The second EIRT Safety of Flight review would be later of the Northrop YF-17s as the planned first flight for the Northrop YF-17 #1 was not until June 1974.

Planning of the scheduled shipment of the completed YF-16 #1 (tail number 72-01567) to Edwards AFB in early-December 1973 for the year long flight test evaluation, allotted a three-weeks block of time in the program schedule to resolve any flight safety issues that might evolve from the YF-16 EIRT flight-readiness review by the Brig. Gen. Sylvester review team.

The YF-16 EIRT review took place during mid-November 1973 at the General Dynamics facilities in Fort Worth. Membership of the working members on the EIRT team were mostly senior engineers from within the Aeronautical System Division and the several Air Force Research Laboratories at Wright-Patterson – all of whom had not been involved directly in the YF-16 program activities at General Dynamics to maintain their independence. The EIRT reviewed the safety of flight and readiness for flight aspects in the following areas:

Functional Interfaces
Flight Control/Airframe
Structures
Propulsion & Secondary Power
Crew Station & Escape
Test & Armament
System Safety

Prior to the EIRT team travel to Fort Worth, Texas, the Author (Project Engineer for the USAF LWF Program and assigned on loan from the Air Force Flight Dynamics Laboratory) had been requested to give a briefing to the members of the EIRT review team who were from the laboratories which had spawned the technology base each laboratory had provided to General Dynamics (and to Northrop) for the LWF Program. The purpose of the briefing was to assist Gen. Sylvester organize the EIRT with the appropriate membership each of whom were knowledgeable and experienced in the flight safety issues which were in need of review.

General Dynamics created an outstanding presentation in each of the emphasis areas to be reviewed by the USAF EIRT team, and in which all of the known risks and issues that in any way related to flight safety at Edwards AFB were factually and candidly presented. Some of the charts presented heretofore in this book were used by General Dynamics in their presentation to the EIRT.

To prioritize the number of issues and their relative importance to flight readiness, the EIRT established four categories as follows:

Category A – Must be resolved prior to first flight
Category B – Must be resolved prior to first in-flight operation
Category C – Does not prevent flight, but imposes flight restrictions until resolved
Category D – May be flown, but operator must be aware of potential safety problem

Within each of the four categories (A, B, C, and D), there were identified numerous items for which tests and/or analyses were planned to be conducted by General Dynamics for the YF-16 (and later those by Northrop for the YF-17) but which were not completed as of the time of the EIRT Review. Not unexpectedly, these constituted the majority of the findings by the EIRT. There were 3 items in Category A that General Dynamics discussed in-depth with the EIRT review team in meetings following the initial introductory informational meeting:

1) Of highest concern to the EIRT was whether there were any critical failure nodes in the fault-tree analysis of the safety-critical flight control computer that would go untested before first flight. The General Dynamics Flight Control Department leader (Tom Collins) identified that the flight control computer in the YF-16s would not have undergone any lightning strike testing, but added that: (a) the YF-16s had been designed for clear weather conditions and that lightning strikes at Edwards AFB in clear weather are a near-zero probability; (b) the flight control computer is a quadraplex analog computer and had been mounted with circuit-breaker fuses at lightning strike-prone mounting points of the flight control computer to all but prevent any extremely high voltage/current surges inherent in lightning strikes from damaging even only one of the four quadraplex branches; and (c) that there could be damage to up to three primary branches (A, B, and C) of the flight control computer but the remaining branch (D) contained stand-by flight control gains enabling the pilot to land safely (on the Rogers Dry Lake Bed if necessary).

2) Of next highest concern to the EIRT was the single piece bubble canopy being lost either due to bird-strike or to mechanical failure of the canopy locking mechanism. General Dynamics responded to mechanical failure aspect of this concern by showing a fault-tree analysis of the canopy locking (and unlocking in the event of commanded ejection) showing that two independent failures in the mechanism would need to occur to prevent the safe locking or unlocking mechanism for release and ejection of the canopy. A single failure potential did exist at one pivot point in the canopy mechanism but that the pivot point had been mechanically activated for over 200 cycles with no wear evidenced in the pivot made of high

strength steel. For the bird-strike concern, analysis did show that the thickness of the original canopy (made of single-piece polycarbonate formed over a heated mandril fixture) could cause the canopy to deform and suffer cracking if at cold temperature when impacted by a five pounds large bird at high speed. There is only negligible frequency of large birds being present at the AFFTC, and that occurs during the rainy season which infrequently occurs at the Edwards AFB vicinity. If the usually dry lake beds in the vicinity flood with rain water, the dormant brine shrimp are activated. Large sea gulls do fly into the Edwards AFB/Rosamond area from coastal California to eat the then-active brine shrimp, but the birds' flight activity is generally very near the surface of the flooded dry lake bed and rarely occurs higher than 200 feet. General Dynamics did concede that the thickness of the front portion of the YF-16 canopy would need to be increased for safer bird-strike capability and probably a laminate of acrylic would need to be added to the exterior of the canopy (polycarbonate is soft and easy to scratch) if that single piece canopy design were to be used in flight operations in an environment having a high presence of large birds.

3) The third item of significant concern to Gen. Sylvester's EIRT team was the loss of power from the single P&W F100 engine with attendant loss of electrical power from the main generator (attached as a gear-driven accessory to the engine). General Dynamics showed the EIRT that a hydrazine-powered emergency power unit (EPU) had been installed for the possible loss of either the engine and with the engine RPM sensor detecting the engine rotating turboshaft spooling down at an abnormal rate (indicating engine loss) and/or the loss of the necessary minimum voltage from the Main Generator (due to self-failure of the generator) to the all-electric FBW FCC. In the YF-16, the hydrazine EPU would be automatically activated in the event of abnormally- high engine spool down RPM rate (in the YF-16s there was no provision for the pilot to command the EPU on or off). The hydrazine propellant for the rocket powered auxiliary generator was contained in a tank sized to provide twelve minutes of EPU operating time – and which General Dynamics had sized to enable an adequately flight-controlled emergency descent from Mach 0.9 at 30,000 feet altitude and controlled landing.

B/G Sylvester approved the EIRT Team final report (issued in late December 1973) for the General Dynamics LWF Prototypes and certified that, with the following restrictions and conditions, the YF-16 #1and YF-16 #2 prototype aircraft were safe for flight during the planned flight test evaluation flights at the Edwards AFB AFFTC: (Author's note: The following sections contain information extracted from internal Program Office Weekly Activity Reports prepared "for the record" by the Author as Project Engineer for the USAF LWF Project Office.)

1) Flight testing of all prototype aircraft shall adhere to all flight safety rules, practices, and regulations established by the Commander of the Air Force Flight Test Center

at Edwards AFB, California. Any aircraft incurring damage to any flight critical subsystem on the test aircraft shall be prohibited from further flight testing until the damage has been certified by a representative of the AFFTC Safety Office to have been successfully repaired and the test aircraft restored to full flight airworthiness.

2) The flight tests of the YF-16 #1 and #2 that have not been lightning strike tested should be constrained to be during only clear weather when the possibility of lightning strike is near zero. Although not mandatory, lightning strike tests are highly recommended to be conducted on the full scale, operational weapons-configured YF-16 #2 prior to any flight testing intended to demonstrate gun-firing with explosive rounds or weapons delivery with explosive ordnance.

3) Flight testing of the YF-16 #1 and #2 at Edwards AFB shall not be conducted when the presence of numerous large birds in the vicinity of flight test operations are known to exist.

4) Personnel on the ground are not be exposed to the toxic fume's effluxion from the hydrazine-fueled EPU if activated while the aircraft is on the ground. The aircraft engine (if running is to be shut-down and the aircraft impounded if the EPU is activated by any activation source. Following safe evacuation of the hydrazine exhaust fumes (if any) from the vicinity of the impounded aircraft, an inspection for cause of the activation is required to be accomplished by the contractor (General Dynamics) and observing all hazardous materials safety handling procedures. Flight test of that aircraft will not resume until the cause of the EPU activation has been determined and preventive measures have been taken to avoid repeat occurrence of the EPU activation.

With almost all no classified new aircraft development programs, a major milestone (for publicity purposes) is a Roll-Out ceremony – typically at the manufacturer's facility and with many dignitaries from the Pentagon, from the state Governor's office, from various organizations of the USAF at Wright-Patterson AFB, from corporate and company upper management, and from most news media organizations. After an orchestra plays the national anthem, the new aircraft is typically towed out to a position in front of the grandstands filled with the attending dignitaries. And the many speeches are delivered. The General Dynamics Roll-Out ceremony was conducted on December 13, 1973 – but to the audience's astonishment, the sound of a jet engine came into hearing and the audience (the Author included) believed the engine sound to be from an aircraft on the Carswell AFB runway taxiing for take-off (to the main runway shared with General Dynamics). As the engine noise grew louder, the YF-16 #1 in its patriotic red, white, and blue color taxied into its spotlight position in front of the grandstand, pilot Phil Oestricher applied the brakes causing the nose of YF-16 #1 to do a beautiful curtsy, and the engine was shut down.

[Authors' note: General Dynamics had invited officials also from Northrop to the Roll-Out ceremony and who were seated in the grandstand one row behind where this Author was seated. In muted voice, the comment overheard by the Author was: "The bastards had to be different and do a taxi-out!". It cannot be confirmed if the observation had been stated by a Northrop attendee, but it sounded much like a sour-grapes comment that would be made by a representative from the Northrop Aircraft Division next to the Hawthorne Airport (in California) where the local Hawthorne community leaders forbid Northrop (or anyone else) from running large noise jet engines at the Hawthorne Airport to annoy local residents.

The photograph next was taken (regrettably in black-and-white photography) at the scene of the YF-16 #1 "in your face" taxi-out (rather than the conventional roll-out process of the new aircraft being towed-out by a ground tug) on December 13th at the General Dynamics/Fort Worth facility (in front of the assembly area configured with a grandstand for the attending dignitaries). The keynote speakers at the Roll Out ceremony included then-Secretary of the Air Force John L. McLucas, General Dynamics President David Jones, and USAF Lt. General James T. Stewart, Commander of the Aeronautical Systems Division.

Taxi/Roll-Out of YF-16 #1 Ft. Worth, TX, December 13, 1973.
(Source: LWF Engineering files)

On the day after the YF-16 #1 Taxi/Roll Out, the YF-16 #1 was then moved back into the assembly area and the wings, stabilators, and engine disassembled from the YF-16 #1 to prepare it for loading into a C-5A Galaxy.

In accordance with the USAF/GD LWF contract plan for delivering the YF-16 aircraft (both YF-16 #1 and #2) to Edwards AFB in California, a USAF C-5A Galaxy flew in to Carswell AFB and a partially-disassembled YF-16 #1 (wings and horizontal stabilators removed) and engine were to be loaded into the C-5A for transfer of YF-16 #1 to Edwards AFB.

A photo of the YF-16 #1 fuselage being loaded in the C-5A Galaxy is shown next:

Loading YF-16 #1 in C-5A to EAFB, CA Carswell AFB, TX, December 13, 1973
(Source: USAF Engineering files)

The EIRT Team, headed by B/G. Sylvester and comprised of the aforementioned technical specialists in the several technical areas to be reviewed for flight safety and readiness for flight, were given briefings at Northrop Corporation in Hawthorne describing the YF-17 design on January 17 and 18, 1974 (introductory familiarization level), and returned status reviews on February 7, February 28, and April 8 1974 to review action items surfaced at the first meeting. The same categorization of action items (Category A, B, C, and D as described earlier) and the same procedure for resolving the numerous items identified by the EIRT at the General Dynamics YF-16 EIRT review were developed for the Northrop YF-17. The YF-17 EIRT final report was issued on 26 April 1974.

Inasmuch as this book is primarily addressing the General Dynamics YF-16 design, and because there were fewer advanced technologies on the Northrop YF-17 than there were on the General Dynamics YF-16 advanced technology demonstrator design, suffice it to be said that the EIRT report for the YF-17 flight readiness (first flight of YF-17 #1 planned to occur June 9, 1974) did not contain as many serious concerns in Category A and Category B in the YF-17 EIRT final report and will not be discussed herein to the same extent as that of the YF-16. The two most major issues the EIRT had with the Northrop YF-17 design were related to the basic design configuration needing consideration if the YF-17 were to eventually be chosen for any further development. These two issues, of which only the first item was a first-flight show-stopper, were:

1) (Extracting from the YF-17 EIRT final report April 26, 1974): "The accessory bays are not isolated from the main landing gear compartment and fire containment is below (safety) standard. Although the potential for fire is primarily associated with catastrophic failure of equipment, and fire detection and extinguishment capabilities are provided in the bays, occurrence of fire has high probability of propagation and sustenance in the unprotected main landing gear compartment. Additional procedural measures are required to minimize or negate the above risk. Retest of the intumescent coating utilizing the FAA standard fire wall burner apparatus to establish fire hardening / containment performance for the engine / accessory bays application is also required."

2) The placement of the nose wheel of the YF-17 is ahead of the propulsion system dual inlets. The location of the nose wheel ahead of the engine inlets of the YF-17 raised a concern of the potential for ingestion of two different types of potential FOD into the inlets:

a) The nose wheel when rolling through puddles of water on the runway surface during high-speed taxi or take-off / landing speeds is highly likely to cause water ingestion into either/both inlets. In order to preclude this potential and the need for subsequent engine blade inspection, planned water taxi tests are recommended to be deferred until the end of the planned flight test program. It is further recommended that the YF-17 be restricted from taxi or take-off operations through puddles until such water ingestion tests are accomplished.

b) The nose wheel when rolling on the runway tossing a piece of FOD (a loose screw/fastener dropped from a previous aircraft being the most common source of FOD) into the inlets which are not separated from the nose wheel center line by a large lateral distance. [Author's note: When Northrop responded to the USAF LWF RFP approximately 18 months prior to the EIRT Review, the P-601 model proposed was configured with the use of a single Pratt & Whitney F100 engine and had a single chin-type inlet directly aft of the nose wheel. That was one of the primary reasons the USAF LWF Proposal Evaluation Team considered the Model P-601 to have a major design flaw for use in an operational environment, and was the basis for preferring the two-engine (GE YJ101-GE-100 prototype engine) Model P-600 configuration proposed as an alternate, and safer, configuration. In the preceding photograph, it can also be seen that a piece of FOD that was not ingested directly into the inlet from the bottom of the nose wheel could also be damaged by a hard object thrown upward by the nose wheel and being deflected off the bottom of the fuselage and with a trajectory somewhat laterally into the inlet.]

In retrospect, the outcome of the EIRT review of both the GD YF-16 design and the Northrop YF-17 design produced a large number of changes and additions that would necessarily be made if the winner of the USAF LWF competition were to be chosen for continued development for inclusion in the USAF fighter aircraft operational inventory.

In contrast to the taxi-out of the General Dynamics YF-16 #1 at its Roll-Out ceremony on 13 December 1973, the Northrop YF-17 #1 was revealed by removing a cover tarpaulin at its Roll-Out ceremony on 4 April 1974 at the adjacent Hawthorne Airport in front of the obligatory grand stand for speakers and other dignitaries at the momentous occasion. The highlight speaker was Dr. John L. McLucas, then-Secretary of the Air Force. The YF-17 #1 LWF prototype (tail number 72-01569) is shown in the photograph next.

This view of the YF-17 better shows the relative location of the nose wheel and the engine inlets that was of concern to the EIRT for having puddle water or hard FOD objects being ingested in the inlet during motion on the runway. (Author's note: During the Lightweight Fighter flight test program, there was only one occurrence of an engine on either the YF-16 or YF-17 being damaged by FOD. This occurred on a takeoff roll by YF-17 #1 early in the flight test program and was to one of the GE YJ101 turbojet engines. A more detailed account will be described in Chapter 9 but a hint of the cause was not as the EIRT had feared – that the nose wheel would toss a piece of FOD from the runway into the YF-17 inlet. The FOD was from within the aircraft itself. Chapter 9 discloses what the source of the FOD was and how it got produced.)

Northrop YF-17 #1 at unveiling, Hawthorne, CA, April 4,1974. (Source: USAF Engineering files)

Unlike the General Dynamics YF-16 #1 being flown aboard a USAF C-5A transport from Fort Worth, Texas, to Edwards AFB in the Rosamond, California area, the YF-17 #1 was transported intact aboard the bed of a truck along the California road system between Hawthorne and Edwards AFB. The Northrop Corporation has a hangar at Edwards AFB which has for a long time been used for the testing of Northrop produced aircraft (most notably the T-38/F-5 series of Northrop aircraft), and it is to the Northrop hangar that the Northrop YF-17 #1 LWF prototype was delivered in late May 1974.

Chapter 9 will describe the year-long flight test program conducted, and the problems uncovered, in 1974 at Edwards AFB of both the General Dynamics YF-16 and Northrop YF-17 LWF aircraft. At the end of Chapter 9 will be presented the decision as to which LWF design had won the USAF Lightweight Fighter fly-off competition.

CHAPTER 9
LWF FLIGHT TESTS, LWF FLY-OFF WINNER ANNOUNCEMENT (1974 TO 1975)

Task 3 of the Statement of Work contained in the USAF LWF RFP (request for proposal) called for the two contractors (General Dynamics and Northrop) to conduct a flight test program to verify the capability of the aircraft to satisfy the design requirements as established by the Contractor considering the Air Force design and performance goals summarized in Attachment A to the RFP (provided earlier in this book). The technologies selected for the YF-16, how those technologies were manifested in the design and manufacture of the YF-16 configuration (external and internal), and the highlights of the ground tests conducted to validate the safety of flight of the YF-16 were described earlier in this book also. In parallel with the ground and preparation for flight activities, the contractors and the flight test planners of the USAF Joint Test Team created a flight-testing plan and schedule for conducting the flight portion of the LWF advanced technology demonstrators. Readers are again reminded that this was not a head-to-head competition wherein the two General Dynamics YF-16s would fly against the two Northrop YF-17s, but rather each design would be flown to evaluate how well each design individually met the USAF design and performance goals.

To conduct the flight tests within the twelve months allowed by the LWF Prototype Project Office per the USAF LWF contract, a Joint Test Team form of management was developed in which test pilots from the two contractors, test pilots and flight test engineers from the Edwards AFB flight test data collection and analysis offices, and pilots from the USAF AFFTC and Tactical Air Command would conduct the flight tests in a phased manner and in adherence to established safety of flight rules of the AFFTC at Edwards AFB. HQ AFSC designated AFFTC at EAFB as the RTO (responsible test organization).

There was a total of ten test pilots comprising the LWF Joint Test Team, two from General Dynamics, two from Northrop, and six from USAF (three from AFFTC and three from TAC).

The USAF pilots were:
Lt. Col. James G. "Jim" Rider (AFFTC, Edwards AFB, CA) Director of the JTT – YF-16 and YF-17 evaluation tests

Major Maurice B. "Duke" Johnston (TAC/AFOTEC, Hill AFB, UT), Deputy Director of the JTT – YF-16 and YF-17 evaluation tests
Major Robert C. "Bob" Ettinger (AFFTC, EAFB, CA) - YF-16 evaluation tests
Major Rutherford D. "Dean" Stickell (TAC) - YF-16 evaluation tests
Major Michael C. "Mike" Clarke (AFFTC, EAFB, CA) - YF-17 evaluation tests
Major Joseph W. "Joe Bill" Dryden (TAC) - YF-17 evaluation tests

General Dynamics pilots were:
Philip F. "Phil" Oestricher, Chief Test Pilot - YF-16 evaluation tests
Neil R. "Neil" Anderson, Test Pilot - YF-16 evaluation tests

Northrop pilots were:
Henry E. "Hank" Chouteau, Chief Test Pilot - YF-17 evaluation tests
Joe Bailey "Joe" Jordan, Test Pilot - YF-17 evaluation tests

Although not in the chronological order of events as they occurred in the LWF flight test program, on Saturday, May 18, 1974, (Armed Forces Day) nine of the ten Joint Test Team had a photograph taken at Edwards AFB in front of the General Dynamics YF-16 #2 and the Northrop YF-17 #1. That occasion when the YF-16 and YF-17 aircraft were that close together during the entire LWF Test Program was rare, as the LWF aircraft were based in two different hangars at Edwards AFB and did not get flown in a head-to-head flight competition. The rare photograph is shown below:

USAF LWF Joint Test Team (with YF-16 #2 & YF-17 #1), May 18, 1974.
(Photograph from USAF AFFTC Museum files)

Missing from the photo is Neil Anderson of General Dynamics who was flying YF-16 #1 on a flight test mission at the time of the photograph. From L-R in the photo: Phil Oestricher (GD), Maj. Bob Ettinger (AFFTC), Maj. Dean Stickell (TAC), L/C Jim Rider (AFFTC), Maj. Duke Johnston (TAC), Maj. Mike Clarke (AFFTC), Maj. Joe Bill Dryden (TAC), Joe Jordan (Northrop), Hank Chouteau (Northrop). There were four pilots who initially flew only the two YF-16s (Oestricher, Anderson, Ettinger, and Stickell) and four pilots who flew only the two YF-17s (Clarke, Dryden, Jordan, and Chouteau). Only two pilots flew both the YF-16s and YF-17s (Rider and Johnston). All ten of the pilots were graduates of the Flight Test Pilots School at the Air Force Flight Test Center (AFFTC).

(Author's note: On December 14, 1959, while in the USAF, then Captain Joe Jordan (second from right in the photograph) became the first person to exceed 100,000 feet altitude in an air-breathing propulsion system aircraft and with only aerodynamic flight controls – in a modified F-104C Starfighter. The 1959 Harmon Trophy was awarded to Joe Jordan for the official altitude record of 103,395.5 feet and he held that record for many years.)

The Joint Test Team pilots designed and wore a patch identifying them as being test pilots on the USAF LWF Flight Test Program. A photograph of that patch is shown below.

Patch Worn by USAF LWF Joint Test Team Pilots
(Source: Author's memorabilia file)

Just as the entire staff of the LWF Prototype Project Office had done, the JTT members pledged to not reveal any test results or any other competition sensitive information about either the General Dynamics YF-16 or the Northrop YF-17 to anyone without the need-to-know about the status of either contractor's efforts and test progress of the other contractor. Otherwise, most of the test results being obtained and problems being encountered were not contained officially with USAF Classified markings (Confidential or Secret). Performance comparisons such as the John Boyd energy-maneuverability overlays comparing the LWF E-M to other country's threat aircraft E-M were originally classified Secret and were protected with all regulations that pertain to the handling of Secret information and/or data. Those original Secret comparisons have long since been declassified and approved for release to the public.

The two General Dynamics YF-16s (tail numbers 01567 and 01568) were transported from the General Dynamics Fort Worth manufacturing/assembly facility aboard a C-5A Galaxy provided by the USAF to the Edwards AFB AF Flight Test Center and to a hangar dedicated to the YF-16 LWF flight test operations at EAFB/AFFTC. The two Northrop YF-17s (tail numbers 01569 and 01570) were transported by truck from the Northrop manufacturing/assembly in Hawthorne, CA, on Los Angeles County roads to the Edwards AFB/AFFTC and to the Northrop hangar on Contractors Row at Edwards AFB. The dates of delivery of the four LWF test aircraft (and their first flight dates) were staggered and occurred on the following dates in 1974:

GD YF-16 #1 Delivered January 8 (January 20 (inadvertent) and February 2 (official)
GD YF-16 #2 Delivered February 27 (9 May)
Northrop YF-17 #1 Delivered April 23 (9 June)
Northrop YF-17 #2 Delivered July 12 (21 August)

All of the four LWF test aircraft were instrumented differently with generally the #1 LWF being instrumented to measure functions of major systems and subsystems, strength/flutter capabilities of structural components, and on a few flights installation of spin chutes during high angle-of-attack flight tests, in gradual expansion of the flight envelope for safety of flight. The #2 LWFs had less safety of flight instrumentation and more test data instrumentation to measure the dynamic performance capabilities of the test aircraft while performing more operational type segments of the flight test program. Also, Contractor pilots flew the initial three or four flights of their respective two LWF prototypes to assure the manufactured aircraft were performing as designed for both safety of flight considerations and performance considerations. After the first three or four safety of flight test missions, USAF pilots assigned to flight test and evaluate the operational capability characteristics of the respective LWF aircraft flew an increasingly higher percentage of the flight tests per the flight test schedule developed by and for the Joint Test Team.

Although there were no detailed performance requirements specified by the USAF LWF Prototype Project Office), the Joint Test Team had developed a flight test program that evaluated both (1) the advanced technologies capabilities embodied in the YF-16s and YF-17s for achieving design goals (design-to-cost and design-to-low weight being the more dominant design drivers), and (2) evaluated the flight capabilities enhanced by the advanced design capabilities to perform operational type combat relevant, multi-role missions (while flight testing the design to performance goals).(Readers are reminded that USAF HQ Commanders had not initially supported the development of any other program that might detract funds needed for completing development of the F-15 Eagle and Pratt & Whitney F100 engine, and therefore no official requirements existed to be verified by flight testing of the USAF LWF advanced technology demonstrators.)

Much emphasis was placed in the initial designs by both General Dynamics and Northrop on the "ilities" (reliability, maintainability, survivability, serviceability, reparability, affordability and availability, etc.) of the designs and chose technologies and equipment items that were of state- ot-art maturity and functional capability comparable to (or better than) that existing for fighter aircraft in the military flight inventory. To evaluate how well the Contractors did in consideration of the "ilities" as a design parameter for their LWF designs, the Air Force Logistics Command (at Wright-Patterson AFB) assigned three very experienced senior maintenance Crew Chiefs to the LWF program and who observed how the "ilities" were actually considered when both contractors were designing-in access panels into their respective designs, selecting high reliability components for installation, and identifying what GFAE (Government Furnished Aeronautical Equipment) would be needed to support the planned flight test program when the assembled aircraft were delivered to Edwards AFB for flight test evaluation. Blue-suit Crew Chiefs were resident at Edwards AFB and divided their time and attention at the two LWF aircraft hangars (General Dynamics YF-16 hangar and Northrop YF-17 hangar) observing how the contractors were maintaining and servicing their respective test aircraft during actual flight test operations.

(Author's note: The purpose of the "fly before buy" concept championed by Deputy Secretary of Defense David Packard included the approach taken for the LWF program that he urged should be taken by the USAF and USN/USMC aircraft developers, to include strong consideration of operational costs when procuring future military aircraft. The assignment of maintenance Crew Chiefs from the very onset of the design process at both General Dynamics and at Northrop Aircraft Division proved to be invaluable to the success of the LWF Program.)

The details of who/what/when/where/how the flight test plan was created is best described in an outstanding historical document recently compiled by the Air Force Test Center History Office at Edwards Air Force Base. That document was written by Wade A. Scrogham and is entitled "Combat Relevant Task: The Test & Evaluation of the Lightweight Fighter Prototypes" and was printed for distribution in September 2014. The

Author absolutely encourages interested Readers to obtain a copy of "Combat Relevant Task" from its source on the internet (by Title and Author) as it is written in much the same style as the Author has written this book. There are a myriad of facts and events described by Wade Scrogham including the transcript of interviews with the pilot members of the Joint Test Team who flew the YF-16s and YF-17s, but also contains interviews with the flight test engineers who processed and analyzed the test data into useable form for input to the Source Selection Team, enabling them to select a winner of the Lightweight Fighter fly-off competition.

So as to avoid great overlap in information between the "Combat Relevant Task" historical document written by Wade Scrogham and the historical contents of the overall USAF Lightweight Fighter Prototype Program within this book, the Author will describe those out-of-expectation occurrences that were generated from the flight tests which caused the Contractors' (General Dynamics and Northrop) back home engineers to conduct some additional analysis, test, redesign, etc. to resolve those safety issues. With those resolutions, the flight test program could be conducted by the JTT with its primary combat relevant task evaluation and with assurance that adherence would be met of all of the flight safety rules imposed by the Air Force Flight Test Center, by the USAF Lightweight Fighter Project Office, as well as being observant of the safety of flight restrictions imposed by the ASD Executive Independent Review Team (B/G Sylvester's EIRT). Unavoidably, some events that took place at AFFTC are identified/contained in both the AFFTC historical "Combat Relevant Task" report and this book. They are discussed from the two different perspectives of flight test evaluation pilots (in the AFFTC historical report), and from Project Office and AFRL Laboratory engineers as to the significance of the problem encountered and how it affected the contracted effort being described in this book.

The General Dynamics YF-16 #1 (tail number 01567) was the first of the four USAF Lightweight Fighters to be delivered to Edwards AFB - on January 8, 1974. The wings were re-installed and the Pratt & Whitney F100 engine re-installed after shipment of the partially-disassembled YF-16 #1 in a C-5A Galaxy from Fort Worth and delivered at Edwards AFB. The planned first test was the Low-Speed taxi test and, if successful, to be followed by the High-Speed taxi test – both scheduled to be accomplished by General Dynamics on January 20, 1974. Per flight safety rules at AFFTC, taxi tests of new aircraft are normally conducted on Runway 04R, a 15,024' long runway with an additional 9,588' runway on the dry lakebed at the NE end of runway 04R, into the normally calm or prevailing wind over the dry lake bed. This direction was chosen in the event a landing gear/braking system failure necessitated a long roll out without brakes to stop the aircraft being taxi-tested. A photograph from above Edwards AFB shows the orientation of the taxiways and runways at Edwards AFB and the Rogers Dry Lake to its east.

Aerial view of the Edwards AFB (near Rosamond, CA)
(Source: USAF Photography files)

The morning of January 20th began with an uneventful Low Speed taxi test on runway 04R – toward the dry lakebed. The weather report for the High-Speed taxi test, planned for the early afternoon, was very unfavorable – a storm front was moving in from the southeast with high wind gusts and lightning producing clouds. The chance of lightning strike of the YF-16 during its high-speed taxi tests was evaluated and except for reversing the planned direction of the high-speed taxi test to be toward the dry lakebed, due to a shift in the wind direction, the decision was made to proceed with the test earlier than planned and to be on Runway 22L – away from the dry lakebed. Because this was a major program event for General Dynamics, a photographic team in a photography van was in position to film the high-speed taxi test. In addition to the JTT Director L/C

Jim Rider in the Operations (Base Ops) tower, the General Dynamics Program Manager Lyman Josephs had arrived to witness the high-speed taxi test event. The LWF Prototype Project Manager Col. Bill Thurman, Col. Del Jacobs, and the Author (the USAF LWF Project Engineer) had also flown in from Dayton to witness the high- speed taxi test with Lyman Josephs. Having a residual concern for the sensitivity of the accelerometers on a rough runway causing feedback to the flight control computer and resulting in high activity of the stabilators, the Author had brought a pair of high-powered binoculars to observe the stabilators activity during the high-speed taxi roll.

Before this event, there was no significant concern that the flight control computer had not yet been qualified for flight in a lightning strike environment as all testing was planned for clear weather conditions. For safety of flight when in a lightning strike environment, protection of critical electronics components (e.g., re-settable over-voltage trips) would have been installed in the appropriate circuits to protect against the high voltage-surge associated with a lightning strike. Additionally, lightning strike tests had not been conducted since the YF-16 flight testing was to be done only in clear weather, VFR conditions. Only rudimentary EMI (electromagnetic interference) analysis and EMI qualification testing had been done due to the nature of the all-electric fly-by-wire flight control system.

The high-speed taxi test had been practiced on the YF-16 flight simulator at Fort Worth many times by the GD test pilot Phil Oestricher, and the test planning called for Phil to retard the throttle to ground idle power shortly prior to the end of the high-speed taxi test - to lighten the aircraft weight on wheels - and to introduce a slight roll command to just get the feel of the aircraft before bringing it to a stop and conclusion of the high-speed taxi test. The initial portion of the high-speed taxi test went very well, and seeing the red, white, and blue aircraft moving rapidly against the very ominous approaching colored cloudy sky was a most impressive sight.

Inadvertent First Flight of YF-16 #1

The historically renowned inadvertent first flight of YF-16 #1 that then occurred at the end of the high speed taxi test on January 20, 1974, is used as the opening flight event to be discussed in this book – from the perspective of the USAF LWF Prototype Program Project Engineer (the Author, who was there as an authorized observer near the edge of Runway 22L at AFFTC and watched the high speed taxi test and its resultant inadvertent first flight through a pair of high magnification binoculars). A montage of photographs extracted from significant frames of the test film sequence on file at the USAF LWF Program Office is shown next. ([Readers who may wish to view a video of the very dynamic YF-16 taxi-test event may find the video on the internet at https://youtu.be/SntFDCx5vkI.)

Inadvertent First Flight of General Dynamics YF-16 #1 (01567)

GD Pilot Phil Oestricher   Edwards AFB, CA   January 20, 1974

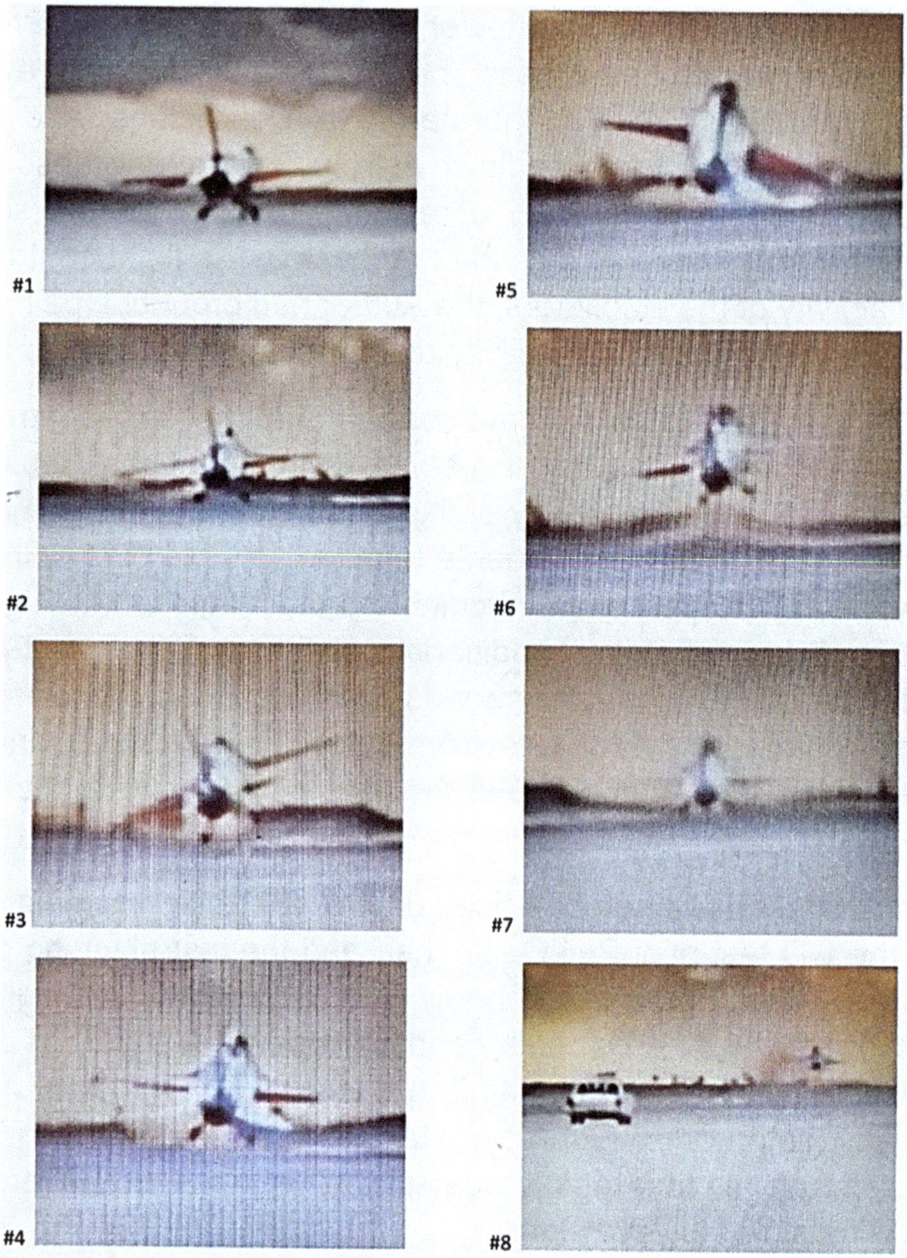

(Source: USAF Photography files)

Total disbelief existed that what was being seen happening could have been planned that way in the JTT flight test plan. It was not widely known that the YF-16 #1 had planned to rotate the YF-16 into a take-off attitude at the end of the high-speed taxi run, but had not planned a take-off of the aircraft. Just after the YF-16 had passed the 3,000 ft runway marker (shown in photo #1 of the montage) where observers were standing, the nose of the aircraft went quite high in angle-of- attack (much above ten degrees), and the three-landing gear all left the runway (shown in photo #2 on the montage). Within a second after the Author called out "She's flying!", the YF-16 made a very sharp left roll (shown in photo #3 of the montage) and the roller on of the AIM-9 missile on the left-wing tip station 1 scraped the runway. Almost instantaneously the aircraft then made a rapid right roll and with the nose angle still very high (shown in photo #4 of the montage), the outboard trailing edge corner of the starboard stabilator also scraped the runway (shown in photo #5 of the montage). This contact was not audible above the engine exhaust roar, but was seen to produce a small puff of white smoke (shown in photo #5 of the montage). It was seen by the Author in the binoculars that a triangular- shaped area of the starboard stabilator had been bent upward with the static discharge fixture pointing upward. Shown next is a clear photo of the YF-16 #1 aircraft just after the outboard trailing edge of the starboard stabilator damage had been incurred, as can be seen upon close examination of the photo. Also, to be noted is that the adverse weather front was almost at the Edwards AFB facility.

YF-16 #1 (GD Pilot Phil Oestricher) in its inadvertent first flight mode, January 20, (Source: USAF Photographic files)

Overheard from the other nearby observers were comments such as “Holy Cow, that wasn’t planned, was it?” (Or more cruder words to that effect). The erratic motion of the YF-16 continued with a roll to the left (as shown in photo #6 of the montage), a roll back to the right, then nose lowered, then nose raised and bouncing on the main gear as it appeared Phil was trying to stop the roll oscillations of the aircraft as well as to use some nose-high aerodynamic braking (shown in the photo #7 of the montage) to try slowing the aircraft down for a safe stop on the runway. Due to the alternating bouncing of the main gear causing the heading of the aircraft to be towards the left edge of the runway, it appeared that Phil was going to be running off the runway. At that point in time, Phil elected to increase the thrust to takeoff power and the aircraft entered into a pogo-stick mode of motion on the thrust column (ground effect) against the runway. Eventually the erratic excursions in roll, pitch and pogo-ing were damped and the aircraft proceeded to take-off somewhat normally (as shown in photo #8 of the montage) near the southwest end of runway 22L as the service road at the end of the runway was rapidly approaching. [The Author turned to the other observers and asked "Have you ever seen a worse take-off than that?"]

Phil flew a pattern to the left (over unpopulated land area) and flew for about six minutes to a perfect landing on runway 22L. The aircraft came to a stop, the canopy was opened, and Phil climbed out of the cockpit taking off his flight gloves - all captured on film by the General Dynamics photography team. Phil obviously had saved the aircraft and elected to stay with it rather than eject which was certainly an option. His recorded comment on film and audio recording was not as might be expected as Phil descended from the cockpit on the access ladder muttering "Aw, (expletive)! I PIO'd it." (Readers unfamiliar with the terminology PIO are informed that PIO stands for pilot-induced oscillation.)

In addition, Phil stated that he was taken by surprise that the engine thrust had not decayed as rapidly as he had expected it would decay based on the simulation flying, he had practiced. It was later observed in the photography that the nozzle of the engine was in a closed position indicating that the engine thrust was being produced at a *flight idle* level rather than less than half that thrust level in a *ground idle* power mode. The simulator also was later found to not have the WOW (weight on wheels) switch function properly built into the simulation such that when the 3-landing gear were off the ground in the inadvertent flight, the WOW switch prevented the thrust level from being commanded to the *ground idle* lower thrust level and instead enabled the *flight idle* higher thrust.

As with all accidents in which damage is done to an aircraft, an Accident Review Board reviewed all of the available information from the instrumentation as well as from the pilot (Phil Oestricher) and concluded that the pilot induced oscillations were caused by extremely sensitive flight control gains particularly in the roll axis and absolved the pilot from any error in omission or commission. To the contrary, the Accident Review Board assessed that the pilot had done extraordinary adaptation to the hair-trigger personality of the first-of-its-kind zero-displacement sidestick and the highly sensitive flight control gains in a first-of-its-kind all-electric fly-by-wire flight control system. Based on this incident, and with other pilots also commenting during the flight testing that the zero-displacement sidestick does not give the pilots a visual indication of how much motion is being commanded, the zero-displacement stick was later structurally re-mounted to have a small amount of deflection. Had Phil Oestricher elected to eject, as a minimum, a very major schedule disruption would have occurred in the LWF Flight Test Program.

As can be seen in the photograph next and on the previous photograph of the inadvertent Flight 1A, the outboard trailing edge triangular section of the starboard stabilator had been damaged and bent upward as well as its underside having been damaged by abrasion against the runway surface.

View of YF-16 #1 Starboard stabilator damaged during taxi test at AFFTC, January 20, 1974 (Source: USAF Engineering files)

The photograph on the previous page shows the degree of angularity to which the triangular shaped outboard trailing edge of the stabilator was bent during the contact with the runway surface. The photograph below shows the laminate on the bottom of the stabilator having been abraded away by contact with the runway surface, acting much like a coarse sandpaper, and exposing the aluminum honeycomb core of the stabilator.

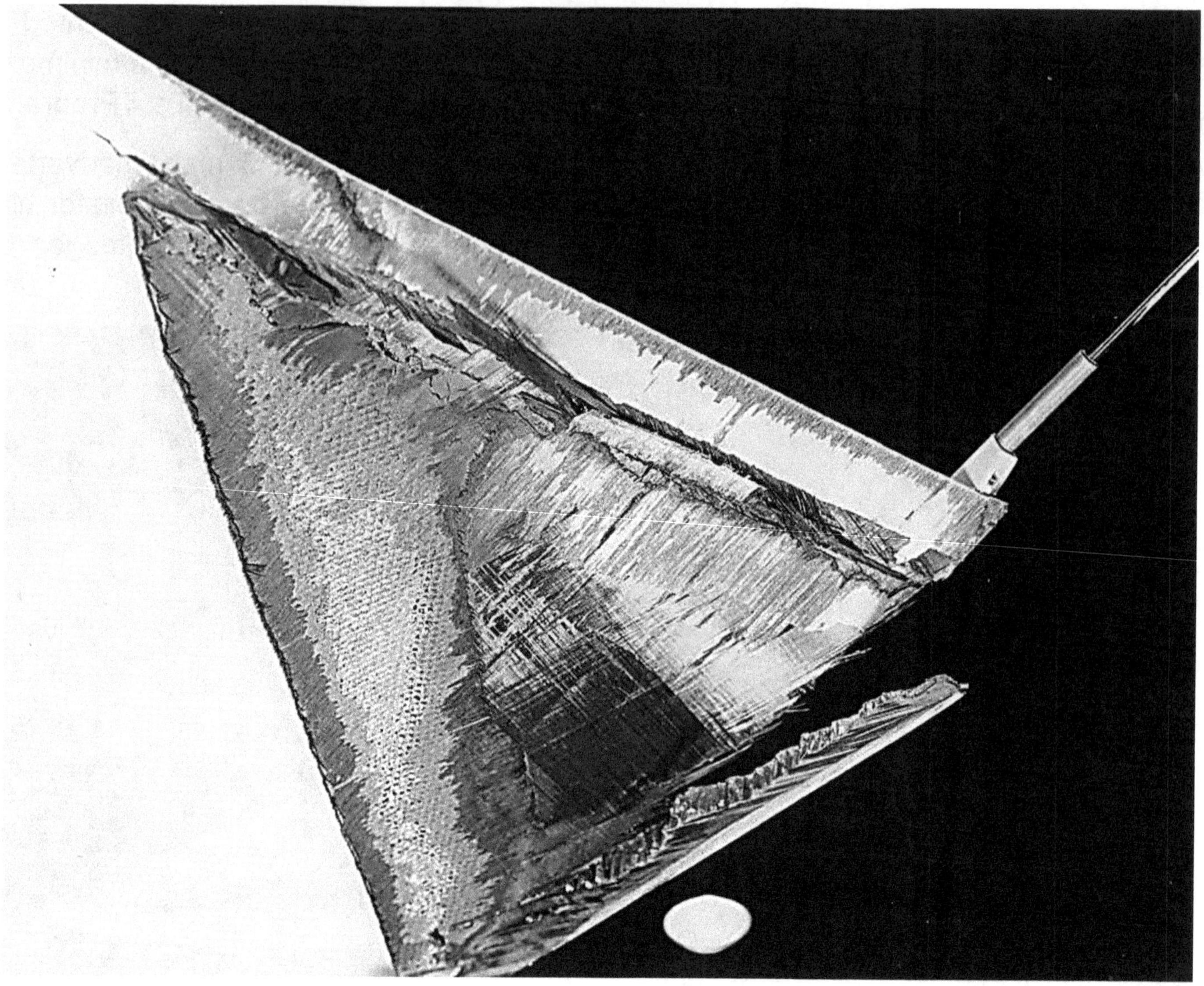

Underside of YF-16 #1 Stabilator damaged in inadvertent takeoff at AFFTC, January 20, 1974 (Source: USAF Engineering files)

In order to adhere to a new re-scheduled first official flight date of February 2, 1974, the right stabilator of YF-16 #2 (nearing completion of assembly at Plant 4) was brought in from Fort Worth while the damaged stabilator shown above was being repaired. The repair of the damaged stabilator, made of an aluminum honeycomb core sandwiched between two laminate material skins, was done quite rapidly and after repair it was replaced on YF-16 #1 for the remainder of the USAF LWF flight test program (ending in December 1974).

The repair of the stabilator damaged during the inadvertent first flight had some positive aspects in that it provided experience in how to repair composite material components of the aircraft that may be damaged in operational use when exposed to the potential for damage by enemy ground gunfire. Much earlier in this book, a discussion of how the battle damage assessment (BDA) of the F-4s in the Vietnam War drove the need for a survivable flight control system that became the driving inspiration for an all-electric, fly-by-wire flight control system in fighter aircraft. That BDA also included the development of damage repair to flight control surfaces of conventional metallic construction, so the repair of the YF-16 #1 stabilator – constructed from a composite outer skin bonded to an aluminum honeycomb core - added a chapter to that battle damage assessment and repair technology. (Readers should appreciate the philosophical tenet that when given too many lemons, a true entrepreneur will make lemonade for sale. Whenever a test has a failure, or if/when an accident occurs, the important lesson(s) learned from the incident/accident will serve to prevent recurrence of the test failure or the accident as well as to add to the technology state-of-the-art data base for future systems applying that technology.)

In celebration of its 30th anniversary, the *Code One* magazine posted an article on January 19, 2015 "What A Wonderful Airplane: YF-16 First Flight" written by Joe Stout. In the article, a description Phil Oestricher gave in 1992 ("one of the thousand times I have been asked to relate the true facts about the inadvertent first flight incident") was printed. That statement by Phil Oestricher in his own words – and no one knows better than the pilot himself what happened on January 20, 1974 – are repeated below.

"We'd progressively taxied the airplane faster and faster and wanted to get a better idea of how it handled. Few people know this, but we actually intended to lift the airplane off the ground that day. Our intention was to move the throttle to military power for a few seconds and let the main gear come up a couple of feet while we went down the runway. At a speed of 130 knots, with the airplane still accelerating slightly, I rotated the nose to about 10 degrees and made small lateral stick inputs to get a feel for the roll response but got nothing, presumably because the main gear was still on the runway which would prevent the airplane from rolling. I then increased the angle of attack a few more degrees to lift the wheels on the main gear off to permit the airplane to roll.

I rapidly encountered two unexpected problems which I had not seen on the simulator. First, the roll control was too sensitive, too much roll rate as a function of stick force. Second, there was too much thrust being produced. We found later that the exhaust nozzle on YF-16 #1 was wired incorrectly. You had to be on the ground for the nozzle to be wide open, so as soon as you took the weight off the wheels, the nozzle closed and essentially doubled the thrust at idle. Immediately upon rotating the second time, the airplane lifted off with the left wing dropping rather abruptly. I rapidly applied a right roll command and the airplane again rolled far more rapidly than I had anticipated. I wasn't aware that the outboard trailing edge tip on the starboard stabilator had impacted the runway until the tower told me it had. Again, I rapidly commanded a left roll and the left roll rate was very high once more, but not as fast as I had first commanded.

After again reacting and commanding a right roll, I realized that I was entering into a PIO (Pilot-Induced Oscillation) *mode. Every time I tried to correct the oscillation; I got a full-rate opposite roll. And the airplane was continuing to accelerate all the while because the nozzle had closed, even though I now had the throttle at Idle power. We had way too much Idle thrust for such an agile, light weight airplane.*

With all the bouncing on alternating left main gear and then right main gear, I couldn't prevent the airplane from wandering to the left of the runway center stripe and I felt that it was impossible to keep the airplane on the runway, even if the nose wheel could be quickly brought down to the runway, as I was getting too close to the left edge. At that point, I instinctively applied Intermediate power and allowed the airplane to climb out slowly in a shallow left turn.

Jim Rider called me from the tower and advised that I not cycle the gear up and to continue the left pattern and return to landing on 22L. Jim said it appeared I had some damage to the starboard stabilator but there were no chase airplanes up – because the local weather was degrading rapidly - to visually confirm how much damage. The airplane flew beautifully for the next five or six minutes and landing was a piece of cake.

*The GD engineers had the problem of control sensitivity solved by that evening. We just put in some logic (*into the flight control computer*) where you selected half-gain for taxiing, takeoff, and for landing. You want the controls to be sensitive for up-and-away, but you don't want that level of sensitivity down in the pattern.*

The nozzle problem took a little longer to square away and wasn't fixed until after the fifth flight – flown by Jim Rider. That nozzle position fix involved putting a change in the switching logic of the WOW (Weight on Wheels) switch to allow the nozzle to be commanded open with the weight off the wheels." (End of Phil's account).

The first official flight of YF-16 #1 occurred as re-scheduled on February 2, 1974, and the photograph next shows YF-16 #1 during the flight:

Official First Flight (Flt 1B) of USAF/General Dynamics YF-16 #1, February 2,
(Source: USAF Photography files)

The official first flight (referred by insiders on the program as Flight 1B) was flown by Phil Oestricher again at the controls. It might be of interest to note the nozzle position in the official first flight photo is in the closed (low exit area) position. A closed nozzle is an indicator that the engine is being called upon to produce a high thrust necessary for flight.

On the official first flight (Flight 1B) with Phil Oestricher at the controls, the YF-16 #1 exhibited almost perfect flight characteristics in all critical systems. The flight lasted for about an hour and attained a speed of 400 knots and an altitude of 30,000 feet. Phil performed a textbook takeoff on the official first flight and, after the test plan for that flight was accomplished (primarily assuring that all flight critical subsystems - flight control, propulsion, fuel system, and control surface deflections - were functioning properly), performed an equally perfect landing.

On the second and third flights, both Phil Oestricher and Neil Anderson reported back to General Dynamics Chief Engineer Bill Dietz that there had been some small but annoying un-commanded motions in the flight control system resembling rudder pulses when rolling at moderate to high roll rate. The YF-16 #1 had not yet been flown by any of the USAF pilots on the LWF Joint Test Team. Bill Dietz invited the LWF Project Engineer (the Author) to come to Fort Worth and to participate in an assessment of the cause(s) for the un-commanded motions of YF-16 #1. The next portion of this chapter describes what took place and what was found during that engineering assessment at Plant 4.

Uncommanded Motion of YF-16s

YF-16 #1 was well instrumented for monitoring flight control computer commands and inputs into flight control surfaces, and the instrumentation traces did show unusual command activity not commanded by the pilot but emanating from the flight control computer during roll maneuvering. A circuit board in the analog flight control computer was suspected to have a manufacturing flaw and had been removed and returned to General Dynamics in Fort Worth for inspection for any manufacturing defects. (A spare board was replaced into the flight computer on YF-16 #1 so as to not impact testing.)

The Flight Control Computers were a "make" product from the several "make or buy" components in the YF-16 all-electric fly-by-wire flight control system and were hand-made in the General Dynamics Electronics Department facility at Fort Worth under the watchful eye of the AFPRO (USAF Air Force Plant Representative Office) at the Plant 4 facility to assure their quality and reliability. The Electronics manufacturing area in General Dynamics Plant 4 was ISO (International Office for Standardization) qualified and certified to meet the strictest quality standards of the US Government military contracts with contractors producing components for military weapons systems.

The suspect circuit board was subjected to a series of stringent electrical tests both on the static bench and on the vibration table to determine if there were any defects in the assembly of the circuit board or in any isolated components populating the board.

Initially, no defects or flaws were found in the board. However, after approximately four hours of the board testing and while on the vibration table, an errant signal was detected in one trace circuit on the board and isolated the source of the errant signal to a set of four glass diodes in the errant circuit trace. Experience with similar type problems suggested that the flaw was related to the operating temperature of the various components on the circuit board rather than being of a mechanical cause (e.g., an improperly soldered connection).

Microscopic visual inspection of the suspect components (four glass diodes) revealed that one of the components was cracked. General Dynamics had made three extra circuit boards as process development boards and as spares for the flight control computer, and these too were inspected for evidence of the same glass diode having any crack in its structure and indeed one of the other boards was found to have the same glass diode with a crack at its mid-body almost exactly where the initial cracked diode was found to be defective. What was visually noticed in the magnification inspection of the two defective diodes was that there was a difference in conformal coating on one-half of the diode than that on the other half of the diode. That led to a review of how the conformal coating material was applied onto the completed, populated circuit board and revealed why there were different thicknesses of conformal coating on the two diodes found to be cracked.

Conformal coatings are specialized film-forming polymers that protect circuit boards and other electronic components from adverse environmental conditions and mechanical abrasion. These coatings conform to the irregular topography of an electronic circuit board, providing increased dielectric resistance, operational and structural integrity, and protection from corrosive elements, moisture, heat, fungus, and airborne contaminants such as dirt and dust. There is also a necessity to provide an additional insulative coating layer between circuits to prevent arcing or current bleed creating sneak circuits. The materials for the most common conformal coatings are acrylic, silicone, or urethane, and are applied either by spraying (manual or robotic on selective areas) on the circuit board or by dipping the assembled board into a heated vat of the conformal coating material. The manual spraying application method is used on small production runs whereas the robotic selective spraying or dipping method is used for larger production runs.

To illustrate what the cause of the problem was determined to be for the cracked diodes on the defective YF-16 #1 flight control computer circuit board, a diagram constructed by the Author is shown next for Reader's information.

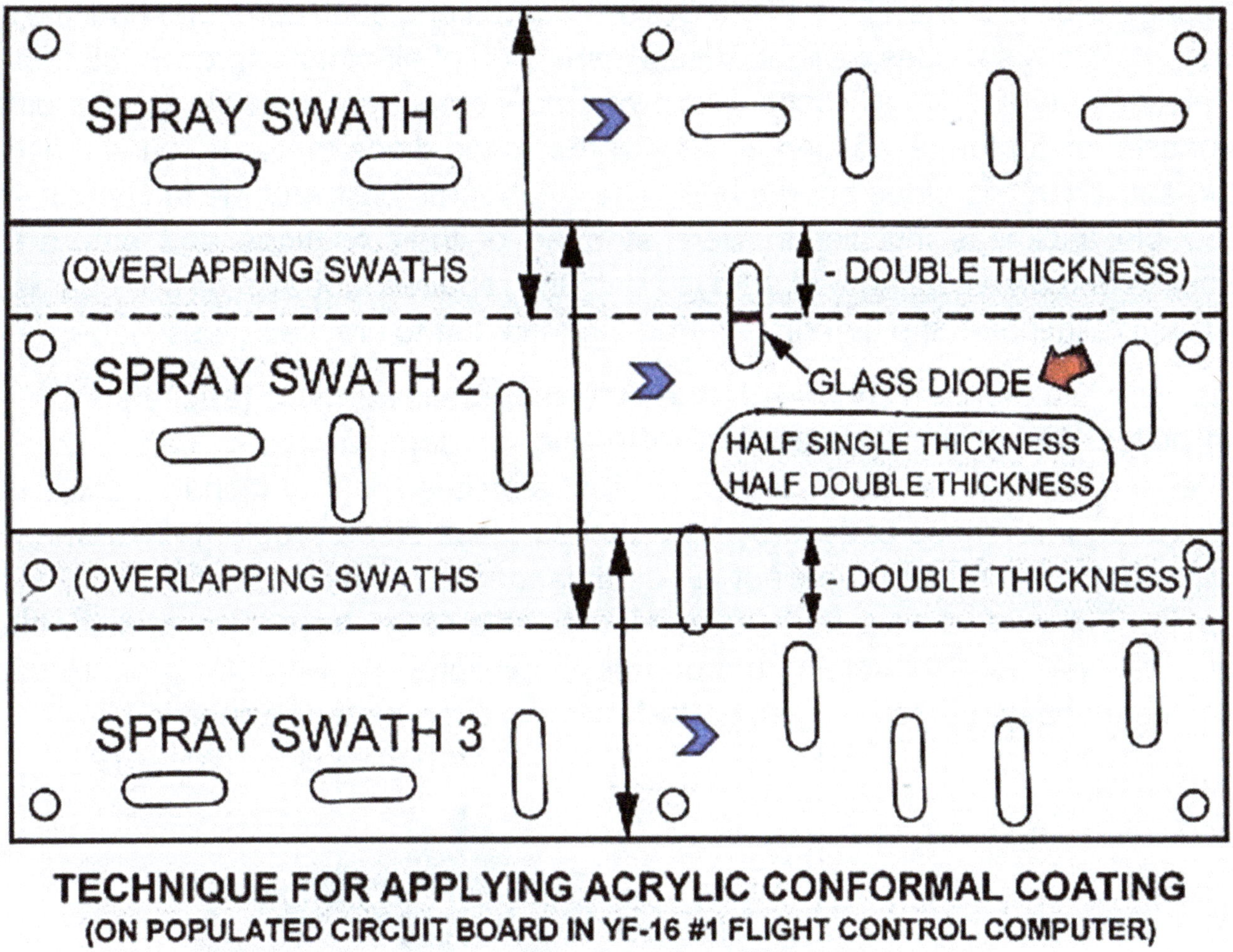

Diagram of conformal coating application
(Source: The Author)

The technician applying the conformal coating used the manual spray method (with a can of acrylic) and applied the spray in three swaths - spray swath 1 from left to right and covering the approximately top third of the circuit board. Spray swath 2 was then applied from left to right over the middle third of the circuit board, and spray swath 3 was sprayed from left to right over the bottom third of the circuit board. That spray pattern assured complete conformal coating of the circuit board. However, there were two horizontal overlaps on the circuit board where swath 1 and swath 2 overlapped and again where swath 2 and swath 3 overlapped. As shown on the diagram, almost all of the glass diodes on the board had received one thickness of coating and one diode had received a double thickness over its entire length. None of these diodes showed any evidence of cracking. The cracked diode, however, was half-way under the double thickness of the overlapping of swath 1 and swath 2, but its other half was only under the one thickness of swath 2. The cause of the crack was attributed to the different coefficient of thermal expansion due to the different thicknesses of the coating and the crack occurring where there was that large difference in coefficient of thermal expansion.

In the original build of the YF-16 Flight Control Computers, each computer had five circuit boards for different modes of flight. Visual inspection of all other boards in the four flight control computers built at Plant 4 showed no components receiving two different thicknesses of conformal coating except for the single diode that was found cracked at its mid-section in two of the boards built. The five boards built with the faulty half-single, half-double thickness installation were stripped of their coatings and only a single thickness conformal coating was sprayed onto the repaired diode on all five boards built. That repair eliminated the un-commanded motions due to cracked glass diodes.

As the flight test program continued, when there were elevated Gs (above about 6-7Gs) being pulled during envelope expansion testing, un-commanded motions again were experienced and were more severe in the amplitude of rotational motion in roll maneuvering than those caused by the cracked glass diode problem. Instrumentation showed that the problem was not un-commanded motions from the Flight Control Computer, but instead were faulty command signals at the bayonet connector into the integrated servo actuator serving the port (left) stabilator of YF-16 #2. A photograph of an Amphenol bayonet-type connector is shown next in a mated connection.

Photo of a typical bayonet connector
(Source: USAF Engineering files)

When access was made to inspect that connector, it was determined that the connector was not locked into its fully mated rotational position. Accordingly, the pins of the male half were intermittently making and losing contact in the female half to cause faulty signals into that port servo actuator and resulting in uncommanded stabilator deflections – not unsafely large, but not insignificant by any means.

When discussing with ground crew maintenance personnel as to the cause for the connector not being in a locked position, several comments were made that resulted in enlargement of the access panel to that connector. The comments of significance made were that the suspect connector was located in a difficult location to access through the original small access panel and could only be reached by one hand, the smaller the hand the better, to make the mating of the two halves (female and male shells).

There is an index stud and cam slot in the connector shells so that the connector pins are mated in the correct relative correspondence. When a bayonet connector is mated, it is recommended that only a torque be applied to the loose half (not a compression force), and the torque applied to that half will rotate the loose half on a built-in cam such that the rotation pulls the pins of the male shell into the female pin receptacles' half of the connector shell. When the two shells are in a locked position, a wavy-washer internal to the female shell will cause an audible snap to indicate the locked connector. In addition, there is a small diameter (about 1/8") hole in the outer shell and on the inner shell is painted a high visibility red dot also about 1/8" diameter. When the connector is properly locked, that red dot can be seen through the hole in the outer shell. In summary then, there are three indicators when a connector is properly mated and locked: (1) the rotation of the outer shell will come to a stop; (2) the wavy washer will produce an audible snap; and (3) the red dot will be visible through the hole in the outer shell.

The maintenance crew members stated that the connector found unlocked was in such an inaccessible location that there was no means to visibly see the red dot when the connector is locked, and if there is other noisy maintenance ongoing at the time or an engine running on an aircraft nearby, the snap of the wavy-washer may not be heard. Third, was that that type connector had an unusual "feel" characteristic unlike other connectors in that the rotational motion of the outer shell did not feel like it had come to a complete stop, but instead acted only like there was increased friction at the final portion of its motion travel. Because of that third comment, the connector (both shell halves) was removed for inspection of its internal assembly.

Rotation of the uninstalled connector shells was not producing the audible snap when the connector halves were fully mated, and the rotation did not physical cease motion – only an increasing friction as it was torqued/cammed toward closure. Internal inspection of the connector by the connector manufacturer (not named herein) revealed that the wavy washer had been installed backwards. The additional friction but no snap was caused by that improperly installed wavy-washer. The connector manufacturer traced to where other connectors made in that same Lot number had been sent and found that about half of that entire Lot of connectors had been improperly assembled (by a new assembler not properly instructed on the connector assembly requirements for the orientation of the wavy-washer). Replacement of the faulty assembled connector with properly assembled ones eliminated the uncommanded motions of the YF-16s – until gun firing tests.

During the initial ground gun-firing tests, there was detected some vibration during the gun-firing entering into the three-axis accelerometer assembly component located near the gun installation. Although the six-barrel Gatling 20-mm gun (Vulcan) when selected to fire at a high rate of 600 rounds per minute (10 rounds per second) does not produce large amplitude vibration, it does produce low-amplitude, high-frequency vibration (much like the "buzz" sound of a chainsaw cutting wood). Installing vibration-absorbing non-metallic mounts under the accelerometer assembly eliminated that potential source of un-commanded motion had it been the source of the un-commanded motion which occurred in flight. However, when subsequent in-flight gun-firing tests were initiated, there was a definite un-commanded yawing (nose-right) motion, which caused gun targeting inaccuracies.

As explained earlier in Chapter 8, when describing the flight control system having a GRI (gun rudder interconnect) function installed, the recoil force of the gun-firing process being 30 inches to the left of the fuselage center line caused a nose-left yawing moment during the short (1.0 – 1.5 second) gun-firing cycle. A reminder of the M61 A1 Vulcan 20 mmm 6-barrel Gatling gun system assembly is shown next:

M61 A1 Vulcan 20 mm barrels offset 30 inches left from fuselage
(Source: USAF Engineering files)

The ammunition drum (large cylinder shown in diagram) was located very close above the YF-16 fuselage center line and the distance of the center of the 6 barrels end clamp was very close to being 30 inches to the left of the fuselage centerline.

The GRI was designed to deflect the rudder an appropriate deflection when firing the gun in the most probable gun-firing arena (Mach 0.9 at 30,000 feet altitude) to cancel out the recoil-induced nose- left yawing moment. In the flight test gun-firing, the GRI function performed as designed at Mach 0.9 and 30,000 feet but still exhibited a small residual nose-right yawing moment. This was initially thought to be an excessive deflection of the rudder to counter the nose-left yawing moment. However, gun-firing at Mach 0.9 at 20,000 feet caused a disturbingly nose- right yawing moment which took the HUD aim-point (sometimes referred to as the Death Dot) far off the target. Gun-firing at 40,000 feet caused a noticeable though less disturbing nose-right yawing moment. Corrections were subsequently made to the GRI in the amount of rudder deflection needed as a function of altitude and Mach number variation when gun-firing at different combinations of Mach and altitude other than the design Mach 0.9 at 30,000 feet; however, there was still an unexplained nose-right wandering at the different gun-firing Mach-Altitude conditions that were not canceled out by the refined GRI rudder deflection correction.

A clue as to what was the additional factor not being sufficiently accounted for is shown in the photograph next.

Gun-firing residue on side of fuselage
(Source: USAF Photography files)

It is readily seen in the photo that there is a residue somewhat elliptical in shape ahead of the gun muzzle blast deflector which is deposited after each gun-firing event. It was rationalized that there had to be a pressure in the gun-blast bubble ahead of the muzzle to deposit that residue on the side of the fuselage - and with that rationalization came the explanation for the additional yawing moment being nose- right being experienced by pilots. The pressure in that elliptical area had a component acting perpendicular to the side of the fuselage and was ahead of the aircraft center of gravity (by a distance of approximately fifty to sixty inches) thereby causing a substantial nose- right yawing moment.

The size of the elliptical shape and the pressure on the side of the fuselage was then realized to be a function of Mach and altitude of the gun-firing event, and was opposite to the correction needed in the rudder deflection imposed by the GRI in the flight control system. That revision was made in the GRI function of the flight control computer and not only does it improve the steadiness of the HUD piper on target but the lesser rudder deflection due to the two opposing effects of gun-firing actually results in a smaller amount of aerodynamic drag due to the lower corrective rudder deflection angle.

Readers should recognize from the three accounts above describing the causes for un-commanded motion that there is an absolute need for high quality components being used in aircraft with all-electric fly-by-wire systems. With easy-to-make revisions in the analog flight control computer in the YF-16, another example is provided of the technologies developed for one aircraft being applied to the benefit of other aircraft currently under development and to be developed in the future. It all further vindicates the Dep. Sec. Def. Packard philosophy of “fly before buy”.

A third problem area arose of un-commanded motion in the first few flights of YF-16 #1 in which pilot commands were not at fault and that was in the engine controls area.

P&W F100 Engine Fuel Control Anomalies

A problem that was experienced in the YF-16 early flight testing was one that was not expected to occur. The YF-16s were equipped with the same engine as in the F-15 Eagle, the Pratt & Whitney F100-PW-100 (2). The first indication that something had possibly not been properly transferred from the F-15 Program to the USAF Lightweight Fighter Project was experienced by Maj. Bob Ettinger when, on his first flight in the YF-16 #1 (late February 1974), the F100 engine throttle control would not respond with any thrust above *ground idle*. Bob taxied YF-16 #1 back to the YF-16 hangar in *ground idle* power and shut the engine down.

Maintenance ground personnel inspected the throttle assembly for the cause of the problem, but did not find any problem with the throttle assembly. Subsequent flights in early March did not experience a repeat in the engine control problem. There was unfortunately an information transfer problem created by the personnel in the Pentagon who were very much against providing any information from the F-15 development program to the LWF Program for fear that somehow it might divert some funds or effort away from development of the F-15 Eagle. If any engine throttle being unresponsive incidents had been occurring in the F-15 program, those events had not been made known to the LWF Project.

On March 25, 1974, Maj. Dean Stickell was flying his initial flight in the YF-16 #1 and was at altitude above Edwards AFB when the engine rolled back into *flight idle* thrust level. Dean landed dead-stick at AFFTC. Again, postflight inspection of the engine control system revealed no cause for the engine rpm to roll back in thrust level.

In the third known instance of the problem, on May 16, 1974, L/C Jim Rider was at about 15,000 feet above AFFTC and the engine rpm rolled back once again into *flight idle* and did not respond to throttle commands. Jim landed the YF-16 #1 on Runway 1A on the dry lake bed. The YF-16 #1 was impounded and a pause in the flight program was declared necessary to determine the cause of the problem. To jump right to the major findings of the flight safety review aspects of the problem, the engine fuel control system was removed from the aircraft and a suspected fuel flow metering valve in the engine fuel control system was internally inspected at the manufacturer's facility. Contamination in the form of aluminum chips (left over from when the valve had been machined) were found to be blocking the metering valve and restricting the fuel flow through the metering valve. Cleaning and replacing the offending valve eliminated that particular problem from recurring during the remainder of YF-16 (#1 and #2) flight testing. The findings were fed back to the F100 JEPO (Joint Engine Program Office) at Wright-Patterson AFB and to the Author's knowledge did not reappear in any YF-16 or F-15 flights during their development flight testing with the P&W F100 engine.

In addition, a control was added to the pilot's cockpit by which the pilot could in effect bypass the engine electronic control and command an engine thrust level directly. This added control was named as being a dial-a-thrust.

Emergency Power Unit Anomaly

Being a single- engine design, the General Dynamics YF-16s had an emergency power unit (EPU) installed from the early conceptual design stage of the program. It should come as no surprise that when a single-engine aircraft loses an engine for any reason, the flight path of the aircraft will take one general direction – down. A pilot of an aircraft with a basic mechanical-type flight control system will more often than not be able to find a suitable emergency landing place and maintain control of the aircraft to make a safe emergency landing. The loss of the engine in the single engine YF-16 design results in a doubly emergency situation in that when the engine rpm spools down it no longer causes a gear-driven main generator to generate sufficient electrical power (low or no voltage) to the flight control computer in the all-electric (no mechanical backup) fly-by-wire flight control system. An EPU was installed in the YF-16 propulsion/electrical power subsystem to provide the pilot with sufficient electrical power into the flight control computer to maintain control of the YF-16 when loss of the engine and/or a failure of the main generator had occurred.

The design of the EPU was basically a rocket-powered motor spinning a rotating armature winding in a magnetic flow field and producing electrical power – until the rocket propellant fuel was consumed. The fuel powering the EPU was hydrazine, an ammonium-based liquid contained in a small tank. In the YF-16 application, the hydrazine tank was sized to hold a volume of hydrazine capable of sustaining the rocket motor spinning and producing sufficient power for the YF-16 flight control computer for 10 minutes of operation. Ten minutes of EPU operation was computed to be the length of time a pilot would have available to descend from Mach 0.9 at 30,000 feet altitude and make a controlled, safe emergency landing. The EPU was designed to be automatically activated only when in flight and by a sensor switch that is closed when the engine rpm was sensed to be spooling down at a rate more rapid than when a normal decrease in engine power (and associated decrease in engine rpm) is commanded by the pilot.

On a routine test flight in YF-16 #1 and while in flight at about Mach 0.9 and about 30,000 feet altitude, the pilot (Maj. Bob Ettinger) was presented with an unusual problem, loss of the flight control computer function. Initially, Maj. Ettinger considered a failure of the engine-driven main generator, but displays did not show any loss of electrical power from the main generator. The EPU of the YF-16 began operating despite the engine rpm not spooling down so rapidly as to initiate activation of the EPU due to rapid decrease of the engine rpm.

Further, the EPU system was not controlled by an on/off switch for the pilot to be able to either activate the EPU or deactivate it at his/her option. As a result of the unusual over-voltage supply, the flight control computer went into a triple failure mode of its quadruple mode capability – and only essential flight control functioning was provided to the pilot. Consulting with the Ops Tower as to what would be the safe procedure to be conducted, Maj. Ettinger was advised to maintain control of the aircraft as best he could for about ten minutes until the hydrazine fuel volume had been expended and then to reset the failed channels of the flight control computer. That procedure was done and the aircraft was brought back to a safe landing with all electrical systems – except the EPU – operating as designed. Inspection of the EPU system revealed that it was the failure of the engine rpm sensor in the propulsion system that had failed. The sensor was replaced and that EPU failure mode did not recur during the remainder of the flight test program.

The rapid post-flight conclusion derived was that the flight control computer was being supplied voltage both from the EPU and from the still-functioning main generator accessory attached to the engine. Flight control system engineers had designed the flight control computer to operate safely but with degraded capability at a low voltage level, and had designed an over-voltage circuit protection system associated with the high voltages of a lightning strike on the aircraft. However, the flight control engineers had not identified any failure mode that would generate a voltage level of twice the normal operating voltage, and, accordingly, had not designed any feature that would mitigate that type of failure mode for the flight control computer.

Valuable lessons learned from the EPU un-commanded on incident were that an EPU on/off switch for the pilot was a mandatory design change for further flights of the YF-16. Redesign of the EPU system in any later operational aircraft would benefit if the EPU could operate on engine bleed-air as a backup for the hydrazine mode of operation, and that adding a battery into the flight control system power circuit would also provide a safety margin.

Once the neonatal early flight tests unusual problems were resolved (un-commanded motion sources – due to the improper conformal coating process on flight control computer circuit boards, the gun-firing vibration isolation from the accelerometer assembly, and the improperly assembled bayonet connectors), and propulsion system related problems - the engine rpm roll backs to *idle* thrust levels, and the quirky single occasion of the EPU being activated by a faulty engine rpm sensor), the remainder of the YF-16 test program went quite smoothly.

YF-16 #2 was rolled out of the Plant 4 assembly facility on February 25,1974. To evaluate the detectability of the small size, lightweight fighter concepts, YF-16 #2 was initially painted in the blue and white camouflage paint scheme shown in the photograph. The paint schemes of YF-16 #2 later were changed to be a gray on gray, and also to red, white, and blue (identical to YF-16 #1). YF-16 #2 was loaded partially disassembled (in the same manner as was YF-16 #1) into a C-5A Galaxy for transport from Fort Worth to Edwards AFB where it was delivered on February 27,1974.

The flight test plan that the Joint Test Team (contractor and USAF pilots and AFFTC flight test engineers) had developed for the 300 flights planned for YF-16 #1 and #2 proceeded – at a slow pace due to the early neonatal problems, but accelerated when YF-16 #2 was delivered to AFFTC on February 27, 1974, and had its first flight on May 9, 1974 (with Neil Anderson piloting YF-16 #2 (tail number 01568). A photograph of YF-16 #2 on the Compass Rose at Edwards AFB, is shown next.

General Dynamics YF-16 #2, May 9, 1974.
(Source: USAF Engineering files)

The first obvious difference between YF-16 #1 and YF-16 #2 was the external paint scheme. Because the YF-16 #1 was to perform flight development tests to evaluate the basic flying qualities and to expand its flight envelope to that which is considered the limits of flight safety – of the pilot, of the airframe, and of the propulsion system, YF-16 #1 was painted in a patriotic red, white, blue which also made it easier to be chased by other aircraft at AFFTC and was instrumented for flight test engineering purposes to evaluate functionality of on-board systems.YF-16 #2 was designated to be the operational factors test aircraft (gun-firing, external stores launching, and maneuvering capabilities) and was instrumented accordingly.

Attention is now turned to the Northrop YF-17. Readers are reminded that Northrop had responded to the USAF RFP with two designs for its Lightweight Fighter – (1) the single engine P610 model that was designed around the two-engine P530 Cobra but substituting one Pratt & Whitney F100 engine, preferred by the USAF as stated in the RFP; and (2) the original P530 Cobra configuration with two General Electric YJ101 engines re-designated as the P-600. The designation for the two-engine P600 was given to the P530 basic design when it was offered to the USAF as the alternate proposal to the Northrop proposal response to the USAF LWF RFP in January 1973. (The easy way for Readers to remember which Model P600 or Model P610 had what number of engines was an intentional designation that the P600 had two holes at the end of the 600 and the P610 only had one hole at its end.)

Both of the Northrop proposal responses were evaluated by the USAF LWF Proposal Review Committee, but the single-engine Model P610 was not pursued for further development as it had a few design flaws resulting from it being a down-modification from its optimized two-engine P530 Cobra basic design. That down-modification placed the nose landing gear directly forward from and in line with the single inlet below the fuselage fore-body. USAF evaluators of the P610 design in the Northrop proposal had unanimously expressed major concern with the nose landing gear kicking up FOD objects from the runway during takeoff and destroying the single engine.

A photograph of the YF-17 #1 on “the blue dot of honor” on its rollout day – which actually was an unveiling, not a rollout - is shown next. The YF-17 #1 had been pre-positioned in front of the visitors stands the night before and was covered with a large cover that was unveiled as the introduction of the YF-17 #1 to the public. Unveiling, rather than rollout, was brought about by the local neighborhood Homeowners Association and Hawthorne Municipal Noise Control Ordinance prohibiting the run-up of turbojet engines at the Hawthorne Airport, which is directly adjacent to the Northrop Aircraft Division facility where the manufacture and assembly of the YF-17 aircraft had been accomplished. The unveiling was done amid the Hawthorne High School band playing a most rousing rendition of “Off we go, into the wild blue yonder”.

Unveiling of Northrop YF-17 #1 Hawthorne Airport, CA, April 4, 1974.
(Source: USAF Engineering files)

Also because of the noise from jet engine prohibition by the City of Hawthorne, Northrop was not allowed to ferry flight the completed YF-17s from Hawthorne to the Edwards AFB for flight testing. As a consequence, Northrop transported each of the two YF-17s as they were completed on the flat bed of large trucks via the Los Angeles public street system and California Highway system to the Edwards AFB facility approximately 50 miles to the north from Hawthorne. The USAF Lightweight Fighter flight test program took on its competitive fly-off atmosphere when the Northrop YF-17 #1 aircraft (tail number 01569) was shipped (by truck from the Northrop Aircraft Division facility in Hawthorne, California – less than two miles from the Los Angeles International Airport to its west). YF-17 #1 arrived at Edwards AFB in mid-April 1974 where it was placed in the Northrop Hangar on Contractors' Row to be readied for its taxi tests and first flight.

The low speed and high-speed taxi tests of YF-17 #1 were in early June at Edwards AFB and were very successful. There were no exciting events occurring such as had been encountered with the General Dynamics YF-16 #1 and the infamous inadvertent first flight. With Northrop Chief Test Pilot Hank Chouteau at the controls, YF-17 #1 made its first flight on June 9, 1974, and the LWF fly-off competition was underway. A photograph of YF-17 #1 lifting off from the runway at Edwards AFB on June 9 for its first flight is shown next.

Northrop YF-17 #1 First Flight Take-Off Edwards AFB, CA (Hank Chouteau) June 9, 1974 (Source: USAF Engineering files)

For those interested in a comparison of the YF-16 cockpit (which was shown earlier) with the cockpit in the YF-17, a photograph of the YF-17 cockpit is shown next. The most obvious difference is the position of the control column "stick". The YF-17 stick was a conventional between the knee's location whereas the YF-16 had a no-displacement sidestick on the right console.

Cockpit of YF-17 showing conventional center stick, 1974
(Source: USAF Engineering files)

The Joint Test Team had created a flight test plan for both the YF-16 and YF-17 to provide the necessary flight test data in approximately 300 flights each over a one-year long flight test program ending in June 1975. Events will be described later which caused the needed flight test completion date forward to December 1974.

In late July, after YF-17 #1 had flown a few of its planned initial flights, the YF-17 #1 experienced an engine FOD incident. Joe Jordan was taxiing out to the active runway in YF-17 #1 when he heard a loud bang and the starboard engine status displays indicated the starboard engine had encountered probably a FOD ingestion. The following explains that while it was a FOD item that had been ingested, it was not a piece of FOD originating from the runway being thrown into the inlet by the nose wheel as the EIRT was concerned about.

General Electric YJ-101 Engine FOD Incident

Northrop had subcontracted with Continental Air Lines to provide experienced maintenance mechanics to maintain the YF-17s during their one-year flight test program at AFFTC. A mechanic had performed some maintenance activity in the nose wheel well of YF-17 #1 which involved removing the fasteners of an access panel. So as to not lose track of the fasteners removed, the mechanic used a bright-colored (rescue- orange) tape to which were attached the several fasteners and stuck the tape in a location above the removed access panel. The mechanic was unable to complete the maintenance action when his shift ended in the Northrop hangar on Contractor's Row, and he left for the day. The next morning a different mechanic was assigned to complete the original maintenance task in the nose wheel well.

Unfortunately, the second mechanic did not observe where the first mechanic had placed the access panel fasteners and went to the maintenance parts crib and drew out the number of fasteners needed to complete the maintenance. Later that morning, when the pilot (Joe Jordan) in YF-17 #1was in a taxi movement to the active runway, a loud bang was heard and the engine instruments indicated that the starboard engine was not operating correctly to continue the takeoff. Joe aborted the takeoff and shut down the starboard engine and taxied back to the Northrop Hangar on the port YJ101 engine only.

Using the engine borescope inspection technique, the damaged YJ101 engine was inspected and a few turbine blades had indeed been physically damaged by an unknown FOD (Foreign Object Damage) object. The inspection however revealed that the indentations on one of the damaged blades were the thread pattern of a fastener. In addition, there was a suspicious bright-orange color piece of what appeared to be tape about ¼" in diameter stuck to one of the other damaged blades. The maintenance crew on duty during that morning did not have any knowledge of how a fastener could have entered the engine and none recognized the small fragment of orange colored tape. However, when the afternoon maintenance crew member came on duty, he immediately recognized the small fragment of orange color tape as being very similar to what he had used in the nose wheel well the night before. Only then did the facts as to the source of the fastener causing the engine damage become known – and another lesson learned entry was made in the maintenance records.

Readers may recall that a major concern of the USAF EIRT when they reviewed the YF-17 design for first safety of flight readiness, the EIRT had identified their safety concern with the design of the nose wheel being a significant distance ahead of the engine inlets of the YF-17. The concern was that the nose wheel might kick up a hard FOD object from the runway and the hard FOD object causing engine damage. Another EIRT concern was that if the nose wheel was running through a wet runway a splash of water might be thrown into the engine inlets and cause a loss of engine power at a critical moment during takeoff. This maintenance error causing FOD of the engine apparently was not caused by the nose wheel picking up the fasteners and tape from down on the runway but rather was caused by the air turbulence within the nose wheel well separating the tape and fasteners from their inconspicuous location and they entered into the inlet on a trajectory from their location in the nose wheel well.

There were three senior USAF maintenance crew members assigned to the Joint Test Team to observe the contractors' maintenance activities from the perspective of what degree of difficulty was encountered during the routine servicing and maintaining of both the General Dynamics YF-16s and Northrop YF-17s and what type of tools were being used.

The need for Ground Support Equipment was also being observed and made a matter of record – not to be used in the sense of being any adverse factor needing to be included in the Flight Test Final Report, but merely to have as information should that design of the test aircraft be chosen to become an aircraft in the USAF inventory.

The damaged engine was replaced in the YF-17 #1 with one of the spare YJ101 engines, and that damaged engine was repaired at the General Electric Engine facility in Lynn, Massachusetts. After repair and certification for safety of flight inspection and testing, the repaired engine was returned to Edwards AFB as a spare. There were essentially no other adverse engine-related anomalies with the General Electric YJ101 engines during the remainder of the USAF LWF flight test program.

In July 1974, as interest in the USAF Lightweight Fighters mushroomed in European NATO Countries as perhaps the USAF LWF winner of the fly-off being considered as a replacement for the aging inventory of F-104 Star Fighters, the Office of Secretary of Defense assigned Dr. Malcolm Currie, The Deputy Director of Research and Engineering (DDR&E) to come up to speed rapidly on how the LWF Flight Tests were progressing. Dr. Currie visited the USAF LWF Joint Test Team flight test facilities at the Edwards AFB and was given familiarization briefings on the technologies embedded in both the General Dynamics F-16 and the Northrop YF-17, and then an at-the-aircraft tour to see the aircraft in the flesh. Shown next is a group of Northrop management officials and AFFTC Lt. Col. Jim Rider describing an important feature of the Northrop YF-17 Lightweight Fighter to Dr. Malcom Currie DDR&E.

JTT Director L/C Jim Rider briefing Dr. Malcom Currie DDR&E on YF-17, September 1974 (Source: USAF Photographic files)

The large keel structure separating the two engines seen in the photograph of the Northrop YF-17 was pointed out to be an ideal structural location for adapting an arresting hook (1) to isolate the two engines in the event one were to catch fire from internal failure and needed to be shut down, leaving the other engine to be capable of returning the aircraft to its base and (2) if either engine were damaged by enemy ground gunfire, the damaged engine would not throw fragments through that keel and destroy the other engine (a lesson learned from the Vietnam War).

Much like the Code of Honor practiced at US military academies, a policy of the USAF Lightweight Fighter Program Office was that no person, government employee (military and civilian) or contractor/subcontractor would intentionally or unintentionally reveal a problem that one of the LWF test aircraft and its design had encountered would be transmitted or revealed in any form to the competitor. To the Author's knowledge, none such offense ever occurred. The two YF-16s were based in a separate hangar (Hangar No. 3) at Edwards AFB and the two YF-17s were based in the Northrop Hangar on Contractors' Row. On only two occasions were the YF-16 and YF-17 aircraft even in close visual proximity to each other. A photograph showing the General Dynamics YF-16 #1 and the Northrop YF-17 #1 in a close formation flight together appears next.

YF-16 #1 and YF-17 #1 in formation over Edwards AFB, July 1974
(Source: AFFTC History Museum files)

The photograph is the only time the two competing aircraft designs were in flight that close, but the USAF History Museum office at Edwards AFB had requested the photograph, taken in July 1974, for part of its "History of Testing at AFFTC" records.

The more challenging flights to establish the full operating flight envelope and capabilities of the aircraft within that safe envelope were left to be accomplished toward the last-half of the planned flight test program as they were the more hazardous to be conducted and increased the risk of loss of the aircraft and pilot. The two best examples of the JTT accepting the hazardous challenge of testing beyond the estimated flight envelope limits (from wind tunnel tests) are those referring to the high angle-of-attack tests (which required the installation of a spin recovery chute) during a 6 weeks period in August and September 1974.

Wind tunnel tests of both the YF-16 and YF-17 in the NASA free-flight tethered testing and vertical (for spin susceptibility and recovery modes) wind tunnel facilities at the Langley Research Center were described earlier in the ground testing section of this book. The high angle-of-attack capabilities shown in the ground test facilities for the YF-16 external configuration were planned for repeat in flight testing at the Edwards AFB AFFTC during the LWF fly-off competition. YF-16 #1 was fitted with a spin recovery parachute assembly in the event the aircraft exceeded the angle-of-attack associated with entry into a spin mode. For the YF-16, the tests in the NASA wind tunnel facilities showed that at angles- of-attack greater than about 30-degrees, the YF-16 had a tendency to depart from controlled flight and to enter a spin mode. (It was for that reason that General Dynamics installed an angle-of-attack limiter at 28-degrees, which limited the YF-16 from pulling higher than the 9Gs at 28^0 angle of attack but which still provided the maximum maneuvering 9Gs desired for pilot safety during air combat engagements as well as for structural considerations). Shown next is the YF-16 #1 with a spin chute system attached behind the aft end of the YF-16 #1 – and above the thrust column from the engine nozzle.

YF-16 #1 with spin recovery parachute system attached, August 1974
(Source: USAF Photography files)

The second highly- hazardous flight test event was accomplished by General Dynamics test pilot Neil Anderson, and that was an attempt to fly an YF-16 to its maximum altitude, above 60,000 feet.

It was no small feat in itself to get a physically large size pilot (as Neil Anderson was) into the very tightly confined YF-16 cockpit, and with Neil wearing a pressure suit instead of the standard flight suit almost required the use of a shoe-horn. (Author's note: With pre-apology to the Alcoholics Anonymous group, professional Human Factors engineers, and readers of Scottish descent, a humorous little analogy overheard among the YF-16 cockpit design group was that "The only thing tighter than the YF-16 cockpit is a drunken Scotsman.")

A photograph next indicates that, for flight safety reasons, General Dynamics test pilot Neil Anderson was required to wear a high-altitude pressure suit when preparing for take-off to attempt achievement of the flight to 62,000 feet altitude. Considering that Northrop test pilot Joe Jordan had set the World altitude record of 103,395.5 feet in 1959 in a modified F-104 Starfighter, Neil posed the question/offer for Joe Jordan to be substituted for the attempt to exceed 60,000 feet in the YF-16. Understandably, Joe Jordan declined.

GD test pilot Neil Anderson readying for flight to 62,000 feet, October 1974 (Source: USAF Photography files).

The urgency of completing the flight test portion of the USAF LWF Program by late December was driven by the desire of then Secretary of Defense Schlesinger to identify to the European Participating Governments (Belgium, Denmark, Netherlands, and Norway) to know which design had been selected by the USAF as its next Air Combat Fighter. That end of December date being moved forward had little impact on the YF-16 test program, but was a major schedule challenge for the Northrop YF-17 which started its flight testing in June (YF-17 #1) and with delivery of YF-17 #2 in late August. The impact was that instead of Northrop having a full year (until June 1975) to complete the original 300 flights planned by the JTT for its combat relevant flight evaluation of the YF-17 capabilities in reality it had only until mid-December. That challenge was well met by the Northrop test team and working closely with the JTT pilots, plus capitalizing on the Air National Guard providing refueling tanker support, the two YF-17s flew 288 flights in the 6 months allotted to them and all significant flight test objectives were met.

Readers should recall that when the USAF LWF contract was awarded to both General Dynamics and to Northrop in 1973, it called for the design and test of two aircraft designs as advanced technology demonstrators. It was described earlier in this chapter that in the early July 1974 time period, the four European countries (Belgium, Denmark, Netherlands, and Norway) had informed Deputy Secretary of Defense David Packard of their high interest in the YF-16 and YF-17 designs as possessing much of the capabilities the four Consortium countries were searching for as a replacement to their aging and not fully capable F-104 Starfighter in the inventory of the Air Forces of the Consortium countries. A string attached to the Consortium countries interest was that it was contingent on the US Air Force proceeding into development of either the YF-16 or the YF-17 becoming the next fighter in the USAF tactical inventory.

The Consortium countries had also informed DepSecDef Packard that since the F-15 Eagle entering the USAF fighter inventory did not have an air-to-ground capability (Readers might recall the mantra of the F-15 Program being "not a pound for air-to-ground"). That factor and the high cost of the F-15 removed the F-15 from consideration by the Consortium as a contender in their F-104 replacement plans. With the capability for carrying external stores on seven external structural hard points built into the LWF prototypes, the Consortium countries saw the USAF Lightweight Fighters as having that multi-mode capability they were seeking.

The external weapons capability of the YF-16 was presented in an earlier chapter describing the design capabilities embodied in the YF-16 design. It is hastened to be added that the USAF LWF flight test evaluation at AFFTC did not include demonstration of those stores identified as ECM, Air-to-Ground Guided Missiles or Special Weapons as the weapons delivery and fire control avionics was very austere and enough to only demonstrate the launching of Mk 82 and Mk 84 bombs, launching of the AIM-9 missiles, and firing of the M61 A1 Vulcan 20-mm six-barrel Gatling cannon.

The air-to-ground stores carrying capability diagram for the YF-16 is repeated to refresh Reader's memory of that capability. The resemblance of the YF-16 in its frontal view to an attacking shark – minus only its teeth – is coincidental.

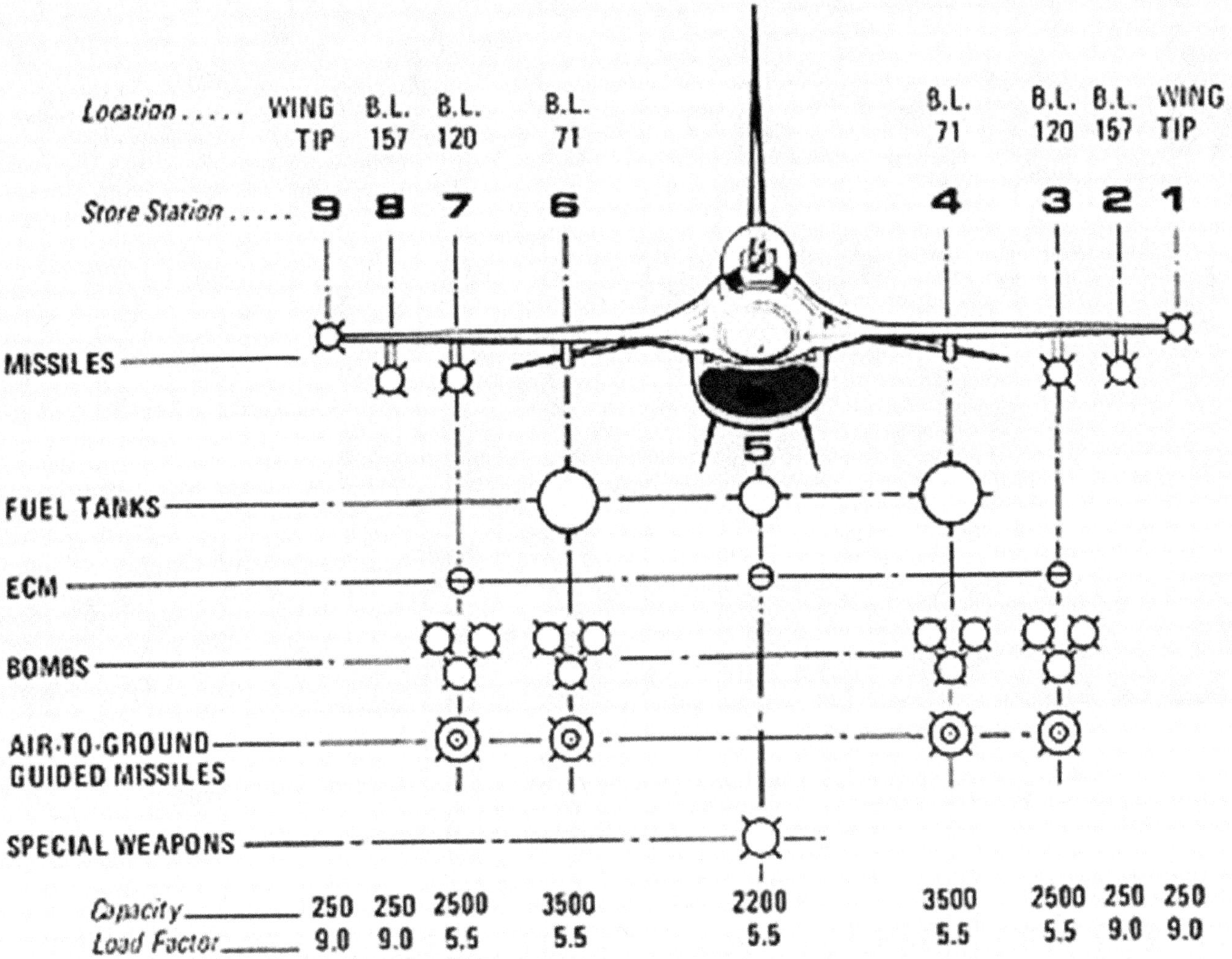

External stores carriage capability designed into the F-16
(Source: USAF Engineering files)

During the LWF flight test program, a substantial portion of the testing was to evaluate the maneuverability of the aircraft throughout the flight envelope established by early flight tests. Later in the flight test evaluation, flight testing was to evaluate the multi-mode capabilities (e.g., its performance while carrying and delivering/launching external stores). of the LWF aircraft. In the photograph shown next, a test was performed to visualize the difference in turn rate of the YF-16 compared to an F-4E Phantom II. Labels have been superimposed on the photograph to explain the conditions of the turn radius comparison.

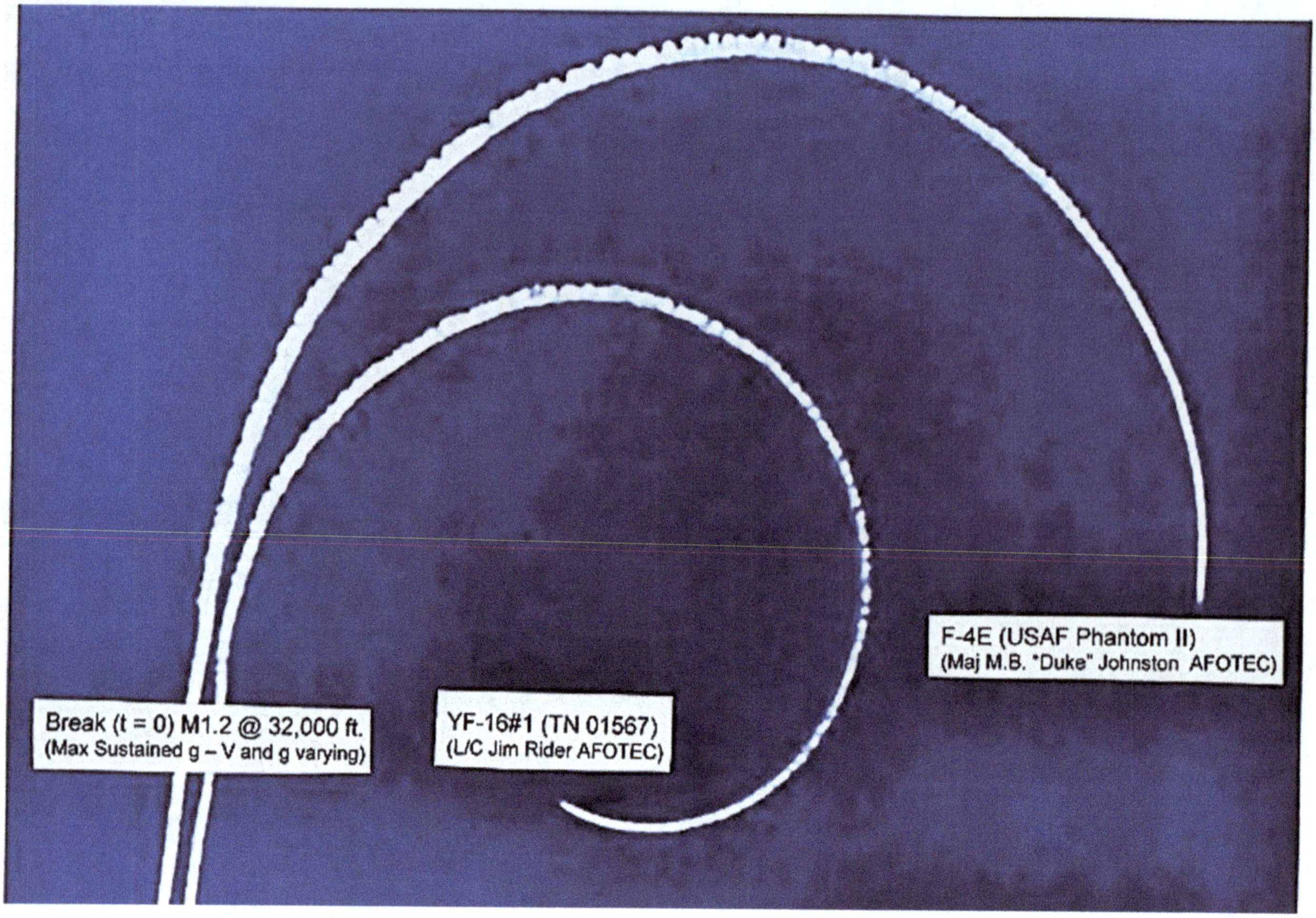

Comparison of turn rate of YF-16 #1 with F-4E Phantom II.
(Source: USAF Engineering files)

There were many combinations of external stores loadings also flight tested. The photograph next shows a heavily- loaded YF-16 #2 preparing for a flight test to evaluate the performance (maneuverability, range, and structural loads) of the YF-16 with the high-drag configuration, and to evaluate the external stores delivery accuracy capabilities of the YF-16 on a weapon testing range.

YF-16 #2 with large complement of external stores
(Source: USAF Engineering files)

Although there was no intention to create the Lightweight Fighter aircraft to have stealth properties (which would be adverse to its maneuvering performance capabilities and destroy its design-to-cost objectives), the YF-16 does have a very low visual detectability in lateral aspects by its small size and in its frontal aspect has a radar cross section (RCS) estimated as 1.2 square meters. In comparison, the F-15 RCS in the frontal aspect is 25 square meters, 20 times larger than that of the YF-16.

At the end of the USAF JTT flight test evaluation of the test aircraft in December 1974, the flight test portion of the contract efforts of both General Dynamics' YF-16s and Northrop's YF-17s were summarized cumulatively as shown next.

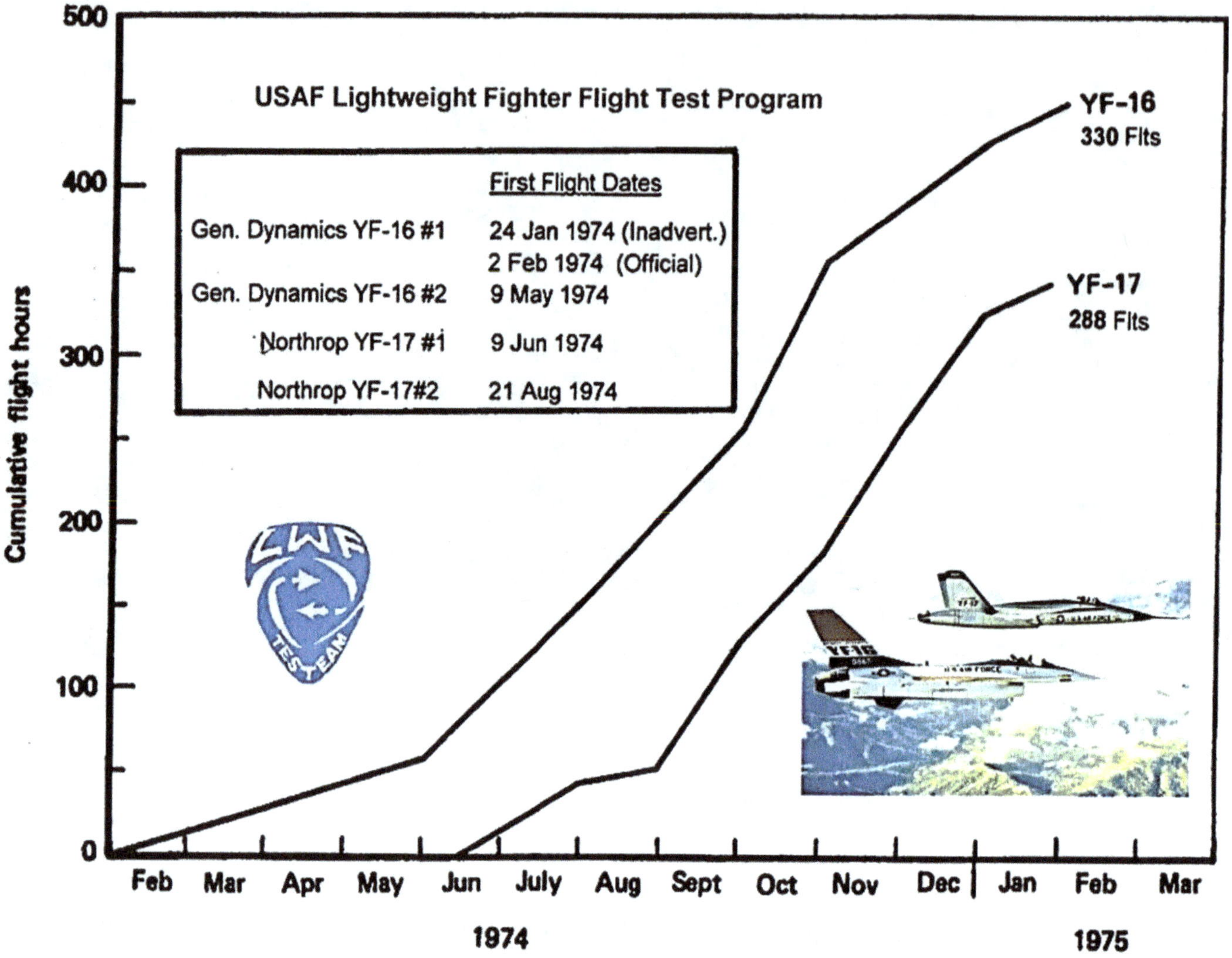

Summary of YF-16 and YF-17 flight test program
(Source: USAF Engineering files)

The chart graphically reveals the rate of accomplishment of the flight tests for both the YF-16s and YF-17s, and it should be noted by Readers that the number of flights per month when only the YF-16 #1 was flying (from early February to early May) were at a relatively constant rate of number of flight hours per month. When YF-16 #2 entered productively into the flight test program in mid-May, the rate of flight hours per month essentially doubled – as expected since two aircraft were producing flight test data - until early October. With YF-17 #1 starting the YF-17 flight test program in early June – 4 months later than the YF-16 – the flight hours per month of the YF-17 #1 from mid-June to late July were at a higher rate than the comparatively same first 6 weeks of the YF-17 #1 in an attempt to make up for the 4 months later start of the YF-17 flight testing.

When YF-17 #2 entered into the flight test program in late August, the rate of flight hours per month increased dramatically. This was due to two factors: (1) there were now two aircraft flying per day, and (2) L/C Rider, the JTT Director, had enlisted the support of the Air National Guard in providing refueling tankers to support primarily the YF-17s. It was based on his experience that once a test aircraft is airborne, the longer it is able to be tested on that same flight, the more useful data will be collected. By mid-December, the two YF-16s had completed the estimated 300 flights of the original test plan snd had accumulated approximately 400 flight test hours. In comparison, by mid-December the two YF-17s had completed approximately 300 flight test hours in about 260 flights. The entire operation at the AFFTC – pilots, flight test engineers, ground crews, administrative staff – had professionally accomplished what was described at the start as being mission impossible, but at the end of the flight test program was re-identified as mission accomplished.

As this book is focused on the YF-16 development, a more detailed flight test accomplishment chart for only the YF-16 part of the USAF LWF Flight Test Program is shown next and with notation of some of the highlight accomplishment as progress milestones during the testing of YF-16 #1 and YF-16 #2 which was completed in December 1974.

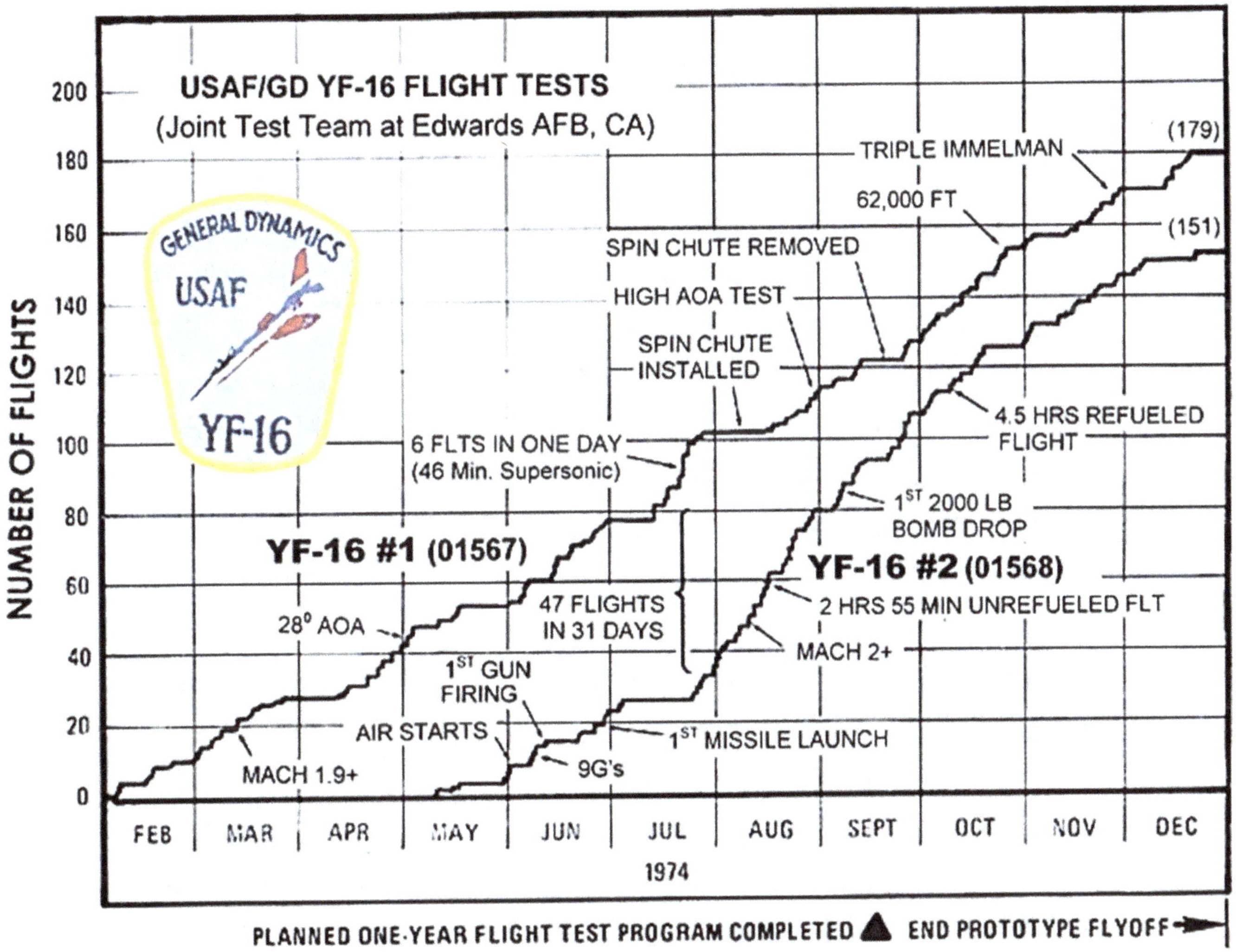

Summary of Both YF-16 Flight Tests
(Source: USAF Engineering files)

Enough credit cannot be given to, and respect expressed for, the ten pilots comprising the USAF LWF Joint Test Team for accomplishing all of the flight testing they did in such a short time period allotted for the test program. They were faced with a whale of a challenge but ate the whale one bite at a time. Not only did the JTT establish the safe flight envelope for the YF-16 (and YF-17) to the safe limits for the aircraft to be flown, but they had to explore the region outside the limit to determine by test just where that limit was in the actual flight regime. As shown on the previous chart, the events considered to be highlights of the test program are indicated by call-outs as they chronologically occurred.

Shown next is a tabulation of the YF-16 accomplishments with numerical statistics about those accomplishments:

YF-16 FLIGHT TEST ACCOMPLISHMENTS

First Flight Prototype No. 1	2 Feb 74
First Flight Prototype No. 2	9 May 74
Total Flights to Date	330
Total Flight Hours to Date	417
Total Supersonic Hours to Date	Nearly 14
Flights With No Discrepancies	169
Most Flights in One Day (YF-16 #1)	6
Quickest Turn Around Time Between Flights	11 Min 50 Sec
Greatest Number of Flights in One Month	47
Maximum Speed Attained	Over Mach 2
Maximum Altitude Attained	Over 60,000 Ft
Maximum "G" Attained	9
Longest Flight Without Refueling	2 Hrs 55 Min
Longest Flight With Refueling	4 Hrs 30 Min
Total Number of Air Refuelings	86
Total Number of Bombs Dropped	10 MK-84's
Total Rounds of Ammo Fired	12,948 Rounds
Total Number of Missiles Fired	7 AIM 9's
Number of Pilots to Fly YF-16	10

Tabulated accomplishments of YF-16 by JTT
(Source: USAF Engineering files)

Somewhat obscure in the earlier charts showing the call-outs of the many accomplishments demonstrated by the General Dynamics YF-16 Lightweight Fighters during their one-year flight testing by the JTT at AFFTC is that their low- maintenance and easy-servicing attributes (just two of the "ilities" important to a weapon system in an operational environment) made the aircraft always available for the flight tests planned for each day. A good fighter aircraft is one that has superior maneuverability, long radius of action, ability to pull 9Gs in turns to convert from being attacked into an air victory with guns or missiles, or to deliver a 2,000-pound weapon on a target with precision accuracy. A better fighter is one that can do all of that in several sorties on the same day instead of a fighter which becomes known as a hangar queen and can only perform one sortie per day at best.

Perhaps Readers might find it interesting to know that before the flight test program had started in early 1974, there was much concern in the LWF Prototype Project Office that there may be perceived that by having a fly-off between two aircraft designs with one design embodying an engine (Pratt & Whitney F-100) that was well along in its development while the other design embodied a prototype engine (General Electric YJ-101) which had almost zero hours of development might appear to many to be a stacking of the cards against the Northrop YF-17 design. Factually, there were more problems experienced with the engine electronic control system in the YF-16 propulsion system with the F100 engine than were experienced with the entire operation of the YF-17 with the prototype GE YJ101 engines. Except for the FOD incident with one of the YJ101 engines, the GE YJ101 performed magnificently well during the entire LWF flight test program.

Earlier in this book, this Author injected descriptions of the attention to "ilities" (reliability, maintainability, survivability, affordability, producibility, accessibility (for maintenance and servicing), etc.) which General Dynamics – Harry Hillaker and Bill Dietz and their engineering teams – required be designed into the YF-16 design concept from the first day of the USAF LWF contract award. The YF-16 was designed with over 100 access panels with quick-opening fasteners and with hinged cover plates that provide absolutely nothing toward improving its maneuverability or long radius of action legs. In fact, they may add up to increase the weight of the aircraft by a few pounds and cost a few dollars more to manufacture the aircraft; however, their existence on the YF-16s made it possible for the tightly scheduled flight test schedule developed by the JTT at AFFTC to accomplish 330 flights in 11 months – an average of 30 flights per month (1 per day), and 6 flights in one day, and 47 flights in 31 days.

The JTT Director (L/C Jim Rider from AFFTC) and JTT Deputy Director (L/C Duke Johnston from TAC) reported the evaluation of both the General Dynamics YF-16 design and the Northrop YF-17 design to the ACF Source Selection Evaluation Committee at Wright-Patterson AFB in late December 1974. The JTT and his Deputy reported that both Contractors LWF designs had been given a thorough and fair flight test evaluation.

On January 12, 1975, Secretary of the Air Force John McLucas announced that the General Dynamics YF-16 Lightweight Fighter had best met the performance goals set by the USAF Lightweight Fighter Program Office at the start of the USAF LWF Program.

Secretary of the Air Force John L. McLucas announced publicly on Monday, January 13, 1975, that the General Dynamics YF-16 LWF design had been selected for continued development as the USAF Air Combat Fighter. On January 14th, Secretary of Defense James Schlesinger held a news media conference and commented on the selection at the time: “It is a happy circumstance that the airplane with the best performance is also the lowest cost.”

Although it was a day before the breaking news of the selection of the General Dynamics YF-16 to be continued on into development as the USAF F-16 Air Combat Fighter, the New York Times on Sunday, January 12, 1975, ran a very well-written page starting A *multibillion-dollar fighter plane decision is due from the Pentagon this week. That decision and the manner in which it is carried out will have a major impact on the United States leadership in the development and production of fighter aircraft for the Free World.* Shown next is a copy of the upper portion of that page in the New York Times chronicling one of the most important events in military aviation history.

THE NEW YORK TIMES, SUNDAY, JANUARY 12, 1975

POINTS OF VIEW

Take-Off of a Crucial Aircraft Program

General Dynamics' YF-16, top, and Northrop's YF-17 are more than just improvements on current fighters.

By ALBERT W. BLACKBURN

A multibillion-dollar fighter plane decision is due from the Pentagon this week. That decision, and the manner in which it is carried out will have a major impact on United States leadership in the development and production of fighter aircraft for the Free World. Knowledgeable observers will be looking for indications that the Pentagon has learned from the

eral Dynamics on a Navy version of the YF-16, while McDonnell Douglas designers in St. Louis plan a Navy model of the Northrop fighter.

Indeed, the program now looms so large that serious consideration has been given to putting both aircraft into production regardless of which wins the Air Force competition.

But there have been prospects of similarly large programs before. In the late 1950's, North American Aviation was the envy of the

TFX go-ahead—10 years earlier in the case of the Skyhawk.

What are the elements of a military aircraft program that determine its success or failure? How is it that on one hand we can spend billions of dollars only to come up with a turkey while on the other, with substantially less investment, we are able to develop a long-lived world-beater? What is to be learned from recent history in this regard?

The Senate subcommittee concluded that

program, Gen. Bellis politely but firmly diverted the visit to his office at Wright Patterson Air Force Base. He made it clear that as long as he was in charge of the program, the team he had chosen to do the work was not going to be disturbed by itinerant probers, however well-intentioned.

A closely related criterion for determining the quality of a program as complex as that for development of a new fighter aircraft is the establishment of well-focused goals. Mul-

Extract of New York Times LWF Competition Article
January 13, 1975

The Author worked closely with the General Dynamics team throughout the entire Lightweight Fighter Program and it is to their honor a photograph (shown next) of the General Dynamics management and design/manufacturing/test team and the LWF YF-16 design that team created.

General Dynamics Management and Design/Manufacturing/Test Team with YF-16 #1
Dec 1973 (Courtesy: Gen Dynamics)

In retrospect, the atmosphere of a "best effort" type of contract without imposition of a multitude of military specifications having to be met allowed every member of the General Dynamics team to use his/her professional talents, experience and creativity to produce the YF-16 Lightweight Fighter. The highest of credit is also given to those in HQ USAF who eventually – after some coercion from the Office of the Secretary of Defense – to approve the unique management approach taken for the Lightweight Fighter Program. It also speaks highly of the value of their being a mutual trust established among all players on the team (whether at the contractor or at the government office sponsoring the contract effort) that a best effort applies to not only the team management but to everyone on the team.

The General Dynamics Corporation published a newspaper monthly to keep its employees informed on what is happening within the world of General Dynamics – and fittingly the name of the newspaper was *GD World*. The above photograph of the General Dynamics team was shown on the lower half of the front page of the February 1975 *GD World*. Shown next is the top half of the front page of the February 1975 issue of *GD*

World informing all employees throughout the General Dynamics Corporation that General Dynamics/Fort Worth had won the USAF Lightweight Fighter competitive fly-off and would be the Prime Contractor for full scale development (FSD) of the USAF F-16 Air Combat Fighter.

GD World

Vol. 5, No. 1 — PUBLISHED BY GENERAL DYNAMICS CORPORATION — February 1975

GD F-16 WINS; USAF CITES AIRCRAFT PERFORMANCE

Secretary of the Air Force John L. McLucas said that "many advantages in performance" and overall cost savings of the F-16 were major factors in selecting the Fort Worth-built aircraft for development as the U.S. Air Combat Fighter.

In announcing the $417.9 million

visibility and better acceleration. The YF-16 met all the performance goals we had established for it."

Secretary of Defense James R. Schlesinger, at a Jan. 14 news conference, called the Air Force selection a "happy circumstance in which the aircraft with higher performance happened to pro-

The developmental planes ordered by the Air Force will be used to evaluate the new air-to-air, air-to-ground fighters under operational conditions, prior to full-scale production. They are expected to be the forerunners of 650, or more, production aircraft to be ordered by the Air Force. Potential orders from

GD World (Company Newspaper) February 1975
(Courtesy: Gen Dynamics)

The YF-16 Program Manager Lyman C. Josephs (5th from left in the team photo) wrote a lengthy article for the February 1975 issue of the *GD World* reflecting on how the unique "adaptive management" and "design-to-goals" approach adopted by the USAF for the Lightweight Fighter Program allowed the General Dynamics design team to use their experience and technical talents to design the YF-16 as an advanced technology demonstrator but with rudimentary operational capabilities – and do so with "best effort" toward achieving challenging design and cost goals identified by the LWF Project Office at Wright-Patterson AFB, Ohio. The next challenge would be to "missionize" that advanced technology lightweight fighter into an operationally capable air combat fighter (ACF) and be effective in a weapon system for use in the inventory of many countries.

Only a part of the article by the GD YF-16 and F-16 ACF Program Manager Lyman Josephs will be extracted from *GD World* for this book, and was as follows (verbatim):

Because of the way in which the Air Force ran the competition, without specifying rigid requirements, and because of the contract latitude to make design tradeoffs and to truly exercise design responsibility, we did not have an overly optimistic proposal which would require us to substantially redesign the airplane. We decided, or rather, I believe the credit really goes to Bill Dietz, that we liked the design we had proposed the way it was; so why not get on with the job of constructing it and flight testing it. This is probably the most important reason that the program has been so successful. Another and perhaps almost as important reason was the way in which we actually managed the program.

During the competition, decision making was delegated to the lowest level possible. No one waited for ratification of a decision but decisions were reviewed after the fact. Because of the high experience level of the managers, few mistakes were made and corrective action even on these was easy. Communication was facilitated by holding a daily meeting at 8 a.m. of all project supervision. This meeting was open to all customer personnel who might be visiting at the time. No problems were unresolved before the personnel involved went home for the night. Either a solution was found or a course of action agreed upon.

Efficient as our management team was, it was certainly matched by the Air Force Prototype Program run by Col. Lyle Cameron (now Brigadier General) and later by Col. Bill Thurman. They ran the office with about only a handful of people reporting to them. Bill Thurman's essential attitude was: 'Our situation is new and different; we'll keep it simple and we'll make it work. You're free to do what your contract said you would and if you have any problems, I'm here to help.' Herb Hutchinson (the Author), *who had Prototype Program Office technical responsibility for the airplanes, and who was Bill Dietz' counterpart, was a pleasure to work with. Fred Wood, who was one-to-one with Blaine Scheideman, certainly met his responsibility to ensure that we fulfilled the terms of our contract.*

Innovation and ingenuity were encouraged. Dan Addison, who managed the tooling effort, invented really austere soft tooling, and, as a matter of fact, he is probably the real hero of the cost control activities. He underran his budget nearly 30 percent. And yet the quality of the fit of the various assemblies and piece parts of the two prototypes was really remarkable. Weldon Bailey, who was in charge of airframe design, and his people came up with novel ideas of using Styrofoam, cardboard, and rope to make accurate mockups of the airplanes various subassemblies and wiring routings. It has been estimated that this technique saved us millions of dollars when compared to the hard mockup technique used on a lot of other airplane programs. Tom Collins admits that the challenges by the purchasing people to some of his design (fly-by-wire flight control system) approaches caused him to redesign in order to meet the purchasing budget without sacrificing quality or performance.

One of the important reasons for the speed of opening up the flight envelope on this program was that in the airplane's structural design it had been decided early on (by Harry Hillaker and Bill Dietz) to incorporate a 125 percent safety margin in the aircraft's critical strength structure. In addition, it was decided to design to a 1.2 factor with respect to the flutter margin instead of the normal 1.15 margin. This permitted the airplane to be tested with confidence that no catastrophic structural failure would occur and also allowed us to go all the way to limit design load without a static test article. We believe the penalty in the structural weight was on the order of only 150 pounds – well worth it.

Included in the February 1975 of the General Dynamics newspaper *GD World* and following the commentary from a management perspective by the GD YF-16 Program Manager Lyman Josephs were articles by Bill Dietz, Director of YF-16 Engineering (commenting on the technical design aspects of the YF-16 program), and by Phil Oestricher, the GD Chief Test Pilot (commenting on the flight test aspects of the YF-16 program).

Excerpted from the Bill Dietz' comments published in the February 1975 issue of *GD World* were his following retrospective observations:

The YF-16 prototype program was the kind of program that an aerospace engineer dreams about participating in, but rarely has the opportunity. The contract required that we design, build, and test the two-prototype aircraft substantially in accordance with the 60-page proposal submitted to the Air Force at the start of the Lightweight Fighter Program. The 'substantially in accordance with' clause allowed complete latitude in making technical decisions without voluminous paperwork as justification.

Most aerospace engineering people are given the opportunity to participate in relatively few first flights of the aircraft they have designed. I can say that the first flight of the YF-16 was the most thrilling of the five I have seen, likely due to my personal close attachment to the program and the fact that the airplane is small enough so that I felt I had a personal acquaintance with most of the nuts and bolts. Most of us have developed the habit of calling the Air Force officials we do business with 'the Customer'. The term 'customer' has a certain impersonal attachment to it and does not represent our true feelings generated during our association with the Air Force program and engineering people on the Lightweight Fighter Program. They are a fine dedicated group, motivated to do the best job they can for the U.S. Air Force and the program. Though they didn't always agree with us, they were always agreeable to discussion and if necessary to consider compromise. They too must be given major credit for bringing the program to fruition and share in its success.

The Maintenance people deserve special commendation. Their first interest was always in ensuring that the airplanes were safe to fly. They never complained when requested to 'take a second look just to be sure.' They took great relish in seeing which crew could accomplish the shortest turnaround. That environment of competition pervaded the entire program externally and internally. The Air Force maintenance people were most constructive in their suggestions and a number of improved maintenance features for the operational airplane can be attributed to their efforts."

Pilot Phil Oestricher's comments in the February 1975 issue of *GD World* were directed toward the flight test aspects of the YF-16 Lightweight Fighter Program, and were cryptic recollections of his and his Joint Test Team associates' experiences during the YF-16 flight test program: *"I credit the success of the overall test program to the Lightweight Fighter Joint Test Force and the entire General Dynamics team. The pride of being associated with the program and the incredibly enjoyable experience of flying the YF-16 stand out in my mind. I remember January 20 (1974) when the high-speed taxi test turned into an unexpected flight.*

(Readers of the Inadvertent First Flight section presented earlier in this book were given both the set of observations made by the Author as to what was externally seen going on as they rapidly occurred, but more interesting was the set of comments by pilot Phil Oestricher himself describing what went through his mind – and what he did with his hands – as the unusual sequence of events unfolded in actual time.

(Continuing with Phil's statements in the *GD World*):
I also recall other incidents and people who participated in the flight test program:
- I heard an Air Force photographer, after filming a YF-16 full-power takeoff and climb vertically to 30,000 feet before even reaching the end of the runway, say 'I haven't seen anything like this since missile shots at Vandenburg AFB.'
- You haven't really heard "The Yellow Rose of Texas" until you have been treated to Lt. Col. "Duke" Johnston's version grunted out while sustaining a 7g maneuver in the YF-16.
– Maj. Dean Stickell flew only 15 minutes on his first YF-16 flight when, in one of the very few instances of serious mechanical problems, the engine fuel control locked in Idle power. Because of his piloting skill and the YF-16's easy handling, the dead-stick landing Dean made on the main runway was smooth and safe.
– W. O. Richardson, a mechanic who doubles as an engine ground-run man, told me 'It was always nice to know that maintenance was not going to hold up the flying. There were just no outstanding problems. The simplicity of operating the engine during our ground testing and trimming made it an enjoyable task.'
– Test pilot Neil Anderson was most impressed with the visibility, commenting: 'You

can see your own contrail in flight'; and with the power of the YF-16: 'Nine seconds on the runway is an exhilarating takeoff'; 'Nine g is a whale of a turn'. "

On January 13, 1975 when the Secretary of the Air Force John L. McLucas announced to the public that the General Dynamics YF-16 Lightweight Fighter design was to be continued on to be developed in a Full-Scale Development Program as the USAF Air Combat Fighter, that decision was based, and stated publicly, as having been based on the Lightweight Fighter fly-off competition during 1974. SAF McLucas said that *there were significant differences between the YF-16 and the competing YF-17 in agility, in acceleration, in turn rate, and in endurance of flight. Other advantages included better tolerance of high g because of the tilt-back seat (30^0 seatback angle in the YF-16 vs 18^0 in the YF-17), better visibility, and had better met all the performance goals we had established for it.*

Readers should note the absence of any cost evaluation numbers in the SAF McLucas announcement, as cost of a production version of either Lightweight Fighter was contractually not a part of the Lightweight Fighter contract effort. The Secretary of Defense James Schlesinger quickly ratified the SAF McLucas decision saying it was "*a happy circumstance in which the aircraft with the higher performance happened to provide the lower cost. The performance of the YF-16 has been remarkable in terms of its agility, acceleration and turn rate. It has proved to be a development of which the Air Force is proud.* The cost information referred to by Secretary of Defense Schlesinger was provided from the Air Force to the Secretary of Defense along with the Transition Contract information showing how the winner of the Lightweight Fighter would look after "missionization" to an Air Combat Fighter configuration – cost, however, was not a major discerning factor in selection of the Lightweight Fighter flight competition winner.

The Author felt it might be of interest to read what the General Dynamics management printed in their General Dynamics Corporation GD World Company newspaper after learning of their being selected as the winner of the LWF fly-off competition.

But the USAF Lightweight Fighter story does not end there.

The General YF-16 design having been publicly announced on January 13, 1975, as the prototype of the new USAF F-16 Air Combat Fighter generated an incredible number of requests to the F-16 Air Combat System Program Office from numerous military organizations for information and/or to allow one or more pilots from those organizations to fly the YF-16 (either #1 or #2). A low number of the requests were honored (for Foreign Technology assessment purposes and for the USAF Fighter Weapons School purposes), but many unfortunately could not be.

Within two weeks after the announcement that the General Dynamics YF-16 had been selected as the winner of the USAF Lightweight Fighter fly-off competition, General Dynamics received an invitation from the sponsors of the 1975 Paris Air Show inviting General Dynamics to participate in the 31st Salon International de l'Aeronautique et de l'Espace at le Bourget Airport in Paris and to conduct flight demonstrations of the newly-named USAF F-16 Air Combat Fighter. Inasmuch as the YF-16 was a government-owned deliverable as part of the USAF Lightweight Fighter original contract, General Dynamics, properly, forwarded the invitation to the USAF F-16 Air Combat Fighter Program Office at Wright-Patterson. The Secretary of the Air Force rapidly approved the request and directed the F-16 Air Combat Fighter Program Office to provide all planning and logistics support to General Dynamics enabling the YF-16 (re-painted with F-16 on its vertical tail and displaying the flag symbols on the left forward fuselage for the USA, Belgium, Denmark, the Netherlands, and Norway. This was in anticipation that the four EPG (European Participating Governments) would formalize their order for 348 of the new USAF F-16 Air Combat Fighters during the Paris Air Show.

In April 1975, the National Society of Professional Engineers selected the YF-16 Lightweight Fighter Program as being one of the ten outstanding engineering accomplishments in the United States for the year 1974. A plaque shown below was presented to General Dynamics as recognition of that award from the NSPE:

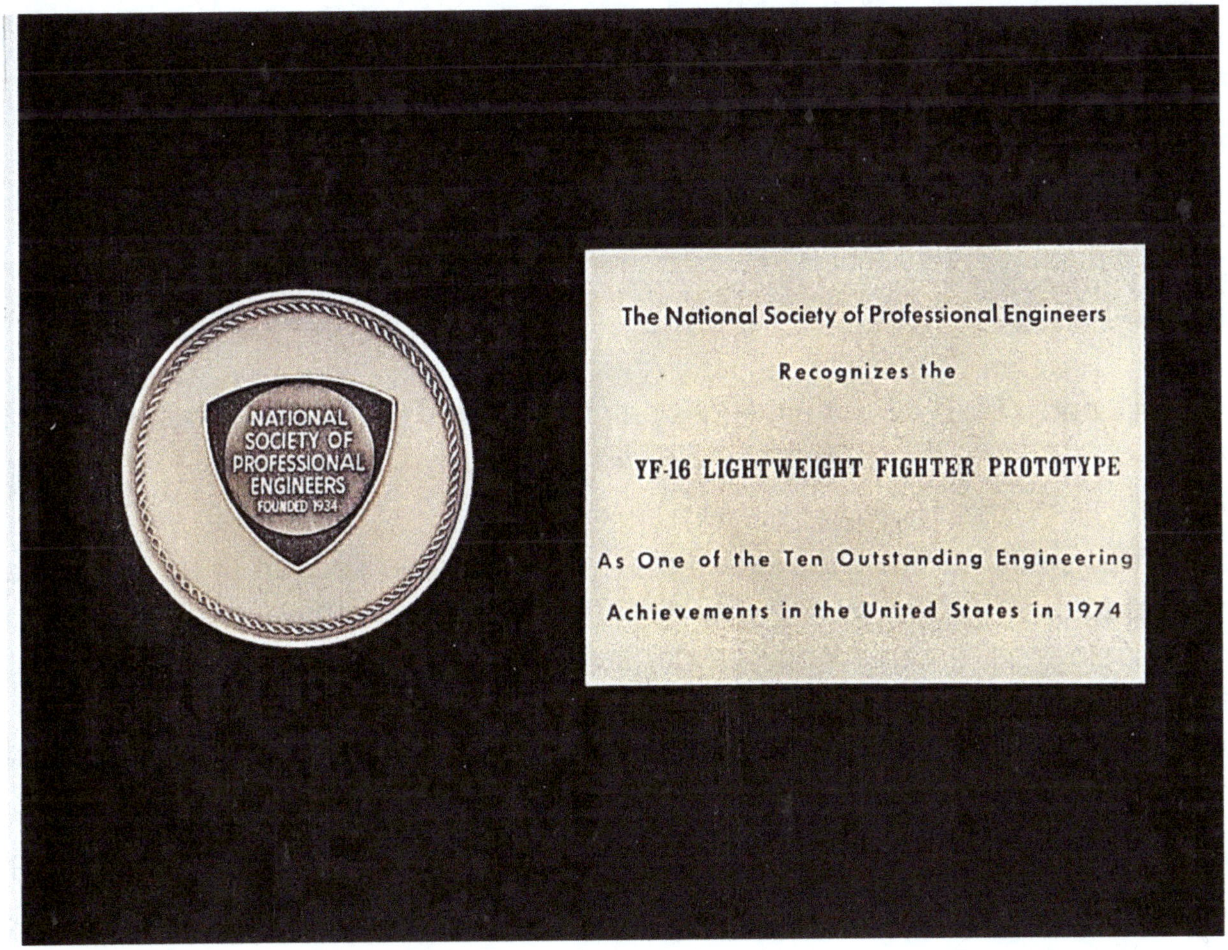

National Society of Professional Engineers award plaque to LWF Program, 1974

In the early-May 1975 time period, YF-16 #2 (tail number 01568) was re-painted in the same red, white, and blue patriotic paint scheme as YF-16 #1 had been and on the vertical tail of YF-16 #2 had been painted "F-16". Because General Dynamics gave the honor of first flight to Phil Oestricher in YF-16 #1 (on February 2, 1974), the honor of flying YF-16 #2 at the Paris Air Show went to Neil Anderson. Neil began practicing the flight demonstration routine for its appearance at the Paris Air Show scheduled to begin on June 7, 1975.

Shown next is a photograph of YF-16 #2 (F-16, piloted by Neil Anderson) in its new patriotic red, white, and blue paint scheme being accompanied (chased) by YF-16 #1 (piloted by Phil Oestricher).

YF-16 #1 and YF-16 #2 both in RWB paint scheme, Ft. Worth, Texas, May 4,1975
(Source: USAF Photography files)

Readers are reminded that the reason that YF-16 #2 was chosen to go to the Paris Air Show instead of YF-16 #1 (which was already painted red, white, and blue) was due to YF-16 #1 being instrumented more extensively for flight testing to define structural air loads, and there was additional testing for air loads planned to be done as part of the F-16 Air Combat Fighter development. YF-16 #2 was configured more for operational type testing/demonstrations as well as for the launching of external weapons.

Shown next is a photograph of the Author (USAF F-16 ACF Program Chief System Engineer) and several of the engineers from the USAF F-16 ACF SPO with General

Dynamics F-16 Chief Engineer Bill Dietz giving an impromptu Safety of Flight Review of YF-16 #2 in preparation for its flight to the Paris Air Show in early June.

Safety of Flight Review Team of YF-16 #2 for Paris Air Show, May 6,
(Courtesy of Gen Dynamics)

(Since it was the year 1975, the common men's style of attire is very obviously seen to be comprised of sporty polyester jackets and trousers with accenting contrast - intentionally).

In the photograph, the access panel on the starboard side of the nose can be seen to have been removed for maintenance of the minimal avionics components that were installed in the nose of YF-16 #2. The nose shape of YF-16 #2 is seen to be a pointy conical shape. That shape and size nose was not adequately configured to accommodate a radar system of the types designed by Hughes and Westinghouse and being tested in a fly-off competition (similar to the manner the YF-16 and YF-17aircraft were tested - in a fly-off competition.) The radar system fly-off flight tests between Hughes and Westinghouse were well underway in May 1975.

While practicing the Paris Air Show flight routine on May 8, 1975, Neil Anderson made a gear-up landing at Carswell AFB. To better understand the causes for a gear-up landing, it must be stated that at the Paris Air Show there is typically a low cloud cover at 500-1,000 feet above the Le Bourget Airfield. Accordingly, the demonstration aircraft perform their flight demonstration routines at relatively low altitudes just below the cloud cover for the spectators. On May 8, 1975, Neil Anderson took off in YF-16 #2 repainted RWB and F-16 on the vertical tail shown in the earlier photograph) from Carswell AFB in Fort Worth. (The immense General Dynamics manufacturing facility is on the opposite side of the main runway from Carswell AFB. The Author was employed at General Dynamics in B-58A Hustler Flight Test Engineering from 1959 to 1961 and knows almost every seam, gap, and bump in the Carswell AFB runway caused by the flight operations of the USAF Strategic Air Command flying heavily loaded B-52s and KC-135 tankers *very* frequently from that runway).

In a high-performance takeoff, Neil rolled the aircraft in a very high-rate roll while the landing gear mechanisms were still in a retraction mode. He had performed most of the demonstration routine when a chase aircraft advised Neil that his right MLG door was not closed. Neil aborted the demonstration routine and tried to lower his landing gear in preparation for landing back at Carswell AFB, but the right MLG would not extend. The nose and left MLG were extending normally, but even with pulling high Gs to coerce the gear down, the right MLG would not extend. Neil was advised by the tower to expend as much fuel as possible and to retract the operable nose and left MLG to clean up the bottom of the aircraft as much as possible before making a gear up landing on the grass strip adjacent to the main runway. With about five gallons of fuel left, Neil made the landing very tangentially on the grass strip. However, early in the slide the aircraft was bumped upward as it transited over a mound of grass-covered gravel/dirt (left over from a prior runway repair project) and then impacted the grass-covered flat surface on the underside of the inlet. The remainder of the slide went smoothly and the aircraft decelerated rapidly (due to the two ventral fins plowing two straight rows that any farmer would be envious of plowing manually). The two ventral fins soon sheared off and the aircraft slid approximately one hundred feet further and stopped.

Neil made a quick egress from the aircraft and, as he ran from the landing site, dropped his flight helmet - which was then run over and demolished by a responding fire truck. He later related *Loss of my helmet was the worst part of the whole incident because I was wearing a specially- painted helmet to wow the spectators at the Paris Air Show.* There was no fire as the aircraft was extremely low on fuel. Shown next is a photograph of the YF-16 #2 being lifted from the grassy surface next to the Carswell AFB runway used for the gear-up landing.

YF-16 #2 (without ventral fins) after gear-up landing, Carswell AFB, Texas, May 8, 1975 (Source: USAF Engineering files)

(Yes, there were comments overheard that the technology of high-lift-devices on fighter aircraft had indeed come a long way in the recent decade.)

The damaged aircraft was inspected and inspection revealed that the right MLG wheel was tightly jammed against the wheel rotation snubber preventing the entire assembly from being extended. A photograph below shows that minimal damage was done to the underside of the inlet, and when the inlet duct impacted the ground after being bumped upward and then down onto the grass–covered hard surface, the inlet had been compressed and bent the thin inlet strap as shown in the photo next.

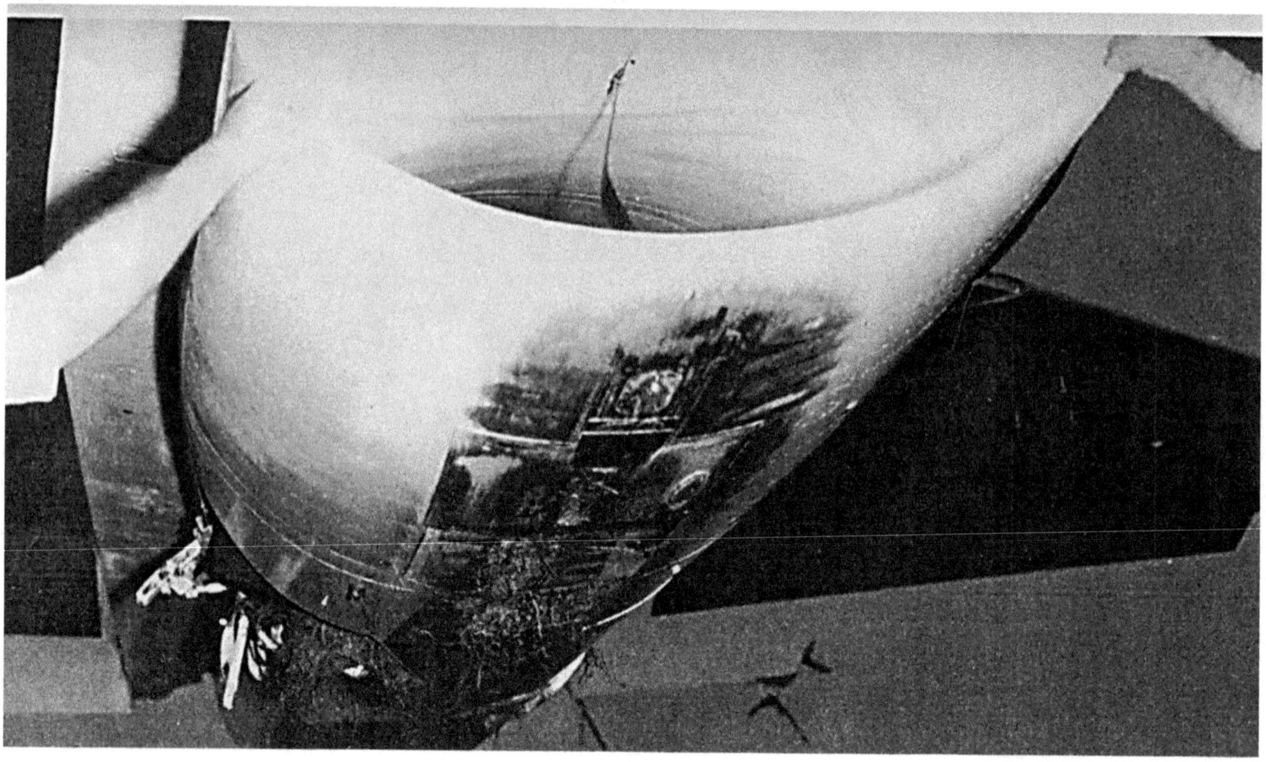

Underside fuselage and inlet damage incurred in gear-up landing, May 8,
(Source: USAF Engineering files)

Not visible in the photograph is the internal damage done to horizontal stringers near the top of the cockpit side rails due to the high stress loads when the under fuselage impacted the ground. There was no debris ingested into the inlet and therefore no engine damage had been incurred.

The root cause for the incident was determined to be the high roll rate of the aircraft being commanded while the landing gear was being retracted. With the right MLG wheel spinning rapidly, gyroscopic forces were induced into the gear retraction linkage and the linkage was deflected sufficiently to cause the spinning wheel to be compressed and jammed into the wheel rotation snubber.

The stress damage done to the stringers was considered to be severe enough to cancel all plans for the YF-16 #2 to risk being flown across the northern Atlantic to Europe for the Paris Air Show and to perform high Gz maneuvers in demonstration flying. As a substitute, YF-16 #1, which was already in the original red, white, and blue paint scheme, was quickly readied for the flight to and at the upcoming Paris Air Show in early June 1975. A technical decision was made, influenced by the marketing Manager at General Dynamics, to not strip out the flight test instrumentation that was aboard YF-16 #1 (and not in YF-16 #2) before it left for the Paris Air Show. This was agreed to as it would preclude the criticism, mostly by the French operators of the Paris Air Show, that the aircraft performing so magnificently for the spectators are well below the operational combat weight status when in the military operational inventory. The marketing ploy was to advertise at the Paris Air Show venue that the instrumentation left in YF-16 #1 is almost exactly the same weight as the operational equipment (radar and other avionics) planned to be integrated in the production version of the F-16 Air Combat Fighter.

General Dynamics also noted that after the gear-up landing (during which the YF-16 #2 had been bounced upward by the grass-covered mound and then impacted the ground with a significant rate of descent), the original (smiley) inlet casting of warped elliptical contour had deformed slightly flatter due to the unusually high compression loading of ground impact and bent the inlet strap during the compression loading on the inlet strap. The same compression effect on the inlet was known possible to happen during a rapid rotation of the nose gear in a hard landing as the nose gear is attached to the bottom side of the inlet duct just aft of the inlet lip casting. The inlet strap was originally designed to take tension loading and maintain the warped-elliptical shape of the inlet when subjected to high air pressure during high-speed flight and inducing the inlet to deform to a more circular shape.

As with almost every accident or incident involving a military item of equipment – and especially with aircraft accidents that tend to draw a disproportionate amount of media attention - once the root cause for the accident/incident has been identified by an Accident Review Board, there are usually several lessons learned documented. These Lessons Learned are fed back to the engineers and military specifications writers at Wright-Patterson AFB for inclusion in future specifications as "What to do to avoid a repeat of the accident or incident". (Specifications do not include "What not to do's.)

A Tee-slot shape attachment fitting for the inlet strap installation was designed (to enable the new strap to react tension loads but to not be subjected to compressive loadings). The inlet strap was made thicker, hollow, and with a symmetrical airfoil cross section. Redesign of the attachment for the inlet strap to the inlet casting led to a fixed attachment with a horn protuberance that inserted inside the new hollow strap designed to accommodate the tension loads when in flight and to accommodate the compression loads into the inlet casting from the nose gear when landing. A heater circuit was also added to the thicker strap for the F-16 Air Combat Fighter development program to avoid ice accretion when in flight and under icing environmental conditions. A photograph of the subsequent re-design of the F-16 inlet strap is shown below:

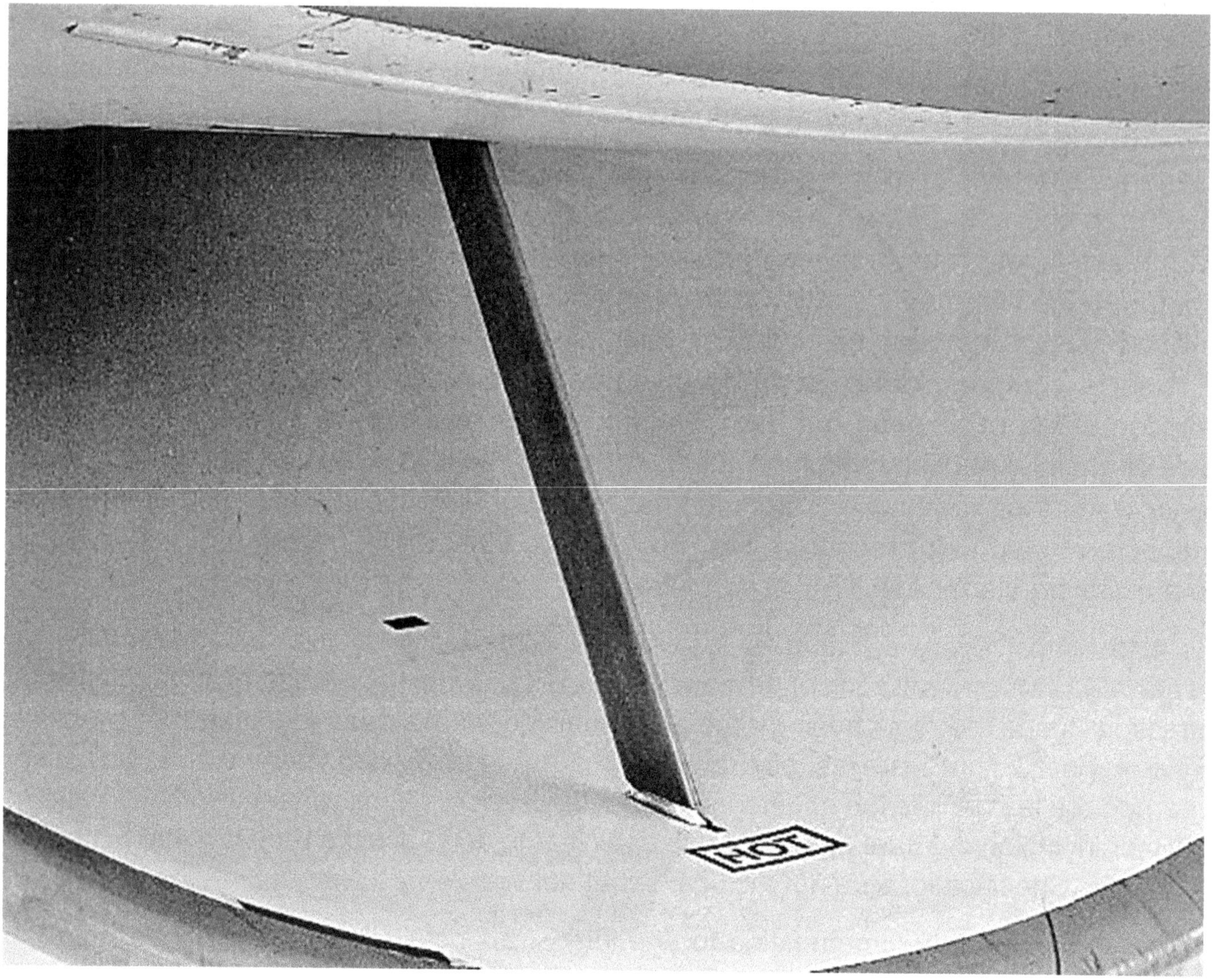

Inlet Tension Strap in F-16A & B Aircraft
(Source: USAF Engineering files)

YF-16 #1 made the trans-Atlantic flight to Paris for the Paris Air Show with no problems and performed exceedingly well, much to the spectator's great delight. In an Aviation Week and Space Technology Editorial ("A Good Show" by Robert Hotz) dated June 16, 1975, the following account was given: *Hit of the show was obviously the General Dynamics F-16, whose flying performance across the Atlantic and at the Paris Air Show certainly reinforced the wisdom of the four-nation NATO consortium (Belgium, Denmark, the Netherlands, and Norway) to standardize on this fighter as their Lockheed F-104 replacement. The flying performance of the U.S. fighters – Grumman F-14, McDonnell Douglas F-15, General Dynamics F-16, and Northrop F-17, if it had been there – showed clearly to the air force chiefs assembled from all over the world that they represent a new generation of burgeoning technology while their European competitors sadly represent the final refinement of an earlier generation. The NATO consortium decision also imposes a heavy responsibility on the U.S. Government to deliver the offset production, program management, and technical excellence they have promised. The F-16 flying was very much a team effort with Lt. Col.*

Jim Rider, commander of the joint flight test task force at Edwards AFB, making the long transatlantic hop from Pease AFB in New Hampshire to Ramstein AFB in Germany with five refueling drinks from a KC-135 tanker, and Lt. Col. M. B. (Duke) Johnston and General Dynamics test pilot Neil Anderson accompanying in an F-4 chase plane and then alternating in flying the F-16 during the show. It is worth noting, in view of some of the French critics' snide comments, that the F-16 flew with 4,500 lb. of fuel on board at a gross weight of 20,000 lb., within about 500 lb. of combat weight, and the 1,800 lb. of test equipment on board equaled the weight of future operational systems such as radar. The wingtip AIM-9 missiles also were full combat weight. The F-16 burned only about 1,800 lb. of fuel per demonstration. USAF wanted to supplement one demonstration with a couple of hours of high-altitude orbit to demonstrate endurance, but the French vetoed the plan. This was a very strong show for the U.S.. Great credit should go to those in the Defense Dept., Commerce Dept. and Federal Aviation Administration who backed the industrial components with full strength. Every contractor who benefitted from this presence at Le Bourget should get this message across to his specific legislators before some dim-witted bean-counters in the Office of Management and Budget kill this strong support for U.S. aerospace exports. – Robert Hotz

(The Author had been invited by General Dynamics to attend the Paris Air Show:

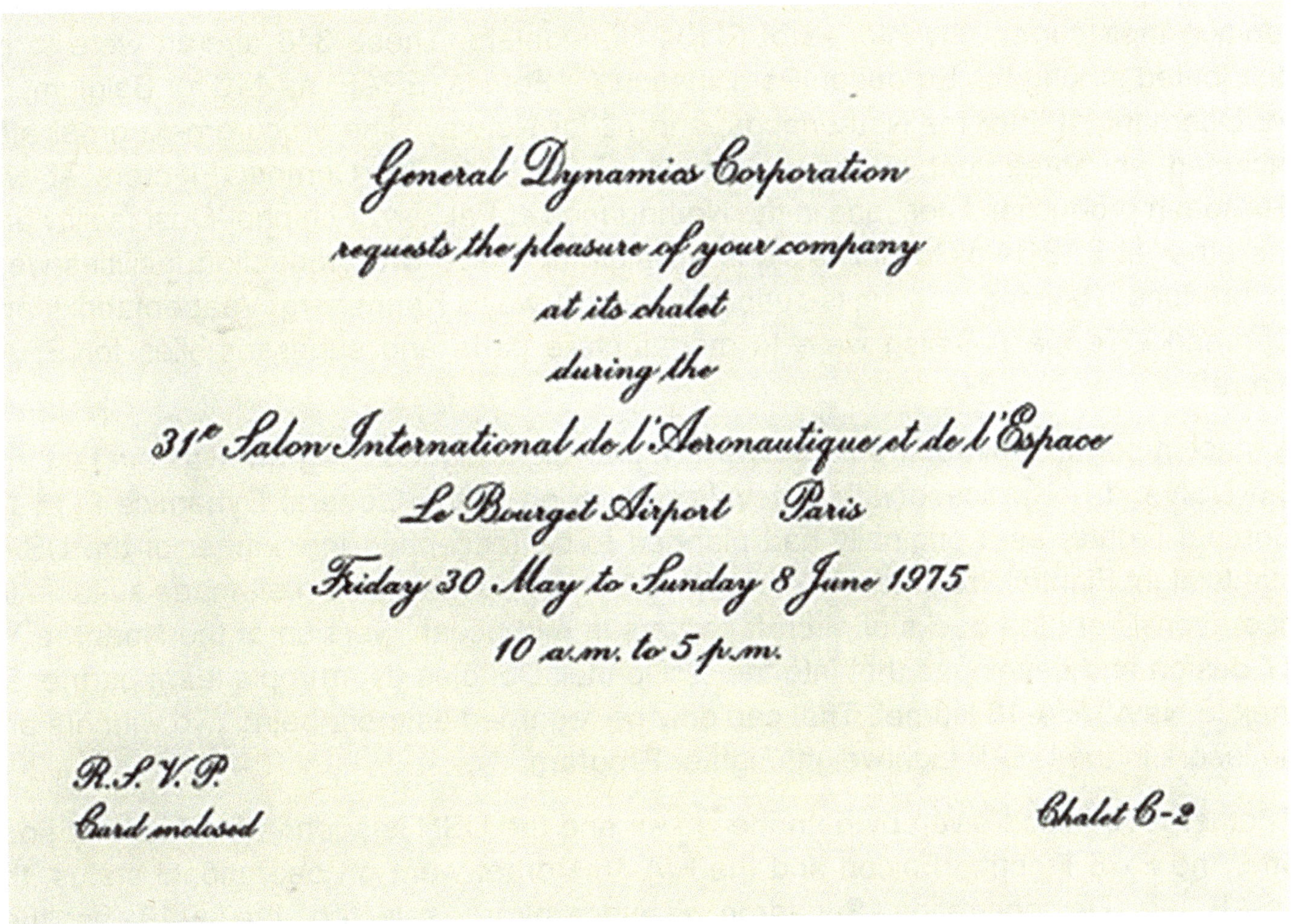

General Dynamics Corporation
requests the pleasure of your company
at its chalet
during the
31^e Salon International de l'Aeronautique et de l'Espace
Le Bourget Airport Paris
Friday 30 May to Sunday 8 June 1975
10 a.m. to 5 p.m.

R.S.V.P.
Card enclosed

Chalet C-2

Invitation to Attend 1975 Paris Air Show
(Author's memorabilia files)

However, this Author chose not to attend, as it was considered a higher priority to attend one of this Author's daughters High School graduation ceremony in Dayton, Ohio, in early June 1975. This comment is made to impress on young Readers who are at the start of their professional careers and perhaps also embarking on the start of a young marriage and raising of children, that there will be a constant need to balance their career activities with their personal life activities. It is often a difficult choice whether to attend a special event which is job-related but not mandatory or to instead attend a family event that can happen only once in a family lifespan together as a family. The art of successfully setting priorities comes primarily with experience and varies with each family situation. It was very troubling to witness the Flight Control System engineer assigned to the USAF F-16 ACF SPO – to be left identified only as JD - undergo a divorce because he was at the General Dynamics engineering facility for 99% of his time during 3 years and devoted 100% of his loyalty and time to development of the magnificent fly-by-wire system on the F-16 ACF. His efforts were extremely valuable to advancing fly-by-wire technology but were at the sacrifice of a normal family life.)

During the closing days (on June 7, 1975) of the Paris Air Show, the EPG countries (Belgium, Denmark, the Netherlands, and Norway) officially announced that they were placing a purchase order to the United States for 348 F-16 Air Combat Fighter aircraft to replace their current inventories of F-104 Starfighters. These 348 aircraft were to be distributed among the European Participation Air Forces (EPAF) as 116 for Belgium, 58 for Denmark, 102 for the Netherlands, and 72 for Norway. The procurement order also included an agreement for co-production of the F-16 Air Combat Fighters at two European production lines, one in the Netherlands at Fokker's Schiphol-Oost facility and the other at SABCA's Gosselies plant in Belgium. These two production facilities were to produce 184 and 164 units respectively. Norway's Kongsberg Vaapenfabrikk and Denmark's Terma A/S also were to manufacture parts and subassemblies for EPAF aircraft.

A most important spin-off from the USAF Lightweight Fighter Program was the fact that the USN did not proceed with a navalized version of the General Dynamics F-16 Air Combat Fighter as it originally had planned to do (accepting the winner of the USAF Lightweight Fighter Program). The US Navy selected a navalized (i.e., made suitable for operations from the decks of aircraft carriers in mid-ocean) version of the Northrop Yf-17 design and developed that into the McDonnell Douglas (Northrop's team partner for the US Navy) F/A-18 Hornet. That can only be regarded as there being two winners and no losers in the USAF Lightweight Fighter Program.

Another recognition given by both the USAF and the USN is seen in the fact that soon after the F-16 Fighting Falcon and the F/A-18 Hornet went on operational status, the USAF Air Demonstration Squadron (Thunderbirds) selected the F-16 as their Thunderbird demonstration aircraft, and the USN Air Demonstration Squadron (Blue Angels) selected the F/A-18 as their Blue Angel demonstration aircraft.

What can be a more convincing statement than the "past" (the Lightweight Fighter Program) having become the "prologue" for the F-16 Fighting Falcon being in the military inventory of twenty-six countries and the F/A-18 Hornet and Super Hornets being in the military inventory of six other countries?

As this book closes, it should now be realized by all those Readers who are embarking on a new career in a field of engineering now can appreciate that it is the human participation in advancing a new technology and applying it properly in a useful application. It is a matter of doing every task in that development of an engineering product that two main factors will determine the success of an engineering venture. First is the paying close attention to the small details involved in the venture and treating every challenge that arises as an opportunity to display your professional talents. Second is that every task should be done with a passion. Following that simple advice will gain you well-deserved recognition for your efforts.

As an example, and not as a form of bragging, the Author was given recognition by the Commander of the USAF Aeronautical Systems Division for his efforts on the USAF Lightweight Fighter Program. A copy of that recognition letter from Lt. Gen James Stewart to the Author for a job well done (an "atta boy") is shown next. Another reward for an engineer doing his best job is to receive some form or award that also puts bread on the table for his/her family. As a result of the Letter of Commendation from the Commander of the Aeronautical System, this Author was requested to remain on the job as the Chief System Engineer for development of the F-16 Air Combat Fighter into full scale development as the F-16 multi-national weapon system. A promotion to Federal Civil Service Grade GS-15 was given to the Author, and the accompanying salary increase reward for doing a good job was greatly appreciated – far better than getting a poke in the eye with a sharp stick.

At the risk of sounding corny, the Author felt it imperative that he do his best to complete the creative genius and thinking of John Boyd and Harry Hillaker from ideas on paper to actual hardware. The best award a person can get is knowing inside that he/she did the best job that could be done.

DEPARTMENT OF THE AIR FORCE
HEADQUARTERS AERONAUTICAL SYSTEMS DIVISION (AFSC)
WRIGHT-PATTERSON AIR FORCE BASE, OHIO 45433

Office of the Commander

2 8 OCT 1975

Mr. Herbert A. Hutchinson
Deputy for Engineering
Aeronautical Systems Division
Wright-Patterson AFB OH 45433

Dear Mr. Hutchinson (Herb)

It is a pleasure to award you the Air Force Systems Command Certificate of Merit. Your performance as Chief Systems Engineer, Deputy for the Air Combat Fighter/F-16 System Program Office from July 1974 to July 1975 is deserving of this recognition.

As a result of your dedicated efforts, the Air Combat Fighter made a successful transition from the prototype program, through source selection, into a rigorous full scale development program. The F-16 Air Combat Fighter has introduced many unique technological innovations into weapon systems development.

I commend you for this outstanding performance and wish you continued success.

Sincerely

J Stewart

JAMES T. STEWART
Lt General, USAF
Commander

Letter of Commendation from Commander of ASD
(Author's memorabilia files)

Ex-F-16 pilots (they refer to themselves as "Viper Drivers") who are close associates of this Author have been reviewing the contents of this book as it has been in the writing process for a few years. In a most honoring manner, those Viper Drivers have bestowed the tactical call name "Stuffer" on the Author in recognition for him having "stuffed" the insides of the F-16 Fighting Falcon (a.k.a. "the Viper") with all the equipment components that allowed them to take the Viper into hostile environments safely, 100% achieve the mission objectives, and with every expectation of returning safely to Happy Hour. One F-16 was shot down by enemy ground fire (a Surface to Air Missile) – and for which Hollywood produced the movie "Basher Five-Two" reenacting the actions taken by the F-16 pilot Capt. Scott O'Grady leading up to his rescue from behind enemy lines.

CLOSING REMARKS

It is hoped that Readers of this book will now be better informed as to what world geopolitical and military events occurring during the time period 1900 to 1975 did in advancing the technologies of flight for USAF fighter aircraft. Equally important is that Readers can now better appreciate how all the myriad of events, decisions, studies, and surfacing of technology advances foreign countries and allies made in fighter aircraft designs in what is known as the technology leap-frog game coalesced into the circumstances of how the USAF Lightweight Fighter program was created, when, and by whom.

This book was/is not intended to mislead any Reader into thinking that the only way a development program gets started by an USAF organization is for one dominant personality within the Air Force chain of command to use devious paths to get some funding authorized for the project by back door methods. John Boyd and his Lightweight Fighter Mafia associates saw that the conventional approach used in filling the USAF needs for a particular capability, such as that being filled by development of the F-15 Eagle air superiority fighter, had cast a shadow on the need for the capabilities of a lower cost – and more affordable – approach to complement the high-end F-15 development with a low-end multi-mode fighter. The F-15 Program was distorting the fact that the F-15 was evidently the best aircraft for accomplishing its specified air superiority capabilities but that it had sacrificed a credible air-to-ground capability to achieve that air superiority capability. Those in the F-15 Program Office who chanted "Not a pound for air-to-ground" were clearly soon identified as having the blinders on by the Boyd Lightweight Fighter Mafia. Ironically, the Lightweight Fighter Mafia had to some degree the same wearing of blinders flaw in trying to develop a fighter aircraft which had only a day, clear-weather operating capability with avionics capability limited to that only for a single mode of operation in order to keep its cost and weight to a low level.

By good fortune, there were a substantial number of other high-ranking officers in the USAF Development Planning and Acquisition Planning offices who saw that there were several advanced flight technologies that were emerging from the USAF Research and Development Laboratories that were of sufficient maturity that they deserved to be flight tested to validate their capability potential for five different flight vehicle concepts. This book contains copies of the more significant charts used by USAF Brig. Gen. Kenneth Chapman in a highly common-sense briefing in late July 1971 to Deputy Secretary of

Defense David Packard. Dep. Sec. Def. Packard totally agreed with the content of the B/G Chapman briefing and caused funding, completely separate from that needed by the F-15 Program, to be authorized for spending on those five-prototype flight vehicle prototype concepts – the Lightweight Fighter being one of those five air vehicle concepts.

In the Preface of this book was shown the two statues at the US National Archives which bore the inscriptions "Study the past" and "What is past is prologue". Readers of this book should consider this book as having studied the past from 1900 to 1975 and which led to the development of the two advanced USAF Lightweight Fighters – the General Dynamics YF-16 and the Northrop YF-17.

It would be incorrect to say that the Northrop YF-17 lost the USAF Lightweight Fighter fly-off competition to the General Dynamics YF-16. Both performed so well that the General Dynamics YF-16 went on to be developed into the USAF F-16 Air Combat Fighter and the Northrop YF-17 underwent navalizing changes to become the USN F-18 NACF and still later went on to be developed as the USN/USMC F/A-18 Hornet and Super Hornet. There were no losers in the USAF Lightweight Fighter Program. What better "prologue" can there be than to witness two military fighter aircraft – the USAF F-16 Fighting Falcon and the USN McDonnell Douglas F/A-18 Hornet and Super Hornet continuously operating in the US military fighter aircraft inventory for over seventy years? The two winners of the USAF Lightweight Fighter Program have been in the USAF and USN/USMC fighter aircraft inventories from 1976 through 2017 and both are projected to undergo service life extensions and upgrades that will see them remaining on operational status into the 2040s.

The "What is past is prologue" part of the story of the USAF Lightweight Fighter program, the USAF ACF (Air Combat Fighter), and USN NACF programs that ensued in 1975, and their production runs (from 1976 through 2017), is contained in a separate book by the Author entitled "The USAF F-16 Multi-National Weapon System (1967-2017)" to be published soon after this book has been published. Both books have been written as stand-alone books; i.e., readers can decide if they wish to learn only how an advanced technology demonstrator aircraft evolved from past historical and technology events. Other readers who are interested only in how the USAF Lightweight Fighter program evolved into becoming a major weapon system in the inventory of 26 countries may elect not to not be concerned about how the lightweight fighter came to be, but only how a prototype aircraft could become a major weapon system.

In fact, the last three chapters of the first book form the basis for the first three chapters of the second book. Readers who may wish to know the full story of the F-16 "from womb to tomb" may elect to acquire both of the books and read the overlapping three chapters only once in the two books.

It is the Author's hope that either or both books will serve to inspire persons newly embarking on a technical career to pay attention to the details of whatever it is they will be working on during their professional career. A parallel hope is that seeing how career paths never evolve the way as first planned will cause readers to be flexible in their career planning and to be in the right place at the right time to take advantage of opportunities that arise. Use common sense, and by no means ever stop thinking and use the flawed logic of "the computer says".

Contact the Author

Comments, questions, or constructive criticism about this book are welcomed. It is also welcomed for any young student contemplating embarking on a career in engineering to ask for career guidance suggestions by this "gray hair". Any and all correspondence is requested to be sent to this Author at herbineer@aol.com

APPENDIX 1

Biographies and/or Obituaries

of Key Individuals in USAF Lightweight Fighter Prototypes

(without whom the USAF F-16 may never have been developed)

John R. Boyd (and Lightweight Fighter Mafia)

and

Harry J. Hillaker

USAF Col. John R. Boyd

Harry J. Hillaker

Col. John Boyd Is Dead at 70; Advanced Air Combat Tactics

By ROBERT MCG. THOMAS JR. MARCH 13, 1997

Col. John R. Boyd, a legendary Air Force fighter pilot whose discovery that quicker is better than faster became the basis of a far-reaching theory that helped revolutionize American military strategy, died on March 9 at a hospital in West Palm Beach, Fla. He was 70 and had lived in Delray Beach.

The cause was cancer, his family said.

To combat pilots of the late 1950's, it was always high noon in the skies above the Nevada desert. A pilot, a crack instructor at Nellis Air Force Base, perhaps, or a hotshot Navy flier passing through would get on the radio to call him out and within minutes Colonel Boyd would have another notch in his belt.

They did not call him 40-second Boyd for nothing. From 1954 to 1960 virtually every combat pilot in the country knew that Colonel Boyd, a former Korean War pilot who helped establish the Fighter Weapons School at Nellis, had a standing offer: take a position on his tail, and 40 twisting, turning seconds later he would have the challenger in his own gun-sights or pay $40. Colonel Boyd never lost the bet in more than 3,000 hours of flying time.

A high school swimming champion who won an athletic scholarship to the University of Iowa, Colonel Boyd, a native of Erie, Pa., had superior reflexes and hand-eye coordination, but what made him invincible in mock combat was something else.

At Nellis he taught himself calculus so he could work out the formulas that produced his repertory of aerial maneuvers and led to his 1960 report, "Aerial Attack Study," the bible of air-to-air combat.

His combat experience was limited to a few missions in Korea, but they were enough to produce a breakthrough insight. Wondering why the comparatively slow and ponderous American F-86's achieved near total domination of the superior MIG-15's, he realized that the F-86 had two crucial advantages: better visibility and a faster roll rate.

This, in turn, led Colonel Boyd to develop what he called the OODA Loop, to denote the repeated cycle of observation, orientation, decision and action that characterized every encounter. The key to victory, he theorized, was not a plane that could climb faster or higher but one that could begin climbing or change course quicker -- to get inside an adversary's "time/cycle loop."

The fast-cycle combat theory, expanded by Colonel Boyd into a lecture he later delivered hundreds of times, has since been widely applied to fields as diverse as weapons procurement, battlefield strategy and business competition.

One implication of the theory was that the best fighter plane was not necessarily the one with the most speed, firepower or range. Colonel Boyd, who enrolled at Georgia Tech after his Nellis tour, was helping a fellow student with his homework over hamburgers and beer one night when he had an insight that led to a way to quantify his ideas. The resulting Energy Maneuverability Theory, which allows precise comparisons of maneuverability, is now a standard measure of aerial performance.

Assigned to the Pentagon in 1964, Colonel Boyd became an important figure in a movement that started in response to $400 hammers and other headline excesses of Defense Department spending and soon expanded to question the need for many hugely expensive weapons systems.

Although he had allies in the Pentagon, Congress and business, Colonel Boyd's ideas often went against the grain of a military-industrial bureaucracy devoted to the procurement of the most advanced, most expensive and (not coincidentally, he felt) most profitable planes.

Although his design ideas helped give the F-15 a big, high-visibility canopy, his major triumph was the F-16, a plane lacking many of the F-15's high-tech, expensive features, but which is far more agile and costs less than half as much, allowing for the purchase of many more of them for a given expenditure.

Top Air Force officers were so opposed to the concept of producing a plane that did not expand on the F-15's cutting edge technology that Colonel Boyd and some civilian allies developed it in secret.

The plane was hailed for its performance in the Persian Gulf war, a war whose very strategy of quick, flexible response was based largely on ideas Colonel Boyd had been promoting for years.

Colonel Boyd, who maintained that the lure of big-money defense contracts invariably perverted weapons assessment, was so personally fastidious that during his years in the Pentagon be became known as the Ghetto Colonel because he lived in a basement apartment.

He carried his notion of propriety to such an extreme that when he retired in 1975 and began some of his most productive work, as a Pentagon consultant, he insisted that his family live on his retirement pay. Initially offering to work full time without pay, he was persuaded to accept one day's pay every two-week pay period, because he had to be on the Pentagon payroll to have access to the building, before retiring in 1988.

When a headstone (such as that of John Boyd) is inscribed with the lifespan of the deceased being "1927 – 1997", that small "–" between the start and end of life years represents a mountain of creativity, accomplishments and failures, opportunities and challenges, sadness and happiness, and every emotion human can sense. Perhaps the technology of headstones will advance someday to have a compartment containing a small booklet in the form of a biography of the person so that the accomplishments (personal and professional) of that person can be available to viewers of the person's headstone – and in which can be read all of the many personal and professional accomplishments that the small "- "represents on the headstone.

A photo is shown below of John Boyd's headstone at the Arlington National Memorial Cemetery, not far from the Pentagon where his foresight and persistence in the need for a Lightweight Fighter for the USAF inventory were so loudly and emphatically made (and will be eternally remembered by those within earshot):

Col. John R. Boyd (1927–1997) Arlington National Memorial Cemetery
(USAF File Photo)

In his book "*Boyd*", author Robert Coram describes John Boyd's personal life and his military career in absolutely factual detail. Every Reader of this book should read the Coram book about "*Boyd*" as it reveals the true John. If John had been given the option, he probably would have asked that his headstone bear the simple epitaph "See, I told you so."

Those of us who were privileged to have known and worked with John Boyd may not have enjoyed the way he spoke (typically ranted and with a cigar stub clenched between his teeth) but there is no one who can say that the meaning behind his words did not command respect for a genius in military fighter aircraft tactics and design requirements. He lived to see his beloved LWF materialize as he had envisioned in the Lightweight Fighter Program – and winced every time a new item of equipment (with weight and cost) was added to the Lightweight Fighter under the guise of "missionization".

Harry J. Hillaker: "Father of the F-16"
by Bob Cox, Fort Worth Star Telegram, February 10, 2009

Harry J. Hillaker at a USAF F-16 Thunderbirds Air Show in 2008

General Dynamics Chief Design Engineer Harry J. Hillaker, died February 8, 2009. It was a chance meeting in 1962 at the Eglin Air Force Base Officers Club bar with a loud-mouthed Air Force fighter pilot that set Harry J. Hillaker on a path that led to the original design of the F-16 fighter jet, arguably the best military airplane of the jet age. Mr.

Hillaker, an Aeronautical Engineer at General Dynamics for 44 years and known to many as the "Father of the F-16," died Sunday at his home in Fort Worth. He was 89. As a Senior Design Engineer at General Dynamics' Fort Worth aircraft plant in the 1960s, Mr. Hillaker led a design team that worked, secretively at first, with a small group of Pentagon insurgents known as the "Lightweight Fighter Mafia", led by USAF Colonel John R. Boyd, to turn a collection of ideas, theories and concepts into what would become the F-16. Their success in convincing HQ USAF to proceed with the advanced technology demonstrator, which rapidly was changed into a light weight fighter prototype of the F-16 is evident in that four decades later, the plant, now part of Lockheed Martin, is still producing F-16s. More than 4,400 have been built and delivered worldwide. At the peak of production in the 1980s, close to 25,000 people were working on the program.

"Harry's legacy is an incredible aircraft that has become the mainstay of 25 nations and continues to be in demand today after 30 years of production," said Ralph Heath, President of Lockheed Martin Aeronautics Co. "The early F-16 versions paved the way for tens of thousands of jobs, over $100 billion in sales and customer relationships that are the cornerstone for Lockheed Martin's transition to the future with our new aircraft programs." Any success has numerous "fathers", and the F-16 is no different. But people close to the F-16 program say Mr. Hillaker's engineering expertise, open-mindedness, and loyalty, to a concept originally known simply as the "lightweight fighter", were vitally critical to the program. "Without Harry, I don't think anything close to the F-16 would have come to fruition," said Jay Miller, an Arlington aviation historian. Mr. Hillaker, born in Flint, Mich., and educated at the University of Michigan, went to work for Consolidated Aircraft in San Diego in 1941. A year later, he was sent to the company's Fort Worth plant. With Consolidated Vultee Aircraft, better known as Convair/Fort Worth Division, and which still later became General Dynamics, Mr. Hillaker worked on most of the company's major projects, including the B-36 and B-58 Strategic Bombers, and the F-111 Aardvark fighter/bomber built in Fort Worth. One night in 1962 at the Eglin Air Force Base officers club, Mr. Hillaker was introduced to Maj. John Boyd, an abrasive and cocky but highly intelligent fighter pilot. Informed that Mr. Hillaker had worked on the USAF F-111, then under development, Boyd launched into an expletive-laden tirade about what a poorly designed, underperforming aircraft the F-111 was fated to be. According to numerous reports of that meeting, Mr. Hillaker quickly realized that Boyd knew far more about airplane design and performance than most pilots and invited him to sit. Soon, the two men were exchanging ideas and formulas on cocktail paper napkins.

In the years that followed, Boyd, assigned to USAF Requirements Office at the Pentagon, argued the cause for a lightweight, highly maneuverable and affordable fighter plane, the polar opposite of the F-111. His magnetic personality attracted a few adherents, notably fellow fighter pilot USAF Col. Everest "Rich" Riccioni, a civilian

Pentagon official named Pierre Sprey, and a high-level civilian mathematician at Eglin Air Force Base, Florida, named Tom Christie. The Fighter Mafia, as the four became known, concocted a scheme to covertly begin work on just such a plane. Covert, because top Air Force brass were spending billions to develop the new USAF F-15 jet and were opposed to any new activity that would potentially draw funds away from the F-15 Program. In 1969, Riccioni wrote a vaguely titled budget request and received $149,000 for performance and design studies. General Dynamics and Northrop were selected to work on competing design concepts, under the guise of them being advanced flight technology demonstrators with no production being planned by the USAF for them.

Mr. Hillaker, who since getting to know Boyd had quietly guided some internal lightweight fighter design work, was General Dynamics' point man for the program. On numerous occasions over the next two years, Harry Hillaker secretly flew to Washington and met with Boyd, Sprey and a few others to hash out theories and share data and design concepts. "We used to stay up all night arguing about fighter design performance and trade-offs," Sprey said. "He gave us a lot of insights both into design and internal General Dynamics internal politics. He was committed to doing it right." Mr. Hillaker, Sprey said, meshed well with the mercurial Boyd and "was very open-minded. Among designers in the military aircraft business, that was very rare."

The lightweight fighter incorporated a number of advanced technologies, in particular fly-by-wire flight controls, all aimed at making it the most agile and lethal aircraft and capable of winning one-on-one dogfights against the best Soviet-bloc aircraft of the day. Top civilian Pentagon officials, most notably Chuck Myers in the Office of the Secretary of Defense, at the urging of Boyd, Sprey and Riccioni, eventually gave their blessing to the program, and contracts were let for each team to design and build two each demonstration prototypes. A fly-off, under stringent representative operational flight test evaluation conditions set as goals by the Fighter Mafia, was held in 1974 at Edwards Air Force Base in California. General Dynamics' YF-16, a close derivative of their Model 401, was selected over Northrop's YF-17, also a close derivative of their P-530 Cobra. The YF-17 was reconfigured to be USN carrier capable, and was produced eventually in large numbers as the USN F/A-18 Hornet. Sprey says Mr. Hillaker and his team were due the largest share of the credit for the F-16 success. "I can practically run down the long list of things that wouldn't have been in the airplane to make it maneuver so effectively, have the first fly-by-wire control system in an operational fighter, and have the advanced cockpit features if it wasn't for Harry," Sprey said.

Mr. Hillaker went on to hold several positions with General Dynamics in Advanced Design concepts, and to further explore advanced flight technology versions of the F-16. He retired in 1985, but remained active with numerous aerospace organizations and advisory groups until he passed away on February 8th, 2009.

ABOUT THE AUTHOR

Herbert A. Hutchinson graduated from the nationally renowned Brooklyn Technical High School (Arista) in New York, and appears as one of a hundred Notables in the BTHS "100 Years of Excellence" who graduated in the first fifty years of Brooklyn Tech. He received a Bachelor of Science degree in Aeronautical Engineering at Georgia Tech in Atlanta, Georgia, in 1953. After serving as a Combat Engineers platoon leader in the U.S. Army Corps of Engineers in the Korean War, he returned to Georgia Tech to earn a Master of Science in Aeronautical Engineering degree (1959).

The initial nine years (1952-1961) of his professional engineering career were as a wind tunnel test engineer (Lockheed C-130) at the Georgia Tech Research Institute, as an aerodynamicist (F-84, F-105B and XF-103) at Republic Aviation in New York, and as a flight test performance engineer (B-58) at General Dynamics in Texas.

The middle half of his career (1961 – 1982) was as the Chief of the Aeromechanics Branch of the Air Force Flight Dynamics Laboratory, and as the Chief System Engineer (GS-15) for several U.S. Air Force weapons systems developments, including the U.S. Air Force Lightweight Fighter program and the U.S. Air Force F-16 Fighting Falcon Multi-National Weapon System, developed by the U.S. Air Force at Wright-Patterson Air Force Base in Dayton, Ohio. He was on the team conducting exploitation testing of foreign flight system technologies at the formerly very highly classified clandestine flight test site now known as Area 51. With extensive applied system engineering experience, he was selected to serve on three Presidential Advisory Committees, participated in numerous aircraft crash investigations (with numerous fatalities), and served three years as Chief of the USAF Flight Technologies Division (Aeronautical Systems Division).

Based on his having been the Chief System Engineer for the USAF YF-16 Lightweight Fighter Prototype Program and the ensuing F-16 Fighting Falcon multinational fighter for the USAF and twenty-eight other countries, former pilots of the advanced technology F-16, used in the several Gulf Wars such as Desert Storm, have awarded him the tactical call name "Stuffer". He has authored two books: "The Inside History of the USAF Lightweight Fighters, 1900 to 1975" and "The F-16 Fighting Falcon Multinational Weapon System, 1972 to 2019", both published by the Xlibris Publishing Company, and which reflect his understanding of the complexities and interactions that exist in applying emerging technologies in "big picture" development of large, complex systems.

The final twelve years (1982 – 1994) of Hutchinson's career were as a Manager of System Engineering at the Northrop Aircraft Division in California managing the system engineering of a new fighter aircraft design for NATO forces in Europe and of the design of stealthy flight vehicles (the Tri-Service Standoff Attack Missile (TSSAM) and the YF-23 Advanced Tactical Fighter (YF-23). He retired from his 41 years professional career at the end of 1994.

After retirement, he applied his system analytical and database research experiences to becoming a world-recognized genealogist and created the extensive family tree histories for his family, of English and British West Indian descendancy (fourteen generations 1595 to 1995), and that of his wife's family, of European descendancy (thirteen generations 1620 to 2000).

In collaboration with the Barbados Museum and Historical Society, he capitalized on the fact his two Barbadian grandfathers were part of the labor force in creating the Panama Canal (in this book as an example of a highly successful early "green" international environmentally-friendly project) to support the Barbados Government and UNESCO in naming the Panama Canal Zone as a Historical World Heritage Site. He invented "An Optronic System for the Measurement of Vehicle Traffic" and holds US Patent 6,750,787, measuring vehicles speeding on neighborhood streets to improve pedestrian safety in neighborhoods. His extensive experience in genealogy and in patenting an invention for improving human safety provided a wealth of knowledge about the human factors involved in everyday living in different regions and migratory habits of humans in different climates.

He is the author of the book: "Dealing with the Real Climate Change" which received the Pacific Book Review Award, and also the contemporary book "Facts and Fallacies of Climate Change on our Planet", both written as a public service and intended to inform the public worldwide that inevitable climate change is happening but at different rates and with different effects and different state of urgency to be mitigated depending on the latitude of the region on the planet. In his books on global warming and climate change, Hutchinson provides data and a system-level, "big picture" analysis concluding that there is no "one-size-fits-all" solution or goal for mitigating the environmental and human problems brought about by global climate changes occurring in different regions of the planet. His fourth book assesses a set of twenty-five energy and transportation technologies being developed to mitigate climate change effects.

Hutchinson is an Associate Fellow Emeritus (seventy-four years continuous membership) in the American Institute of Aeronautics and Astronautics (AIAA), is a life member of the Veterans of Foreign Wars, is a member of the Air Force Association, and is a member of Mensa. His name is on the Smithsonian Institute Wall of Honor. He and his wife Arline reside in Durham, North Carolina. Herb continues to be active in the aeronautical engineering profession, passing along his extensive experiences to any group (educational, industrial, or Governmental) pursuing the application of new technologies. A visual indicator of his dedication and passion for describing STEM activities applied in the "real world" is shown as follows:

(From Author's Memorabilia – at Lockheed Martin, Greenville, SC)

Contact the Author:

The author of this book, *The Inside History of the USAF Lightweight Fighters 1900 to 1975*, as well as the four books *The F-16 Fighting Falcon Multinational Weapon System 1963 to 2019*, *Facts and Fallacies of Climate Change on Our Planet*, *The Big Picture of Climate Change on Planet Earth*, and *Dealing with the Real Climate Change*, may be contacted with questions or constructive criticism at: **herbineer@aol.com**.

The Collier Trophy is an annual aviation award administered by the U.S. National Aeronautic Association (NAA), presented to those who have made "the greatest achievement in aeronautics or astronautics in America, with respect to improving the performance, efficiency, and safety of air or space vehicles, the value of which has been thoroughly demonstrated by actual use during the preceding year."

The Robert T. Collier Trophy

Awarded to the General Dynamics F-16 Team, for year 1975

VERITY PUBLICATIONS

ISBN